Birkhäuser

Vita Mathematica

Volume 21

Edited by
Martin Mattmüller

The series Vita Mathematica traces the professional lives of many of the world's great mathematicians – from antiquity to the present day. These books are written broadly following the same pattern and take into consideration the research carried out on the history of science in recent decades.

Egbert Brieskorn • Walter Purkert

Felix Hausdorff

Mathematician, Philosopher, Man of Letters

Translated by David Rowe

Birkhäuser

Egbert Brieskorn
Eitorf, Germany

Walter Purkert
Mathematics Institute
Bonn University
Bonn, Germany

Translated by
David Rowe
Mathematics Institute
Mainz University
Mainz, Germany

ISSN 1013-0330 ISSN 2504-3706 (electronic)
Vita Mathematica
ISBN 978-3-031-52134-8 ISBN 978-3-031-52135-5 (eBook)
https://doi.org/10.1007/978-3-031-52135-5

Mathematics Subject Classification (2020): 01A55, 01A60, 01A70

Cover illustration: Felix Hausdorff at Work in his Home Office.

This book is published under the imprint Birkhäuser, www.birkhauser-science.com by the registered company Springer Nature Switzerland AG
The registered company address is: Gewerbestrasse 11, 6330 Cham, Switzerland

Preface

The present biography of Felix Hausdorff grew out of a much larger undertaking dating back to the early 1990s. At that time, Egbert Brieskorn, Friedrich Hirzebruch, Walter Purkert, Reinhold Remmert, and Erhard Scholz founded the Hausdorff editorial project, which between 2001 and 2020 published *Felix Hausdorff · Gesammelte Werke* (*HGW*) in ten volumes.[1] Volume IB, published in 2018, contains the original German biography by Brieskorn and Purkert, a detailed study of 1,130 pages. The abridged edition, *Felix Hausdorff: Mathematiker, Philosoph und Literat* [Brieskorn/Purkert 2021], is roughly half as long as the original and has served as the basis for this somewhat adapted English version. A few words about Hausdorff's intellectual accomplishments will help illuminate why the authors were moved to document his life in such detail.

Felix Hausdorff was an extraordinarily versatile mathematician whose creative achievements left a lasting mark on a number of mathematical fields. Probably the strongest influence came from his magnum opus, the monograph *Grundzüge der Mengenlehre* (1914), which exerted a major impact on the development of mathematics in the first half of the twentieth century. In this work, he introduced the theory of topological spaces based on point sets satisfying various properties, including his second separation axiom (such spaces are called Hausdorff spaces today). On this basis, he launched general topology as an independent mathematical discipline. A particularly important case of Hausdorff spaces are the metric spaces, which he dealt with systematically based on the three defining properties of a metric. He later also introduced a special metric that measures the distance between two closed and bounded subsets of a metric space, today known as the Hausdorff metric. This triggered numerous investigations of hyperspaces in infinite-dimensional topology. Hausdorff spaces, but especially metric spaces are, so to speak, omnipresent in today's mathematics.

Hausdorff was a leading early representative of a new orientation in mathematical research that later came to be seen as the modern approach of the twentieth century. For mathematical modernity, the chapters on topology in Hausdorff's *Grundzüge der Mengenlehre* constitute one of the central texts illustrating this approach, which can thus be counted as a pioneering methodological achievement. Characteristic for this new conception of mathematics was the presentation of fundamental structures as abstract axiomatic theories. This approach aims to extract a common structural core from concrete examples on the basis of suitably chosen axioms. Proceeding deductively, a well-chosen axiom system leads to general theorems in an overarching theory that embraces various examples as special fields, but which also affords many possibilities for further applications. The construction of such abstract theories leads to greater generality and transparency, while producing improved clarity, uniformity, and ultimately economy.

[1]The acronym *HGW* for *Hausdorff · Gesammelte Werke* is itself an abbreviation for *Felix Hausdorff, Gesammelte Werke einschließlich der unter dem Pseudonym Paul Mongré erschienenen philosophischen und literarischen Schriften und ausgewählter Texte aus dem Nachlaß.*

Whereas Hausdorff's *Grundzüge* exerted a strong influence on set theory and general topology already in the 1920s, his innovative ideas in measure theory – familiar today in the notions of Hausdorff measure and Hausdorff dimension – only gained widespread currency during the last third of the twentieth century. Since then, they have profoundly influenced developments in geometric measure theory. At the same time, using representations of fractals by means of computer graphics, these ideas have stimulated research on a broad array of scientific problems. Within mathematics proper, Hausdorff's measure-theoretic concepts have found applications in a variety of fields, including number theory, harmonic analysis, potential theory, nonlinear partial differential equations, dynamical systems, and stochastics, as well as in fields closely tied to physical research, such as ergodic theory, turbulent flows, statistical mechanics, and meteorology. One of Hausdorff's most famous findings – the paradox bearing his name and later generalized by Stefan Banach and Alfred Tarski – has since led to developments that today fill an entire monograph.

Although Hausdorff followed and appreciated the efforts of Ernst Zermelo, Abraham A. Fraenkel, and others who sought to avoid the antinomies of set theory by placing the subject on axiomatic foundations, he had no burning desire to participate in this research himself. Instead, he took up the field of ordered sets; indeed, as Arthur Schoenflies emphasized in 1913, practically all progress achieved in this branch of set theory since Georg Cantor's time was due exclusively to Hausdorff. He introduced the concepts of *cofinality* and *coinitiality*, which became fundamental for set theory, and he was the first to distinguish between *regular* and *singular* cardinal numbers. By noting that regular cardinals with a limit index – if they were to exist at all – would have to be "of exorbitant size," he initiated the investigation of inaccessible cardinals. In the theory of ordered sets, he showed that gaps in dense ordered sets can be uniquely characterized by pairs of regular cardinal numbers. Hausdorff also introduced partially ordered sets and showed that any ordered subset of such a set can be extended maximally (Hausdorff's maximal chain theorem). With his general product and power notions for order types, he created and classified an enormous arsenal of new order structures, among them the η_α-sets, which are important in model theory today. Hausdorff found the fundamental recursion formula for aleph exponentiation which bears his name. He also undertook important work on Cantor's continuum hypothesis (CH) within the context of descriptive set theory, proving that CH holds for Borel sets in Polish spaces. He was also the first to formulate the generalized continuum hypothesis (GCH).

Beyond all these impressive achievements in the broader realm of set theory, Hausdorff was also deeply engaged in analytic research and made particularly noteworthy contributions to functional analysis. He originated a large class of matrix methods for limiting divergent series, today called Hausdorff methods (the earlier classical methods of Otto Hölder and Ernesto Cesàro are special cases). The basis for this theory arose from his solution of the so-called Hausdorff moment problem. He enriched the theory of L^p spaces through the Hausdorff-Young inequality, which generalizes a corresponding result connected with the classical theorem of Riesz and Fischer. Beyond functional analysis, Hausdorff published noteworthy results in four other fields: Lie algebras (the Baker-Campbell-Hausdorff formula), number theory (an essential simplification of David Hilbert's solution of Waring's problem), mathematical statistics (introduction of cumulants), and actuarial mathematics (individual risk theory).

Hausdorff's mathematical work and its influence on the development of mathematics represent, however, only one side of his fascinating personality. Indeed, for almost two decades he led a double life, so to speak. On the one hand, he struggled to find his way as a scientist, starting from less than successful work in optical astronomy to his breakthrough as a leading expert in set theory; on the other, he made his mark as a philosopher, man of letters, and critical essayist, writing under the pseudonym Paul Mongré. In this respect, he occupied a singular place in the cultural life of Germany during the two decades that preceded the outbreak of the Great War. Already during his years as a student in Leipzig, Hausdorff pursued wide-ranging interests in the sciences as well as the humanities. For many years, he was deeply occupied with problems concerned with the epistemological nature of space and time, taking inspiration from Immanuel Kant while distancing himself from him and his followers. In developing his own ideas, he was above all influenced by the philosophy of Friedrich Nietzsche, whose radical views took on existential importance for the young Hausdorff. It was through a critical examination of Nietzsche's "proof" of the eternal recurrence of the same that Hausdorff was led to *his* field in mathematics: set theory and its applications. Under his pseudonym, he published the aphorism volume *Sant' Ilario. Gedanken aus der Landschaft Zarathustras*, the epistemological work *Das Chaos in kosmischer Auslese*, the book of poems *Ekstasen*, and the very successful one-act theatrical piece *Der Arzt seiner Ehre*, as well as a series of essays in leading literary journals.

These dual aspects of Hausdorff's intellectual life and personality provided a strong motive for undertaking an edition of his Collected Works. There was another very different motivation, however, and that was his tragic end as a victim of the anti-Semitic racism of the National Socialist regime. Into his old age, Hausdorff was increasingly marginalized and harassed as a Jew. Finally, faced with imminent deportation to a concentration camp, he took his own life on January 26, 1942, together with his wife and sister-in-law. All contributors to the Hausdorff edition were deeply aware of this moral dimension and felt it their duty to research his life's work in order to bring him closer to posterity, thereby honoring him in a special way.

As mentioned above, the comprehensive biography in Volume IB of the Hausdorff edition was published in 2018; this volume, however, had a long prehistory. To mark the 50th anniversary of Hausdorff's death, Egbert Brieskorn organized a memorial colloquium on January 26, 1992, held at the Mathematics Institute of the University of Bonn. That event led to the anthology *Felix Hausdorff zum Gedächtnis – Aspekte seines Werkes* [Brieskorn 1996]. Brieskorn also organized an exhibition, "Felix Hausdorff – Paul Mongré" and prepared a catalog for it.

In the course of organizing these activities, Brieskorn had years earlier begun to collect biographical material documenting Hausdorff's life. He had numerous conversations with Hausdorff's daughter, Lenore König, and he established contacts with Charlotte Hausdorff's niece, Else Pappenheim, and other contemporary witnesses. He also contacted numerous archives and was able to collect a great deal of valuable material through extensive research over the years.

After funding for the Hausdorff edition was secured, Brieskorn took on the task of preparing a detailed biography as part of this project. In addition to describing Hausdorff's extraordinarily influential mathematical work, he also wanted to take into account all his other major fields of interest. His aim thereby was to give an accurate and comprehensive picture of Hausdorff's unusually complex and convoluted intellectual journey. Philosophy formed one key component, especially the ideas of Kant, Schopenhauer, and Nietzsche, which eventually led to Hausdorff's interactions with leading figures associated with the Nietzsche Archive in Weimar. At the same time, Hausdorff's interests in epistemology spilled over into reflections on language criticism as well as on space and time; his literary work put him in touch with leading German writers; his deep knowledge of and passion for music ran from Richard Wagner to Max Reger; his love for the fine arts found expression in his friendship with the sculptor Max Klinger. Brieskorn explored all these corners of interest, but also the family history in the context of Jewish history, in particular the history of antisemitism as background to Hausdorff's tragic last years under the Nazi dictatorship.

During the many years Egbert Brieskorn worked on the biography, he continually collected new material that led to new drafts dealing with Hausdorff's early years. This was a story that had barely been touched beforehand; he had to write it from scratch. There were also longer interruptions, mainly due to his deep interest in and commitment to environmental issues and species conservation. As a result, he began writing the final version relatively late in his life; he worked on it very intensively, especially during a period when he was seriously ill. During that time, he faced death with philosophical calmness, though one of his worries was what would happen were he unable to finish the biography. I promised him that I would complete the work as best I could and that everything he wrote would go into the biography, unabridged and unaltered. Three weeks before his death, he sent me the last section he was able to finish, a particularly difficult part, in fact, that concerned Hausdorff's relations with the philosopher, mystic, and anarchist Gustav Landauer. Sensing that he would not be able to do more, he proposed that we meet to discuss how he envisioned the further course of the biography. We set July 12, 2013 as the date for that meeting, but on the evening of July 11, 2013 he passed away.

At the time of his death, Brieskorn had completed 546 pages of the biography; they have been included, as promised, in Volume IB of the edition, unabridged and unaltered. For my work on the remainder of the book, Brieskorn's posthumous papers were an indispensable aid. After her husband's death, Heidrun Brieskorn made these documents available to me, for which I am very grateful. Furthermore, I must acknowledge my debt to the other volume editors of the Hausdorff edition, most of whom contributed to the project on a voluntary basis, often over a period of several years. These contributors are (were) Josef Bemelmans, Christa Binder, Egbert Brieskorn, Srishti D. Chatterji, Moritz Epple, Ulrich Felgner, Horst Herrlich, Stefan Hildebrandt, Mirek Hušek, Vladimir Kanovei, Peter Koepke, Gerhard Preuss, Reinhold Remmert, Udo Roth, Winfried Scharlau, Felix Schmeidler, Erhard Scholz, Wolfgang Schwarz, Werner Stegmaier and Friedrich Vollhardt. I especially wish to remember with great gratitude all those colleagues who did not live to see publication of the original biography published in Volume IB or the abridged edition [Brieskorn/Purkert 2021].

To produce a new abridged version of a book written by two authors when one of these died nearly eight years ago is an unusual and problematic undertaking. On the other hand, it would have been impossible to publish a book of over 1,000 pages in a series such as Vita Mathematica. Frau Brieskorn encouraged me to take this unusual step, thereby enabling an international readership to have access to the results of her husband's many years of work on Hausdorff's biography, as fully documented in Volume IB of the edition. In the present version, the following parts are almost completely due to Egbert Brieskorn: Chapters 1 and 4 as well as Sections 2.1, 2.3, 5.1–5.4. For other parts – 7.1, 7.2, 7.5, 8.1, 8.2, 8.7, 9.1, 9.4 and 9.5 – I relied heavily on the wide-ranging research results he left behind in his estate. The remaining parts of the book constitute in a revised form my own contributions from Volume IB.

I would like to express my sincere thanks to the German Research Foundation and the North Rhine-Westphalian Academy of Sciences and Humanities, whose financial support made the edition and thus the biography possible in the first place. I would like to thank the Hausdorff Center for Mathematics at the University of Bonn and the Manuscript Department of the University and State Library in Bonn for their continuous support of my work.

I would also like to thank my friends and colleagues Reinhard Siegmund-Schultze and Erhard Scholz, who accompanied my work on the biography with advice and support from the beginning. During the end phase, the final text benefited greatly from Martin Mattmüller's careful editing; the help he has lent this project is most deeply appreciated.

My dear friend and colleague David Rowe took on the difficult and time-consuming task of translating [Brieskorn/Purkert 2021] into English. This was an especially fortunate circumstance, since he combines broad mathematical knowledge and philosophical understanding with an intimate feeling for the language. He also contributed a number of suggestions for improving the text, not only with regard to its structure and readability but also its content and visual illustrations. For all this I am especially grateful for his support.

Bonn, November 2023 Walter Purkert

Notice for the Reader

Many footnotes and references in [Brieskorn/Purkert 2021] have been omitted from the present English edition, since these are generally only relevant for those with proficiency in German. In citing works published in the volumes of Hausdorff's Collected Works, we employ the abbreviation *HGW* together with the volume number. These volumes are (along with the year of publication):

Volume IA	*Allgemeine Mengenlehre* (2013)
Volume IB	*Biographie* (2018)
Volume II	*Grundzüge der Mengenlehre* (2002)
Volume III	*Deskriptive Mengenlehre und Topologie* (2008)
Volume IV	*Analysis, Algebra und Zahlentheorie* (2001)
Volume V	*Astronomie, Optik und Wahrscheinlichkeitstheorie* (2006)
Volume VI	*Geometrie, Raum und Zeit* (2020)
Volume VII	*Philosophisches Werk* (2004)
Volume VIII	*Literarisches Werk* (2010)
Volume IX	*Korrespondenz* (2012)

The first part of the bibliography gives a complete list of Hausdorff's writings, including those published under the pseudonym Paul Mongré. These works are referred to by the symbol H, thus [H 1897b] for *Sant' Ilario* or [H 1914a] for *Grundzüge der Mengenlehre*. When citing passages from his published works, we first give the pages in the original and then those for the reprinted text in the edition. For example, if a passage is quoted from pages 211–212 in *Sant' Ilario*, the reference is given in the form [H 1897b, 211–212; VII: 305–306]; the reprint of this passage is thus found in volume VII of the edition on pages 305–306.

The second part of the bibliography presents all other writings referenced in the book, except for those by Hausdorff. The biography also quotes from numerous letters written to and from Hausdorff. For these, the only source given is the corresponding page in *HGW* Volume IX, since that volume provides full reference details, including information about the location of the original letters.

Walter Purkert cataloged Hausdorff's extensive estate of about 26,000 pages in preparation for subsequent work on the edition. Only a few printed copies of this finding aid are available, but it can be accessed electronically by searching for the *Findbuch* at www.sammlungen.ulb.uni-bonn.de.

Contents

Part I

Family Background and Early Intellectual Interests

Chapter 1

Hausdorff's Childhood and Youth

1.1 Ancestry

Felix Hausdorff was born on November 8, 1868 in Breslau in Silesia, today the city of Wroclaw in western Poland. His father, Louis Hausdorff, was a Jewish merchant, whereas his mother Hedwig (Johanna)[1] descended from the Tietz family in Breslau. The ancestors on the father's side can be traced back to Felix Hausdorff's great grandfather, Michoel Shoul Hausdorff. He lived in the town of Rawitsch, which was only 60 kilometers from Breslau, but in the Prussian province of Posen rather than Silesia. Rawitsch was a regional center for cloth weaving with a large Jewish population. A Torah school was founded there in the 1770s, and some members of its Jewish community undertook Talmudic studies diligently.

Nothing is known about how Michoel Shoul Hausdorff made his living, nor do we possess any information about his wife's family. Even the family name "Hausdorff" is shrouded in mystery. By the sixteenth century, the Christian population throughout central and eastern Europe had acquired family names, unlike the Jews, who usually only acquired such names in the nineteenth century. Before then, a Jewish child carried the father's name, for example, Jacob ben Salomon (Jacob, son of Salomon). Toward the end of the eighteenth century, members of the Jewish community in both Austria and Prussia were compelled to take on permanent family names. For the governmental district of Breslau this began already in 1791, whereas most parts of Prussia introduced this requirement following the Edict of Emancipation of 1812. Unlike in Austria, Prussian Jews were allowed to choose a family name based on their geographic locality. In the region surrounding Rawitsch und Breslau there were four places with the name "Hausdorf," one larger village and three smaller ones. In all likelihood, Michoel Shoul Hausdorff or another ancestor came from one of those villages, thus the family name.

[1] Johanna was her formal name registered with the synagogue; in daily life she was known as Hedwig, the name that appears on her gravestone.

E. Brieskorn, W. Purkert, *Felix Hausdorff*, Vita Mathematica 21,
https://doi.org/10.1007/978-3-031-52135-5_1

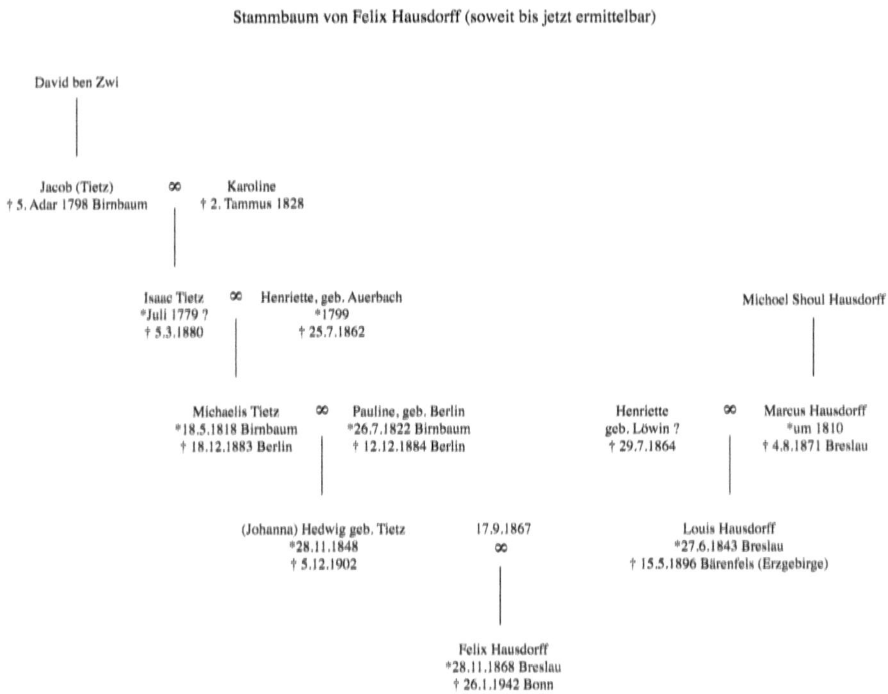

Figure 1.1: Felix Hausdorff's Ancestral Tree

Michoel Shoul Hausdorff had three sons and a daughter: Marcus, David, Julius und Faigele.[2] Marcus, a merchant who settled in Breslau some time before 1847, was Felix Hausdorff's paternal grandfather. In that year, he opened a business on Carlsstrasse, where he sold canvases, draperies, and dry goods up until his death in 1871. Marcus Hausdorff and his wife Henriette, *née* Löwin, had three sons: Siegfried, born in 1841, Louis, born two years later, and Moritz, who died in childhood. The boys grew up in a deeply religious home, as their father placed great value on the spiritual development of his sons.

Marcus Hausdorff was, in fact, not just a merchant; he was also a prominent Talmudic scholar. What is more, his second son, Louis – who would become Felix Hausdorff's father – showed true scholarly talent at an early age. He evidently realized this potential, as his later accomplishments were remembered in an obituary article from 1896:

> Born in Breslau on 27 June 1843, [Louis] Hausdorff was educated in Judaic studies by his father, a well-known Talmudic scholar. He devoted himself to his studies with such burning passion that he was already able to obtain the Morenu title at the age of 13. By studying on his own, he also acquired knowledge of secular subjects. Not only was he adept at learning, he was also active as a freelance writer. A lively and original mind enabled him to impress the public – especially after 1870 when he settled in Leipzig – with his independent and original viewpoints on a variety of different areas of knowledge. (*Mittheilungen vom Deutsch-Israelitischen Gemeindebunde*, Nr. 44, Leipzig 1896, 36.)

The Morenu title was normally awarded to those who passed an examination required in order to become a rabbi. Candidates were tested not only on their knowledge of the Talmud but also on later decisions passed down by rabbinical authorities.

As a young man, Louis Hausdorff had relatively little time to devote to his studies. By 1866, at the latest, he was working with his brother Siegfried as a salesman in his father's business. Both also married around this time, taking as their brides two sisters from the Tietz family. Siegfried married 19-year-old Natalie Tietz, and on 17 September 1867 Louis took her younger sister Hedwig as his wife. The latter young couple moved into an apartment not far from the family store on Carlsstrasse. After the birth of Felix came their daughter, Martha, born on 19 December, 1869, followed by a third child, Vally, who was born on 21 June, 1874 when the family was already living in Leipzig.

The family history of Felix Hausdorff's mother is far better known than that of his father. According to oral tradition, the ancestors of the Tietz family left southern France for the Rhineland during the 12th century, settling afterward in the Netherlands. The earliest documented record of a family member concerns

[2] Among the descendants of David belongs Mr. William (Bill) Hausdorff (Brussels), to whom I owe the information about Michoel Shoul Hausdorff.

David ben Zwi, who in 1772 was living in the newly formed Prussian district of the Netze, established after the first division of Poland. There he was given land near the village Tütz (today Tuczno) as well as certain privileges as a Jewish settler. In this period, during which the Prussian government ceded property in newly occupied territories to promote economic development, it relied in part on Jewish merchants and settlers. As with the family name Hausdorff, one may assume that the family name Tietz dates back to the village of Tütz.

David ben Zwi's son, Jacob, later settled in Birnbaum in the district of Posen, a region that fell to Prussia after the second divsion of Poland in 1793. He attained considerable wealth in Birnbaum, which he passed on to his three sons: Hirsch, David, and Isaac. One of Hirsch Tietz's grandchildren was Hermann Tietz, who together with his nephew Oscar founded the first department stores in Germany, including the famous chain in Berlin "Hermann Tietz OHG." The store on Alexanderplatz was known for its spectacular fassade, some 250 meters long. By the time the Nazi government came to power, the company had stores in many cities throughout Germany. With the help of a consortium of banks, these were appropriated ("aryanized") during the Third Reich and reopened under the name HERTIE, a chain of stores that continued to do business after the war in West Germany.

David and Isaac Tietz continued to live in Birnbaum, where they worked as merchants. The eldest son of Isaac, Michaelis, was born in 1818; he would become Felix Hausdorff's maternal grandfather. Michaelis Tietz married Pauline Berlin, and together they had nine children.

The founding of the German Empire under Prussian hegemony in 1871 brought about a uniformization of the legal system in all German states. As a consequence, Jewish citizens were guaranteed the same rights before the law as all others. In fact, such a law had already gone into effect in July 1869 for states in the North German Confederation, and this was adopted in some of the southern states as well in 1870. However promising these measures may have seemed as a sign of progress for German Jews, this equality de jure by no means signified true equality de facto. Latent forms of antisemitism continued to exist just as before, though traditional religious antisemitism was increasingly giving way to racial antisemitism, which played a role, for example, when Jewish candidates came under consideration for university appointments. Nevertheless, these new legal rights represented decisive progress compared with earlier times, especially for Jews who participated in the larger economy and the free professions. As a consequence, an increasing number of Jewish families living in the eastern border regions of Prussia began to migrate into the larger cities.

Michaelis Tietz, his wife Pauline, and their son Heinrich settled in Berlin. Siegfried Hausdorff brought his wife Natalie and children there, too, following a lengthy stay in Leipzig. Another son, Adolf Tietz, opened a business in Schwerin that sold all manner of luxury articles from porcelain and chandeliers to walking sticks, clocks, and gifts for weddings and other special occasions. In later years, Felix Hausdorff and his family maintained close relations with their relatives in

Schwerin, with whom they remained in contact into the period of the Nazi dictatorship. Michaelis Tietz died in 1883 in Berlin, followed a year later by the death of his wife Pauline. Their children afterward established a family foundation in their honor that offered financial assistance to members of the extended family who had fallen into debt through no fault of their own.

Louis and Hedwig Hausdorff also decided to move westward with their two children. Before departing from Silesia, Louis obtained a certificate from the Prussian government in Breslau certifying – for purposes of residency in the Kingdom of Saxony – that he possessed Prussian citizenship as a birth right. By November 1870, the Hausdorffs had registered as residents in Leipzig, where they began a new life, especially for two-year-old Felix, who would spend the next forty years there. Leipzig was a vibrant modern city with a long and rich history, much of which still remained visible during that time.

1.2 Louis Hausdorff in Leipzig

In 1165 Margrave Otto von Meissen granted Leipzig, originally a Slavic settlement called "Libzi," the legal status of a city with the right to a regular weekly market. Soon thereafter, Leipzig grew to become a major medieval city, though its historic importance derives from another royal benefit bestowed more than three centuries later. In 1497 Emperor Maximilian I granted Leipzig the status of an Imperial Trade Fair City, a watershed event that decisively shaped its future development. The trade fairs held there profited, in turn, from Leipzig's favorable geographical location at the crossing of two important and venerable trade routes, the *Via regia* and the *Via imperii.* Prior to the onset of industrial mass production, the Leipzig Messe was always a commodities fair that handled goods from all over the world. Foreign merchants brought their wares to the fair and those that went unsold were stored in vaults, which were owned by Leipzig citizens, until the next fair came.

The street "Brühl" – known mostly as "the Leipzig Brühl" – was the main center for these sales and settlement trades and, in particular, a transshipment point for wool, all kinds of textiles, leather goods, and especially furs. By around 1700, Leipzig had surpassed Augsburg, Nuremberg, and also Frankfurt as a trading and trade fair city. In addition, it developed into a center for the book trade and publishing. The increasing wealth of the city found expression in magnificent Baroque and early Rococo buildings. Leipzig's most prominent boulevard, Katharinenstrasse, led from the Brühl to the market square with its beautiful Renaissance town hall. One side of this boulevard survived the destruction from bombings during the Second World War, so that one can still view the carefully restored buildings along it with pleasure.

In 1870, when the Hausdorff family moved to Leipzig, the city was a major urban center with around 100,000 inhabitants. It was around this time that the Leipzig Trade Fair began to evolve from a commodities fair to a commercial

Figure 1.2: Louis Hausdorff (1843–1896)
Source: Hausdorff Estate, Capsule 65, Nr. 1.

Figure 1.3: Katharinenstraße 29 in Leipzig, Site of Louis Hausdorff's Company.
Source: Sammlung historischer Fotos des Stadtgeschichtlichen Museums Leipzig.

samples fair, and beginning in 1873 the first magnificent trade fair passages were built in the city center. The Brühl was no longer the heart of the fair; it was now increasingly occupied by merchants' shops, company offices, and, as before, numerous hostels and restaurants. Louis Hausdorff had probably already prepared his business activities in Leipzig while he was still living in Breslau. In any event, by 1872 he was co-owner of a linen and cotton goods store originally founded in Leipzig by the merchant Salomon Mosler; the address was Brühl 25. On old photos one can recognize the company sign for Hausdorff & Mosler located there. In 1878, its founder left the firm, and Louis Hausdorff's brother Siegfried took over Mosler's place. In the meantime, Siegfried's family had also moved from Breslau to Leipzig. By the end of 1885 Louis, too, left the company, letting his brother run it for another year before it then closed. Siegfried continued working as a merchant in Leipzig until 1892, at which time he moved to Berlin with his family.

Nothing is known about whether or to what degree the company Hausdorff & Mosler proved to be profitable. The following circumstances, however, reveal that Louis Hausdorff had already acquired considerable capital by 1880; they also suggest that he was not averse to taking risks as an entrepreneur. In that year he bought a plot of land located near the Bayerischer Bahnhof that was owned by Adolf Moritz Paufler. A year earlier, Paufler had entered into a contract with Metzger & Wittig, a well-known printing company. This contract obligated him to build a four-story building on said property within one year, to equip it with a steam engine and transmission system, and afterward to rent it out to Metzger & Wittig. Apparently Paufler had difficulties financing this project. He had to take out several loans, so when Hausdorff bought this property at the end of 1879 the mortgages on it totaled 135,000 marks, against a final purchase price of 180,000 marks. Louis paid 23,000 in cash, while assuming 120,000 marks from the mortgage debts. The remaining 37,000 marks was paid through a claim Siegfried Hausdorff had against Paufler for the sale of another property in Breslau, a claim Siegfried ceded to his brother. To get an idea of the relative value these sums of money had at the time, the average annual income of a worker in the metal industry around 1880 was roughly 850 marks. After concluding the purchase, Louis Hausdorff signed a lease contract with Metzger & Wittig. The yearly rent was 9,000 marks, income Louis could use to make mortgage payments that carried interest rates between 4.5 and 5%. The business relationship between the Hausdorff family and Metzger & Wittig continued even after Louis' death; in fact, Felix Hausdorff's classic book, *Grundzüge der Mengenlehre*, was printed and bound by Metzger & Wittig in 1914.

Louis was also involved with a publishing business. Around 1885 he became a partner in the company Leipzig Court Newspaper Werner & Co. This newspaper, which initially appeared on a weekly basis, later twice a week, dealt with legal questions and reported on court cases of general interest. Louis' dealings with the company lasted only until 1891, and nothing more is known about his involvement with it. He took over a second publishing venture at the end of 1885, however, and in his hands it grew to be highly successful.

In 1883 the businessman Siegfried Hepner had founded an advertising paper, *Der Spinner und Weber*, for the textile industry. On December 31, 1885, he and Louis Hausdorff appeared before the district court in Leipzig to file a declaration stating that they were forming a limited partnership under the company name "Spinner & Weber Hausdorff & Co." at the location Brühl 25, thus Louis Hausdorff's business address. Hepner was only a limited partner in this venture with an investment of 6,000 marks. In the beginning, *Der Spinner und Weber* was purely an advertising platform for textile companies across Europe and even beyond. The advertisements related to machines, raw materials, and accessories of all kinds for the textile industry; there were also job advertisements as well as notices about changes of ownership, bankruptcies, auctions, etc. The publishing house Spinner and Weber, Hausdorff & Co. grew splendidly, so that by 1893 circulation had already reached 10,000 copies. As early as 1887, Louis moved the company headquarters from the old building at Brühl 25 to more luxurious quarters located at Katharinenstrasse 29. Louis' brother-in-law Heinrich Tietz had already acquired another business in this building in 1885, and Louis purchased this business from him three years later.

Der Spinner und Weber gradually evolved from an advertising pamphlet into a reputable trade magazine for the entire textile industry. After his father's death in 1896, Felix Hausdorff became co-owner of the publishing house. In Section 9.4, we describe the circumstances surrounding the forced sale of the company during the Nazi era.

1.3 Leipzig's Jewish Community

The pulsating activity at the Leipzig trade fairs and business life on the Brühl were inextricably linked with the history of the city's Jewish population. The first reports of Jewish settlements in Leipzig date from the end of the twelfth and beginning of the thirteenth centuries, after which treatment of Jews alternated between toleration and violent expulsion. From the mid-fifteenth until the early eighteenth century, Jews were barely a tolerated presence at the fairs. Although the trade fairs depended heavily on the participation of Jewish merchants, especially fur traders from the East, these so-called "fair Jews" were harassed with special levies and forced to accept degrading working conditions. These included carrying a "gelbes Flecklein" (yellow spot) on their clothing at all times or face a fine of 20 thalers, a considerable sum. Furthermore, they could only stay in quarters along the "Judengasse"; only later, after the "Judengasse" was demolished, were they allowed to lodge in hostels on the Brühl.

These conditions improved somewhat during the Napoleonic period and also after the Battle of the Nations of 1813, which was fought just outside Leipzig. Not until 1831, however, when Saxony enacted a constitution, did substantial changes begin to take place. This constitution guaranteed equality, freedom of movement, and individual freedom, including freedom of religious belief. Leipzig, though, only slowly implemented these conditions, and in 1836 its city council rejected granting full equality to Jews. One year later, the state passed a law guaranteeing residential and civil rights to Jews, including the right to build synagogues and to conduct free trade. Yet, once again, Leipzig's civic leaders delayed implementing this law; not until 1839 were Jews granted these civil rights for the first time. In 1846 the city officially approved an Israelite religious community, which one year later had 33 members. In 1855 the community inaugurated its newly built synagogue, though its rabbi, Adolf Jellinek, left already in 1857 to serve in Vienna. He was succeeded by Dr. Abraham Meyer Goldschmidt, previously from Warsaw. Goldschmidt was a representative of liberal Judaism who, following the Enlightenment spirit of Moses Mendelssohn and Gottfried Ephraim Lessing, advocated mutual understanding and respect between Jewish and non-Jewish citizens. His career will be discussed later in Section 2.3, as he also happened to be the grandfather of Charlotte Goldschmidt, who became Felix Hausdorff's wife.

Among the Leipzig Jews, a large number were orthodox immigrants from the east, whose religious inclinations were hardly satisfied by the liberal community, a situation that led to intermittent conflicts. Louis Hausdorff soon played an important role in the affairs of the Leipzig Jewish community, and his influence even extended far beyond it. With regard to the tensions between liberal and orthodox members, he sympathized with the latter as a conservative, though without bowing to orthodox demands calling for a complete rejection of liberal Judaism. In fact, he worked closely with leading liberals in efforts to promote the rights of Jewish communities throughout the German states. This movement began only shortly before Louis Hausdorff's arrival in Leipzig in 1870.

In June 1869, the Israelite religious community in Leipzig, led by Moritz Kohner and Rabbi Abraham Meyer Goldschmidt, took an important initiative toward founding an umbrella association representing Jewish communities throughout Germany. At that time, a synod of liberal rabbis and lay people convened, in parallel with another meeting called by the board of Leipzig's Jewish community. At the suggestion of the Dresden lawyer Emil Lehmann, the Leipzig board also invited representatives from other Jewish communities to discuss their common interests. Lehmann, a member of the local board in Dresden, advocated radical reforms in the sense of liberal Judaism, which promoted integration into German civil society.

Those who gathered in Leipzig forged a general plan for creating an organization representing Jewish communities across Germany. They resolved that such an association would be officially founded if 100 or more communities were to pledge at least 2,000 thalers in support of this undertaking. A provisional board was elected, whose chairman was Moritz Kohner. After many debates about its tasks and goals, the *Deutsch-Israelitischer Gemeindebund* (German-Israelite Association) (DIG) was finally founded in Leipzig on April 1872. Initially, 113 communities declared their willingness to join. An executive committee, chaired by Kohner, formulated the DIG's goals. These included promoting the cohesion of Jewish communities, supporting weaker communities, initiating social projects, and improving the training of religious teachers. Above all, the DIG aimed to represent the interests of German Jewish vis-à-vis the states to which they belonged. From its inception, the membership was dominated by representatives of Reform Judaism.

Despite this auspicious start, with moral and financial support from over 100 local communities, the DIG lost momentum almost immediately after its founding. Less than half of the member congregations sent delegates to the inaugural meeting, and only 35 delegates attended the second meeting in 1875. In 1877, the year Moritz Kohner died, only 25 delegates came to the meeting. Thus, just five years after its founding, the DIG already found itself facing a crisis situation. Kohner's successor as head of the Leipzig synagogue community, the banker Jacob Nachod, was determined to reverse these ill fortunes. Louis Hausdorff had befriended Nachod after settling in Leipzig, and together with the Dresden lawyer Emil Lehmann they began hatching plans to reform the DIG.

In September 1877, Jacob Nachod called an extraordinary meeting of the DIG's executive committee in Leipzig to discuss badly needed organizational reforms. Those present resolved that the executive committee should be reconstituted with a permanent office based in Leipzig to ensure stable working conditions. Nachod was appointed as chairman, and other new committee members were then chosen, including the banker Alexander Werthauer and Abraham Meyer Goldschmidt from Leipzig, Emil Lehmann from Dresden, Moritz Lazarus from Berlin, and four other municipal representatives. The DIG's office remained in Leipzig until 1882, but afterward relocated to Berlin, where the physician Samuel Kristeller took over as committee chair.

The new statutes of 1877 also gave the DIG's executive committee the right to appoint new members. Nachod, as chair, made use of this right at a meeting held on January 2, 1878 by nominating Louis Hausdorff for membership. In doing so, he argued that the committee would benefit from allowing "the decidedly conservative direction to which Hausdorff belongs" to be represented. This motion passed, and Louis Hausdorff remained a member of the DIG executive committee up until his death in 1896. In the committee's obituary, he was remembered as "one of its oldest, most loyal and enthusiastic members." A man of vitality and originality, he worked tirelessly on behalf of the DIG, both by promoting its work in public appearances as well as through his lively engagement in all aspects of the committee's work.

This impression of Louis Hausdorff's important role in the DIG is confirmed by the protocols from committee meetings, which were published every six months. Some of the issues he addressed as a committee member were particularly noteworthy, one being Jewish religious instruction, which was very important to him. Louis contributed to the Herxheimer Foundation, which awarded scholarships to future teachers of religion. Against the misgivings of fellow committee members, he campaigned to support the Ez chajim Jewish School in Breslau. He was personally familiar with the school and its program of religious instruction, which taught children from poorer families free of charge. Social problems were particularly close to Hausdorff's heart. He was also involved in efforts to fund retirement programs for Jewish community officials as well as to memorialize Jewish culture. Together with Abraham Meyer Goldschmidt, he campaigned to purchase the house in which Moses Mendelssohn was born in Dessau. The building was later damaged from bombing during World War II and finally torn down in 1957.

The committee also tasked Hausdorff with overseeing and further developing organizations that offered financial assistance to foreign-born Jews, especially Eastern European immigrants. These funding efforts by German Jews were mainly undertaken as a way to counter negative images of Judaism by reducing public begging and other forms of undesirable social behavior. In 1880, roughly five million Jews were living in the far western part of the Russian Empire known as the Pale of Settlement, which stretched from the Baltic to the Black Sea. Most were Yiddish-speaking Ashkenazi Jews who gathered in *shtetls* (small towns) surrounded by poverty and hostility. After the assassination of Czar Alexander II in 1881, these Jewish communities suffered a barrage of pogroms that caused many to flee westward. A large number of them eventually found their way to the New World (Canada, Argentina, and especially the United States), whereas others settled in Western European countries. Before the Nazis came to power, over 500,000 Jews were living in Germany, though many more lived in Poland and Rumania. The wealthier or better integrated German Jews generally evinced no interest in mingling with their newly arrived Eastern European brethren.

In his work for the DIG, one particular question especially interested Louis Hausdorff. He was keen to ascertain whether Jewish students attending public schools had the opportunity to observe the rites and laws of their religion on the Sabbath and the High Holy Days. He considered this an important question of freedom of conscience, but he was also personally affected as a result of his son Felix's school experiences. On behalf of the committee, Louis conducted an inquiry to determine whether, how, and on what legal basis dispensation from school work was granted to Jewish students on the Sabbath and religious holidays in the various German states; he also asked whether steps might be taken to improve this situation. After conducting this survey, he published the results in the Notices of the DIG, together with his personal recommendations. The committee, as it turned out, did not entirely concur with Hausdorff's views and also expressed reservations when it came to his proposals for improving the situation.

1.4 Confronting Antisemitism

One overriding problem deeply concerned the DIG's executive committee, namely how to combat growing and increasingly aggressive antisemitism in the German Reich. In his study of this problem, Helmut Berding has described the situation the Jewish minority then faced:

> In the decades before the founding of the German Empire, antisemitism, while decreasing on the surface of political events, took root in the

> deeper layers of peoples' mentalities. Hypostasis and demonization marked the emergence of a new image of Jews, an image widely disseminated through works of literature. The outlines for what would become a coherent racial ideology were already evident. It blamed the Jew for all conceivable crises in economy, politics and culture, while channeling aggression toward the Jewish minority. A dangerous hostility built up that could suddenly be activated in times of crisis. [Berding 1988, 84]

Those times of crisis were not long in coming: after the euphoria of the founding era (Gründerzeit) came the sudden fall in 1873, the beginning of a nine-year-long economic depression. This, however, was not caused by the stock market crash but rather by a boom in overproduction. At the height of the crisis, in January 1878, the Protestant court preacher Adolf Stoecker founded the Christian Social Workers' Party. The centerpiece for his broad-based agitation was political antisemitism that made Jews responsible not only for all the negative effects of capitalism but also for the socialist counter-movement. This alleged "Jewish being" was, in Stoecker's words "a drop of poison in the heart of our German people. If we want to be healthy, we have to get rid of the poisonous drop of the Jews from our blood" [Berding 1988, 84].

In 1879, the journalist and radical anti-Semite Wilhelm Marr published a pamphlet entitled *The Victory of Judaism over Germanism. From a Non-denominational Point of View* [Marr 1879]. Marr's hostility towards Jews had nothing to do with religious differences, as was partly the case with Stoecker, but rather he based this on alleged qualities of a specific Semitic race. Members of this race, according to Marr, possess all the negative characteristics that were later over time attributed to Jews. He ended his tirade by painting a vision of a future world dominated by the Jews – and this meant the end of "Germania." In September 1879, Marr and other activists founded the League of Anti-Semites, the first political organization based on antisemitism. Later Marr became an increasingly marginal figure in this movement, although in 1888 he worked on a new edition of the *Antisemiten-Katechismus* (Anti-Semite Catechism). The first edition was published one year earlier in Leipzig by the antisemitic author and publisher Theodor Fritsch, and the book later met with stunning success. In 1907 the 26th edition came out under the new title *Handbuch der Judenfrage* [Fritsch 1907]. Two readers who especially admired its malicious propaganda were Adolf Hitler and Heinrich Himmler; in 1944 the 49th edition rolled off the press.

Among the numerous public figures who rode the waves of German antisemitism during the 1870s and 1880s, a special place was occupied by Heinrich von Treitschke, a distinguished and well-known historian and political publicist, who also held a seat in the Reichstag. In 1879 he published an essay entitled *Unsere Aussichten* (Our Prospects) [Treitschke 1879], which began with reflections on foreign policy before discussing the domestic situation in the German Reich. Treitschke pointed to "economic hardship," "many false hopes" and "the sins of the early days (Gründerzeiten)" as factors that had brought forth a "wonderfully powerful excitement in the depths of our people." As a symptom of this excitement he noted "the passionate movement against Judaism," and he then raised the question whether "these outbursts of deep, long-lasting anger are only a fleeting swell." His answer:

> No, the instinct of the masses has indeed correctly recognized a serious danger, a highly worrisome damage to new German life; it is no empty phrase if one speaks today of a German Jewish question.

What Treitschke demanded of the Jews was that they abandon their Jewish identity and completely assimilate, a step the majority had so far refused to take:

> What we have to ask of our Israelite fellow citizens is simple: they should become Germans, and simply feel like Germans [· · ·] because we do not want millennia of German civilization to be followed by an age of mixed German-Jewish culture.

He went on to list typical symptoms of non-Germanic manners and mixed German-Jewish culture: a "dangerous spirit of arrogance in Jewish circles," "heavy complicity in the crass materialism of our day," and the "over-representation of Judaism in the daily press." Treitschke was not a conventional racial anti-Semite like Marr or later Fritsch and Eugen Dühring; he fought with a finer blade and distanced himself from more brutal reactions, for which he nevertheless showed understanding. For him, it was important to make antisemitism acceptable in broad circles of the bourgeoisie, especially among its highly educated members:

> If one gazes over all these relationships – and how much more could be said! – it appears that the loud agitation of the moment is only a brutal and hateful, but natural reaction of Germanic feeling toward a foreign element that has acquired too much space in our lives. [· · ·] From circles of the highest education, among men who would reject any thought of clerical intolerance or national arrogance, there sounds today as if from a single mouth: the Jews are our misfortune!

Treitschke achieved his goal, and far more. His saying "The Jews are our misfortune" ("Die Juden sind unser Unglück") later became the subtitle for the notorious Nazi propaganda newspaper *Der Stürmer*. The longstanding debates over his essay, which went down in history as the "Berlin antisemitism dispute" ("Berliner Antisemitismusstreit"), animated broad circles of the bourgeois public and contributed to the fact that antisemitism became a *cultural code* in the German Empire. Treitschke's views especially fanned the flames of antisemitism that swept through student organizations at this time.

The DIG took on the task of defending itself against this growing and highly aggressive antisemitism. The minutes of the monthly committee meetings during the years 1878–1880 show that the committee dealt with the "defense question" at almost every meeting. Since the authorities – especially after the conservative political turn of 1878/79 – often gave the anti-Semite movement free rein, the

DIG found it questionable to take the previously tried and tested methods of defense, namely through petitions or complaints to the legislative assemblies of the empire and the individual states. In April 1881, a report on the committee's activities against the anti-Semitic movement acknowledged that its options were rather limited. It counseled avoiding public discussions of the problem in the press so as not to stir up even more anti-Semitic agitation. Nor were there many successful attempts to take legal action against particularly aggressive anti-Semitic brochures or newspaper articles.

In 1879, in an effort to "refute the slanderous charges on the part of the anti-Semitic agitators," the DIG distributed two books that dealt with Marr's writings. They also issued a pamphlet in June of that year entitled "Has Judaism promoted Usury?" From the positive side, members were urged to participate in community life and to discipline themselves through learning and studying the history of Judaism as a means toward overcoming religious and racial hatred. The organization attached great importance to writings that extolled the history of the Jewish people, including works by the famous botanist and co-founder of cell theory, Matthias Jacob Schleiden. In one such work, Schleiden gave a detailed description of the martyrdom suffered for centuries by the Jews in Christian Europe during the Middle Ages. Finally, in response to ongoing anti-Semitic agitation, the community encouraged all fellow believers to adopt exemplary civil and civic-minded behavior, acting with modesty and restraint, while conducting business affairs with the utmost conscientiousness.

Within the executive committee of the DIG, a variety of views were expressed about the stance it should take in this increasingly difficult situation. Minutes from these meetings reflect that Louis Hausdorff often took the opportunity to voice dissenting opinions. Thus, at a meeting on September 3, 1878, the committee chairman urged decisive action against anti-Semitic riots and abuse inculcated by Stoecker's Christian Social Party during the election campaigns for the Reichstag. To this, Hausdorff responded that the DIG should simply ignore "attacks against our fellow believers [···] at meetings of the Christian-Social Party [···] because this party stands outside the realm of any serious discussions." Hausdorff's comments at another meeting, held on November 4, 1879, went in a similar direction. The question under discussion was how best to counter the anti-Semitic agitation of hate-mongers like Marr. Two possible actions were proposed: either to submit a protest to the Saxon and Prussian Ministries of Justice or to send a memorandum to the Emperor or Chancellor. Hausdorff thought both proposals were inappropriate:

> Mr. Hausdorff warns against overly hasty steps. It seems as if one is now expecting a great and brilliant deed from the DIG. Yet if the matter were gauged by its true value, then this request from outside could well be calmly rejected. The elections are over and the calm will return, which is why he recommends "doing nothing."

Hausdorff finally agreed to support submitting a complaint to the Saxon Ministry of Justice, but this was voted down in favor of sending a memorandum to the Chancellor, which Hausdorff opposed:

> Herr Hausdorff sees no great danger in all the anti-Jewish agitation and consequently cannot attach to it the same significance found by the committee. Calling on the Chancellor or even the Emperor himself will only exaggerate the matter and give it an importance it does not deserve. The existing laws protect us sufficiently; he is therefore against taking such an approach.

1.5 Louis Hausdorff's Religious Views

If from 1878 to 1880 the position of the DIG regarding the Jewish question was increasingly cautious, the position of Louis Hausdorff was even more careful. On many occasions, he advised waiting, exercising caution and restraint, and relying on confidential channels. In his publications, on the other hand, he revealed himself to be a stubbornly argumentative conservative Talmudic scholar. For example, he took a strong position with regard to burial practices.

In 1883 the liberal Jewish community in Berlin decided to allow the burial of non-Jewish spouses from mixed marriages in the Jewish cemetery. The Jewish community in Leipzig followed suit with a similar decision at the beginning of 1884. In response, the ultra-conservative Jewish magazine *Jeschurun* published a letter heavily criticizing the Leipzig decision. During the 1880s, there was an extensive *halachic* discussion about the funeral question in which Louis Hausdorff also participated. In 1884, the co-founder and co-editor of the magazine *Beth Talmud*, the distinguished Talmudic scholar Meir Friedmann, published an article in Hebrew entitled (in English translation): "The Funeral of a non-Jew on a Jewish Field of God. Allowed according to Talmudic Law." As the title suggests, Friedmann came to the conclusion that it is permitted to bury a gentile in a Jewish cemetery. He even claimed that this was regular practice at the time of the emergence of Mishna, and that Jews and Christians were buried in the same cemetery, if not in a common grave. Hausdorff responded sharply (also in Hebrew) to Friedmann's article, provoking a counter-response by Friedmann in *Beth Talmud*. He accused Hausdorff of allowing the forbidden and forbidding the permitted. Hausdorff complained that Friedmann had insulted him while pretending to hold the wisdom of the world.

It is noteworthy that the discussion about the funeral question continues to this day. Orthodox rabbis strictly reject the burial of non-Jews in Jewish cemeteries, sometimes expressly with the intention of preventing the increase in mixed marriages. Reformed rabbis take a differentiated position. It is interesting to see that a relevant response from American rabbis in 1983 recapitulated the *halachic* discussion of the 1880s, especially the exchange between Friedmann and Hausdorff. Leo Baeck, the famous rabbi and namesake of the Leo Baeck Institute, wrote in a letter to the liberal rabbi Peter Levinson in 1951 about the funeral question:

> For decades the rabbinate in Berlin had approved non-Jewish spouses, men and women who had had such a wish, to be buried next to the Jewish spouse in our cemetery in Weissensee. The cases in which husbands bravely and willingly showed devotion to their Jewish wives in difficult times are particularly to be assessed in this sense, and the funeral should hardly be refused. (Quoted in N. Peter Levinson: Die liberale Halacha, *UDIM, Zeitschrift der Rabbinerkonferenz in Deutschland*, XIX (1999), 139–150, p. 143.)

Louis Hausdorff could not have foreseen that his son Felix would marry a baptized Christian whose grandfather was a rabbi. Their only child Lenore married a non-Jew, who stood by her "bravely" and with great sacrifice, saving her from being murdered during the most trying of times.

Louis Hausdorff not only contributed to the *halachic* discussion through his writings but also by taking part in lively debates concerning questions of biblical translation. In 1894, his comprehensive treatise on the history of the Targumim, according to Talmudic sources ("Zur Geschichte der Targumim nach talmudischen Quellen"), appeared in *Monatsschrift für Geschichte und Wissenschaft des Judentums* (Monthly Journal for the History and Science of Judaism). A Targum (plural Targumim) is a translation of a biblical text, mainly into Aramaic. Since the Persian period, Hebrew had been supplanted by Aramaic in many parts of the Near East, so that during synagogue services after the reading of the Torah in Hebrew there followed an oral translation in Aramaic. Hausdorff was primarily concerned with the Targum Onkelos as well as a Greek biblical translation attributed to a proselyte named Aquilas.

In studying Talmudic texts in the Targumim, it had long been observed that the stories told about Aquilas in the Palestinian Talmud were similar to those about Onkelos found in the Babylonian Talmud. Already in the Middle Ages, this led to the identification of these two figures. In the 1570s, the eminent Jewish humanist Azaria de Rossi undertook to resolve this question. He came to the conclusion that the Greek translation attributed to Aquilas and the Aramaic Targum attributed to Onkelos were translations made by two different men. Long afterward, during the eighteenth and nineteenth centuries, a whole range of Christian and Jewish authors debated this problem anew, arguing more or less convincingly either for or against the identity of Aquilas and Onkelos. In his contribution to this debate, Louis Hausdorff came to the conclusion that Aquilas and Onkelos were, indeed, one and the same person. This result, he emphasized, was "in complete contrast to de Rossi's view, which has been adopted by all men of science." In the end, Hausdorff thought he could even sketch a curriculum vitae of Aquilas that enabled him to date the completion of the translations in question: the Greek translation of Aquilas between AD 55 and 60, the Onkelos Targum between AD 75 and 80.

Despite Hausdorff's article, the debates over Aquilas and Onkelos continued and they have not yet come to a complete rest. Hausdorff's analysis was based almost exclusively on the interpretation of Talmudic texts, which in his opinion were related to Targumim. His rival, Meir Friedmann, also discussed the problem examined by Hausdorff. In a long article, also based on Talmudic sources, he confirmed de Rossi's results, though "with minor deviations." Both Hausdorff and Friedmann ignored the historical-critical method that had already been developed by various Old Testament scholars. This proved to be problematic, as shown by the fact that these two excellent Talmud scholars reached precisely opposite conclusions regarding the identity of Onkelos and Aquilas. Today, a multitude of different sources and methods are used for historical research into the origin, translation and traditions of Old Testament texts; the one-sided limitation of studies confined to the Talmud was already scientifically unsustainable back then, and even more so today. However, it remains in the nature of such things that, even if all available modern methods are used, uncertainty in answering important questions, e.g. on dating, cannot be avoided.

Louis Hausdorff thought that his method, based on the ingenious analysis of rabbinical tradition, would lead to certain results. He presented his insights with great self-assurance, calling them "evident" or "clearly proven" and using words like "absolutely," "undoubtedly," "without question," and "impossible." He was a quarrelsome private scholar in the tradition of Talmudic discourse who mixed high demands with high self-esteem. His son had an altogether different personality.

1.6 School Years in Leipzig

In 1874, when Felix Hausdorff turned six, he entered the local elementary school, which pupils attended for four years. Afterward they could either pursue their further education at a six-year high school that offered instruction in modern foreign languages or else attend a traditional Gymnasium, where heavy doses of Latin and Greek were taught. The building of this elementary school, erected between 1837 and 1839, was of classicist design and located where Pfaffendorfer Strasse and Lortzingstrasse meet the inner city ring. The Hausdorffs happened to live on Pfaffendorfer Strasse, which meant little Felix only had a short walk to reach the school. The fact that in 1874 neither he nor his parents held Leipzig or Saxon citizenship apparently posed no problem for school entry. In any event, in January 1875 Louis Hausdorff sought Leipzig citizenship rights as a merchant, while simultaneously applying for himself, his wife and their three children to become subjects of the state of Saxony. His application was granted in the same month, and in March Louis Hausdorff took the official oath to become a citizen of Leipzig.

While we have no concrete information about how Felix Hausdorff was raised as a child, it can be safely assumed that he had a strict religious upbringing. In a presentation to the executive committee of the German-Israelite Community

(DIG), his father stated that the focus of Judaism was not on preaching or church services but rather on religious family life. The following entries from his son's book of aphorisms, published in 1897, no doubt reflect how Felix Hausdorff as an adult looked back on this own upbringing:

> Whoever invented the fable of childhood bliss forgot three things: religion, schooling, and adolescent development. [H 1897b, 254, VII: 348]

And with regard to contemporary pedagogy, he wrote:

> But the method still means today: exterminate, curb, cut, prevent, restrict, forbid – it is a fundamentally negative, depriving, and prohibitive method of educating, pressuring, punishing, and abolishing instead of creating, amputating in place of healing. [H 1897b, 62, VII: 156]

In terms of his religious upbringing, this experience produced precisely the opposite effect as that which his father had hoped for; as an adult, his son no longer practiced the Jewish faith. Felix Hausdorff became an agnostic, critical of Judaism as well as Christianity; however, he always remained a member of the Jewish community and met his pecuniary obligations conscientiously. He also never considered baptism, though this would have helped him to advance his career.

On March 1, 1878, he was admitted to the Nicolai Gymnasium, the former Schola Nicolaitana, a humanistic high school with a long and proud tradition as the oldest urban secular school in Germany. Its history was closely connected with the founding of Leipzig University in the early fifteenth century. Founded on St. Nicholas' Day in 1512, the Nicolai school gained new quarters in 1568 after a fire destroyed the original. The new building was redesigned in Renaissance style in 1596/97. Today it stands near the St. Nicholas Church (Nikolaikirche) as one of the touristic attractions of the old city. The Nicolai Gymnasium moved in 1872 to a new building located to the southeast in Johannistal at the crossing of Königstrasse and Stephanstrasse, about a half-hour's walk from Felix Hausdorff's home. (Today Königstrasse is called Goldschmidtstrasse in honor of the feminist and social pedagogue Henriette Goldschmidt, grandmother of Felix Hausdorff's wife Charlotte Goldschmidt.)

Up until 1716, instruction at the Nicolai school had been exclusively in Latin; only afterward did the German language find its way into the curriculum. After the turn of the nineteenth century, neohumanism had a growing influence on the school. Neohumanism aimed to reveal the intellectual treasures of classical antiquity to the younger generation, instilling in them a love for classical cultures. This was still the case at the time Felix Hausdorff attended the school, although the predominant role of traditional language teaching had by now been questioned in public discussions. At the newer secondary schools, the curriculum was weighted toward mathematics, natural sciences, and contemporary languages.

At the Nicolai Gymnasium, teachers remained steadfastly committed to neohumanist ideals. For example, the new rector, Karl Mayhoff, dedicated his inaugural speech on October 25, 1884 to defending the traditional ideals of education grounded in classical languages. According to Mayhoff, this tradition "sets the firm foundation for an *aesthetic* education, which is all the more valuable the more the restless confusion and the spreading aberrations of taste in modern production are likely to disturb judgment." Classical literary taste creates "a firm hold on the style of the ancients, who are rightly credited with sober simplicity, plastic clarity, and calm grandeur in their artistic revelations of the true and beautiful." Beyond that, traditional language instruction promotes "the educational task of the school, since it can purify and ennoble love of the fatherland, consolidating and deepening the monarchical spirit, that firm haven of legal order and civil freedom, in a way that supports the cultivation of religious and Christian sensibilities." Such teaching also promotes "a truly historical sense, ... the unmistakable characteristic of all genuine *humanitas*," as opposed to the superficiality of learning based exclusively on the acquisition of knowledge. The classics thus strengthen the sense of scientific truth in a morally ennobling manner.

The upper classes of the Nicolai Gymnasium always attended such festive events, like the introduction of a new rector, which makes it likely that Felix Hausdorff, then nearly 16 years old, heard Mayhoff's speech. What might have gone through his mind back then? What we know is that the immense tension between such institutionally conveyed ideals and values, on the one hand, and the complex social, cultural, and intellectual upheaval at the turn of the century on the other, found deep individual and consciously contradictory expression in his biography and especially in his literary works.

The venerable tradition of the Nicolai Gymnasium was a source of considerable pride, and that for good reason. In the course of its long history its pupils included such famous names as the universal scholar Gottfried Wilhelm Leibniz, the philosopher Christian Thomasius, and the composer Richard Wagner. Besides Felix Hausdorff, the school also produced a second eminent mathematician, Hermann Hankel, whose work had importance for Georg Cantor's theory of point sets and thus indirectly for set-theoretical topology, the field associated with Hausdorff's most important mathematical achievements.

Before taking a closer look at Hausdorff's experiences as a student, first a few words about the overall structure and organization of the Nicolai Gymnasium. In modern times, the German Gymnasium emerged as the most advanced form of three types of secondary schools, followed by the Realschule and the Hauptschule. All were exclusively for boys, whereas special secondary schools for girls terminated with the tenth grade. Boys who successfully completed their education at a Gymnasium, which ended with the Oberprima (corresponding to thirteenth grade), qualified to attend a German university, already a high distinction in Hausdorff's time. The German Gymnasium thus aimed to promote academic learning, like the British grammar school system or private prep schools in the United States.

By the end of the 1860s, the number of students at the Nicolai Gymnasium grew sharply, so that only a decade later the student body had already doubled. Throughout the 1880s it consistently exceeded 500 students, making it necessary to introduce parallel classes in 1878. In his inaugural speech, Rector Mayhoff remarked that he had been tasked with running an institution as large as two normal-sized schools. In order to gain an impression of its curriculum, one can consider the requirements for the senior class as set forth in the new teaching regulations from 1882. These called for the following distribution of hours per week for these subjects: Latin 8, Greek 6–7, German 3, history 3, religion 2, French 2, Sport 2, Mathematics 4, Physics 2; there were also these elective courses: Hebrew 2, English 1–2, singing 1, drawing 2, stenography 1. Classical language instruction thus took up about 45% of the total compulsory teaching time, and the level of difficulty was correspondingly high. Students read Greek works by Herodotus, Xenophon, Thucydides, Demosthenes, Plato, Homer, Hesiod, Aeschylus, Sophocles, and Sappho as well as Latin texts by Caesar, Livy, Sallust, Tacitus, Ovid, Vergil, Catullus, and Horace.

The teachers at the Nicolai Gymnasium were exceedingly well trained. Many of them had attended an elite Gymnasium in Saxony, such as the Kreuzschule in Dresden, the Johanneum in Zittau, or even Leipzig's Nicolaitana. Several also held doctoral degrees and participated in academic life through their own publications. Those who were in charge of a class held the title of *Ordinarius*, just like full professors at a German university. This group taught four principal subjects: Latin, Greek, German and history, again reflecting the neohumanist educational tradition at the Nicolai school. In fact, there was a strict rank order within the teaching staff, as documented in the yearly programs issued by school. A teacher's place within the hierarchy depended not only on the subjects taught, but also on his suitability, qualifications, and performance. The rector stood at the top of this pyramid, followed by the senior teachers, listed by rank, then came the provisional senior teachers, followed by those who taught gymnastics, singing, and stenography, the assistant teachers, and finally candidate teachers. Those in the upper ranks often held the title of professor. In the school year 1886/87, when Hausdorff was in the Oberprima, his class teacher (Ordinarius) was the Rector, Prof. Dr. Mayhoff. His mathematics teacher was ranked third in the school's hierarchy as the "1. Mathematicus," Prof. Dr. Adelbert Gebhardt, whom he also had for two earlier school years. Gebhardt was surely an excellent teacher, as mathematics instruction at the Nicolai Gymnasium was considered to be among the best available in Saxony. In 1895, the King of Saxony bestowed the Albrecht's Medal of the Knight's Cross First Class on Gebhardt for his services to the Nicolai Gymnasium.

From the school's student directories one can also gain an idea of the social standing and religious affiliation of the families who sent their sons to the Nicolaitana. In 1879, Hausdorff's class was composed of 45 students, from which some 40% had fathers engaged in the private sector as merchants, booksellers, publishers, manufacturers, or bankers. About 20% held professional positions as teachers, lawyers, doctors, or theologians. Another larger group were those whose fathers were white-collar workers or minor officials, and none came from a working-class background. Of the 511 students who attended the Nicolai Gymnasium in 1885,

462 were Lutheran, 19 Protestant-Reformed, 22 Jewish, and 6 Catholic. There were only two non-denominational students (referred to as dissidents): Karl and Theodor Liebknecht, the sons of the socialist politician Wilhelm Liebknecht.

The strict ranking of the teaching staff was also mirrored in the ranking of the students in each individual class. This was based on overall grades, but also on conduct, as measured by the extent to which a student's behavior complied with the strict rules of the school. Its regulations from 1878 contained eighteen paragraphs devoted to disciplinary matters, as for example, §11 stated:

> If a student believes that a teacher has wronged him, he must present his justification in the teacher's home, but always in a modest tone. Any kind of objection or pleading in front of the assembled class, or even in front of witnesses, is strictly forbidden.

Student behavior outside the school was also controlled. Participation in public dance entertainments was "absolutely prohibited to students," and they were only allowed to visit restaurants, cafés, and entertainment locales in the company of their parents or other persons of respect. Pupils caught disobeying the school's rules were punished by physical detention.

Felix Hausdorff left no personal testimonials with reflections on his school days. Nevertheless, information about him can be found in the school's annual reports, which record some of his early academic achievements. The Nicolai Gymnasium had a tradition of honoring between 5 and 10% of its best students with special awards, either as books or in monetary form. This took place twice each year, once at Easter and then in September, either on Sedan Day (celebrated on the 2nd of September to commemorate Germany's victory over France in the 1870 Battle of Sedan) or at the Feast of St. Michael on the 29th. From the school's reports, we know that Hausdorff received a book award every year for five consecutive years at the Nicolai Gymnasium, followed by a book premium with a cash bonus of 40 marks during 1886–87, his final year as an *Oberprimaner.* School records of student rankings for the last six years, starting with the *Untertertia,* are also available. Starting then, Hausdorff was ranked second in his class during his first two years, and afterward he rose to the top. When he left the school in Easter 1887, he was the only one among the 33 students in the two graduating classes to receive the highest overall grade of 1.

During his school days, Felix already distinguished himself as a literary talent. One of the special occasions celebrated at the school was the birthday of King Albert of Saxony. This took place on April 23, 1885 in the auditorium of the Nicolai Gymnasium, as reported in a local newspaper. On such occasions, it was a special honor for a student to recite a poem for the assembled audience. On that day, after the singing of a motet by Haydn, Hausdorff read a poem he had composed in Latin verse. The next year, he twice took the stage to recite poetry, first in March at a graduation ceremony that also celebrated the birthday of his Majesty the Emperor Wilhelm, and then in September for the celebration of the Day of Sedan. The school's annual report contains an account of the latter event:

Figure 1.4: Class Photo showing Felix Hausdorff standing on the far right, 1887
Source: Stadtarchiv Leipzig, Nicolaischule, Nr. 280.

> The ceremonial act ... in the school auditorium opened with Luther's prayer "Give us peace graciously." There followed readings of poems in German and Latin composed by the seniors Alfr. Marschner, Isaak Belmonte, and Felix Hausdorff, declamations by some younger pupils celebrating this day, and finally a speech by Dr. Krieger, who linked the uplifting contemplation of the unforgettable deeds of 1870 with an instructive account of prehistoric Germanic life, especially the role of horses, those faithful comrades in war of the ancient Germans.

One of the requirements of the final exam before graduation was to compose an essay in German. Hausdorff's topic was the following question: "What were the main factors that shaped the young Goethe's views?" In Latin, students were required not only to translate but also to write a Latin essay in class. The topic that Hausdorff had to write about was: "Cupidius quam verius Cicero dicit res urbanas bellicis rebus anteponendas esse" (freely translated: "it reflects Cicero's interest more than the truth when he asserts that matters of public welfare take precedence over those of warfare.")

Compared with such high standards in Latin, those in mathematics were relatively modest. During Hausdorff's final year, the topics covered included combinatorial theory and the binomial theorem for exponents with natural numbers, continuation of solid geometry, the synthetic treatment of the parabola and ellipse, and a general review, to be covered in four hours each week. It would appear unlikely that a student of Hausdorff's abilities would have felt at all challenged or stimulated by topics such as these. On the other hand, his mathematics teacher, Adelbert Gebhardt, lectured and published not only on pedagogical matters. His interests extended to problems in mathematics, physics, and epistemology, which suggests that he may well have encouraged his best student to study mathematical and scientific literature that went well beyond the subject matter he taught. If so, then Hausdorff's epistemological interests may also have received an early impulse from Gebhardt. At the school's graduation ceremony on April 23, 1884, Gebhardt gave a speech in which, citing Hermann von Helmholtz, he described the physical properties of optical instruments, which impose limitations on the accuracy of our visual perceptions when using a microscope or a telescope. He drew the conclusion that, "the deeper one dives into science, the more one becomes conscious of human barriers and boundaries." Felix Hausdorff surely heard this lecture at that time; perhaps it even prompted him to think about the space problem for the first time.

Chapter 2

Hausdorff's Student Years and Short-lived Astronomical Career

During his senior year at the Nicolai Gymnasium, Felix Hausdorff probably had a difficult time deciding which course of study he should pursue at the university. His literary and linguistic talents were well nurtured, thanks to the school's strong neohumanistic curriculum, but these fields were perhaps already eclipsed by his passionate interest in mathematics and the natural sciences, despite having had less exposure to these subjects during his school years. Alongside this broad spectrum of possibilities, however, stood another, namely the world of music. For Felix was a talented pianist and a fervent devotee of the operatic works of Richard Wagner, whose cult of worshipers celebrated his memory each year at festival performances in Bayreuth.

Many years later, one of Hausdorff's students, Magda Dierkesmann, was aware of his musical talents and she later made mention of his youthful dreams in this direction. Dierkesmann visited the Hausdorffs often when she was a student in Bonn from 1926 to 1932, and in 1967 she reported that her former teacher had once contemplated a career in music. "His multifaceted musical talent was so great," she wrote, "that it was only due to his father's urging that he gave up his plan to study music and become a composer. He then studied mathematics in Leipzig, Freiburg and Berlin."

In all likelihood, Hausdorff's dreams of a life as a composer quietly passed away toward the end of his days at the Nicolai Gymnasium. In the school's annual report for the year 1886–87, one reads under Hausdorff's name that he planned to study natural sciences in the future. Probably he had no clear direction in mind at that age, as his early university studies suggest that it took some time before he settled on physical astronomy as his principal field of research. In following his early years at the university, we will gain an impression of the circumstances that led Hausdorff into that particular scientific arena.

E. Brieskorn, W. Purkert, *Felix Hausdorff*, Vita Mathematica 21,
https://doi.org/10.1007/978-3-031-52135-5_2

There is, however, one surviving clue of his early interest in astronomy, the subject that he eventually chose as his main field of study during his last university semesters. By a happy coincidence, Egbert Brieskorn was able to acquire a book Hausdorff probably heard about during his school days. This was a set of lectures on astronomy by Professor H.W. Brandes, published in two volumes by the Leipzig firm Göschen in 1827. The copies Brieskorn found bear the inscription "F. Hausdorff" in pencil with lettering that deviates somewhat from Hausdorff's later handwriting. This may well be an indication that he signed these two volumes as a Gymnasium student; surely, he will have read them as well.

2.1 University Studies

On April 18, 1887, Felix Hausdorff enrolled to study mathematics and natural sciences at Leipzig University. At this time, Leipzig probably had more mathematicians on its faculty than any other university in Germany with three full professors, one honorary professorship, and two associate professors. Leipzig's fame was long associated with August Ferdinand Möbius, an astronomer and mathematician who became director of the observatory in 1848. Although he did pioneering work in geometry, his name is mainly remembered only in connection with a special topological object, the Möbius band. Leipzig's senior professor was Wilhelm Scheibner, a versatile scholar whose interests spanned from theoretical astronomy, mechanics, optics, potential theory, and elliptic functions to invariant theory and number theory. He was mainly known as a textbook author, whereas his research work had little lasting impact, nor did he exert any lasting influence on Hausdorff. Scheibner's main importance, however, came from his contributions to building the mathematics program in Leipzig, which benefited greatly from his clever appointment policy.

When Möbius died in 1868, Scheibner succeeded in appointing Carl Neumann, son of the famous Königsberg physicist Franz Neumann and at the time professor of mathematics in Tübingen. In that same year, Neumann made a lasting contribution to higher mathematics by co-founding *Mathematische Annalen* with his friend Alfred Clebsch. As natives of Königsberg, both were eager to create a new journal that could compete with the older *Journal für die reine und angewandte Mathematik*, founded in 1826 by August Leopold Crelle and long dominated by mathematicians in Berlin. Clebsch was appointed to Bernhard Riemann's chair in Göttingen in 1868, whereas Leipzig was home to the publishing firm of B.G. Teubner, making these two cities natural allies in this undertaking. The *Annalen* began to flourish as a leading international journal in the late 1870s, when Felix Klein, a protégé of Alfred Clebsch, served as its principal editor. Soon afterward, Leipzig became an important station in Klein's ascending career trajectory.

Wilhelm Scheibner had long hoped to gain funding for a new professorship in geometry in Leipzig. This field had developed rapidly over the course of the century and, following the death of Möbius, the faculty no longer had a competent authority representing this discipline. For years, Scheibner lobbied to fill that gap before he finally succeeded. In 1880, the Ministry in Dresden created a chair in geometry and quickly succeeded in recruiting Felix Klein to occupy it. One year later, Klein founded the mathematics seminar, which soon drew a number of talented young mathematicians to Leipzig. Its reputation as a leading center for geometry continued after Klein's departure for Göttingen in 1886, as he was able to arrange for the appointment of his Norwegian friend, Sophus Lie, to succeed him. Since foreigners were rarely ever offered positions at German universities, Klein's coup stirred up considerable controversy at that time. Lie eventually became the most influential member of the mathematics faculty in Leipzig up until he resigned in 1898.

These three mathematicians – Scheibner, Neumann, and Lie – were the full professors, though Hausdorff only took courses offered by the latter two. One might have imagined that the novelty of Lie's theory of continuous groups would have led young Felix to pursue his studies in this direction, but possibly owing to his strong interest in the natural sciences this was not the case. Carl Neumann – an expert in potential theory, mathematical physics, and complex functions – may have exerted a stronger influence on him. Neumann founded a tradition of excellence in mathematical physics that has continued in Leipzig up to the present day. One of his principal concerns was clarifying the principles underlying physical theories, as in his inaugural lecture, held on November 3, 1869. On that occasion he spoke about the fundamental principles of classical mechanics, such as the Galilean principle of inertia, which Newton had attempted to establish by appealing to the notion of an absolute space. Neumann instead famously postulated the existence of an absolute inertial frame of reference corresponding to a hypothetical rigid body, called the body Alpha. Hausdorff chose a related topic for his own inaugural lecture, "The Space Problem," delivered on July 4, 1903, when he joined the Leipzig faculty. In the meantime, he had carefully studied Neumann's earlier inaugural lecture, which he referred to in his presentation.

Still more important for Hausdorff's early career was Leipzig's honorary professor, Adolf Mayer, with whom he established friendly relations almost from the beginning of his studies. Although seldom remembered after his death in 1908, Mayer was an important mathematician, not only for his work as an analyst but also as a key figure in promoting the work of Sophus Lie long before his arrival in Leipzig. Moreover, after the sudden and wholly unexpected death of Clebsch in 1872 left the future of *Mathematische Annalen* in doubt, Mayer played a major role in ensuring its long-term survival. After 1875, when Klein succeeded Otto Hesse in Munich, he and Mayer took the *Annalen* under their wings and within little time made it one of the leading mathematics journals worldwide.

Adolf Mayer grew up in Leipzig as the son of a wealthy businessman, circumstances that perhaps account for the fact that he held only an honorary professorship during the 1880s and only became a regular full professor in 1890. Mayer's main fields of expertise were the theory of first-order partial differential equations, calculus of variations, and mechanics. C.G.J. Jacobi had shown that the integration of any system of first-order partial differential equations can be reduced to the integration of complete systems of linear homogeneous equations. Mayer went still further by showing that such complete systems could be reduced to integrating a system of ordinary first order differential equations. With this reduction, the Jacobian method was considerably simplified, so that today one speaks of the Jacobi-Mayer integration method for systems of partial differential equations of the first order.

Mayer's research in the calculus of variations was also partly inspired by Jacobi's studies of the second variation in Lagrangian problems with single integrals. He found a particularly simple normal form for the second variation and derived necessary and sufficient criteria for optimality using the so-called Mayer determinant. This marks the beginning of the theory of conjugate points on extremals. Mayer was also one of the first to pursue Hilbert's program for developing a theory of extremal fields (Mayer extremal groups, Mayer field), and to publish work that anticipated ideas in the important theory of optimal control (Kuhn-Tucker conditions in optimization theory). Mayer's work in mechanics was also closely connected with the calculus of variations, as witnessed by his inaugural lecture on the history of the principle of least action. Adolf Mayer was a very popular teacher with a reputation for the clarity of his lectures. He also took a personal interest in his students, as became evident in Hausdorff's case already in his second semester.

Mayer's expertise in differential equations made him a natural ally of the Norwegian mathematician Sophus Lie, one of the most important figures of the era. Lie came to Leipzig as Klein's successor in 1886. They had first met in Berlin during the winter of 1869/70 when Lie was on a travel grant. Both had broad interests, but their mutual interest in Julius Plücker's line geometry proved to be a decisive factor cementing their friendship. In the spring of 1870 they met again in Paris, parting ways suddenly in July when the Franco-Prussian War broke out. Lie returned to Christiania (today Oslo) in the fall and completed his doctoral thesis in 1871; the following year he became a professor there.

One often hears the names of Klein and Lie in connection with the theory of continuous groups, although this was hardly the focus of their collaborative interests. Much of their research dealt with geometrical transformations, particularly mappings that preserve tangency relations, which became the basis for Lie's theory of contact transformations. He discovered a particularly famous example, the line-to-sphere transformation, which established a deep link between Plücker's line geometry and French sphere geometry. Lie's mapping has the interesting property that it takes the asymptotic curves on surfaces onto lines of curvature of the image surface. In a joint work, Lie and Klein used this transformation to determine the asymptotic curves of Kummer surfaces.

Contact transformations and their associated infinitesimal generators were fundamental tools for Lie's integration theory for partial differential equations of the first order. This theory reduces the integration of such a differential equation to the integration of a system of ordinary differential equations (characteristic equations of Lie). This work led to the theory of "finite continuous transformation groups," which he presented with the help of Friedrich Engel in the large three-volume study *Theorie der Transformationsgruppen* (1888, 1890, 1893). These are called Lie groups today, though the modern concept of a Lie group is somewhat more general than the original idea. To such groups Lie associated a purely algebraic structure – known today as a Lie algebra – which is formed by its infinitesimal transformations. The core of Lie's theory is the following fundamental theorem, which in modern wording reads: Every finite-dimensional Lie algebra over $\mathbb{R}$ (or $\mathbb{C}$) is the Lie algebra of a suitable analytical group. Lie himself gave interesting applications of his theory to differential equations as well as to the Riemann-Helmholtz-Lie space problem. The combination of algebraic and topological structures created by Lie has proven to be extremely fruitful for nearly all branches of mathematics, and today it is also of fundamental importance for theoretical physics as well.

When Felix Klein received a call to Göttingen at the end of 1885 (which he followed by Easter of 1886), he tried to win Lie as his successor for Leipzig. In recommending his appointment to the Ministry, Klein summarized Lie's most significant scientific results and then added in conclusion:

> There are other geometers, who, with their diverse training, have found important results in one direction, as well as the other, and know how to present the current state of geometric science well in lectures. But Lie is the only one who is able to establish an independent geometric school, thanks to his powerful personality and the originality of his thinking. [Beckert/Purkert 1987, 220-225]

Lie accepted the call to Leipzig, and on May 29, 1886 he gave his inaugural lecture on the influence of geometry on the development of mathematics. He developed a lively teaching activity in Leipzig, above all on the theories he had created, and drew a whole host of talented students from home and abroad. Of the 56 mathematicians who received their doctorates in Leipzig between 1887 and 1898 (Lie's last year), 26 were students of Lie.

Three other mathematicians on the Leipzig faculty played a role in Hausdorff's studies. The first was the geometer Friedrich Schur, who habilitated in 1881 under Felix Klein before he was promoted in 1885 to associate professor. Schur was known for his contributions to Riemannian geometry and Lie groups as well as to the foundations of geometry. Especially noteworthy was his book *Grundlagen der Geometrie* (1909), which was awarded the Lobachevsky Prize. Schur offered exercise courses for students in their early semesters, which Hausdorff attended, but in 1888 he already left Leipzig to accept a full professorship in Dorpat (Tartu). In 1885, two other mathematicians habilitated in Leipzig: Eduard Study and Friedrich Engel. Both played an important role, as friends as well as supporters, in Hausdorff's later life. Study worked as a private lecturer in Leipzig until he left for Marburg in 1888. Before then, he was not much involved in Hausdorff's educational development.

Friedrich Engel earned his place in the history of mathematics primarily through his many contributions to the development of Lie's world of ideas. Lie was guided by an ingenious intuition, but he lacked the ability to present his ideas in such a way that they could be understood by a broad mathematical public, much less meet the standards of rigor of the time. Mayer and Klein realized this, too. Engel had recently completed his doctorate under Mayer with a thesis on contact transformations, so Mayer and Klein now considered sending Engel to Lie in order to help him develop his theory of transformation groups. Lie welcomed the idea as well, and so Engel set off for Christiania in September 1884. There he spent nine months working with Lie on the first volume of his treatise on transformation groups. When Lie joined the faculty in Leipzig in 1886, Engel became his assistant; he continued his work on Lie's project until all three volumes were published. In 1889 Engel was appointed associate professor and in 1899 he finally became a full honorary professor at the University of Leipzig. Five years later, in 1904, Engel was called to Greifswald as Study's successor. Study and Engel had been good friends since their time together as private lecturers in Leipzig, and their friendship lasted up until Study's death. Unfortunately, Lie and Engel became, little by little, deeply estranged (see [Purkert 1984] and [Fritzsche 1991]). Lie later learned that he was suffering from pernicious anemia, which led to psychological changes that presumably played a role here. Many years after Lie's death in 1899, Engel began working on his truly impressive edition of Lie's Collected Works (*Gesammelte Abhandlungen*), which eventually was published in seven volumes with masterful commentaries. We will encounter Engel often in the course of this book.

From documents issued by the University of Leipzig, we have records of the courses Hausdorff attended during his first two semesters (summer semester 1887 and winter semester 1887/88). Furthermore, additional documentation of his activity from this time can be found in his estate, which contains handwritten transcriptions or elaborations from the lecture courses and seminars in which he participated. During his first semester, Hausdorff took Lie's course on plane and solid analytical geometry supplemented by exercises in a seminar conducted by Friedrich Schur. He also attended Mayer's lectures on algebra and the theory of determinants, from which he produced a careful elaboration (Hausdorff estate, Capsule 56, Fascicle 1170, 59 pages). One can assume that he was particularly interested in those lectures that he worked out carefully from the transcripts.

During his first semester, Hausdorff also took two courses in the natural sciences, one with Gustav Heinrich Wiedemann, a leading authority in the field of electricity. This introductory course dealt with topics in general physics, acoustics, and optics. The other course was in mathematical geography, taught by the astronomer Heinrich Bruns. Mathematical geography was understood at this time to encompass mathematical methods for determining the shape and size of the earth. This included calculating the coordinates of a point on the earth's surface and determining the current location and position of the earth in the solar system.

Heinrich Bruns soon emerged as a figure of decisive importance at the beginning of Felix Hausdorff's scientific career. Bruns had studied mathematics, astronomy and physics in Berlin from 1866 to 1871, where his principal teachers were Ernst Eduard Kummer and Karl Weierstrass as well as the astronomers Arthur von Auwers and Wilhelm Foerster. An excellent student, he twice received a special monetary award from the Berlin Mathematics Seminar [Biermann 1988, 108]. After taking his doctorate in 1871, Bruns spent two years at the Pulkowo observatory working as a calculator. Afterward, he was employed at the observatory in Dorpat (Tartu) from 1873 to 1876, during which time he completed his habilitation. With the support of Kummer and Weiertrass, he was then appointed associate professor of mathematics at the University of Berlin. In addition to introductory courses, he also taught differential equations, mathematical geography, and the application of elliptic functions, while working on the side at the Military Academy and the Geodetic Institute [Biermann 1988, 132–134, 137]. In 1882 Bruns accepted a call to Leipzig as professor of astronomy and director of the observatory, a position he held up until his death.

Apart from the two years he spent in Dorpat, Bruns was not engaged in observational astronomy. His main areas of work were celestial mechanics, geometric optics, geodesy, and numerical mathematics, as well as probability theory and mathematical statistics. Bruns' most important contribution to celestial mechanics was a theorem on the n-body problem that he proved in 1887. A classical result dating back to Euler and Lagrange showed that the differential equations in such systems lead to 10 linearly independent algebraic integrals, thus algebraic functions of position, velocity, and time. Bruns proved that these are the only linearly independent algebraic integrals that can arise, a result that opened up new insights into the n-body problem as developed, above all, through the works of Henri Poincaré. He also published important results on geodesy (Bruns' polyhedron) in his work *Die Figur der Erde* (1878).

Bruns' most important contributions to geometrical optics appeared in his comprehensive study *Das Eikonal* (1895). In it he considered optical images in which the source and image spaces are homogeneous and isotropic so that light rays will be straight lines. He then treated an optical mapping as one which satisfies Malus' condition in geometrical optics: the manifold M_1 of lines mapped to lines in the manifold M_2 must have the property that those forming normals to a surface in M_1 will be mapped to lines normal to a surface in M_2. Bruns found an elegant way to represent such a mapping $\Phi : M_1 \to M_2$, which initially requires four component functions, by means of a single function that he called the *Eikonal*. In 1901, Felix Klein showed how this was closely connected with Hamilton's characteristic function, which the latter had developed based on Fermat's principle in geometrical optics. A series of other researchers soon developed Bruns' work further, including Karl Schwarzschild, who used it to calculate higher-order image defects for the construction of telescope lenses. As discussed in Section 2.2, Hausdorff began his career as a private lecturer with a paper on geometric optics that was directly related to Bruns' *Eikonal*.

The evaluation of geodetic and astronomical data lead to problems in numerical mathematics, a field closely linked with probability theory and mathematical statistics (error analysis). Under Bruns' direction, the Leipzig Astronomical Observatory employed mathematicians for purposes of practical data analysis, which, as we will see, was one of Hausdorff early activities. Bruns himself taught numerical methods – e.g. interpolation procedures or methods for numerical integration. He presented these in lectures and seminars and eventually published a compendium of such techniques in [Bruns 1906]. In the same year Bruns published a book on probability and mathematical statistics (*Wahrscheinlichkeitsrechnung und Kollektivmasslehre*) that was also based on his lecture courses. The terminology "collective measure theory" in place of mathematical statistics was introduced by Gustav Fechner, the founder of psychophysics.

Statistical error theory in astronomy and other fields was long based on the properties of normal distributions. The influential Belgian astronomer and statistician Adolphe Quetelet made this into something akin to a dogma by suggesting that every reasonable statistical entity was normally distributed. Fechner was among the first to cast doubt on Quetelet's paradigm. After decades of collecting statistical data in various fields, he emphasized that many distributions were not normal and he noted, in particular, the importance of asymmetric distributions. He did not go so far, however, as to accept the notion of arbitrary distributions for a statistical population, as Bruns was the first to do in [Bruns 1897]. This study dealt with series expansions by means of Hermitian polynomials, which led to an analytical representation of arbitrary distribution functions. These were at first called Brunsian series, but today are known as Gram-Charlier series. Hausdorff would later also take up this idea of Bruns, whose approach to probability theory was based on the concept of relative frequency, which he called "quote calculation" (*Quotenrechnung*). Richard von Mises later promoted this framework by means of his axiom system for this approach to probability theory.

During his first semester in Leipzig, Felix Hausdorff was interested not only in his main subject fields, mathematics and the natural sciences; he also took courses in the humanities. He attend a lecture course on the Faust saga given by Carl Biedermann, a distinguished scholar, politician and publicist. He also attended the lectures of Conrad Hermann in his course "General Grammar and Language Philosophy." These were based on his book *Die Sprachwissenschaft nach ihrem Zusammenhange mit Logik, menschlicher Geistesbildung und Philosophie* (*The Science of Language in Connection with Logic, Mental Construction, and Philosophy*). The focus of Hermann's philosophy was the relationship between language and thought. Although Hausdorff never referred to Hermann later, one can easily imagine that these lectures might have given him an impulse to study this problem, which he eventually did.

The same might be said about another course he took on the fundamental problems of philosophy that was taught by Hermann Wolff, director of a local vocational school (*Bürgerschule*) but also a private lecturer at the University of Leipzig. Wolff's best known publication was his *Kosmos*, which appeared in two volumes in 1890. Like Gustav Theodor Fechner and Friedrich Albert Lange before him, Wolff was concerned with the problem of reconciling the foreign, cold mechanistic-materialistic worldview dominated by the natural sciences with the image of a lively, spiritual world. These two worldviews were conceptualized or imagined very differently. Hausdorff's later philosophical writings, published under the pseudonym Paul Mongré, reflect his deep interest in these debates over worldviews and related questions concerning the relationship between philosophy and science. Surely he thought about these matters already as a student, though no evidence survives that suggests an influence tracing back to Wolff's lectures.

In Hausdorff's second semester, the winter of 1887/88, he took several offerings in mathematics, including Lie's course on spatial projective geometry and Mayer's introductory calculus with supplemental tutorial. He also attended Engel's theory of algebraic equations, in which he learned the fundamentals of Galois theory. This topic evidently appealed to him, as he worked out his own elaboration based on these lectures, though he later completely forgot which courses he had taken with Engel. Finally, he enrolled in Friedrich Schur's supplemental tutorial on quadric surfaces. He also continued his physics program under Gustav Wiedemann with part two of his course on experimental physics (heat and electricity).

In the area of humanities, Hausdorff was even more active during this second semester with five courses in a variety of subjects. He heard lectures by the theologian Gustav Adolf Fricke, pastor of the St. Peter's Church in Leipzig, on the scientific foundations of belief in a personal God. He also took in the lectures offered by the philosopher and theologian Rudolf Seydel in his course "On the present task of philosophy in relation to theology and natural science." Seydel was a representative of "speculative theism," a philosophical trend that emerged in reaction to the pantheistic or atheistic radicalism of the Young Hegelians as well as the materialism of the natural scientists. Its main goal was to reinterpret parts of the Christian doctrine of salvation, such as the creation story, speculatively. Seydel interpreted this as a "fusion and reconciliation of Christianity with modern culture."

Hausdorff was also very interested in social issues. He took a course on the history of socialism offered by the economist Otto Warschauer and attended another given by the political scientist Karl Walcker on "The Workers' Question." This term had been current since the middle of the nineteenth century, when problems arising from the economic and social lives of the newly emergent class of industrial workers first gained widespread attention. Finally, he studied the history of music in a course taught by the musicologist Oscar Paul, associate professor for music theory and its history. Paul also taught piano and musical theory at the Leipzig Conservatory, where his students included musicologists, performing musicians, and the composer Leoš Janáček. For many years Paul also edited the music section of the *Leipziger Tageblatt.* His lectures dealt with a number of different topics: music history, harmony, musical instruments, and the recent theory of tonal sensations set forth by Helmholtz.

During Hausdorff's time, there were no general regulations governing university studies throughout the states of the German Reich, except for the field of medicine. One enrolled in a certain subject area, such as in his case mathematics and natural sciences. However, there were no uniform study plans, so students were completely free to choose their courses, just as Hausdorff did. Those with sufficient financial means were also encouraged to transfer at least once to another university. In this way, students could broaden their horizons, become acquainted with other academic teachers, and acquire new viewpoints and research interests, not to mention the chance to become familiar with other cities, landscapes and social circles. Moreover, moving from place to place was a simple matter because, once admitted, universities required no examinations or other hurdles as part of the course of studies. The only formal requirements for a Ph.D. came at the very end: the successful completion of a doctoral degree required that candidates submit a thesis containing original research results. If accepted, they were then admitted to the final oral examinations, usually in three disciplines. This final phase was often quite arduous and demanded intense concentration on one's field of study and related subjects. Those contemplating a career as a Gymnasium teacher had to pass a state examination for qualification to teach certain special subjects at various levels; in many cases students first acquired a doctorate before taking this examination for future teachers.

Following these traditional customs, Hausdorff planned to spend his third semester and possibly also his fourth at another university. It is quite remarkable, however, that already during his first year of study he was able to develop a close personal relationship with one of his mathematics professors, Adolf Mayer. Presumably Mayer took notice of him as particularly talented student when he taught his seminar on differential and integral calculus. The following letter from Hausdorff to Mayer testifies to their relationship; incidentally, this is the oldest surviving letter written by Felix Hausdorff (*HGW* IX, 493):[1]

> Dear Professor!
> Since, to my deepest regret, I did not find you at home during my visit today, I hereby take the liberty of writing to you to express my heartfelt thanks for the abundance of instruction and stimulation that I have received and will continue to receive from your lectures during the last two semesters. That it was granted to me to become personally known to you gives me heartfelt joy, though also a sense of regret that I did not make use of this privilege properly in my own favor, i.e. for my scientific advancement; however, given the perfect clarity and comprehensibility of your lectures, I really did not find any opportunity to make use of your private help. – Since I intend to leave Leipzig for one or two semesters at the most, I commend myself to your benevolence and am grateful to you with the assurance of sincerest devotion. Your
> Felix Hausdorff
>
> March 23, 1888

[1]For all correspondence to and from Hausdorff, full bibliographic information is given in *HGW* IX; see "Notice for the Reader."

Hausdorff spent the summer semester of 1888 in Freiburg in Breisgau. No concrete documentation of his studies during this period can be found in his estate, but the names of three of his instructors are given in the curriculum vitae he wrote when he submitted his doctoral dissertation. These three professors were Eugen Baumann, Hugo Münsterberg and Ludwig Stickelberger. That semester Baumann offered two courses: Inorganic Experimental Chemistry and Physiological Chemistry, though which of these Hausdorff attended is unknown. Ludwig Stickelberger, a student of Weierstrass, was best known as an algebraist. That semester he taught a calculus course and another on descriptive and synthetic geometry along with a mathematics seminar. Hausdorff attended his lectures on integral calculus: his estate contains a transcript entitled "Integralrechnung," which can hardly be ascribed to a course he might have taken elsewhere. Whether he attended Stickelberger's geometry lectures remains unclear.

In the case of Hugo Münsterberg's course on experimental psychology, we can give some details about his teaching and research activity. This course had real importance for Hausdorff, as the experiments he carried out mark the beginning of his decades-long interest in philosophical and epistemological reflections relating to time, including psychological experiences of time or timelessness. In 1882 Münsterberg had begun studying medicine in Leipzig, where he soon began attending the lectures of Wilhelm Wundt, the founder of experimental psychology. This experience set the course for Münsterberg's future career, as he began working in Wundt's laboratory and then obtained his doctorate under him in 1885. He afterward studied philosophy and medicine in Heidelberg before habilitating in Freiburg in 1887, the year prior to Hausdorff's arrival; indeed, Münsterberg was only five years older than Hausdorff. As a young private lecturer, though, he had gained financial support from the University of Freiburg to set up a private laboratory for psychological experiments in his home. For several semesters, he conducted experimental psychological work at this facility.

Münsterberg published the methods and results of his experimental work in the periodical *Contributions to Experimental Psychology*. For his experiments on the "sense of time," carried out during the summer semester of 1888, he reported being assisted by two students, Hausdorff and Kalchthaler. They used a special apparatus, first constructed by Wundt, for measuring time intervals. Münsterberg hoped to find experimental evidence supporting his "tension theory," which conjectured that the ability to estimate the lengths of time intervals was related to the inner perception of sensations of tension in the body generated over time by external stimuli or by physical processes, in particular by breathing. To prove this, he carried out two series of experiments.

Using Wundt's apparatus, his assistants generated two acoustic signals by striking a hammer at preset time intervals of between 6 and 60 seconds. Münsterberg sat at a certain distance in an armchair and tried to reproduce the time interval as accurately as possible. To do this, he would take the second hammer blow as marking the start of a time interval that would end with a third signal, which he triggered by an electrical contact held in his hands. The percentage error for the reproduced time interval was then recorded. In the first series of experiments, the assistants did not take account of Münsterberg's breathing. In the second, however, they coordinated the two signals so that they fell within the same breathing phase. The results, based on hundreds of individual experiments, showed that in the first case the average error was 10.7%, whereas in the second the errors only averaged 2.9%. Münsterberg regarded this marked difference as confirming the validity of his tension theory. For Hugo Münsterberg, these studies stood at the beginning of his life's work in psychology, which took a dramatic new turn in 1891, when he met William James at the first International Congress for Psychology. Soon thereafter, he emigrated to the United States, where he became one of the pioneers of applied psychology. He taught at Harvard University until his sudden death on 16 December 1916 in the middle of a lecture.

At the beginning of September 1888, Hausdorff returned home to Leipzig, but he left soon thereafter to continue his studies in Berlin. Documentation of the courses he attended there during the winter semester 1888/89 can be found in the certificate he received on leaving the university. For a number of these courses, one can find notes and transcriptions in Hausdorff's estate. During this stay, he attended lectures offered by the Director of the Berlin Observatory, Wilhelm Foerster, on general astronomy and astrophysics as well as on error theory and measurement. In these fields as well as in geodesy, Foerster was an influential figure both within Prussia and beyond. In 1870 he initiated the founding of the Royal Prussian Geodetic Institute in Berlin, and one year later he proposed building an astrophysical observatory, a plan that was finally realized when the Potsdam Observatory opened in 1879. He also founded an institute for astronomical calculations in Berlin, which specialized in charting the orbits of asteroids. As a researcher, Foerster was best known for his investigations of phenomena associated with luminous night clouds. Finally, in 1888, he helped launch the private company *Urania*, which promoted popular science and built the first public astronomical observatory.

Hausdorff also took a course on celestial mechanics offered by Friedrich Tietjen, director of the Astronomical Computing Institute and professor of astronomy. Tietjen's most important achievements were in the field of geodesy (surveying Northern Germany) and in the field of computational astronomy. He published numerous articles on methods for calculating the orbits of objects in the solar system. The distinguished geodesist Friedrich Robert Helmert also taught at Berlin University. Helmert held a professorship in Aachen from 1870 to 1886 before becoming director of the Geodetic Institute in Potsdam, which became an important international center under his leadership. His two-volume study, *Die mathematischen und physikalischen Theorien der höheren Geodäsie* (1880, 1884), was a standard work for decades. Hausdorff heard his lectures on the method of least squares, which was based on his textbook [Helmert 1872]. He also took a course entitled "Hansen's Method for calculating absolute perturbations" offered by the private lecturer Rudolf Lehmann-Filhés. This dealt with a method introduced in 1859 by Peter Andreas Hansen for calculating perturbations of small planets.

Hausdorff's choice of these five courses strongly suggests that he planned to pursue a career in the broad field of research connected with computational astronomy. Since Berlin was an important center for these related fields, he certainly went to the right place for this purpose, but he also studied other subjects during his stay there. He took two courses in pure mathematics, both offered by Lazarus Fuchs: one on the theory of elliptic functions, the other on analytical mechanics. Fuchs was an eminent mathematician, having first studied in Berlin under Kummer and Weierstrass before completing his habilitation in 1865. He went on to make fundamental contributions to the theory of ordinary linear differential equations in one complex variable [Gray 1984]. In 1866 he introduced the class of "regular singularities" for such differential equations, and he considered the associated local monodromy transformations for the system of solutions. These ideas are still of great importance today for the theory of Gauss-Manin connections [Deligne 1970].

Fuchs grew up in a Jewish family in Posen. In 1858, when he obtained his doctorate in Berlin, there were no full professors of Jewish faith at any of the German universities. Afterward, these conditions gradually liberalized, beginning with the elevation of the Göttingen mathematician Moritz Abraham Stern to *Ordinarius* in 1859. He remained for some time the only full professor from a Jewish family who had not been baptized as a Christian. It took Stern thirty years to gain this appointment after habilitating in 1829; indeed, had he not come from a wealthy family he never would have been able to pursue a university career. His unique success illustrates the difficulties young Jewish scholars faced during the period when Hausdorff was studying; in many cases, they were faced with the decision of having to forgo a university career or convert to Christianity. The mathematician Leo Koenigsberger, a good friend of Lazarus Fuchs, wrote about the latter's situation in his memoirs: "Thus, Fuchs and I had to face the question whether we should sacrifice our scientific careers and entire existence because of the dominant narrow-minded views of the government or, after having long since stripped away religious prejudices, we should adopt Christianity" [Koenigsberger 1919, 31]. Koenigsberger and Fuchs decided to undergo baptism and become Lutherans, the latter in 1860.

Felix Hausdorff – as mentioned earlier – chose not to convert to Christianity. Nevertheless, the question of conversion could well have crossed his mind. During his semester in Berlin, he probably read a novel published a short time before in 1885, in which this topos plays a role: *Die Sebalds* by Wilhelm Jordan. The heroes in this utopian work are two brothers: Arnulf Sebald becomes an astronomer, a profession symbolizing a modern scientific worldview. His brother Ulrich, a Protestant theologian, represents a new religiosity, a Deification of Man, in opposition to Lutheran Orthodoxy. At the end of the novel, after all sorts of misunderstandings and confusion, both brothers gain the two women whom they love. Ulrich wins Cäcilie, the daughter of a Jewish banker, against the resistance of an Orthodox rabbi, whereas Arnulf marries Hildegard, daughter of a Catholic Count, by overcoming the intrigues of the Catholic clergy. None of these four, however, converts to another religion, a major point of the novel. As evidence that Hausdorff probably read it while in Berlin, one finds his copy of the poem "Niagara" from the second volume of the novel in his notes from that time.

As before, in Berlin Hausdorff also signed up for several other courses outside his major fields of study. One of these concerned reflections on recent findings in the natural sciences, a series of lectures offered by the now 70-year-old Emil du Bois-Reymond, Director of the Physiological Institute at Berlin University. These were popular lectures that du Bois-Reymond had been giving for many years, as he himself noted in a footnote to a well-known address from 1872, when he spoke about the limits of scientific knowledge at the annual meeting of the Society of German Natural Scientists and Physicians. This was the famous speech that triggered what came to be called the "ignorabimus dispute," remembered from his closing words: "Regarding the question, however, as to what matter and force are and how we should conceive of them, [the researcher] must once and for all arrive at the much more profound truth: *ignorabimus* (we will not know)."

Du Bois-Reymond's skepticism was greeted by some, but others objected to it violently, in particular monists such as Ernst Haeckel. Many were alarmed that such an important and highly respected natural scientist would voice the opinion that mankind can never find answers to key questions and will always suffer from a fundamental lack of knowledge. Yet the distinguished physiologist sincerely doubted whether the subjective qualities of human feeling and thinking could be traced back to mechanical or physical states found by studying the matter in the brain. Hausdorff, too, later viewed this type of reductionism critically. The philosophical debate about the scope of the natural sciences for understanding human feelings continues to the present day (see the anthology [Bayertz et al. 2007]). Given ongoing discussions about the capabilities and future applications of artificial intelligence, this debate is more topical than ever.

David Hilbert also alluded to the *ignorabimus* dispute in his famous lecture "Mathematical Problems," delivered in 1900 at the International Congress of Mathematicians in Paris. There he underscored his conviction that in mathematics there are no questions the human mind cannot answer in principle. The passage in question ends with the following emphatic pronouncement: "This conviction that every mathematical problem can be solved is a powerful incentive for us during our work; we hear the constant call within us: *There is the problem, look for the solution. You can find it by pure thinking; because in mathematics there is no ignorabimus!*" [Hilbert 1900, 266].

Hausdorff also attended lectures on the physiology of language given by Johannes Gad, a neuro-physiologist who had earlier worked as an assistant to Du Bois-Reymond before becoming associate professor in Berlin. Since 1885 he headed the experimental department within the Berlin Physiological Institute. The fact that Hausdorff chose to attend these lecture courses reveals his continued interest in the phenomenon of language, a theme we will discuss in 5.7 in connection with Fritz Mauthner's "Language Critique." Hausdorff also attended lectures on the relationship between mental disorders and crime in a course taught by the psychiatrist and neurologist Carl Moeli, whose main area of expertise was forensic psychiatry. His study of mentally deranged criminals [Moeli 1888] served as the basis for these lectures.

While in Berlin, Hausdorff also took two courses taught by philosophers. One was Edmund König's lecture course on the epistemological and logical foundations of modern conceptions of nature. König taught at a Gymnasium while serving as its principal. He was associated with the journal *Kant-Studien* and published a considerable number of philosophical books and essays. Hausdorff also took part in a tutorial based on Kant's *Critique of Pure Reason*, which was taught by the philosopher and educator Friedrich Paulsen. Paulsen had completed his habilitation in Berlin in 1875 with a thesis on the history of Kantian epistemology. In 1878 he was appointed associate professor for philosophy and education in Berlin, and he later taught these subjects as a full professor there. Eventually, he became known far beyond the borders of Germany for his writings on philosophy as well as pedagogy. Hausdorff's knowledge of Kantian philosophy was, of course, by no means restricted to what he learned in this tutorial or by reading other neo-Kantian writers, such as Friedrich Albert Lange, Kuno Fischer, or Otto Liebmann. Far more decisive was his own thorough study of Kant's writings. Still, it is noteworthy that Hausdorff's notes from Paulsen's course contain the seeds for his own epistemological essay *Das Chaos in kosmischer Auslese*, as discussed in Section 4.4.

Following this semester in Berlin, Hausdorff returned to Leipzig in March 1889 to complete his studies. As we have seen, during his first two semesters there he took a very wide range of courses, many far beyond his major fields, mathematics and natural sciences. Now, however, he focused increasingly on mathematics and astronomy. In fact, during his last five semesters he showed a strong preference for courses taught by just two professors – Heinrich Bruns and Adolph Mayer. He took the following lecture courses taught by Bruns: mathematical geography (SS 1889), orbit determination and special perturbations (WS 1889/90), practical astronomy and probability theory (SS 1890), celestial mechanics (WS 1890/91), mathematical geography and general astronomy (SS 1891). For nearly all of these courses, one can find Hausdorff's transcriptions and notes in his estate. He also took part in Bruns' astronomical seminar three times (SS 1889, WS 1889/90 and SS 1890), but chose to bypass his seminar on scientific calculations, which was offered twice. As for the courses taught by Mayer, Hausdorff took: calculus of variations (SS 1889), dynamical differential equations (WS 1889/90), ordinary differential equations (SS 1890), analytical mechanics (WS 1890/91), and first-order partial differential equations (SS 1891). He also attended Mayer's seminar every semester except for the last.

When he first returned to Leipzig in the summer semester 1889, Hausdorff also took two lecture courses with Sophus Lie. One dealt with applications of contact transformations in geometry and mechanics, the other with Lie's introduction to the theory of transformation groups. Lie offered a continuation of the first course the following semester, which Hausdorff attended up until 11 November 1889. At that time Lie had to stop teaching due to a nervous condition. He took sick leave and entered a sanatorium, appointing Friedrich Engel to take his place, by which time Hausdorff had dropped out. Lie returned and taught again during Felix Hausdorff's last two semesters, but Hausdorff appears not to have taken any of these courses. He did take two lecture courses taught by Carl Neumann: analytical mechanics (WS 1889/90) and the mechanical theory of heat (WS 1890/91), and he also took a physics lab course under Gustav Wiedemann in the summer of 1890.

During this final period at Leipzig University, Hausdorff took only one course in the humanities. This was a philosophical tutorial on Kant's metaphysics of morals taught in the winter semester 1889/90 by Professor Max Heinze, whose expertise was in the field of history of philosophy. After taking his doctorate, Heinze had initially been employed as a teacher at Pforta, the famous classical Gymnasium. One of his pupils there was Friedrich Nietzsche, who was evidently fond of his teacher. After Heinze married an earlier acquaintance of Nietzsche, a personal relationship developed between them. Nietzsche's letters to other friends reveal, however, that he thought little of Heinze as a philosopher, as he complained that the latter did not understand him at all. A student like Hausdorff, who was seriously interested in Nietzsche's writings, could hardly have learned anything about them from Heinze, though the same could be said about all German philosophy professors at this time.

2.2 Hausdorff as an Astronomer

During his final semester in Leipzig, Hausdorff had to concentrate on preparing his dissertation for Heinrich Bruns, so he quite naturally decided to take only a few courses at that time. No doubt he also wanted to ready himself for his oral exams as well. As described in the previous section, Bruns' astronomical research was mainly focused on his related interests in geometrical optics, and in this respect the astronomer Felix Hausdorff followed closely in his footsteps. In 1891, Bruns published an extensive work on the theory of astronomical refraction. As his student, Hausdorff had carried out special calculations for this work, as Bruns expressly acknowledged [Bruns 1891, 190]. Certain essential points were still open, however, so he entrusted these matters to his talented student as the topic for his dissertation.

Before describing this work, some information about the physical background should be briefly given. Due to the low, but not negligible density of the earth's atmosphere, a ray of light reaching the earth from space will be refracted, thereby making the angle of elevation at which a star is seen above the horizon of the observer appear greater than the "true" angle of elevation, i.e. what one would see without such an atmospheric effect. The difference between these two angles is called the astronomical refraction. At the zenith this measure is zero, whereas at the horizon it reaches a maximum value, somewhere between 34 and 39 minutes of arc. Normally one works with the zenith distance, i.e. with $\vartheta = 90° -$ height angle. The observed zenith distance is therefore smaller than the "true" one. The phenomenon of atmospheric refraction was already known in antiquity; astronomers in Islamic cultural regions also dealt with it. Johannes Kepler constructed one of the first refraction tables (values of the refraction depending on the apparent zenith distance). The mathematical treatment of refraction began with the emergence and development of infinitesimal calculus in the 18th century. Here Newton, Euler, Laplace and Oriani should be mentioned, in particular.

The mathematical theory of refraction leads to what is called the refraction integral in astronomical literature. Let P_0, P_1 be two points of the light path (or refraction curve) characterized by their "heights", i.e. their distances r_0, r_1 from the center of the earth ($r_0 < r_1$). Let $\mu(r)$ be the refractive index at height r (for the sake of simplicity, one assumes μ only depends on the distance from the center of the earth). In most cases, one takes P_0 to be the observation site and P_1 as the point where the light beam enters the atmosphere. Now imagine the atmosphere as broken into thin spherical layers, each of which makes a small contribution to the refraction; the sum of all these then approximates the total refraction. By passing to infinitely thin layers and thus to the integral, by the law of refraction we obtain the refraction for the apparent zenith distance ϑ:

$$R(\vartheta) = \mu_0 r_0 \sin\vartheta \int_{\mu_1}^{\mu_0} \frac{1}{\sqrt{\mu^2 r^2 - \mu_0^2 r_0^2 \sin^2\vartheta}} \frac{d\mu}{\mu}\,.$$

Here μ_0, μ_1 are the values of $\mu(r)$ at r_0 and r_1, respectively. The refractive index $\mu(r)$ is a function of the density of the air and consequently of the temperature and air pressure.

If one uses a barometric formula for altitude, then one can specify $\mu(r)$ as a function of altitude if one also knows the course of temperature T as a function of altitude, i.e. the function $T(r)$. The measurement of the temperature change through the atmosphere, however, was not yet possible around 1890; measurements for moderate heights in the mountains were falsified by the influence of meteorological conditions in and around the mountainous landscape. This shortcoming was not very serious for practical astronomy. Around 1800, Laplace and Oriani had proven independently of one another that up to zenith distances of 75° $R(\vartheta)$ was independent of $\mu(r)$, which meant that it sufficed to know the air pressure and temperature at the observation site. Even in the range of $75° < \vartheta < 85°$, the uncertainty of $T(r)$ only causes errors no greater than the uncertainty of the astronomical measurements around 1900. Thus it was only in the range $85° < \vartheta < 90°$, i.e. the area very close to the horizon, that one needed to employ the theory. Before the time of Bruns and Hausdorff, the standard procedure was to assume a plausible functional relationship between the state variables of the air as a function of altitude. One then calculated $\mu(r)$ with this hypothesis, and then determined $R(\vartheta)$ from the refraction integral. During the second half of the nineteenth century,

Carl Maximilian von Bauernfeind and Hugo Gyldén developed theories based on various hypothetical assumptions about the temperature distribution, which still contained certain parameters that were then determined by temperature observations at low altitudes.

Bruns took up the "inverse problem" as it were: his goal was to determine the change in temperature throughout the atmosphere, which, as already noted, could not be found by direct measurements. This type of inversion of the refraction problem had once been considered by Friedrich Wilhelm Bessel, who wrote about this in 1823 in a letter to Heinrich Christian Schumacher. Bruns first altered the refraction integral by introducing $\nu = r \cdot \mu$ as the new variable of integration and by letting the observation point P_0 vary on the refraction curve. He wrote ν at the point P_0 as σ and called τ (initially also viewed as a variable) the value of ν on the boundary of the atmosphere. The refraction integral then takes the form

$$R(\sigma, \vartheta, \tau) = \sigma \sin \vartheta \int_{\sigma}^{\tau} \frac{m(\nu)}{\sqrt{\nu^2 - \sigma^2 \sin^2 \vartheta}} d\nu \tag{2.1}$$

with $m(\nu) = -\frac{d \ln \mu}{d\nu}$. Bruns now introduced a hypothetical analytical function $R(\sigma, \vartheta)$ by means of an "interpolation formula" (containing $\frac{\sigma}{\tau}$ as a parameter), such that $R(\sigma_0, \vartheta_0)$ approximates a function $R^*(\vartheta_0)$ reasonably well. The values $R^*(\vartheta_0)$ are measured or calculated on the basis of other theories and data collections for refraction values on the surface of the earth. From this interpolation formula one then calculated $\mu(r)$ and ultimately the temperature $T(r)$. In short, Bruns' inversion consisted in the following: whereas previously one calculated R from a hypothetical μ, now one sought to determine μ from a hypothetical R.

In order to narrow the class of interpolation formulas to be considered, Bruns obtained the following partial differential equation by differentiating the refraction integral (2.1):

$$\cot \vartheta \frac{\partial R}{\partial \ln \sigma} - \frac{\partial R}{\partial \vartheta} = \frac{d \ln \mu(\sigma)}{d \ln \sigma} \tag{2.2}$$

For each solution of (2.2), i.e, for any $R(\sigma, \vartheta)$ that comes in question, the associated μ or $m(\nu)$ can be uniquely determined. This thus enabled Bruns to invert the hitherto usual procedure for dealing with atmospheric refraction.

The interpolation formula hypothesized for R is an analytical function of ϑ that contains functions of σ which still need to be determined. Inserting this hypothetical expression for R in (2.2) leads to a system of ordinary first order differential equations for the functions σ to be determined. By choosing a fixed σ_0 and adjusting $R(\sigma_0, \vartheta)$ to given observed or known values $R^*(\vartheta)$, one obtains initial conditions for this system of ordinary differential equations. Their solutions then yield a largely fixed expression for R, from which one can determine the integrand of the refraction integral using (2.2). If, after choosing appropriate constants, this calculated integral agrees with the expression for R, then this expression, according to Bruns, serves as a suitable interpolation formula for the refraction.

The example that Bruns carried out in detail involves the so-called partial fraction approach. Using this, one alters (2.2) by means of the coordinate transformation $\xi = \cot \vartheta$, $\eta = \ln \sigma$, which with $R(\sigma(\eta), \vartheta(\xi)) =: R(\eta, \xi)$ results in

$$\xi \frac{\partial R}{\partial \eta} + (\xi^2 + 1)\frac{\partial R}{\partial \xi} = \frac{d \ln \mu(\eta)}{d\eta}.$$

One then takes $R(\eta, \xi)$ to be rational in ξ, and since $\xi = \cot \vartheta \to \infty$ for $\vartheta \to 0$ and $R = 0$ at the zenith ($\vartheta = 0$), this assumption leads to

$$R(\eta, \xi) = \sum_{j=1}^{q} \frac{1}{a_j(\eta)\xi + b_j(\eta)}.$$

Bruns then showed that, assuming sufficiently large q, one can replicate the observations (or a table of values) for any number of prescribed positions (in the sense of an exact matching). As he explained further:

> for practical purposes, it is of course essential that a complete determination be achieved with only a small number of terms. We will treat the cases $q = 1$ and $q = 2$ and use the following refraction tables instead of actual observations, which is sufficient for our purpose. [Bruns 1891, 189]

The tables Bruns used for this purpose were those published by Bessel (B), Oppolzer (O), and Gyldén (G). For $q = 2$, one had five disposable parameters, which Bruns was able to reduce to four based on certain physical properties.

With Bruns' method, the type of hypothetically adopted refraction formula plays a decisive role. One could try finding different formulas that fit the same observational data, but these might then lead to quite different temperature extrapolations. If such were the case, then his approach would have no value for meteorological studies. Bruns had used the partial fraction method to calculate atmospheric temperatures from the data in (B) and (G), and in this connection he formulated the still open problem:

> according to the method developed here, to derive the vertical temperature distribution from observed refractions one must first ask whether the result depends on the interpolation formula used to represent and adjust the observations. [Bruns 1891, 210]

In order to examine this question, it was necessary to find different types of interpolation formulas for the refraction and to adapt these to Bruns' method. This was precisely the task Bruns assigned to his student Hausdorff as the challenge for his doctoral thesis. In his report on Hausdorff's dissertation from 12 July 1891, Bruns wrote:

> In the meeting reports of the Royal Saxon Scientific Society, I recently published a method for deriving the constitution of the atmosphere

> from observed refractions by reversing the previously used method in the theory of atmospheric refraction. The examples I gave were merely intended to explain the method, while a major question, namely the question of the accuracy that needed to be achieved, remained unanswered. For this it was necessary to solve the following tasks: 1) to find more sufficiently accurate and practically useful interpolation formulas for the refraction, 2) to complete the computations for the found formulas, 3) to derive the curves for the vertical temperature change. (Archiv der Universität Leipzig, Phil. Fak., Promotionsakte Nr. 742, p. 1)

If one can prove that "[the method's] success depends only to a subordinate degree on the formula, but instead is significantly influenced by the observational data," there still would remain a serious astronomical problem. As Bruns emphasized, for the method to be truly useful in practice one would need to extend the observations to cover large zenith distances over longer periods of time [Bruns 1891, 210]. These types of observations, however, had never been undertaken. Moreover, the tables that did exist, such as (G), used values for large zenith distances based on a hypothetical temperature distribution extrapolated from the known data. This then posed a major hurdle for the applicability of Bruns' method for determining the true temperature throughout the atmosphere, as will be discussed further below.

Hausdorff submitted his dissertation on July 8, 1891. He entitled it – just like Bruns' work – "On the Theory of Astronomical Refraction" [H 1891, V: 5–90]. This work deals with three hypothetical approaches to refraction, and it related these to the data found in refraction tables, particularly Gyldén's table (G). At first Hausdorff attempted to modify Bruns' partial fraction approach by dropping the condition that required an exact correspondence between the hypothetical values and the table data for zenith distances near the horizon, replacing this with a correlation based on averaged values calculated using the method of least squares.

Hausdorff introduced two new hypothetical refraction formulas, the second of which is briefly presented here:

$$R(\sigma, \vartheta) = \sum_{k=0}^{n} a_{2k+1} \chi^{2k+1} \tag{2.3}$$

with $\chi = \tan(\frac{1}{2}(\arctan(\sin\varphi \tan\vartheta)))$. Here $\sin\varphi$ and the coefficients a_{2k+1} are functions of σ. The method, mentioned above, that uses the partial differential equation (2.2) requires ingenious and extremely extensive calculations. For these calculations, Hausdorff made use of the extensive apparatus of formulas from the theory of special functions (gamma, beta, and spherical functions), which he initially provided. For φ he obtained the result $\cos\varphi = \frac{\sigma}{\tau}$. The coefficients are then linear combinations of odd powers of $\tan\varphi$ with constant coefficients. These coefficients and φ can be chosen in such a way that for a certain σ_0 the values fit a given refraction table. Hausdorff showed this for three different tables, noting that the first four terms of (2.3) already indicate good agreement with the values of these tables for $85° < \vartheta < 90°$, the zenith distances of interest. At the end of his work, he set out the enormous formula apparatus needed for all three approaches, each of which should then lead to finding the meteorological results. Hausdorff left the numerical calculations open, however, so he gave no specific temperature distributions for the atmosphere.

In his report Bruns wrote with regard to the three tasks he had set:

> The present document solves the first and second problem and provides the mathematical preparations necessary for the third, whereas the author reserves the purely arithmetic elaboration of the third problem for a later publication. [...] Since he follows essentially known roads in the purely arithmetical parts, it suffices to remark that this side of the subject has been treated with skill and care. For task 1), on the other hand, the author had to rely on his own ingenuity, and he dealt with the difficulties that arose in a way that testifies to excellent dexterity in the use of mathematical tools. (Archiv der Universität Leipzig, Phil. Fak., Promotionsakte Nr. 742, pp. 1–2)

Both referees, Heinrich Bruns and Adolph Mayer, accorded this work the highest grade: *egregia*. Hausdorff's oral examination in the subjects of astronomy and mathematics with Bruns and Mayer, respectively, took place on July 30, 1891; here, too, both examiners awarded the highest possible grade. On the same day, Felix Hausdorff signed the document acknowledging the formal Latin doctoral vow, and thus after nine semesters of study he was now a doctor of philosophy.

This happy event was then followed by a phase in his life that certainly did little to further his scientific work: on October 1, 1891, Hausdorff began his military service as a one-year volunteer. The legal basis of this form of service in the German Empire was §8.1 of the German Army Order of November 22, 1888, based on older Prussian military laws:

> Young people of education who dress, equip, and care for themselves during their service, and who have presented the knowledge they have gained to the prescribed extent, become after one year of active service in the standing army – counted from the day of commencement of service – officers on reserve leave. [John 1981, 54]

Shortening active service to one year was a privilege. Normally, compulsory active service in the standing army lasted three years, though in 1893 this was shortened to only two. The prerequisite for obtaining this privilege was a certificate of eligibility confirming that a young man met the educational standard, the *Abitur* being sufficient for this purpose.

Eligibility to serve as a one-year volunteer was not only desirable because of the shortening of the active service period, but for several others interrelated reasons. For example, the certificate of eligibility gave its holder the right to choose his unit and location. Felix Hausdorff served in the 7th Infantry Regiment

Prince George, Number 106 in the 4th Company. The regiment Prince George was stationed in a barracks complex in the Leipzig suburb of Möckern, about three kilometers northwest of the city center. Further privileges of the one-year volunteer included being allowed to live and sleep outside the barracks and, to the extent his service permitted, to pursue studies or cultural interests. After a tough four-week period for new recruits, they were freed from performing drudgerous duties such as cleaning, looking after equipment and clothing, etc. They thus received preferential treatment from their superiors and wore green and white cords (the Saxon colors) around the epaulettes of his uniform that set him apart from ordinary soldiers. So Hausdorff did not have to sleep in these barracks, as his home was only about one-half kilometer away from them.

The decisive privilege of the one-year volunteer, however, was that his service offered him the prospect of a career as a reserve officer and thus a respectable position in the upper class society of the Empire. A number of stages were envisioned in this selection process. A widely used textbook for one-year volunteers describes the first stages as follows:

> Those one-year volunteers who have managed themselves well and have acquired sufficient service knowledge can be promoted to irregular corporals (*überzähligen Gefreiten*) after at least six months of service and those among the latter who are particularly distinguished by diligence and knowledge can be promoted to irregular NCOs after at least nine months of service. [Menzel 1897, 43]

The adjective irregular or surplus (*überzählig*) refers to the fact that these were not regular promotions, i.e. they took place outside the funding structure for permanent military positions. In accordance with these regulations, Felix Hausdorff was promoted to irregular corporal on April 1, 1892 and to irregular NCO on July 1, 1892.

The next level took place shortly before the service period ended. This is when regular officers and the troop commissioner took note of those one-year volunteers whom they regarded as suitable aspirants for the reserve officer corps. This group then underwent a practical and theoretical examination. Those who passed were then released into the reserves with a certificate stating that they had qualified as a reserve officer aspirant. Hausdorff was clearly assessed positively by his supervisors and then passed this exam, because he was released on October 1, 1892 with this certificate.

The next steps toward becoming a reserve officer involved completing two eight-week training courses, "course A" in the first year and "course B" in the second after completing active service. Following "course A" candidates took an exam and those who passed were promoted to vice-sergeant; Hausdorff successfully completed "course A" and obtained that title on 23 April 1893. The following year, he also completed "course B," at which point he met the formal qualifications for selection as a reserve officer. To be selected, however, required the recommendation of the local commander and then the approval of the district commander, after which the aspirant had then to be elected by the officer corps of the district. This selection mechanism was designed to ensure that the reserve officers chosen would be socially suitable for this position. For example, someone who sympathized with social democratic ideas would certainly not be elected. Since racially motivated antisemitism was endemic in the officer corps, Jews were nearly always excluded as well. Since the 1880s, in fact, there were no Jewish officers in the German army, except for a few in Bavaria [John 1981, 150–220]. Little wonder, then, that Hausdorff decided to withdraw his name from the election process. This decision was recorded on his military passport, dated 11 September, 1894: "Deleted at own request from the list of officer aspirants."

Immediately after his active service, Hausdorff began working on the continuation of his dissertation, as Bruns had already noted in his report. In fact, already in early 1893 Bruns was able to present Hausdorff's results to the Royal Saxon Society of Sciences [H 1893, V: 93-135]. Hausdorff first developed another hypothetical refraction formula, a kind of dual of formula (2.3). Since ϑ enters into (2.3) via the tangent, he designated this the tangent refraction, whereas he called the new formula the secant refraction, since ϑ appears via the secant. This has the form

$$R(\sigma, \vartheta) = \sin \vartheta \sum_{k=0}^{n-1} b_{2k+1} \psi^{2k+1} \tag{2.4}$$

with $\psi = \tan(\frac{1}{2}(\arctan(\tan \varphi \sec \vartheta)))$. At the end of this work, Hausdorff calculated the temperature curves for the three hypothetical refraction formulas used in his dissertation. For these calculations, he limited himself to the refraction data in Gyldén's table (G). Hausdorff considered the correspondence of the temperature curves provided by the three refraction formulas to be sufficiently good that "Herr Bruns' inverse treatment of the refraction problem meets the demands made of it."

From February 1893 to February 1895, Hausdorff worked as a calculator at the Leipzig observatory. Bruns had succeeded in getting the Saxon Ministry of Culture to set up two rooms there for this purpose. These calculators typically received 60 marks a month for their work, usually carried out daily from 8 a.m. to 1 p.m. 60 marks was roughly the monthly wage paid to port workers in Hamburg. In a letter to Sophus Lie, dated March 28, 1894, Hausdorff wrote that "the morning work in the computer room, which I was obliged to undertake for Prof. Bruns, was extremely strenuous in the beginning" (*HGW* IX, 436).

In addition to work for the zone catalog of the Astronomical Society, the Leipzig observatory undertook important selenographic studies and also carried out parallax determinations. Hausdorff certainly contributed to the extensive computational work, though nothing is known about what he did in detail. Probably he worked on the second supplement to his dissertation [H 1893, V: 136–182] along with his other duties. This work – like the dissertation itself and the first supplement to it – required a huge amount of computation; some formulas are so long that the typesetter had to place them across the page, reading from bottom to top. This study dealt with the special problem of the influence of the flattening of the earth. As a rule, when calculating refraction, one assumes that the earth is a perfect sphere so that the physical state variables of the atmosphere only depend on the distance from the center of the earth, an assumption that is not completely rigorous. Hausdorff treated this as a perturbation problem by calculating the refraction in terms of powers of the earth's flattening, for which one can neglect terms of higher order than the first power. It then turns out that the flattening only has a noticeable influence on refraction effects at zenith distances of more than 80°. In the immediate vicinity of the horizon, the flattening of the earth also leads to a small refractive influence in the azimuthal direction.

As noted earlier, precise observations of refraction near the horizon were lacking, so these meteorologically relevant data were needed in order to pursue the path envisaged by Bruns. Hausdorff noted this fact, after he had calculated the temperature curves, when he wrote:

> For an actual application of the method, now that its applicability has been shown by means of pure calculated examples, only the data is missing. One may hope for its procurement in the not too distant future by one of the more conveniently located observatories. [H 1893, 162, V: 135]

This hope was, in fact, realized a few years later, but with the result that all of Hausdorff's efforts in his three studies on atmospheric refraction became superfluous. Up until 1896, when he was still in Munich, Julius Bauschinger carried out refraction observations close to the horizon over a long period of time using what was then an ultra-modern instrument. Although he published these results in [Bauschinger 1898], Bruns and his students ignored these findings in their calculations for temperature curves using the Bruns-Hausdorff method. This line of investigation turned out to be unpromising, however, because Bauschinger showed that quite different assumptions about the temperature profiles at high altitude – namely the various hypotheses of Radau, Ivory and Gyldén – were compatible with the refraction observations. His résumé, which made no mention of the work of Bruns and Hausdorff, reads as follows:

> One often encounters the opinion that information on the temperature distribution in the top layers of the atmosphere can be obtained by astronomical refraction observations. This is only to a very limited degree correct. For the influence of the temperature distribution on refraction is far outweighed by other factors. ... The series of observations that I undertook was partly in the expectation of obtaining a more secure finding than before. ... Numerous observations in zenith distances of 85° – 88°, carried out with high precision and with an exquisitely constructed instrument, led me to expect this. This hope was not realized.
> ...

> It is doubtful that one could obtain a more favorable result if observations had been available at lower heights than 2°. For here, as many observations demonstrated, the local disturbances in the lowermost air layers so predominate as to completely obscure the influence of the temperature decrease in the upper layers. In relation to the law of falling temperature one will therefore always be dependent on meteorological observations, mainly those conducted in an air balloon. [Bauschinger 1898, 212, 215]

Owing to the development of balloon technology and aviation after 1900, it soon became possible to take measurements of the upper atmosphere. That meant that one could abandon the path pursued by Bruns and Hausdorff of laboriously calculating temperature curves based on refraction observations, although this method might still have been considered reasonable before the turn of the century.

The most detailed theory of astronomical refraction based on measurements of temperature and pressure in the upper layers of the atmosphere was published in 1924 by Paul Harzer [Harzer 1922–1924]. Many subsequent investigations on refraction built on this study, for example, the compilation *Refraction Tables of Pulkowo Observatory*, 3rd ed., Moscow 1930 (the first edition had presented Gyldén's tables). Harzer not only relied on extensive meteorological data pertaining to the so-called normal state of the atmosphere, he also considered deviations in temperature and barometric pressure, moisture levels, wind speeds, light color, and the role of the annual and daily periods. This direct method, exploiting the meteorological data, was for him easily justified:

> The fact that our self-evident approach for constructing tables for the deflection of rays was here carried out for the first time is due to the fact that the necessary knowledge regarding properties of the atmosphere was unavailable and therefore had to be replaced by plausible assumptions. [Harzer 1922–1924, 4]

The alternative approach of Bruns and Hausdorff was not even mentioned by Harzer. Hausdorff himself never commented on his scientific debut, but it must have been frustrating for him to see that the huge effort he had invested in this special problem only took him down a dead end road.

At the beginning of 1894, Hausdorff began working on his post-doctoral thesis for *Habilitation*. In a letter to Sophus Lie, dated 28 March 1894, he wrote in this regard: "At the beginning of the year, I received a topic from Prof. Bruns related to the preliminary work I had done thus far." The idea for this study came, however, from Hausdorff himself, as Bruns described in his final report:

> Some time ago I showed that in the treatment of astronomical refraction it was appropriate to reverse the ordinary train of thought and to derive the relationship between refraction and elevation by viewing refraction as a given function of zenith distance. At that time, Herr Dr. Hausdorff continued to pursue my approach and in that connection

> remarked that the same approach was appropriate for studying extinction, the knowledge of which is a necessary prerequisite for measuring astronomical light intensity. (Universitätsarchiv Leipzig, PA 547, p. 6)

The theoretical study of atmospheric extinction (today called absorption) treats the atmosphere as composed of homogeneous thin layers. Monochromatic light coming from celestial bodies will be weakened as it passes through these atmospheric layers in which the density of the air depends on the height of the layer. One obtains an approximation for the total amount of absorption by summing the contributions over all the layers or, taking these to be infinitesimally small, by integrating over the entire light path from the observer's location to the boundary of the atmosphere. The extinction so calculated then depends on the apparent zenith distance at which the observed celestial body is located. This method neglects changes in temperature, air pressure, density and moisture content, and similar meteorological factors, which have a non-negligible effect on measurements of extinction, just as one also disregards localized pollution of the atmosphere, such as dust or smoke. The fundamental ideas behind this mathematical theory can be summarized as follows.

Let the arc length along the light path (i.e., the refraction curve) be denoted by s and the intensity of the light by $J(s)$. A basic assumption of the absorption theory takes the intensity change dJ to be proportional to the density $D(s)$, the intensity $J(s)$, and the path length ds:

$$dJ = Q\,J\,D\,ds\,, \quad Q = \text{ const.}$$

Next let s_1 denote the boundary of the atmosphere and $J_1 = J(s_1)$ the initial intensity of the light coming from space. We assign the value $s = 0$ to the observer's location on the earth's surface. Then for observations on the earth's surface at the apparent zenith distance ϑ and intensity J_ϑ ($\vartheta = 0$ is the light path perpendicular to the earth's surface):

$$\ln \frac{J_1}{J_\vartheta} = Q \int_0^{s_1} D(s)\,ds\,. \tag{2.5}$$

Since J_1 is not known, one must seek a measurable quantity that can used to determine it. For the zenith, equation (2.5) yields:

$$\ln \frac{J_1}{J_0} = Q \int_a^{a+H} D(r)\,dr\,.$$

Here a is the radius of the earth, H the height of the atmosphere, and J_0 the apparent brightness of the observed object at the zenith. The fraction J_0/J_1 (ratio of apparent brightness at the zenith to the initial brightness) is called the transmission coefficient. It can be determined by adjustments taken from photometric observations.

The integral in (2.5) can then be related to the refraction theory. First, one has $ds = \frac{dr}{\cos\kappa}$, where κ is the apparent zenith distance at the location s. From the known invariance relation in refraction theory

$$\mu r \sin\kappa = \mu_0 a \sin\vartheta = \text{ const } =: N$$

it follows that

$$\ln\frac{J_1}{J_\vartheta} = Q\int_a^{a+H} \frac{\mu\, r\, D\, dr}{\sqrt{\mu^2 r^2 - \mu_0^2 a^2 \sin^2\vartheta}}\,. \tag{2.6}$$

Here $\mu(r)$ is the refractive index at height r and $\mu_0 = \mu(a)$ the refractive index on the earth's surface. The integral (2.6) plays the same role for the absorption theory as the refraction integral (2.1) for refraction theory. If one assumes that there is a direct relationship between the effects of refraction and absorption in the atmosphere, i.e., if one postulates a functional connection between density and refractive index, for example

$$D = c\,(\mu^\alpha - 1)\,, \qquad c = \text{ const. }, \tag{2.7}$$

then the evaluation of the absorption integral requires knowledge of changes in the refractive index μ as a function of the height, just as in the refraction theory.

The analogy between the problems of refraction and absorption of light in the atmosphere served as the basis for the optical theory of Pierre Simon Laplace, who assumed the proportionality of the refractive power and absorbent effect for all layers of air in the atmosphere. Isaac Newton had already assumed this condition and he adopted $D = c\,(\mu^2 - 1)$ (Eq. (2.7) for $\alpha = 2$). In 1805, Laplace took up this Newtonian formula in close approximation using $\mu^2 \cdot \frac{\mu r}{\mu_0 a} = 1$, and finally derived the formula

$$\ln\frac{J_1}{J_\vartheta} = \gamma\,\frac{R(\vartheta)}{\sin\vartheta}, \qquad \gamma = \text{ const.}, \tag{2.8}$$

in which the absorption is expressed directly by the refraction $R(\vartheta)$.

Hausdorff presented his post-doctoral thesis no later than in April 1895, the month he applied for the *venia legendi* in astronomy and mathematics. This certificate had to be approved by the entire Philosophical Faculty at the University of Leipzig, since it authorized Hausdorff to offer courses on these subjects. In the first part of his thesis, which was published that same year as [H 1895, V: 215–296], he dealt with the theory of Laplace, who had derived his absorption formula under the assumption of constant atmospheric temperature, thus an exponential decrease in density.

Hausdorff replaced this overly simple assumption by a flexible formula for the refractive index μ that still made it possible to calculate the refraction integral. The absorption integral then yields a series expansion for $\frac{J_1}{J_\vartheta}$ in powers of $\ln\mu$, whose coefficients depend on ϑ and whose third-order terms can be neglected. The first-order term gives the Laplacian approximation, which had been criticized by the Munich astronomer Hugo von Seeliger, who noted its failure to correspond accurately with the absorption values measured by Friedrich Max Müller. This new result thus opened the possibility that by taking the second-order term in Hausdorff's series into account one might reach an improved correlation with Müller's findings. This, however, proved not to be the case; on the contrary, the differences between the theoretical and observed values became even greater than before. From this Hausdorff concluded that the theory of Laplace was basically unsuitable for describing the absorption of light in the atmosphere as a function of ϑ. He therefore proposed to develop a new approach that was wholly independent of measurements of atmospheric refraction.

Hausdorff thus dropped the hypothetical equation (2.7) for the relationship between refractive index and density, and in place of $dJ = Q\,J\,Dds$ put

$$d\ln J = f(s)ds$$

with an unknown "absorbent force" $f(s)$. He then introduced $\nu = \mu\,r$ as independent variables and letting ν be its value at a fixed but arbitrary point on a light path, ν_1 the value at a point on the boundary of the atmosphere, and denoting by ν' a variable point between ν and ν_1, then $\ln \frac{J_1}{J_\vartheta}$ will be a function of ν and ϑ for which

$$\ln \frac{J_1}{J_\vartheta} = \int_\nu^{\nu_1} \frac{F(\nu')\,\nu'\,d\nu'}{\sqrt{\nu'^2 - \nu^2 \sin^2 \vartheta}} \tag{2.9}$$

with an unknown function $F(\nu)$ that depends on $f(s)$ in a somewhat entangled manner. This $F(\nu)$ can also be interpreted as a kind of "absorbent force." The analogy with refraction theory becomes clear if one writes the refraction integral in the Brunsian form (2.1), where ν was designated as σ and ν_1 as τ:

$$R(\nu, \vartheta) = \nu \sin \vartheta \int_\nu^{\nu_1} \frac{m(\nu')\,d\nu'}{\sqrt{\nu'^2 - \nu^2 \sin^2 \vartheta}} \tag{2.10}$$

with $m(\nu)\,d\nu = -d\ln\mu$.

The analogy between (2.9) and (2.10) can now be used to derive a functional relationship between $\ln \frac{J_1}{J_\vartheta}$ and $R(\nu, \vartheta)$ by means of suitable conditions on $F(\nu)$. This leads to

$$\frac{R(\nu, \vartheta)}{\nu \sin \vartheta} = \int_\nu^{\nu_1} -\frac{d\ln\mu'}{\nu'\,d\nu'}\, \frac{\nu'\,d\nu'}{\sqrt{\nu'^2 - \nu^2 \sin^2 \vartheta}}\,.$$

Thus by setting $F(\nu) = -\frac{1}{\nu}\,\frac{d\ln\mu}{d\nu}$, one obtains

$$\ln \frac{J_1}{J_\vartheta} = \frac{R(\nu, \vartheta)}{\nu \sin \vartheta}\,. \tag{2.11}$$

This looks much like the formula (2.8) of Laplace. In order to clarify the difference with the usual Laplacian formula, Hausdorff called (2.11) a "functional relationship" or a "heuristic principle." This means that one can exploit the Bruns-Hausdorff approach to refraction by using its interpolation formulas for $R(\nu, \vartheta)$. These contain free parameters that can then be determined by connecting them with the values in an absorption table. In the usual application of the Laplacian formula, one took the refraction values $R(\nu_0, \vartheta_0)$ at ground level from a table and then used the formula to calculate the numerical absorption values.

Hausdorff's "heuristic principle" thus showed that one could use any permissible refraction formula as a corresponding absorption formula. Just as one could calculate $m(\nu)$ by following the Brunsian method, so could the "absorbent force" be found from a suitable absorption formula for $\ln \frac{J_1}{J_\vartheta}$. In effect, Hausdorff's "heuristic principle" enabled one to draw conclusions about the absorbent force from refraction formulas. However, certain cases led to absurd consequences, including some that corresponded very well with Müller's absorption tables, as at certain elevations they predicted a physically senseless "absorbent force" that actually amplified the light intensity. Hausdorff thus drew the following negative conclusion:

> The numbers in the usable extinction tables cannot be precisely represented by a physically useful absorption formula, namely, one in which the absorbent power of the layers of air is always positive. [H 1895, 422; V: 236]

Bruns' principle of inversion – the possibility of inferring data about the physics of the atmosphere from observed astronomical data – obviously failed in the case of absorption.

Ultimately, Hausdorff recommended the following formula for the zenith reduction as one of the main results of his work:

$$\ln \frac{J_0}{J_\vartheta} = \frac{a \tan^2 \frac{\vartheta}{2}}{1 + b \tan^2 \frac{\vartheta}{2}} . \tag{2.12}$$

Here $a > 0$ and $b > -1$ are suitably chosen constants.

A year after Hausdorff's work on the absorption of light in the atmosphere was published, Paul Kempf discussed his results in an extraordinarily detailed review [Kempf 1896]. In doing so, he severely criticized several of the results found by Hausdorff. At that time Kempf was professor and observer at the Astrophysical Observatory in Potsdam and one of the leading experts on problems concerning the photometry of celestial bodies.

The criticisms put forward by Kempf were primarily directed against Hausdorff's view that Laplace's theory was fundamentally unsuitable for correctly representing atmospheric absorption as a function of the apparent zenith distance ϑ. As mentioned, Hausdorff held this opinion because of inexplicable differences between the observational data in Gustav Müller's Potsdam absorption table and the values he calculated according to Laplace's theory. He had overlooked, however, that the Potsdam table was not drawn from homogeneous observational material but rather consisted of two distinct parts, which had been obtained in completely different and independent ways. The first part extended up to a zenith distance of 80° and was derived by comparing five fixed stars with the pole star at all possible zenith distances between 0° and 80°. For zenith distances beyond 80° Müller's procedure was based instead on planetary data. During a succession of particularly clear nights, he tracked large planets located between 80° and 90° zenith distance and found absorption rates for the respective zenith distances by determining the differences in two successive measurements. This two-fold division was unavoidable due to the latitude of Potsdam, which ruled out observations of stars near the zenith that move all the way to the horizon. If Laplace's formula was applied separately to these two parts of the Potsdam table, however, a perfectly satisfactory agreement resulted. Kempf also pointed out that Müller had obtained his observational data on the summit of the Säntis mountain under perfect weather conditions, and that his findings corresponded perfectly with Laplace's theory. Summing up, he concluded: "Laplace's theory of absorption corresponds fully with the actual findings within the accuracy currently achievable" [Kempf 1896, 16].

It is hardly surprising that a young researcher like Hausdorff, who was more theoretically oriented, was not familiar with the special character of Müller's extinction tables. What does seem astonishing, though, is that an experienced astronomer like Bruns could have made such a mistake. In his report on Hausdorff's *Habilitation*, Bruns did not take note of Müller's results at all, but rather – in agreement with Hausdorff – wrote:

> First it is shown that the till now preferred methods of Laplace, even when elaborated further, do not conform with the observations. The reason lies in the assumption that a simple relationship obtains between the refraction exponent and the absorption coefficient, when in reality this does not exist.

Furthermore, Kempf criticized the way in which Hausdorff had connected interpolation formulas with the data in tables, a procedure he called "absolutely inadmissible":

> From the series of observed values he takes as many as the formula has numerical parameters, determines the latter in such a way that those numbers are represented in absolute terms, and then uses these constants to calculate the entire course of the formula. That this method can under certain circumstances lead to very significant deviations in parts of the curve somewhat distant from the connecting points should be altogether evident. [Kempf 1896, 18]

As an example of the sensitivity of the formulas in this regard, Kempf focused on formula (2.12), which Hausdorff had recommended for practice. Hausdorff had determined the two constants a and b by requiring an exact match at $\vartheta = 85°$ and $\vartheta = 87,5°$ with the data in Müller's table. Kempf calculated a and b using the method of least squares, also on the basis of Müller's table, thereby obtaining a qualitatively different dependence on the zenith distance from the one Hausdorff had calculated; the two curves thus came out quite differently.

Kempf also connected formula (2.12) with the Säntis observations, again using the least squares method. This comparison provided another impressive example showing why a good fit with the observed data hardly meant that one could draw sound conclusions about the constitution of the atmosphere. Although the adapted formula (2.12) reproduced the data in both tables well, it provided the transmission coefficient 0.707 for Potsdam and 0.79 for the Säntis, an absurd result, since it implied that the light intensity was greater at the former location than on top of a mountain with an elevation of 2500 meters. As Kempf concluded:

> From this example one sees that a formula can very well satisfy the requirements set out by Bruns and yet be absolutely unsuitable for inferring properties of the atmosphere from it. The question thus arises as to when it is justified to assume that the physical conclusions drawn from the expressions for absorption resp. refraction can be regarded as realistic. [Kempf 1896, 21]

This question could only be answered on the basis of known meteorological data, which put the entire Bruns-Hausdorff procedure in doubt. "But then," wrote Kempf, "we revert to the understanding that the meteorological information constitutes the basis for the theory, an assumption that Bruns' method sought to avoid" [Kempf 1896, 22].

Hausdorff's *Habilitation* opened on May 3, 1895 with the formation of a commission, which included Bruns, Lie and Mayer. Since the candidate's qualifications were long known and well recognized, the outcome was surely a forgone conclusion. On May 20, Bruns submitted his report, which summarized the content of the thesis on a single page without noting criticism or praise. Only the last two sentences of his report contain an evaluation: "In my opinion, the treatise certainly meets the requirements to be placed on a habilitation thesis. Thus, for admission to the other requirements." Sophus Lie wrote under this: "I agree with the conclusion," and Adolph Mayer signed "likewise." These "other requirements" consisted of a colloquium followed by a trial lecture, the first for members of the commission and the second for the full faculty.

Hausdorff's collqouium took place on June 18, and his performance gained "the full satisfaction of the commission." For this public lecture, Hausdorff proposed the following three topics:

1) The Gaussian law of errors
2) The variation of the pole heights
3) The determination of solar parallax.

The commission decided unanimously on the first topic, and on July 25th the trial lecture took place, "before the dean and the commission [and] to general satisfaction." On the same day Hausdorff signed a pre-printed declaration stating that he had taken note of the fact that the *venia legendi* granted did not entitle him to support in the form of gratuities, fixed salary, or future acquisition of an extraordinary professorship. With its acquisition, though, Hausdorff became a private lecturer for astronomy and mathematics at the University of Leipzig, which opened up the possibility of a future scientific career.

In his first semester as a private lecturer (*Privatdozent*), the winter semester of 1895/96, Hausdorff offered two lectures courses: "Figure and Rotation of the Celestial Bodies" and "Map Projection." The first, which attracted just one auditor, touched only indirectly on astronomical topics. Instead, Hausdorff dealt mainly with theoretical mechanics, including gyroscopic theory, hydrostatics, and hydrodynamics, drawing on potential theory and even branching theory for nonlinear integral equations. The second lecture touched on geodesy, another research area pursued by Bruns. It was also mathematically demanding and only two auditors attended. Hausdorff offered it again during the winter semester 1900/01, when the course drew an audience of six.

During this time, Hausdorff published a paper on geometrical optics [H 1896, V: 315–366], his final publication in this field of research. This study, which followed directly on Bruns' monograph *Das Eikonal* [Bruns 1895], was still related to Hausdorff's work at the observatory, insofar as the theory and practice of optical instruments were of great importance for astronomy. Here, he examined anastigmatic images by studying certain single-parameter groups of contact transformations. These can be understood mathematically as images formed when the rays emanating from a point P in the object space (so-called centric bundles) are transformed into rays of the image space that converge in points P'. The sets M and M' of such points P and P' form the stigmatic sets. Ideal optical instruments would be those with three-dimensional stigmatic sets that are as extensive as possible, for example, if M is the entire object space then M' would be the entire image space.

Hausdorff restricted himself to purely dioptric (i.e. refractive) systems. His main result was that telescopic aplanasia – freedom from spherical aberration – cannot be realized dioptrically, i.e. an ideal telescope lens cannot be produced dioptrically. Since Hausdorff's paper was ignored in the contemporary literature and no references to it were made in more recent works on geometrical optics, we add only a few remarks. There was, however, one singular reaction to this paper, namely, a three-page note by Lie that appeared immediately following Hausdorff's article in the *Leipziger Berichte.* Lie's note merely wished to underscore his priority with respect to the application of the theory of contact transformations in optics. Although no names were mentioned, many readers would have understood this as a reproach directed against Bruns. Hausdorff referred to Lie several times in his article, whereas Bruns never mentioned him in *Das Eikonal.* Nevertheless, the fact that Lie's paper appeared directly after Hausdorff's was surely an unpleasant experience for him.

At the latest by 1896, when Kempf's criticism of his habilitation thesis appeared, Hausdorff must have realized that the path he had taken into science by following Bruns' ideas had led to a dead end. Not surprisingly, he began to turn away from astronomy in order to concentrate his energies on new areas of interest. During the years to follow, he explored various topics in mathematics while pursuing a range of ideas and interests in philosophy and literature. This intellectual shift took place at the same time as did other major events in his personal life.

2.3 Family Life

An event of great import in Felix Hausdorff's life was the early death of his father. Louis Hausdorff died on May 15, 1896 at the age of 53, while staying in the spa town of Bärenfels, about 30 km south of Dresden. In all likelihood, he was seeking recovery from an illness there, though nothing more is known about the circumstances of his death. His grave, located in the Jewish cemetery along the Berliner Strasse in Leipzig, has a tombstone marker that remains in excellent condition. In addition to the German inscription, another in Hebrew reads: "But the teachers will shine like the heaven's gleam, and those who led many to justice will shine always and forever like the stars." This text comes from Daniel 12.3 and is part of the prayer *El Male Rachamim* (God full of mercy), which is spoken for the soul of the deceased at the burial and each year on anniversaries marking the day of death. The selection of precisely this verse was perhaps also intended to honor the deceased as a Talmud scholar and leading personality in the circle of his fellow believers.

As his father's only male descendant, Felix Hausdorff had the task of dealing with all financial matters relating to the family. His sister Martha, who had been living in Prague since her marriage to the merchant Anton Brandeis, had already given him power of attorney, and so did his mother Hedwig and his sister Vally.

Louis Hausdorff's widow took possession of the firm "August Dietze & Co." located on Katharinenstrasse. The property at Hohe Strasse 1 initially went to all four of his heirs: Hedwig, Felix, Martha and Vally, but on December 1, 1896 the younger three agreed to give their shares to their mother, so Hedwig became the sole owner. The decision to transfer ownership of this property probably resulted from Vally's marriage. On December 8, she wed Anton Glaser, owner of a factory for iron constructions in Prague. Vally had been living in Leipzig, but after her marriage she, like her sister, lived in Prague. Felix now succeeded his father and became co-owner of the company Spinner and Weber, Hausdorff & Co. He was joined at the same time by Siegfried Hepner from Schöneberg, who had previously been only a limited partner, but now became a full co-owner with the right to sign contracts. The private lecturer Felix Hausdorff thus became financially secure as co-owner of a flourishing publishing house; its emergence has already been briefly described in Section 1.2.

Three years after Louis Hausdorff's death, a new chapter in his son's life began. At the end of May 1899, Felix sent Heinrich Köselitz a printed card with the following message:

Figure 2.1: The Graves of Hausdorff's Parents in Leipzig's Jewish Cemetery
Foto: Walter Purkert

> I have the honor to announce the engagement of my daughter Charlotte to Dr. Felix Hausdorff, Privatdocent at the University of Leipzig.
> Reichenhall, May 22, 1899 Dr. Sigismund Goldschmidt
> (*HGW* IX, 399)

On June 25,1899, Hausdorff then married Charlotte Sophie Friederike Goldschmidt in Bad Reichenhall. The following year, on February 1, 1900, a daughter was born to the young couple in Leipzig, their first and only child. The girl was named Lenore (usually later called Nora) and she was baptized on April 19, 1900 in Reichenhall according to the Protestant confession.

The three first names of Hausdorff's wife trace back to a noblewoman among her ancestors. One of her maternal great-grandmothers was Charlotta Sophia Friederika von Lützow, who was born in 1786 and came to occupy a special place in the genealogy of the Hausdorff and Goldschmidt families.[2] Prior to her marriage to the physician Baron von Eckoldtstein, she was court lady to Princess Therese von Thurn und Taxis and then lady-in-waiting to Queen Luise of Prussia. Luise was a famously popular queen, remembered for her negotiations with Napoleon at Tilsit, where she tried to obtain milder peace terms for Prussia. Charlotta von Lützow was, in turn, the daughter of the Württemberg Oberjägermeister Julius

[2]The genealogy of Hausdorff's wife is printed in (*HGW* IB, 451–453). Further information comes from Egbert Brieskorn's conversations with Hausdorff's daughter Lenore König.

Friedrich von Lützow and his wife Charlotte, a born Freiin von Franquemont. This *Freiin* (Baroness) von Franquemont was an illegitimate child of Duke Karl Eugen von Württemberg. At that time the illegitimate children of a reigning prince commonly received a noble title, in this case von Franquemont. Karl Eugen, by the way, had been a student of Euler's at Berlin in 1741/42 and was the duke against whose absolutist rule young Friedrich Schiller was moved to rebel. His ancestors can be traced over sixteen generations through a strictly male line leading back to Ulrich I., Count of Württemberg, who lived in the thirteenth century. Thus, this line of ancestry of Felix Hausdorff's wife had its roots in the high nobility of the Middle Ages.

The family of Charlotte Hausdorff's father, Sigismund Goldschmidt, also had interesting historical roots. Sigismund was born in Warsaw on March 13, 1845 as the second of three sons of the Jewish preacher Abraham Meyer Goldschmidt, one of Leipzig's first rabbis (see Section 1.3). However, he barely knew his biological mother Pauline, *née* Sternfeld, who died already in 1846 at the age of 26. For him, his mother was Henriette Goldschmidt, born Benas, who married Abraham Meyer Goldschmidt in 1853. Sigismund Goldschmidt studied medicine in Leipzig from 1864 to 1869. He must have served as a doctor in the Franco-Prussian War of 1870/71, because he later occasionally used the title "Stabsarzt a.D." (Staff Doctor, retired). From 1872 onward, he worked in Berlin as a doctor for the poor. These were physicians employed by the city of Berlin with the task of treating the needy free of charge.

This experience led him to devote himself to the fight against pulmonary tuberculosis, which was especially widespread among the poor. Goldschmidt believed that scientific progress in medicine would eventually succeed in defeating this insidious enemy of the human race. In 1894 he published a book about this disease. In it, he decried above all how tuberculosis claimed the lives of so many children. His compassion for the lot of the poor led him to criticize contemporary social conditions, and his book was one of the first that called for large public sanatoriums. His critique underscored how cramped and unsanitary housing conditions contributed to the spread of tuberculosis, a situation the state and society had an obligation to overcome in his view.

Alongside tuberculosis, Sigismund Goldschmidt also dealt intensively with asthma. He held the view that asthma was a type of neurosis, and he attached to this his hope that neurophysiology would eventually bring progress in treating this disease. He considered it not merely justifiable, but even altogether necessary to relieve the suffering experienced by patients from severe asthma attacks. These attacks often produced a mortal fear of suffocation, which could be reduced by administering drugs, such as morphine or opium; he, of course, recognized the dangers associated with this approach.

During their time in Berlin, four children were born to the couple Coelestine and Sigismund Goldschmidt: Charlotte, Sitta, Hannelore, and Edith, though Hannelore died of diphtheria at the age of six. Soon after the birth of their last child, the family moved to Bad Reichenhall, where Dr. Goldschmidt practiced as a spa physician. The Bavarian spa and health resort of Reichenhall was well suited for the treatment of most respiratory diseases, including tuberculosis, when such cases were not too severe. However, in Goldschmidt's opinion the spa facilities were less suitable for those suffering from asthma.

As a bath and spa physician, Sigismund Goldschmidt became an institution in Bad Reichenhall. Although a Jew, he was well integrated into the bourgeois society of his time; his entry in the list of the fourteen physicians on the spa staff differed little from that of his colleagues: "Dr. Sigismund Goldschmidt, general practitioner, staff physician a.D. (ret.), knight, etc., consulting physician and honorary member of the *Poliambulanza Internationale* in Venice, lives at Gewerkenstrasse No. 261 (consulting hours from 11–12 and 4–5 o'clock)." Sigismund Goldschmidt shared many values commonly held within the German *Bildungsbürgertum*. Not only did he firmly believe in science and progress, he also held a deep appreciation for German, and more generally European culture, literature, music and architecture. These views are evident from his handbook for visitors to Bad Reichenhall, which describes its history and the mentality of the people, but also describes touristic sights and cultural events conveyed by way of his own personal experiences [Goldschmidt 1892].

It was particularly tragic for the physician that his own wife Coelestine fell ill with tuberculosis not long after moving to Reichenhall. She was therefore sent to the then world-famous spa town of Meran in Southern Tyrol, a place considered particularly favorable for treatment of this disease because of its dry climate, which was mild and nearly windless, even in winter. She remained there until shortly before her death in Innsbruck on April 7, 1886. When she died, her children were not yet teenagers, and the youngest, Edith, who had lived with her mother in Meran, was only three years old, whereas Charlotte was twelve and Sitta eleven. Thus, Charlotte gradually assumed the role of caretaker for her youngest sister; this affected their relationship in later years, when Charlotte's dominance persisted to some extent. During their time together in Bad Reichenhall, the family household was in the hands of an Italian housekeeper.

The girls received their education at a higher girls' school that belonged to the religious order "Institutum Beatae Mariae Virginis." It was run by nuns (the English Misses, as they were popularly known) from the convent St. Zeno in Bad Reichenhall. This school at St. Zeno imparted the typical educational values deemed important for "daughters from good families" during the nineteenth century. Later, the English Misses also established an elementary school for girls that Lenore Hausdorff attended from May 1909 to October 1910, when her parents were in the process of moving from Leipzig to Bonn. The Hausdorffs often sent their daughter to stay with Grandpa Sigismund and Aunt Sitta in Bad Reichenhall when they were traveling abroad.

It was another stroke of bad luck for the physician, who had dealt so intensively with asthmatic cases, that his daughter Sitta also suffered from this disease. He was occasionally forced to treat her with morphine when she was suffering from a severe attack. Sigismund Goldschmidt believed, after all, that with proper care morphine addiction could be avoided. In any case, Sitta became addicted, though possibly only after her father's death in 1914. Sitta continued to live in Bad Reichenhall; she remained unmarried and died in 1930. Around 1900, Edith Goldschmidt moved to Leipzig, where she began her training to become a kindergarten teacher under her grandmother Henriette Goldschmidt. However, before she could complete her training, her father called her back to Bad Reichenhall so that she could care for her sister Sitta, who had fallen ill.

In 1908 Edith married Martin Pappenheim, a Jewish neurologist and psychiatrist born in Pressburg (today Bratislava). He had converted to Protestantism in 1908, but left the church in 1918, whereas Edith had been baptized as an infant. Her older sisters Sitta and Charlotte also converted to the Protestant faith; they were baptized in 1896 in the church of the French Reformed congregation in Berlin-Friedrichstadt. In 1911, Else Pappenheim, the couple's only child, was born in Salzburg; the family then moved to Vienna one year later. Martin Pappenheim eventually left his wife for another partner and divorced her in 1919. Edith remained with her daughter in their apartment in Vienna. To earn money, she worked in a communal kitchen and later in a nursery for working-class children, but she always had financial worries. Edith and her daughter Else often spent several weeks during the summers with Aunt Sitta in Bad Reichenhall or with the Hausdorffs in Greifswald or Bonn. Else eventually became a psychiatrist and later emigrated to the USA after the Nazis came to power in Austria. In October 1938, Edith was forced to vacate her apartment in Vienna, after which she went to live with the Hausdorffs in Bonn. This part of her life and her final fate are taken up in detail in 9.4.4.

Finally, we turn to the best known personalities from the family of Hausdorff's wife: Sigismund Goldschmidt's parents, Rabbi Abraham Meyer Goldschmidt and especially his second wife Henriette, who had raised Sigismund and his two brothers in their mother's place. Abraham Meyer Goldschmidt was a child of Eastern Jewry, having been born in Krotoschin in 1812. After attending the Gymnasium in Breslau, he worked as a tutor in Krakow and Warsaw for several years. Between 1839 to 1857 he served as a rabbi for the German-Israelite congregation in Warsaw, after which he assumed the same position in Leipzig up until his death in 1889. Although he came from an orthodox tradition, Meyer Goldschmidt was firmly committed to the ideals of the Enlightenment. He thus became a leading representative of moderate-liberal Judaism. During his more than 31 years in Leipzig, he worked steadfastly to promote mutual understanding and respect between Jews and Christians in the tradition of Lessing and Moses Mendelssohn. He thus urged Jews to integrate into the German nation and culture without abandoning the core of their religious identity. A characteristic statement comes from his ceremonial address at a memorial service in 1861 honoring Mendelssohn, where he said:

> We have entered the stream of German national and cultural development and promote it according to all the forces and means at our

Figure 2.2: Henriette Goldschmidt at Age 94
Frontispiece Photo from *Henriette Goldschmidt. Ihr Leben und ihr Schaffen* [Siebe/Prüfer 1922], Project Gutenberg Ebook Nr. 42651

> disposal. Not only have our personal relationships taken on the character of German life and German customs: our religion and lifestyle, too, have, so far as their essence permits, surrounded themselves with forms adapted to German culture and customs. Nevertheless, our religion has lost nothing by this; indeed, it has gained! (*Festrede bei der am 3. Januar 1861 vom Verein zur Förderung israelitischer Interessen in Leipzig veranstalteten Gedächtnisfeier Moses Mendelssohn's.*)

Louis Hausdorff and Abraham Meyer Goldschmidt knew one another well; for many years they worked together in the governing body of the German-Israelite Community Federation (DIB). Louis Hausdorff, as a representative of Conservative Judaism, surely disagreed with some of Rabbi Goldschmidt's teachings and the opinions he advocated. Nevertheless, they achieved a number of important things together, including the purchase of the birthplace of Moses Mendelssohn in Dessau (see 1.3). The young Felix Hausdorff surely heard many of Goldschmidt's sermons during his school years and perhaps afterward as well, and we can safely assume that the Goldschmidt and Hausdorff families occasionally came together privately. In fact, it is more than likely that this connection enabled Felix Hausdorff to meet his future wife, Charlotte, as well as her famous grandmother, Henriette Goldschmidt.

Already as a young woman, thus well before her marriage to Abraham Meyer Goldschmidt in 1853, Henriette was a bright and curious personality with remarkable energy (on Henriette Goldschmidt see [Siebe/Prüfer 1922], [Rapp 2011]). When measured against the then common image of a wife and mother, her life was, indeed, singularly eventful. Beginning with a very modest education in the provincial town of Krotoschin, she went on to found a college for women in Leipzig in 1911. Looking backward, she saw the revolutionary year of 1848 as having planted the seeds for her later involvement in the women's movement. Before her political activism, however, she spent her first years in Leipzig settling into the cultural milieu of the city, while caring for her three growing stepsons. Over time, though, she no longer felt fulfilled by her life as a housewife. During this period, she met Luise Otto-Peters and Auguste Schmidt, with whom she founded a women's educational association in 1865. That year they also invited women to attend a conference in Leipzig, at which a General German Women's Association (*Allgemeiner Deutscher Frauenverein*, or ADF) was founded. Peters was appointed chairwoman and Schmidt vice-chair, and in 1867 Goldschmidt was elected to the board, a position she retained until 1906. The founding of the ADF marks the beginning of the organized women's movement in Germany, which primarily fought for the rights of women to pursue education and employment in the workplace.

Henriette Goldschmidt promoted these goals through a busy lecture schedule and other activities. Particularly noteworthy were her efforts to prepare women for careers in early childhood education. By chance, she had visited a Leipzig kindergarten and thereby gained acquaintance with Friedrich Wilhelm Fröbel's methods for engaging young children by means of physical activities, games, and songs. After studying Fröbel's writings and gaining inspiration from his ideas, she developed a vision of women as caretakers with a "human-caring destiny." This led to her realization that careers in early childhood education could provide women a new and highly significant social task and position. Moreover, this vision merged with her desire to enable women to freely participate in cultural and social life. Under the impression of Fröbel's ideas, she decided "to make the educational profession the cultural profession of women," and in the following decades she worked relentlessly on this task.

In December 1871, she founded the Association for Family and Public Education (*Verein für Familien- und Volkserziehung*) with about 150 members; by 1913, it had around 500. Some were distinguished Leipzig citizens, university professors, and dignitaries, such as the mathematician Adolf Mayer. Hausdorff's mother Hedwig, her sister Natalie, and later he and his wife were also paying members. Natalie Hausdorff was one of the women on the board responsible for supervision and administration. In the fall of 1872, the association opened its first public kindergarten; others followed in 1874, 1876, 1889, and 1911. Already in 1872, as a second stage, a seminar was established for training future kindergarten teachers. Two years later, a third stage emerged when the association initiated a *Lyceum für Damen*, which served as a forum for holding scientific lectures. Finally, in 1911, the *College for Women* opened in Leipzig, the first institution of its kind in Germany. The financial means for its founding were provided by the Jewish music publisher Henri Hinrichsen.[3]

The work of Henriette Goldschmidt was increasingly appreciated by the city of Leipzig as well as the state. Shortly before the outbreak of the Great War in August 1914, Saxony's King Friedrich August awarded her the Order of Maria Anna for her services to education. By the time she died of old age on January 30, 1920, the College for Women had already been placed under the supervision of the Saxon Ministry of Culture and had thus acquired the character of a state educational institution, achieving national as well as international recognition. Since 1947, a street in Leipzig bears the name of Henriette Goldschmidt, and in the German Democratic Republic the college she founded became the *Fachschule für Sozialpädagogik 'Henriette Goldschmidt'*, today known as the *Berufliches Schulzentrum für Sozialwesen Leipzig, Henriette-Goldschmidt-Schule.*

In her later years, many guests, friends, and relatives came to see Henriette Goldschmidt in her home, among them the young Hausdorff couple with their daughter Nora. The Hausdorffs visited regularly on Sundays to greet the woman they called *Großjettchen.* During those visits they usually had coffee and cake, often together with other guests who were often interesting or colorful people. An inconspicuous sign of Felix Hausdorff's esteem for his wife's grandmother managed

[3]Hinrichsen, privy councilor of commerce, commercial judge and city councilor, was one of Leipzig's most generous benefactors. In 1926, he donated the gigantic sum of 200,000 Reichsmarks to the university for purchase of a famous collection of musical instruments, for which he was later awarded an honorary doctorate. His death came on September 17, 1942, when he was murdered in Auschwitz.

to survive the turmoil of two world wars. This can be found in a copy of his poetry volume *Ekstasen* from 1900. On the title page, in Hausdorff's small handwriting in the upper right corner, one finds this dedication: "To the most excellent of all grandmothers Frau Dr. H. Goldschmidt presented by the author."

Chapter 3

Hausdorff's Early Mathematical Works

From 1897 onward, Felix Hausdorff presented himself to the public in two completely different guises. Within the mathematical world he published the results of his research under his own name, whereas in broader intellectual circles he became known under the pseudonym Paul Mongré as a talented poet, stimulating essayist, and original philosophical thinker. During these years, sometimes the one side predominated, sometimes the other. Thus, from 1897 to about 1904, Mongré played the principal role, after which the mathematician Hausdorff gradually assumed center stage. This theme of Hausdorff's dual existence is discussed in detail in Part II of our book. The present chapter provides a prelude to that theme by describing how he took his first steps into the realm of mathematical research, a path which itself was most unusual for a mathematician who would later attain international fame. Indeed, as discussed in Chapter 2, he began his career with optical investigations in astronomy, a field far removed from the mathematical topics he took up later.

3.1 From Astronomy to Mathematics

Already during the time he was engaged in astronomical work, a possibility opened for Hausdorff to move into mathematics when Sophus Lie approached the talented young man. Lie was very interested in working with younger mathematicians who could help him make his ideas accessible to the mathematical public in the form of monographs. He had previously worked closely together with Friedrich Engel, who during the decade from 1884 to 1893 drafted the three volumes of Lie's fundamental work *Theorie der Transformationsgruppen.* A second such selfless supporter was Georg Scheffers, who took his doctorate in 1890 and then habilitated in Leipzig one year later. Scheffers was responsible for writ-

E. Brieskorn, W. Purkert, *Felix Hausdorff*, Vita Mathematica 21,
https://doi.org/10.1007/978-3-031-52135-5_3

ing up three extensive monographs based on Lie's lecture courses, including the lectures Lie delivered on the theory of contact transformations (*Geometrie der Berührungstransformationen* (1896)).[1]

Lie arranged to have Hausdorff prepare a similar book dealing with his integration theory for first-order partial differential equations. This probably took place before the beginning of Hausdorff's military service, which was roughly half over when Hausdorff wrote to Lie on 27 March 1892:

> First, I apologize for responding so late to your kind offer concerning the publication of your theory of partial differential equations; I was initially hindered, due to many duties and then by a six-week illness, from requesting an appointment to discuss this with Prof. Bruns, as I had earlier indicated that my answer depended on the outcome of that discussion. Recently I spoke with him, and I am very pleased to report that my astronomical work places no obstacle in my way regarding your honorable proposal. I could take up this project after completing my year of military service – before then any serious work would be unthinkable – along with work on the second part of my dissertation, which requires essentially only computations, assuming that you still as before wish to honor me with this position. (*HGW* IX, 435)

At the end of February 1893, Hausdorff had a final conversation with Lie about this matter, as can be seen from a letter, written by Hausdorff on 28 March 1894, in which he recapitulated earlier events. After that meeting, Lie must have considered the planned cooperation promising in view of other contemporaneous events. In 1892, Lie's former student Elling Holst began an earnest effort to bring Lie back to his native country, supported by leading Norwegian figures such as the Arctic explorer Fridtjof Nansen and the famous poet Bjørnstjerne Bjørnson. The latter wrote to Lie on 26 February 1893 to ask whether he would be inclined to return to Christiania if offered a professorship with a salary between 8,000 and 10,000 crowns. In his reply, Lie described the scientific conditions that especially tied him in Leipzig, including its excellent library and outstanding collection of mathematical models, as well as the university's financial support for his two assistants, Engel and Scheffers. Still more important, however, was the circumstance that in Leipzig he had "three unusually excellent young assistants (Engel, Scheffers, Hausdorff)," whereas he would hardly find comparable talents in Oslo [Stubhaug 2003, 420].

A little more than one year later, on 28 March 1894, Hausdorff wrote a long letter to Lie that reflects how their relations had changed in the meantime:

> Dear Professor!
>
> For a long time, I believe I notice your deep displeasure with me,

[1] The other two volumes were Lie's *Vorlesungen über Differentialgleichungen mit bekannten infinitesimalen Transformationen* (1891), *Vorlesungen über continuirliche Gruppen mit geometrischen und anderen Anwendungen* (1893).

> which I am not aware of deserving to such an extent, although I do not feel free of fault. As your former pupil, whom you honored in the highest way in both scientific and personal relations, I am much too bound to you not to be pained by anything troubling those relations. I beg of you, highly honored professor – assuming that your dissatisfaction stems quite rightfully from my inadequate efforts to further the work undertaken at your request – to permit me this, albeit very late, attempt at explanation (*Rechtfertigungsversuch*) in the form of a brief presentation of the facts.

Hausdorff then described a series of professional as well as personal circumstances that prevented him from dedicating himself to work on Lie's theory of partial differential equations, after which he continued:

> I have by no means abandoned my strong interest, highly honored Professor Lie, in the theories that carry your name, just as I hope never to lose the prospect of pursuing pure mathematics, whereas I am less inclined toward or capable in practical astronomy. Of course, I realize that you may have since found another, more appropriate person to do this work, but if this is not the case, I would be extremely glad if you would still count on me, despite the meager services I have performed thus far. Would you then, dear professor, through your kindness and patience be willing to grant me an extension that I might first complete my habilitation? Based on my assessment of the situation at this moment, I hope this will not involve an unusually long period of time. After my habilitation – this applies as a binding promise – I will turn directly to your partial differential equations. (*HGW* IX, 436–437)

Nothing came of these plans, however, as there is no evidence in Hausdorff's papers that he ever did any work on the book Lie had planned. One can therefore only speculate as to the reasons why this project failed. Lie's displeasure with Hausdorff would surely be understandable given that the latter had for two years apparently done virtually nothing after first accepting this work in March 1892. On the other hand, other factors may well have played a role in shaping Lie's attitude at this time. Lie's general behavior was probably affected by pernicious anemia, the disease that eventually took his life, though this diagnosis only occurred some years later. In the fall of 1893, however, he and his most loyal supporter, Friedrich Engel, had a major falling out [Purkert 1984], [Fritzsche 1991]. Hausdorff was surely to some extent aware that Lie's behavior toward Engel had changed radically. Apart from this, it would seem highly doubtful that Hausdorff would have agreed to work for Lie in the same selfless manner as Engel had long done. After completing his postdoctoral thesis, he was seeking a new orientation, one in which he could break out on his own and not only in mathematics. He had the same longing for creative work in philosophy and literature.

As we have seen, during Hausdorff's first semester as a private lecturer he taught two courses on topics in astronomy and geodesy. Beginning with his second semester in the summer of 1896, though, his teaching was almost exclusively confined to mathematics. Moreover, unusually for a private lecturer, he often taught introductory courses for younger students, such as analytic geometry, projective geometry, differential and integral calculus, or ordinary differential equations. These courses typically drew more than 20 students, whose fee payments provided Hausdorff with a modest additional income. At most universities, the senior professors taught these more lucrative courses, but in Leipzig the elderly Wilhelm Scheibner had reduced his teaching before retiring in 1898/99 for health reasons. His colleague Carl Neumann mainly taught courses in complex analysis and analytical mechanics.

Hausdorff also offered a broad spectrum of courses for more advanced students. Several of these reflected his own personal ambition to familiarize himself with areas of mathematics that he had not studied deeply as a student. Among the advanced courses that he taught, sometimes more than once, were: conformal transformations, theory of curves and surfaces, higher geometry (including non-Euclidean geometry), differential geometry, complex numbers and vectors (including hypercomplex number systems), probability theory, set theory, non-Euclidean geometry, analytical mechanics, theory of transformation groups, number theory, algebraic equations, and algebraic numbers. Beyond these fields stretching across the broader domains of geometry and algebra at that time, Hausdorff also enriched the Leipzig curriculum with courses he taught in a special applied field, namely, insurance and finance mathematics.

The situation that led up to this was the following. Since the so-called founding days, the insurance industry had developed into an important economic factor in the German Reich. In connection with this, efforts were made in the 1990s to set up university training for future specialists in this field, including the basics of actuarial mathematics. In the winter semester of 1895/96, a seminar for insurance science was founded in Göttingen, offering courses taught by the private lecturer Georg Bohlmann. Even before this, the Dresden Polytechnic offered regular lectures on actuarial mathematics, which led to the founding of a corresponding seminar in 1896. Leipzig was, however, Saxony's most important financial and commercial center, and so it was only logical that the government made an inquiry with the city's business community. In the summer of 1895, state officials asked the Leipzig Chamber of Commerce whether establishing a curriculum for actuaries along with a corresponding state examination might be a "significant state interest." The Chamber of Commerce decidedly rejected a separate course of study, but warmly welcomed lectures on matters relating to insurance. In response to this, the Saxon Minister of Education Paul von Seydewitz turned to the Philosophical Faculty of the University of Leipzig for further advice.

The faculty's response came in March 1896:

> the question of training for the insurance field had long ago attracted the attention of the local mathematicians and economists at whose instigation the private lecturer Dr. Hausdorff has decided to teach a relevant lecture course – at first to meet the needs of students of economics and statistics – in the coming summer semester: Introduction to the mathematical theory of insurance, [...] and next winter another on mathematical statistics. [Girlich 1996, 36]

In connection with this "instigation" (*Veranlassung*) of certain faculty members, it is interesting to read what Heinrich Bruns wrote to Felix Klein about this matter. His letter was occasioned by a vacancy in Göttingen, for which Hausdorff was one of the candidates under consideration. About him, Bruns wrote:

> H. habilitated here in astronomy and mathematics. He holds his fair share of the purely mathematical lectures, and we have also assigned him the theoretical lectures on insurance mathematics. For the latter task, the specific dispositions of his race (unbaptized) undoubtedly come in handy. (Bruns to Klein, 7 July 1897, SUB Göttingen, Cod. Ms. Klein 8, Nr. 131)

Bruns was a liberal man who had always supported Hausdorff, so he was by no means an anti-Semite; nevertheless, he obviously accepted this standard cliché.

In the summer semester of 1896, Hausdorff taught a course entitled "Mathematical Introduction to Insurance Practice." This dealt with compound interest and annuity calculations as well as basic probability theory needed for the construction of mortality tables and analysis of related problems, but above all for calculating the cash values (and thus the net premiums) of various pensions and insurances. The concept of risk was only briefly discussed. In his preliminary remarks, Hausdorff emphasized the yawning gap between theoretical insights and procedures that were useful in practice: "there is nothing to be warned against more than theoretical over-confidence, the mistake of translating from paper to real life." He also noted there that the course was prompted by "ministerial recommendation." In further fulfillment of this "ministerial recommendation," he twice offered a course on "Political Arithmetic" (summer 1897 and winter 1898/99). At the time, this notion meant interest calculations as well as repayments and annuities together with elements of probability and life insurance mathematics. Beyond this introductory course, Hausdorff taught "Mathematical Statistics" in 1897/98 and "Insurance Mathematics" during the summer semesters of 1898 and 1900.

In April 1898 the public commercial school (later commercial college) was founded in Leipzig. Some students from this business school, following its recommendation, attended the relevant lecture courses taught by Hausdorff. He must left a positive impression on them, because afterward he regularly taught courses there, beginning with "Political Arithmetic" in the winter semester of 1901/02.

This outside teaching went on until the winter semester of 1909/10 when he left Leipzig to accept his first professorship in Bonn.

3.2 First Publications

In retrospect, one can say that until the end of 1901, when he published his first works on set theory, Hausdorff had not yet found *his* field in mathematical research. Until that turning point arrived, though, he produced four mathematical works in four very distinct fields: actuarial mathematics, non-Euclidean geometry, theory of algebras, and probability theory. One could well imagine that he was testing these fields out, still looking for a new orientation. After 1901, he began to focus on set theory and its applications in topology and measure theory, the areas of his greatest success. As we shall later see, he was led to set theory primarily through his philosophical interests.

Hausdorff prepared manuscripts for all his lecture courses, written elegantly in his own hand. They are available for study today in his estate, and many are of almost printable quality. A comparison of the relevant manuscripts with Hausdorff's first four publications reveals that all were closely tied to his teaching activities: two resulted directly from courses, a third was inspired by a relevant lecture, and the fourth was motivated by a course he had planned to teach. Here we briefly discuss this work and its connection with the relevant lectures.

3.2.1 Risk in Random Games

Hausdorff's paper "Risk in Random Games," published in [H 1897a, V: 445-496], was anticipated to a large extent by a topic covered in his course on political arithmetic, which he taught during the summer semester 1897. This course already reflected some typical traits of the mathematician Hausdorff as can be seen in all his later work. He typically studied previous work critically, correcting errors and inaccuracies, while looking for possible generalizations, which often led to new fundamental concepts that he placed at the center of his own contributions.

In the present case, Hausdorff recognized that various applications of probability theory to finance and insurance could be described by one and the same mathematical model, namely, by discrete random variables that assume finitely many possible values. He refers to such a random variable as a *random game*. The most important application in his paper deals with an insurance policy running over n years for which the possible values of the random variable X are the differences between the payments of the insurance company and the policyholder in the individual years, each discounted from the beginning of the policy. The probabilities for each value are then based on certain probabilities for death or survival. The equivalence principle of actuarial mathematics then states that the random variable X, which can also be described as the loss to the insurer, has the expected value of zero, or – as Hausdorff expresses this – that the random game is a "fair game." The gross premiums, which the policyholder actually has to pay are, of course, higher than the amount calculated according to the equivalence principle.

Hausdorff's lasting contribution to actuarial mathematics was the introduction of the variance of X as a characteristic parameter. He considered the variance to be a far more appropriate measure of the risk to an insurance company than the usual one, the *average risk* $D = \frac{1}{2}\sum_{i=1}^{n} p_i|x_i|$ (x_i are the possible values of X and p_i their probabilities), and he designated var X as the *median risk.* He also uncovered a series of errors in the literature that had occurred in previous risk theory because of failure to distinguish carefully between dependent and independent events, resp. random variables. With the introduction and examination of the properties of the median risk, Hausdorff created one of the foundational concepts of so-called individual risk theory.

Hausdorff also calculated the median risk as well as the conditional median risk for a series of insurance policies, assuming that the policyholder survives during the first k years. All of Hausdorff's results were soon taken up in the specialized literature, especially in influential textbooks by Czuber, Küttner, Broggi and Berger [Czuber 1903], [Küttner 1906], [Broggi 1911], and [Berger 1939]. In his five-part study on the development and status of risk theory, the Munich mathematician Friedrich Boehm devoted no less than 23 pages to the discussion of Hausdorff's paper! [Boehm 1933–1939, 917–939].

Individual risk theory never gained much importance in practice, and after around 1930 it was gradually replaced by so-called collective risk theory. However, Hausdorff's concept of the variance of loss to an insurer remained an important quantity for evaluating a life insurance policy with a given performance and premium plan, for example when comparing two policies with the same payouts but different premiums. Although forgotten today, Hausdorff also gave the first correct proof of the theorem of Hattendorf, which is important in the theory of reserve capital. A history of topology or set theory or measure theory that failed to mention Hausdorff's name would be inconceivable. Yet in a history of actuarial mathematics – were one to be written – the name Felix Hausdorff should also not be missing.

3.2.2 Analytical Contributions to Non-Euclidean Geometry

At a meeting of the Royal Saxon Society of Sciences held on 6 March 1899, Hausdorff's former mentor Heinrich Bruns submitted his paper "Analytical Contributions to Non-Euclidean Geometry" [H 1899a, VI: 213–266].[2] This was Felix Hausdorff's only mathematical publication dealing with this subject, although his epistemological-critical writings were deeply concerned with non-Euclidean geometry, as can be seen from Section 4.9.

[2] The following remarks about this work were communicated to me by Moritz Epple, editor of *HGW*, Volume VI, to whom I express my sincere thanks.

Hausdorff used so-called Weierstrassian coordinates as his main mathematical tool both in this paper as well as in the corresponding lecture course. Wilhelm Killing used these in several publications, as well as in his monograph *Die nichteuklidischen Raumformen in analytischer Behandlung* (Leipzig 1885), to coordinatize the hyperbolic plane. First, one chooses an initial point O and perpendicular axes OX and OY. Then, from an arbitrary point P, drop normals PA und PB to the two axes to obtain

$$x = \sinh BP, \quad y = \sinh AP, \quad p = \cosh OP,$$

which are the Weierstrassian coordinates of the point P with respect to the given system of axes. Here x, y and p satisfy the equation $p^2 - x^2 - y^2 = 1$. Using these coordinates, one can easily explain some of the analytical relationships connecting the geometry of the hyperbolic plane with planar and spherical trigonometry. Hausdorff discusses how simple transformations of the hyperbolic plane in Weierstrassian coordinates are realized in the Beltrami and conformal image spaces. He also discusses, among other things, the geometric interpretation both for linear homogeneous substitutions of the Weierstrassian coordinates as well as for Möbius transformations of the hyperbolic plane, where he considers the analytical relationships between the Weierstrassian, Beltramian and conformal images.

Already at the outset of his paper, Hausdorff indicated how this train of thought could be developed in a lecture course that began by introducing Weierstrassian coordinates in non-Euclidean geometry, but then used Beltrami's mapping to pass over to the Cayley-Klein metric as a crowning completion of the theory [H 1899a, 161; VI: 213]. Soon afterward, during the winter semester 1899/1900, Hausdorff took exactly this approach in his lecture course on higher geometry.

Heinrich Liebmann mentions Hausdorff's work in two places in his well-known book *Nichteuklidische Geometrie* [Liebmann 1905/1912]. One concerns his treatment of Möbius transformations in the hyperbolic plane, the other the role of Weierstrassian coordinates. With regard to the latter, he quoted directly from the introduction to Hausdorff's paper, where he wrote that these coordinates are "the prescribed analytical instrument for introducing a non-Euclidean metric without projective [coordinates]".

3.2.3 On the Theory of Complex Number Systems

Hausdorff's course in the summer of 1899 on "Complex Numbers and Vectors" dealt with so-called hypercomplex numbers, such as the Hamiltonian quaternions and Grassmann's alternating numbers. His manuscript shows that he was familiar with most of the relevant literature, including [Weierstrass 1894], [Study 1898], and [Scheffers 1891] as well as Hermann Hankel's book [Hankel 1867]. In his lectures, Hausdorff presented none of the new ideas found in his paper "On the Theory of Complex Number Systems" [H 1900b, IV: 407–425], but one can assume these naturally arose in direct connection with his preparations for teaching this course.

In that paper, Hausdorff began with an associative $\mathbb{C}$-algebra A with identity of dimension n. To each such A he associated a new algebra called the index system of A, and along with it the algebra $End(A)$ of all $\mathbb{C}$-endomorphisms of A. The index system of A is the subalgebra of $End(A)$ generated by right and left multiplication by elements from A. Its dimension is called the index p of A, for which the inequalities $n \leq p \leq n^2$ hold. The case $p = n$ occurs exactly when A is commutative, whereas $p = n^2$ (i.e., the index system is identical with $End(A)$) precisely when A is a central-simple algebra. Hausdorff furthermore proved the following theorems: If the given algebra A is a direct resp. tensor product of two algebras A_1, A_2 with corresponding index systems I, I_1, I_2, then I is a direct resp. tensor product of I_1 and I_2. In the first case, $p = p_1 + p_2$, in the second, $p = p_1 \cdot p_2$.

In his commentary on [H 1900b], Winfried Scharlau conjectured that Hausdorff hoped by investigating index systems to find a new way to approach the classification problem for finite dimensional algebras. His paper was based on the classification of Study and Scheffers. Hausdorff determined the index systems for all three- and four-dimensional algebras, for which he found that only a five-dimensional non-commutative algebra occurs as index system for a three-dimensional algebra. From this, he concluded "that among the numerous types of complex number systems (with 5 units there are, according to Scheffers, already 33) very few have the character of index systems; it would be interesting to find simple criteria for this characterization" [H 1900b, 56; IV: 420].

Hausdorff evidently never pursued this idea further, however. His construction of index systems associated to an algebra A was later reintroduced under the name multiplication algebras. Nathan Jacobson dealt with this in Chapter V of his *Structure of Rings* (AMS Coll. Publ. 37, Providence 1956), though with no mention of Hausdorff. Scharlau closed his commentary with the comment that an in-depth examination of "multiplication algebras" using methods available today has yet to be undertaken, adding that "it could perhaps be rewarding to pick up this idea of Hausdorff's."

3.2.4 Contributions to Probability Theory

By far and away the most influential of the four publications described here was Hausdorff's paper on probability theory [H 1901a, V: 529–555]. This gave results he had presented in a lecture course taught during the winter semester 1900/1901. These lectures were for more advanced students and drew nine auditors, two of whom were, like Hausdorff, private lecturers: Gerhard Kowalewski and Heinrich Liebmann. Kowalewski mentioned in his memoirs that he and Liebmann "stood on the best of terms [with Hausdorff] and attended his higher lectures, for example, a wonderful course on probability" [Kowalewski 1950, 100].

Hausdorff's paper [H 1901a] consists of three specific investigations relating to completely different topics. The first concerns conditional probability, a concept soon recognized in the following years as truly fundamental, not least owing to Hausdorff's work. In order to illustrate the important progress made after

Hausdorff clarified this notion in 1901, one can compare his paper with the article on probability published by Emanuel Czuber in the *Enzyklopädie der Mathematischen Wissenschaften* [Czuber 1900]. This comprehensive survey of the existing literature up until roughly 1900 does not even explicitly mention the concept of conditional (or in Hausdorff's terminology relative) probability. Czuber implicitly discusses this in two different places, once under the heading "composite probability" (multiplication property for non-independent events), and elsewhere in a very different context under the heading "posterior probability" or "probability of causes."

In the latter case, Czuber's notion of contingent probabilities relates to the Bayesian formula (then usually known as Bayes' rule) for calculating *a posteriori* probabilities. He formulates this rule as follows [Czuber 1900, 759–760]: If an observed event E results from n independent causes C_i $(i = 1, 2, \dots n)$ for which the *a priori* probabilities are ω_i, then its *a posteriori* probability is

$$P_i = \frac{\omega_i p_i}{\sum_1^n \omega_i p_i},$$

where p_i is the probability of E given C_i. Here Czuber even makes a standard mistake, criticized by Hausdorff, of conflating independent events with those that are mutually exclusive [H 1901a, 158]. This presentation, often linked with Poisson's explanation by means of drawings from an urn, was standard in the textbook literature from around 1900, including even the notation p_i and ω_i (sometimes also $\overline{\omega}_i$). One still finds this in some textbooks published in the 1920s.

Hausdorff was thus the first to highlight the fundamental meaning of the concept of conditional probability and to introduce the symbol $P_F(E)$ (today usually written $P(E|F)$) for the probability of E given that the event F has occurred. Furthermore, he stressed that the Bayesian formula in the form

$$P_E(C_k) = \frac{P(C_k)\, P_{C_k}(E)}{\sum_i P(C_i)\, P_{C_i}(E)} \tag{3.1}$$

"is nothing but an elementary consequence of the first fundamental concepts and in no way subject to any particular criticism" [H 1901a, 161; V: 538], a viewpoint that appears to us today as self-evident.

The criticism of the Bayesian principle that Hausdorff had in mind appeared at the time in numerous philosophical debates with regard to the calculus of probabilities. This concerned the following problem: consider a Bernoulli experiment in which the event A has the unknown probability p. Suppose that after n trials the event A occurs k times: then based on this observation we are to determine the probability $P_{\alpha,\beta}(k,n)$ that $\alpha \leq p \leq \beta$, where $0 < \alpha < \beta < 1$.

The "Bayesian principle" is given by the following formula

$$P_{\alpha,\beta}(k,n) = \frac{\int_\alpha^\beta x^k\,(1-x)^{n-k}\,dx}{\int_0^1 x^k\,(1-x)^{n-k}\,dx}\,. \tag{3.2}$$

This is obtained from the Bayesian formula (3.1) by a transition to a continuous scheme of causes $C_x : p = x$ $(0 \le x \le 1)$ under the assumption the a priori distribution of p is a uniform distribution on $[0, 1]$. Hausdorff clearly recognized that no criticism could be directed against the Bayesian formula (3.1) or its continuous extension (3.2), which are proven within the mathematical theory. If problems arose, these could only pertain to the modeling behind the mathematics, i.e. to "freely disposing over the unknown a priori probabilities." In the present case, the criticism applies to modeling with a uniform distribution, which illustrates the "tendency, despite absolutely no knowledge and a lack of all numerical data, nevertheless to apply theorems from mathematical probability" [H 1901a, 161; V: 538]. With this critical remark, Hausdorff anticipated to a certain extent the debates that broke out decades later between different schools of mathematical statistics, in particular, between "Bayesians" and "frequentists."

Hausdorff did not intend to set forth any new results with his remarks about the concept of conditional probability. Instead, his aim was to elaborate clearly a fundamental concept – just as he did later in other areas of mathematics, for example topology, with great effectiveness. A side result of this effort was his definition of stochastic dependence or independence without recourse to content-related assumptions as to whether or not one event influences the occurrence of another, i.e. without recourse to non-mathematical realities. His hint at the end of §1, where he asserted that probability theory can dispense with all causal and temporal (i.e. non-mathematical) auxiliary representations and be grounded on a single logical function (later written $\omega \in E$), pointed far into the future.

Hausdorff's comments on the concept of conditional (relative) probability were taken up in a number of textbooks. Already in 1901, Bruns emphasized the importance of this work in recommending that Hausdorff be promoted to a titular associate professorship. His report mentions the paper "Risk in Random Games" [H 1897a] and then goes on to praise "Contributions to Probability Theory" [H 1901a]. Bruns regarded it as "even more important for treating the much disputed Bayesian principle, so fundamental in previous presentations of the discipline, in such a way that it eliminates the essential part of the difficulties." In his textbook from 1906, Bruns emphasizes that the Bayesian principle "appears as a simple corollary of purely arithmetic theorems" if one follows Hausdorff and "introduces the concept of 'relative' probability" [Bruns 1906, 31].

Hausdorff's remarks on conditional probability were widely disseminated through Czuber's textbook, a standard work in this field for decades. However, the first edition [Czuber 1903] contains no reference to Hausdorff – there Czuber's presentation follows the same line he took in his encyclopedia article [Czuber 1900].

Only with the second edition were the corresponding sections completely revised, and in the preface Czuber called attention to the "use of the concept of relative probability" [Czuber 1908, IV]. He also adopted Hausdorff's notation, in which the conditional event is written as an index, and he followed Hausdorff's presentation of the multiplication theorem. In a footnote, Czuber acknowledged this as the starting point for his revision:

> F. Hausdorff pointed out the important role that can be assigned to the concept of relative probability in laying the foundations of probability calculus. ... In fact, this concept permits a clear formulation and a transparent symbolic representation for many of the theorems that follow. [Czuber 1908, 45].

At first, Czuber writes the Bayesian formula as he did in his encyclopedia article. Afterward, however, he writes: "the whole presentation becomes more transparent when one makes use of the concept of relative probability" [Czuber 1908, 197]. He then writes down the Bayesian formula again, this time in the form (3.1) given by Hausdorff. Ugo Broggi also adopted Hausdorff's terminology and symbolism in his textbook on insurance mathematics, where the Bayesian formula appears with reference to Hausdorff as in (3.1) [Broggi 1911]. Independent of Hausdorff, the Russian mathematician A.A. Markov introduced a special form of conditional probabilities (though with no special term for it) in his famous textbook on probability theory [Markoff 1912]. He denotes the conditional probability $P(B|A)$ by (B, A), "the probability of the event B if the facts A are known" [Markoff 1912, 15].

The modern concept of conditional probability (more generally conditional expectation) is due to Andrei Kolmogorov. He first presented this in his book *Fundamental Concepts of Probability Theory* [Kolmogoroff 1933], a work generally considered as marking the birth of modern axiomatic probability theory. In the preface, Kolmogorov emphasizes that the conception of probability as a normalized measure had been known in certain mathematical circles for some time, and the same was true for expectation value as expressed by a Lebesgue integral. The fact that Hausdorff belonged to those circles is shown particularly impressively in his Bonn lecture on probability from the summer semester of 1923 (*HGW* V, 595–722); commentary by S.D.Chatterji, pp. 723–756). What Kolmogorov characterized as particularly new was the theorem establishing the existence of measures in sequence and function spaces, "but above all the theory of conditional probabilities and expectations" [Kolmogoroff 1933, III]. Conditional probability appears in this theory as a random variable; the classical conditional probability

$$P(A|B) = \frac{P(A \cap B)}{P(B)}$$

(Kolmogorov denotes this like Hausdorff with $P_B(A)$) then results from the general concept as a very special case.

The second special topic appears in §2 of [H 1901a], where Hausdorff provides a robust estimate for the "precision measure" of the normal distribution. Carl Friedrich Gauss wrote the density of the standard normal distribution in the form

$$\varphi(x) = \frac{h}{\sqrt{\pi}}\, e^{-h^2x^2},$$

where h refers to the precision measure. This formulation remained standard until around 1920, whereas today one obtains the usual form by means of the substitution

$$h = \frac{1}{\sqrt{2}\,\sigma},$$

where σ^2 is the variance.

For a sample $(X_1, \ldots, X_n)$, Hausdorff uses the ranking sizes of $(|X_1|, \ldots, |X_n|)$ to determine an estimate H of h. In obtaining H, the resulting size of $E(H-h)^2$ turns out to be about one and one-half times larger than the corresponding value for the maximum-likelihood estimate. However, Hausdorff emphasized the advantage of his estimate over the latter due to the ease in calculation, since it only requires knowing the rank ordering rather than the actual sizes of the observed values.

This already points to another peculiarity of Hausdorff's estimate, namely, that it is significantly less influenced by extreme values (so-called outliers) than is the case with the maximum-likelihood estimate; this makes it a more robust estimate than the latter. In this respect, Hausdorff touched on a central problem that has played a major role in modern statistical studies since the 1960s. In the modern literature on robust statistics, Hausdorff's estimate plays no direct role, though one would place it under the robust estimates in the realm of rank-ordered samples.

Nevertheless, Hausdorff's estimate soon found a place in the textbook literature. It appears in the second edition of the classic work of Friedrich Robert Helmert on *Ausgleichsrechnung* (curve fitting) and also in the second edition of Czuber's textbook, which refers to Hausdorff as well as Helmert [Czuber 1908, 270]. Also in *The Calculus of Observations* [Whittaker/Robinson 1924], a book especially well-suited for practitioners, the authors refer to Hausdorff's estimate in a detailed discussion of his method for calculating the precision measure.

The third and final paragraph of [H 1901a] is mainly concerned with the Central Limit Theorem (CLT). Hausdorff's principal goal is to account for how the mysterious Gaussian Law arises as a collective error from the interaction of numerous individual errors. For this purpose, he begins with some comments on the concept of randomness, a topic which deserves special attention from a historical point of view.

Random variables were studied in implicit form in various branches of stochastics, for example in the investigation of discrete distributions, error theory, and in statistics. But these ideas were only taken up systematically around 1900 in the Russian school of probability theory. Drawing on the earlier developments, its members established the notion of a random variable as a central and fundamental concept. It is particularly noteworthy, and shows Hausdorff's strong sense for fundamental concepts, that already in 1901 he formulated the concept of randomness independently; his "observation mode" (*Beobachtungsmodus*) [H 1901a, 167; V: 544] is nothing but the general concept of the distribution of a random variable. From the outset, Hausdorff deals with discrete as well as continuous random variables, and he even takes note of the general case in which the distribution function could have discrete jumps in addition to continuous sections. The modern definition of a random variable $X(\omega)$ on a probability space $(\Omega, \mathcal{A}, \mathcal{P})$ as a $\mathcal{A}$-measurable function $X : \Omega \to \mathbb{R}$ was introduced by Kolmogorov in 1933, though even into the 1950s the concept of a random variable was very often simply identified with the concept of a distribution.

Hausdorff introduces the expectation value operator, which he immediately applies to functions $f(X)$ of random variables X; for $f(X) = e^{uX}$ he obtains the generating function $\Phi_X(u) = \mathrm{E}\, e^{uX}$. He takes the important inverse formula for the Fourier transform from Gauss' unpublished papers and uses this to derive the relation $\Phi_{X+Y}(u) = \Phi_X(u)\,\Phi_Y(u)$ for independent random variables X, Y. All the derivations are so straightforward and clear that part of this Section 3 was included under the heading "Generating and Characteristic Functions" in a collection of classical texts on probability theory (alongside works by Laplace, de Moivre, and Lévy) [Schneider 1989, 432–434].

In addition to moments, the notion of semi-invariants (also called cumulants) plays an important role in probability theory and especially in mathematical statistics. Hausdorff was led to these parameters

$$M_k = \frac{d^k}{d\,u^k} \ln \Phi_X(u) \bigg|_{u=0},$$

which he called canonical parameters, by expanding the logarithm of the generating function in a power series. He evidently thought he was the first to introduce this idea, since he speaks of the "newly introduced concept of *canonical parameters*," but he was anticipated by the Danish astronomer, statistician and actuarial mathematician Thorvald Nicolai Thiele. In the original Danish edition of his book from 1889, he introduced semi-invariants recursively and called them "Halvinvarianter." Thiele's book later became widely known after an English translation appeared in 1903, so that most later authors refer to the Thiele semi-invariants. An exception is the Japanese mathematician T. Kameda, who wrote a lengthy work on generating functions which includes a section on "Canonical Parameters," where he cites Hausdorff's paper [Kameda 1915/16, 357–359]. One can assume that Hausdorff did not know Thiele's book when he wrote [H 1901a]; he therefore rediscovered the semi-invariants independently of him.

With the help of semi-invariants, Hausdorff was also able to formulate a sufficient condition for the validity of the Central Limit Theorem. Compared to

Hausdorff's version of the CLT, the versions published by Alexander Liapounov are much more general [Liapounoff 1900], [Liapounoff 1901]. These, too, were surely unknown to Hausdorff when he prepared his lectures for the winter semester 1900/01, which were the basis for his subsequent publication [H 1901a]. Later, Hausdorff carefully studied the works of Liapounov, which were little noticed outside of Russia. After doing so, Hausdorff helped draw the attention of other mathematicians to these important results. His references to them, for example, significantly influenced the work of Harald Cramér, who later wrote in his memoirs: "The work of Liapounov was very little known outside Russia, but I had the good luck to be allowed to see some notes on his work made by the German mathematician Hausdorff, and these had a great influence on my subsequent work in the field" [Cramér 1976, 512].

The intermediary between Hausdorff and Cramér was very likely Cramér's academic teacher Marcel Riesz, with whom Hausdorff was well acquainted. In a letter dated 2 November 1919, Hausdorff also referred Richard von Mises to Liapounov's results (*HGW* IX, 518).

Hausdorff's noteworthy remark "that the interaction of a large number of partial errors does not always lead to Gauss' Law" [H 1901a, 170; V: 547] played a special role in the reception history of his paper. He constructed the following example to illustrate this: If X_1, $X_2, \ldots$ are independent, identically distributed random variables with the densities $\varphi(x) = \frac{1}{2}\, e^{-|x|}$, then the distributions of

$$Z_n = \sum_{k=1}^{n} a_k\, X_k, \quad \text{where} \quad a_k = \frac{1}{(k+\frac{1}{2})\pi}$$

do not converge to a normal distribution but rather to a distribution with the density

$$1 : (\exp(\frac{\pi}{2}\, x) + \exp(-\frac{\pi}{2}\, x)) = \frac{1}{\cosh \frac{\pi}{2} x}.$$

Richard von Mises referred to this example in [Mises 1919]. There von Mises considered the behavior of distributions $W_n(x)$, which arise as a convolution product from n distributions $V_1(x), \ldots, V_n(x)$, for $n \to \infty$, for which he formulated a central limit theorem, i.e. sufficient conditions for the convergence of $W_n(x)$ to the normal distribution $\Phi(x)$. In a footnote he added: "For an example of a case where W_n does not converge to Φ, see F. Hausdorff, Leipz. Ber. 1901, p.166" [Mises 1919, 54]. If Hausdorff's paper might have escaped notice by earlier researchers, by around 1919 experts would have become aware of Hausdorff's example, since all were familiar with von Mises' works.

In 1928 Maurice Fréchet dealt with the question of how observational errors can be described mathematically [Fréchet 1928b]. He suggested broadening the study of experimental error laws beyond the Gaussian case and comparing other error laws with empirical findings. Fréchet referred to Hausdorff's example in order to argue that, even with additivity, the error for $n \to \infty$ does not have to lead to a Gaussian distribution. Paul Lévy interpreted Fréchet's article as a criticism of his book [Lévy 1925] and thought, in particular, that his reference to Hausdorff's example was intended in order to cast doubt on Lévy's version of the Central Limit Theorem. This led to an exchange of letters between Fréchet and Lévy [Barbut et al. 2004, 129, 134] and then in 1929 to a public dispute in which Hausdorff's example also played a role (*HGW* V, 581–582).[3] Lévy apparently considered this whole matter so important that he took it up again more than 40 years later in his autobiography [Lévy 1970, 82–83, 111].

The debate with Fréchet over Hausdorff's example led Lévy to an interesting conjecture. He had already shown that in the class of infinitely divisible distributions, the decomposition of a normal distribution into two independent components was only possible if the components themselves were normally distributed. The question he now raised was whether other distributions for the components might be possible if one no longer restricted the focus of study to the class of infinitely divisible distributions. In articles from 1934 and 1935, Lévy conjectured that this was not the case [Lévy 1934], [Lévy 1935]. Just one year later, Cramér succeeded in proving this conjecture [Cramér 1936]. From the correspondence Lévy–Fréchet it becomes clear that Lévy understood the true foundations of the Central Limit Theorem far better than Fréchet and many other researchers at the time. Today, there are many known examples of sequences $\{X_i\}$ of independent random variables with $\mathrm{E}\,X_i = 0$, X_i bounded, and where for no choice of the sequence $\{a_n\}$ will the distribution of

$$\frac{S_n}{a_n} = \frac{1}{a_n}(X_1 + \cdots + X_n)$$

converge to a normal distribution (see, for example, [Feller 1966, 502], problem 20).

Hausdorff's article [H 1901a] was nothing more than a casual piece of work, but it undoubtedly left its mark on the history of stochastics.

[3]Srishti D. Chatterji called my attention to the dispute between Fréchet and Lévy.

Part II

A Double Life: the Mathematician Felix Hausdorff and the Philosopher/Man of Letters Paul Mongré

Chapter 4

Paul Mongré as a Philosopher

4.1 Hausdorff's Path to Philosophy

As a student at Leipzig University, Hausdorff pursued very wide-ranging interests in his studies, which extended far beyond the standard courses in mathematics and natural sciences. In like manner, his path to philosophy was anything but straight and narrow. In this respect, he was an autodidact, even if he probably picked up ideas and influences from some of his academic courses. The only surviving autobiographical statement written by Hausdorff regarding his own individual road to philosophy appears in a letter written on 22 February 1904, thus more than ten years after he began that journey. He wrote about this in answer to an inquiry from a young librarian in Jena by the name of Franz Meyer. Already as a schoolboy, Meyer had discovered the writings of Paul Mongré and struck up an epistolary exchange with Hausdorff. Unfortunately, his letters are no longer extant, though some of the letters from Hausdorff to Meyer have survived.[1]

Meyer must have asked Hausdorff for suggestions about how to become a philosopher, as he received the following reply:

> What advice should I give you? You want to become a philosopher. If I take this literally and think of the scientific philosophers, the professional philosophers, I can recommend myself only as a deterrent counterexample. A tendency or developmental line that meant a lot for me personally and even appeared necessary, but seen from the outside was really coincidental, determined my sporadic philosophical knowledge: from Wagner to Schopenhauer, from there back to Kant and forward to Nietzsche, along with some contact through mathematics with philosophy of the exact sciences – that's all! [$\cdots$]

[1]These letters were discovered by the historian of science Uwe Dathe in the Meyer estate at the Jena University Library; they are published in [Dathe 2007].

E. Brieskorn, W. Purkert, *Felix Hausdorff*, Vita Mathematica 21,
https://doi.org/10.1007/978-3-031-52135-5_4

> But maybe I misunderstood you: You want to *become* a philosopher – that could well mean that you already *are* one in the deepest sense, and that you are already feeling the fantasy (*Flügelbrausen*) of a great dominating thought that you want to expound. Philosophy as your own answer to the question of the world, not as knowledge of the answers other minds have given before: that could be the goal that draws you [···] . If this is the case, if you are seeking to express in your personal way a response to the world – which, of course, does not even mean you have become aware of this – I could then offer you some advice. For that, though, I would have to know you somewhat better. (*HGW* IX, 503–504)

We already mentioned that during his first two semesters as a student in Leipzig, Hausdorff attended courses on the history of music taught by the musicologist Oscar Paul. Very likely Paul lectured on the works of Richard Wagner since he was a member of the Leipzig's Wagner Club and regarded him as "the greatest dramatic tonal poet of the present era" [Paul 1873, 570]. Not that Hausdorff needed to hear Paul sing such praises in order to be inspired by Wagner's musical achievements. In Hausdorff's literary estate one finds abundant evidence of such enthusiasm along with a quite extraordinary familiarity with Wagner's musical dramas. This interest sometimes surfaced in unlikely places, such as in the notes he took during some of his mathematics courses. He seems to have been bored at times during the geometry courses taught by Sophus Lie and Friedrich Schur. Mixed in with his mathematical notes, one finds numerous caricatures, small drawings and other doodling. Most striking of all, between the mathematical formulas and stenographic notes Hausdorff inserted musical motifs from all of Wagner's later operas: *Lohengrin, Rheingold, Walküre, Siegfried, Tristan, Meistersinger, Götterdämmerung* und *Parsifal.* Overall, he wrote down more than 30 motifs, given in a few melodic bars along with their harmonies.[2] Sometimes a single page contains motifs from several different operas, and in one case there are seven motifs from all the operas of the *Ring* and from *Tristan.*

The Leipzig musical season of 1887/88 reflected the city's intense interest in Wagner to a remarkable degree. Between January 1, 1887 and June 30, 1888 the Leipzig Opera staged 10 of his works, which were performed 89 times over the course of those 18 months, which represents 27 percent of all opera performances during that period. During the winter semester of 1887/88, for example, Felix Hausdorff had the opportunity to take in all of Wagner's operas with the exception of *Parsifal,* thereby providing him plenty of material for the musical motifs that he interspersed in his mathematical notebooks.

Friedrich Nietzsche admired Wagner as a young man, when he sometimes visited him and his mistress Cosima von Bülow in Tribschen, their country house near Lucerne. He eventually distanced himself from Wagner, who died in 1883.

[2]Assisted by some outstanding experts on Wagner's music, Egbert Brieskorn identified all these motifs; the results can be found in (*HGW* IB, 107–108).

Five years later, in *The Case of Wagner*, Nietzsche attacked him as a seducer and a "crafty rattle snake." We know nothing about the kind of seductive influence Wagner's music exerted on Felix Hausdorff. Certainly, though, the Wagnerian experience during his first two semesters at the university had a powerful effect on him. Wagner's elixir was definitely stronger than anything served up in the lectures he heard, and his music may have provided the young student an orientation within the field of tensions that lie between philosophy, religion and a general scientific worldview. For as the passage from his letter cited above makes clear, Hausdorff was not led to philosophy by the lectures of a university philosopher but rather by Wagner's works; it was *not* by reading a systematic highly scientific book, but rather his interest resulted from a rousing emotional experience.

Despite his enthusiasm for the multi-faceted Wagnerian "Gesamtkunstwerk," though, Hausdorff must have had ambivalent feelings regarding the personality of Richard Wagner, whom he probably knew to be anti-Semitic. Already in 1850, Wagner wrote a first form of his essay *Das Judenthum in der Musik*, which he published under a pseudonym. Then, in 1869, he published an expanded version of it as a brochure, this time under his own name [Fischer 2000]. Wagner's perhaps more latent anti-Semitism became much stronger once Cosima von Bülow, an extreme anti-Semite, entered his life and then became his wife in 1870. Afterward, this attitude became the unquestioned consensus view within the inner-circles of Wagner's admirers, who later made Bayreuth into a shrine for defenders of a radically racist Germanic cult. One of the pages in Hausdorff's notebook from Lie's course on projective geometry suggests, in fact, a more distanced view of Wagner. It shows a motif from the *Vorspiel* to *Götterdämmerung* together with the monogram **RW**, Richard Wagner, around which Hausdorff drew a snake (see *HGW* IB, 104). The idea of linking Wagner with a snake, the Old Testament symbol of everything insidious and devilish in this world, was surely Hausdorff's alone; he could not have known that Nietzsche would later call Wagner a "clever rattlesnake."

"From Wagner to Schopenhauer" – this path was not as unusual as it might at first seem. Ever since *Tristan und Isolde*, which was first performed in 1859, the influence of Schopenhauer on Wagner was unmistakable, and thereafter all his music reflected the philosopher's gloomy pessimism. After reading Schopenhauer's famous book, *Die Welt als Wille und Vorstellung* (*The World as Will and Representation*), Wagner wrote to Franz Liszt on 16 December 1854:

> His main idea, the final renunciation of the will to life, is of terrifying consequence, but ultimately redemptive. For me, of course, the idea was not new, and no one can conceive of it at all who does not already live it. Still, it was this philosopher who awoke in me real clarity. When I think back on the storms of my heart, the terrible struggle, with which it – against its will – clings on to the hope of life, yes, when they often swell into a hurricane – so I have indeed found a sedative that finally helps me sleep during sleepless nights; it is the warm and innermost longing

> for death: full unconsciousness, complete non-existence, disappearance of all dreams – the only final salvation! [Kloss 1919, 42]

Nietzsche later wrote in the fourth Section of *The Case of Wagner* that Wagner's ship was initially quite merrily on an optimistic course toward the Golden Age that would throw conventional morality overboard, but then it struck a reef and went aground; "the reef was the Schopenhauerian philosophy." Wagner himself felt great respect for Schopenhauer and sent him a copy of his *Ring of the Nibelungen* along with a letter expressing his deep appreciation.

Among nineteenth-century philosophers, Arthur Schopenhauer was the first to become famous as an outsider. Very early on, he gave up all attempts to pursue a university career. Instead, he lived as a bachelor from his inheritance. His principal work, *The World as Will and Representation*, was first published in 1819. By 1835, sales remained poor, forcing the publisher to sell the remainder as waste paper. In 1844, Schopenhauer published an expanded new edition, but the wider reception of this work only began around 1850. In 1859, a third and still larger edition appeared. All of Schopenhauer's other works elaborate on the central idea of his *magnum opus*, which begins with this sentence:

> "The world is my representation" – this is a truth that applies to every living and comprehending being. Yet only humans can bring this truth into reflective abstract consciousness: and one who really does this will enter into a state of philosophical calm. [Schopenhauer 1999, 31]

This is essentially Kant's position that the things of this world are given to us only as phenomena. Thus, "philosophical calm (*Besonnenheit*)" will occur for those who accept this point in Kant's philosophy. Schopenhauer himself called this Kantian theory the entrance gate to his own philosophy. "Kant's greatest merit is the distinction between appearances and the thing in itself," as he wrote in his "Critique of Kantian Philosophy," the appendix to Volume I of his book [Schopenhauer 1999, 534]. The thing in itself (*Ding an sich*), according to Kant, lay beyond the realm of all possible human experience. Thus, metaphysics based on knowledge of the thing in itself could never, in Kant's view, be a scientific philosophy.

Schopenhauer did not want to accept Kant's views regarding the "thing in itself," i.e., his rejection of metaphysics. If one wants to go beyond ideas (*Vorstellungen*), then one is confronted "with the question of the thing itself, and to answer that is the subject of my entire work, just as it is for all metaphysics altogether" [Schopenhauer 1999, 567]. For Schopenhauer, the formula "thing in itself " was synonymous with "innermost being." Thus, one cannot approach the thing in itself from the outside, as all philosophers before him had attempted [Schopenhauer 1999, 150]. Access to the interior of the world lies *in ourselves*:

> The recognizing subject, which occurs through its identity with the body as an individual, is given in two very different ways: once as a representation in an understandable image, as an object among objects

> that obey certain laws; but then at the same time in a completely different way, namely as that directly known given, which is designated by the word will. [Schopenhauer 1999, 151]

Will is here not only intentional conscious desire, but above all an inner urge that Schopenhauer also referred to as the "will to life," and which expresses itself in unconscious processes such as breathing, digestion, and the beating of the heart. This will to life is characteristic of all living things. Human beings are thus part of nature as a special type of animal with a bodily nature that, like any other animal, wants to maintain its life and must do so, and – to use Darwinian language – in the struggle for survival brings suffering to other beings and to his fellow human beings. As Schopenhauer expressed this:

> This general struggle attains the clearest visibility in the animal world, which nourishes itself on the plant world, and in which every animal becomes the prey and food of another, [· · ·] until in the end the human race, because it overwhelms all the others and regards nature as a product for its use, but also its fellow humans [· · ·] reveals in itself in the most terrifying clarity that struggle, that self-division of the will, and becomes *homo homini lupus* [man is a wolf to man] [Schopenhauer 1999, 208]

Little wonder that Schopenhauer has sometimes been called a Darwinist before Darwin.

One of the dominant ideas in philosophy in the wake of German idealism, especially the Hegelian variety, was the idea of progress. This provided a goal for all human actions and ultimately for the development of mankind; humanity was destined to attain a state that would lead to some kind of ideal life in the future. The ruthless honesty with which Schopenhauer portrayed the position of mankind in nature, after two millennia that saw him as the crown of God's creation, was naturally unacceptable, and not only for academic philosophers but for liberal and progressive thinkers in general. His writings were, therefore, largely ignored by university philosophers. For his part, Schopenhauer often scolded the "university philosophers," those who profess their ideas to earn a living, with very drastic words.

During the latter half of the nineteenth century, the idea of progress was increasingly called into question along with its associated optimism. In Germany, particularly after the revolution of 1848 had failed, Schopenhauer's work drew increasingly strong attention. Especially among writers, artists, and wide circles of educated laypeople he became the philosopher of choice. Those who accepted his honest appraisal of the human condition felt disturbed and shaken by his knowing description of the sufferings inextricably linked with life, by his rejection of all meaning deduced through reason, and by his dark pessimism. One of Schopenhauer's oft-quoted sayings comes from his *Senilia*: "The world is hell, and the people are, on the one hand, its tormented souls and, on the other, its devils."

Nevertheless, his followers and probably he himself could hardly have imagined what the coming century would bring in destruction of human and other forms of life at the hands of human beings.

It seems likely that Hausdorff studied Schopenhauer intensively during his first semesters at the university. In a booklet containing his transcription of a mathematical lecture that he attended during his fourth semester in Berlin, one finds a drawing that looks like an illustration for the sentence cited above. It shows a figure bound to a wheel that rotates over a fire, a shape. A human skeleton serves as the mechanism that drives the rotation of the wheel, whereas below this scene a larger skeleton beats a smaller one. A second indirect indication from this time suggesting that Hausdorff dealt with Schopenhauer's philosophy of suffering and his ethics of compassion can be found in another booklet. There, he listed keywords from the *Veda* and relating to the Hindu priesthood and various aspects of Hindu philosophy. Schopenhauer had connected his own philosophy with such elements, informing his readers that those already consecrated in ancient Indian wisdom had the very best preparation for understanding what he had to tell them. Schopenhauer was especially fond of repeating the Sanskrit saying from Vedantic Hinduism *Tat Twam Asi* ("You are That"). Those who are capable of expressing to themselves this thought "with clear knowledge and firm inner conviction for every being with which they come into contact, such persons are assured of all virtue and bliss on the straight path to salvation" [Schopenhauer 1999, 483]. Because those who practice works of love see through the veil of Maya and the deception of the principle of individuality. They recognize themselves in every being and consequently also in their suffering. This love (*caritas*) is compassion, and for Schopenhauer the straight path to redemption leads to the annihilation of the will to live.

This philosophy, presented in suggestive language and appealing to elements of experience and self-consciousness accessible to everyone, created a deep shock for many who were receptive to it. For some, however, including Friedrich Nietzsche, this initial attraction to Schopenhauer's teachings eventually led to a strong sense of repulsion. As a young man, Nietzsche praised him in "Schopenhauer as Educator," the third of his essays *Untimely Meditations.* Later, in *Nietzsche contra Wagner*, he wrote "Wagner, like Schopenhauer – they deny life, they slander it, so they are my antipodes" [Nietzsche KSA, 425]. Something similar must have occurred with Hausdorff, though only a few scant sources report on the early receptive phase. In particular, some passages in *Sant' Ilario* and in *Chaos in Cosmic Selection* reveal the strong effect Schopenhauer's metaphysics exerted on the young Hausdorff, though mainly by showing how he attempted to overcome this pessimistic worldview. We will return to this briefly when discussing these works, noting here that most of what Hausdorff wrote about Schopenhauer stems from a later time when he had absorbed Nietzsche's sharp critique of the older philosopher's pessimism.

In many ways, Hausdorff's path "from Wagner to Schopenhauer" and then "back to Kant and forward to Nietzsche" was a natural one. In fact, Schopenhauer's book *The World as Will and Representation* contains an appendix with the title "Criticism of the Kantian Philosophy." This begins with exuberant praise of Kant's work and closes with these words: "Incidentally, Kant's works have no need for my weak praise, as they themselves will forever praise their master, and they will, if perhaps not to the letter yet nevertheless in his spirit, continue to live on earth" [Schopenhauer 1999, 532]. Indeed, Schopenhauer wrote this appendix in order to justify his own theory,

> ... since this does not agree with the Kantian philosophy in many points and, indeed, stands in contradiction to it. A discussion about this is nevertheless necessary since my chain of thought, however different from the Kantian, stands under the influence of it, necessarily assumes it, and I confess that the best of my own ideas arising from impressions of the world are indebted to Kant's works and the divine Hindu writings and those of Plato. [Schopenhauer 1999, 533]

Thus, this part of Hausdorff's intellectual journey, which took him from Schopenhauer to Kant, was practically foreordained. As for university philosophy, the situation in relation to Kant was completely different than with Schopenhauer. In the wake of the neo-Kantian movement of the 1870s and 1880s, Kant's philosophy became the focus of quite varied discussions and many German philosophy professors offered courses on his works and ideas. Hausdorff surely gained exposure to Kantian philosophy in some of the courses he took in Leipzig, but also during his semester in Berlin, where he attended Friedrich Paulsen's tutorial concerned with Kant's *Kritik der reinen Vernunft* (*Critique of Pure Reason*). Hausdorff's knowledge of Kantian philosophy was, of course, by no means confined to what he learned from Paulsen or by reading other neo-Kantians, such as Friedrich Albert Lange, Kuno Fischer or Otto Liebmann, all of whom appear in his notes from Paulsen's course. As Hausdorff's later philosophical writings make clear, he studied Kant's own works very closely. Nevertheless, one might well imagine that Hausdorff's later rejection of a priori arguments and his suspicion of transcendental philosophy were influenced by Paulsen's views. At any rate, one finds in his notes from that course an original idea that he later developed in *Chaos in Cosmic Selection*, namely a new argument to replace Kant's proof for the transcendental ideality of time.

Despite his later criticisms, Hausdorff considered the Kantian philosophy in one essential point secure, namely regarding the "unconditional dualism separating appearance from the thing in itself (*Ding an sich*)." For in his view, "here we can save the tuition that the older philosophy has already spent and connect with knowledge that has slipped into our thinking, though without drawing expressly on the results of previous intellectual work, namely the realization that the world given to us through experience is the phenomenon of our consciousness, but that – possibly – a world 'in itself' also exists *independent of our consciousness*" [H 1898a, 1–2; VII: 595–596]. In particular, there is no "transition strip" between the two worlds, no "protrusion of one world into the other." We will later discuss how

Hausdorff drew on Kant when he wrote his *Chaos in Cosmic Selection*, but we next consider the vital next step, "forward to Nietzsche," which led to his mature philosophical views.

4.2 To Know and Honor Nietzsche's Works

After Schopenhauer, Friedrich Nietzsche was the second great outsider among the German philosophers of the nineteenth century. However, his influence, which continues to the present day, far exceeded that of Schopenhauer. The son of a Protestant pastor, Nietzsche began studying classical philology and theology in Bonn during the winter semester of 1864/65. Impressed by the writings of the Young Hegelians – Bruno Bauer, Ludwig Feuerbach and David Friedrich Strauss – he gave up theology and after just one semester left Bonn for Leipzig, where he devoted himself to studying classical philology. On the basis of his publications and the recommendation of his Leipzig mentor Friedrich Ritschl, the 25-year-old Nietzsche was appointed to a professorship in classical philology at the University of Basel in 1869. A decade later, however, he gave up this position, partly for health reasons, and began living as a freelance writer, supported financially by his modest pension from Basel. He tried to mitigate his chronic suffering through long stays in places with a pleasant climate, especially northern Italy, southern France, and Switzerland. While vacationing in Turin, however, he suffered a severe mental breakdown in January 1889 and, after shorter hospital stays, lived thereafter under the care of his mother and later his sister. He never recovered, though, and became increasingly mentally deranged up until his death on August 25, 1900.

Ever since 1894, when Lou Andreas-Salomé published her book about Nietzsche, most interpreters of his philosophical works have divided these into three periods. The first or early period, from 1872 to 1876, reflects the influence of Wagner and Schopenhauer. It was during those years that he wrote *The Birth of Tragedy from the Spirit of Music* (1872) and the four *Untimely Meditations*: *David Strauss, the Confessor and Writer* (1873), *On the Use and Abuse of History for Life* (1874), *Schopenhauer as Educator* (1874) and *Richard Wagner in Bayreuth* (1876). During the second (middle) period, starting at the end of 1876, Nietzsche increasingly distanced himself from Wagner while overcoming Schopenhauerian pessimism. He now emerges as the "free spirit" who no longer writes conventional essays, but rather adopts an aphoristic style instead, as in *Menschliches, Allzumenschliches* (*Human, All Too Human*) (1878–1880), *Morgenröte* (*The Dawn*) (1881) and *Die fröhliche Wissenschaft* (*The Gay Science*) (1882).

At the end of this middle period, Nietzsche produced one of his central works: *Thus Spoke Zarathustra*, a book published in four parts between 1883 and 1885. It was written in hymn-like prose or, as he described it, in halcyon tones that conveyed his idyllically happy and peaceful state of mind. In it, a narrator reports on the ideas and the life of a fictional thinker who bears the same name as the Persian religious figure. One finds in this work several important ideas associated with Nietzsche's philosophy: the death of God, already proclaimed in *The Gay Science*, "Will to Power," the vision of a future "Übermenschen," and the oppressive nihilistic idea of "Eternal Return of the Same." One often reads of these things as elements of Nietzsche's teachings, though he never wanted to teach philosophy. At the very end of his *Zarathustra*, he describes the inevitable failure of all teaching, which can never form a person's individuality or influence someone's own experiences and their specific view of the world. On the contrary, teaching requires general norms that treat everyone the same way. Almost all of the shocking ideas in Nietzsche's philosophy, such as his critique of moral principles, arose from the contrast he drew between individuality and the larger educated public.

The works from the late period, 1886 to 1888, develop many earlier ideas and radicalize them, often with unprecedented polemical sharpness and aggressiveness. (Some authors believe that the last works written in 1888 already reveal signs of his latent mental disorder.) *Jenseits von Gut und Böse* (*Beyond Good and Evil*) (1886), a work consisting of longer aphorisms, belongs to this period, as well as the treatises *Zur Genealogie der Moral* (*On the Genealogy of Morality*) (1887) and *Der Fall Wagner* (*The Case of Wagner*) (1888). Also included in this period are writings that only appeared in print after his mental collapse: *Götzen-Dämmerung* (*Twilight of the Idols* (1889), *Der Antichrist* (*The Antichrist*) (1895) and *Nietzsche contra Wagner* (*Nietzsche contra Wagner*) (1895). His late work also includes the autobiographical sketch *Ecce homo* from 1888, which appeared in 1908 in a falsified edition published by his sister Elisabeth Förster-Nietzsche.

As a philosophical thinker, a poet and writer, and as a brilliant stylist and biting critic, gifted with great psychological empathy, Nietzsche was a singular phenomenon in European intellectual life and afterward well beyond it. However, his fame came only after considerable delay and took hold essentially only after his mental collapse; he himself experienced only the very beginnings of this acclamation. Nietzsche never wanted to develop a philosophical system: "The will to systematize derives from a lack of integrity", he says in *Twilight of the Idols*. Above all, he was concerned with critically questioning traditional moral concepts, religion, philosophy, science and, to some extent, art as well, while offering food for thought for a new orientation. After he had overcome Schopenhauer's pessimism, the focus of this reorientation was a radical affirmation of life. His "deepest conception," that of the eternal recurrence of the same, can only be endured if life is affirmed, no matter how difficult that may be:

> If that thought got hold of you, it would transform you and perhaps crush you; the ever-recurrent all-embracing question 'do you want to experience this again and countless times more?' would weigh as the heaviest burden on your actions! Or how would you have to become good to yourself and life in order *to long for nothing more* than this last eternal confirmation and seal?[3]

[3]Friedrich Nietzsche: *Die fröhliche Wissenschaft* (*The Gay Science*), No. 341. [Nietzsche KSA, Volume 3, 570].

We shall see that Hausdorff's thorough preoccupation with the idea of eternal recurrence was an essential point in his reception of Nietzsche.

Nietzsche's critique of traditional ideas of morality, as for example in *On the Genealogy of Morality*, was based on his detailed examination of the origin and development of moral norms and value judgments. He saw the key to liberation from traditional morality as based above all on overcoming the precepts of the Christian religion. Thus, he writes in the fifth book of *The Gay Science*: "The greatest recent event – that 'God is dead', that belief in the Christian God has become no longer believable – has already begun to cast the first shadow over Europe." The "free spirits" do not fear the "long fullness and succession of demolition, destruction, downfall, overthrow that is now imminent," on the contrary:

> In fact, we philosophers and "free spirits" feel, upon receiving the news that the "old god is dead" as though bathing in the warmth of a new morning sun; our hearts overflow with gratitude, astonishment, foreboding, expectation – finally the horizon appears clear to us again, even if not yet bright, finally our ships may set sail again, sailing to meet any danger, every risk to the knowledge-seeker is again allowed, the sea, *our* sea is open again, perhaps there has never been such an "open sea." [Nietzsche KSA, Volume 3, 573f.]

For Nietzsche, world history had no goal, no determination. It is an endless process of becoming and passing away, of creating and destroying. "God is dead" also means: In addition to this world of ours, there is no second, higher, ideal world, no eternal ideas, no hereafter, no thing in itself that would concern us. Any form of metaphysics is to be rejected. There is no higher destiny to which the individual should subordinate his individuality.

Based on these assumptions, one could now reconsider the human condition from the standpoint of an individual's position in society and their actions within it. Thus, after turning his back on Schopenhauer, Nietzsche chose to face the consequences of his life-affirming philosophy, beginning with his critique of compassion as a principle of moral action, an argument that runs through his entire work. Compassion, for him, was not a virtue but a weakness; it harms those who act on it by weakening their life forces. It also harms the afflicted because it reveals to them their powerlessness and weakness, thereby increasing their suffering and humiliation.[4] Consequently, individuals should follow their own law, independent of traditional morality.

As far as philosophy in general was concerned, in his late work *Twilight of the Idols* Nietzsche sought to smash it altogether. In particular, he criticized the neo-Kantianism of his time, which had further developed Kant's critique of knowledge into a formal epistemology. In *The Dawn*, Nietzsche wrote that just as a tool cannot "assess its own excellence and suitability," so can an epistemology, in which

[4]See Gudrun von Tevenar: *Nietzsche's Objections to Pity and Compassion.* [Tevenar 2007, 263– 281].

knowledge is subject and object at the same time, only produce "incomprehensible nonsense." Nietzsche's work went largely unnoticed for many years. He even had to pay for the printing of the fourth part of *Also Sprach Zarathustra* out of his own pocket because his publisher refused to continue publishing the book. Not until the end of the 1880s, shortly before his mental collapse, did his work begin to receive a noteworthy positive reception. The first university lectures on Nietzsche's thought were delivered in Copenhagen in 1888 by the philosopher Georg Brandes, who came from a Jewish home, more than a decade before philosophers at any of the German universities dealt with his works. Still, this did nothing to detract from his rapidly rising fame in the 1880s, since those who read and admired him moved largely outside academic circles. Most were young intellectuals and iconoclasts who rebelled against oppressive regulations and who questioned everything hitherto considered untouchable in politics, state, religion, morals, science or art. Added to this was the fact that Nietzsche's dense and suggestive, often poetic language exerted a great attraction for those who appreciated his literary talent.

These factors make it hardly surprising that a considerable number of Nietzsche's early followers were of Jewish origin. Having long suffered from exclusion and discrimination, Jewish intellectuals had good reason to identify with his radical rejection of traditional values [Golomb 1997], [Stegmaier/Krochmalnik 1997]. Felix Hausdorff was, in this sense, by no means alone. As a student, he certainly heard nothing about Nietzsche in the philosophical courses he attended, except perhaps for a few derogatory remarks, so he must have come to Nietzsche independently. In any event, a shared enthusiasm for Nietzsche's writings brought him into contact with a number of young artists and writers, a story that can be partially reconstructed on the basis of fragmentary sources.

When Hausdorff began his studies in Leipzig in the summer semester of 1887 the university already had a large number of student associations with a long tradition. In addition to traditional fraternities, there were clubs for gymnastics, choral groups, and different religious denominations, but also clubs restricted to German students and also a few scientific clubs. Many of these clubs belonged to larger cartels or associations, but since nearly all of them were decidedly anti-Semitic only a few came into question for Hausdorff.

The only documented case of Hausdorff's participation in student organizations concerned activities of the Academic-Philosophical Association in Leipzig. This club was established in 1866/67 by the then 23-year-old Richard Avenarius, today remembered as the founder of the philosophical school known as empirical criticism. According to the organizational statutes, which were revised in 1880, any student enrolled at the University of Leipzig could become a regular member of this club. As this shows, the Academic-Philosophical Association was not anti-Semitic, and it stayed that way for a long time. Many of its members were Jewish and the organization took a critical stance regarding antisemitism within the larger student body, distancing itself from the extremely anti-Semitic Association of German Students. In the summer of 1886, for example, they turned down an invitation to attend the latter association's festival celebrating its founding. From around 1892, however, antisemitism also began to take hold within the Academic-Philosophical Association of Leipzig, as it was in that year that a Jewish student was refused admission for the first time. By the end of the 1890s, it had fallen in line with the general situation in Leipzig, as described by a member of the Kyffhäuserverband in 1902:

> The social isolation of the Jewish student has been largely carried out today. All of the respectable color-bearing fraternities – *Korps, Burschenschaften, Landsmannschaften* – and color-bearing gymnasts as well as most of the other associations – the academic gymnasts, singing and scientific clubs – today exclude Jews from membership. [Kampe 1988, 203]

In all likelihood, Hausdorff was never a member of the Academic-Philosophical Association, though he did participate at some of its meetings. His name appears in protocol books on five occasions when he took part in the discussions that followed a lecture. The topics of these lectures were quite varied: "Two Diseases of Philosophy" (15 May 1889), "On the Concept of Student Honor" (13 May 1890), "On Realism in Modern German Literature"(17 June 1890), "The Claims of the Anthropocentric over the Heliocentric Worldview" (23 December 1890), "The Psychological Degeneration of Humanity under Modern Culture and the Possibility of Change" (27 January 1891). This last lecture was delivered by Dr. Wilhelm Schallmayer, a former member and one of the first advocates of eugenics in Germany. Later, Schallmayer wrote popular books on this subject.

Far more important for Felix Hausdorff than such lectures was the friendship he gained through the Academic-Philosophical Association with the future lawyer Kurt Hezel, the son of a merchant from Marienberg in Saxony. Hezel studied law and public administration in Leipzig from the summer semester of 1884, but he was also interested in philosophy, art and literature. As a new member in 1885/86, he gave a lecture on the subject of "The Limits of State Power on the Sphere of Freedom of the individual," a topic considered so explosive that a police commissioner was compelled to attend. Not long afterward, Hezel was elected as chair of the Academic-Philosophical Association. After receiving his doctorate in 1890, he eventually became a legal clerk in Leipzig, and in 1896 he joined the renowned Leipzig law firm of Dr. Felix Zehme, where he worked on matters of interest for commercial firms, in particular copyright laws. This work brought him into contact with the founder of the Nietzsche Archive in Weimar, Elisabeth Förster-Nietzsche and Heinrich Köselitz (Peter Gast). Köselitz, together with his friend Franz Overbeck, had managed Nietzsche's estate during the first few years after his collapse, at which time they began work on the first edition of Nietzsche's collected works. Hezel would later become a significant figure in Hausdorff's life. It was through him that Hausdorff came to know Heinrich Köselitz, a circumstance that brought him into contact with others in Nietzsche's closest circle.

Köselitz, who had studied harmony and composition theory at the Leipzig Conservatory since 1872, became so fascinated by Nietzsche's *Birth of Tragedy from the Spirit of Music* that in 1875 he decided to leave Leipzig for Basel in order to attend his lectures. A friendly relationship soon developed, in which Köselitz served not only as discussion partner and friend but also as a kind of secretary. He prepared transcriptions of Nietzsche's writings for the printed works, read page proofs, and made suggestions for improving certain formulations. Starting in April 1878, he continued this work in Venice, where he lived for more than 10 years. During a stay together at the Recoaro thermal bath, Nietzsche gave Köselitz the name Peter Gast, the pseudonym he adopted from then on. Hausdorff read Heinrich Köselitz's very positive review of Nietzsche's *The Case of Wagner. A Musician's Problem* and was greatly impressed. Since he was still a young student, Hausdorff did not immediately approach Köselitz, who was fourteen years older, so he waited until he felt he had something interesting to say about Nietzsche's ideas.

Evidently, Hezel helped to establish the initial contact with Köselitz three or four years later, by describing Hausdorff as a great admirer and connoisseur of Nietzsche. This motivated Köselitz to send Hausdorff a no longer extant letter, to which the latter replied on 17 October 1893:

> Since 1889, when I encountered your name for the first time in your review of the "The Case of Wagner," making your acquaintance one day became part of my personal program. That wish would probably have remained unrealized, except for that fascinating man of splendor, Kurt Hezel, our "genius of the heart," who brought this about for both him and me – though the favorable, all-too-favorable assessment he gave of me, from which your letter testifies, is one I certainly still need to grow into very much. (*HGW* IX, 377)

In order to show Köselitz that he did have something new to contribute, Hausdorff took a critical stance regarding Nietzsche's attempt to justify scientifically his notion of the eternal return of the same, an argument found only in Nietzsche's unpublished writings. Köselitz valued this idea highly and made it public in the preface to *Human, all too Human*, which appeared in his first edition of Nietzsche's works. In his letter to Köselitz, Hausdorff indicated how Nietzsche's proof could be refuted mathematically, though this in no way affected the great fascination associated with the idea of eternal return. Hausdorff presented this refutation in detail in his volume of aphorisms (see, however, Section 6.1).

In addition to Hezel, another Nietzsche enthusiast wanted to bring Hausdorff and Köselitz together. This was Paul Lauterbach, who in a letter to Heinrich Köselitz, dated 30 December 1893, wrote:

> I just read some manuscripts that allow me to signal a really new author to you – Dr. Felix Hausdorff. He has taken up a surprising amount of Nietzsche and has assimilated him excellently. I read aphorisms reminiscent of *The Gay Science* as well as rhymes that attain the same height. A Dionysian mathematician, that sounds strange, but have

> him send you something, and you can bet with me that there is something to experience about him. (Nachlaß Heinrich Köselitz, Goethe- und Schiller-Archiv Weimar, Bestand Gast, Signatur GSA 102/417.)

Lauterbach studied chemistry at the Fresenius laboratory in Wiesbaden and then at the Zurich Polytechnic, where he also studied philosophy at the university. After acquiring an inheritance, he ended his studies without taking a degree and went abroad. He stayed in England for a long time and then in Geneva. There he married a widow who had three sons from her first marriage. The family moved to Leipzig, where Lauterbach, who had exhausted his inherited assets, managed to make a living by doing translation work and by producing new editions for the firm of Philipp Reclam jun.; he also offered language lessons and public lectures. Lauterbach prepared an influential new Reclam edition of Max Stirner's *The Ego and Its Own*, containing his 8-page "short introduction" (1893). After Stirner's work (first edition 1844) had been almost entirely forgotten, this new edition brought forth a resurgence of interest, which later elicited a reaction from Hausdorff (see 4.6).

Lauterbach's sole published work, *Aegineten. Gedanke und Spruch* (1891), was a collection of aphorisms produced by Constantin Georg Naumann, the publisher of Nietzsche's works. This collection contains the dedication: "To the master of the Zarathustra." As one of Nietzsche's early admirers, Lauterbach made plans to translate his works into French. However, this project came to naught after Nietzsche's sister Elisabeth Förster-Nietzsche awarded the translation rights to Henri Albert. Lauterbach had full command of an impressive number of languages: in addition to Greek and Latin he was also fluent in English, French, Italian and Russian. After Hausdorff made his acquaintance, he decided to take English lessons with him. Lauterbach died on March 24, 1895 as the result of a spinal cord disorder. The nephew of Constantin Georg Naumann, Gustav Naumann, who was also well known to Hausdorff, wrote that he and Felix Hausdorff were the only non-relatives who attended Lauterbach's funeral (Gustav Naumann, *Paul Lauterbach zum Gedächtnis*, Universitätsbibliothek Basel, NL Naumann II,4; Zitat dort S.7.)

Hausdorff and Lauterbach probably first met one another around the end of 1893; as Gustav Naumann reported, they soon became good friends. Only a few items of correspondence (a letter and three cards) from Hausdorff to Lauterbach survive, since they mainly maintained personal contact in Leipzig. From these few sources, as well as from Lauterbach's letter to Köselitz, it emerges that Hausdorff was working on things he probably published later (at least partially) in his aphorisms and poems. The close ties to Nietzsche's thought also resound again and again, as in Hausdorff's postcard from March 6,1894:

> this time the hoped-for sequel to the butterfly flights [the first part of Hausdorff's volume of poems *Ecstasies* is entitled "butterfly flights"] did not flutter to me, but instead only two dozen rondels that I wrote with solemn doggedness, some little things, unfortunately always with erotic coloring. [···]

Figure 4.1: Sils Maria in the Engadin, ca. 1890
www.nietzschehaus.ch/friedrich-nietzsche-in-sils-maria/thema-1

> Before my trip, I will surely see you: this promise I shall hold. Probably this time I'll travel to Eastern Switzerland (Engadin and perhaps Ortler), in which case, I will not fail to take in Sils-Maria, a confrontation that must still move and affect one the way it did with our nearly unique one. – Did you read Lou's book? I did. [This refers to her study of Nietzsche's life and work [Andreas-Salomé 1894].] (*HGW* IX, 422–423)

Nietzsche loved Sils Maria in the Engadin and stayed there no fewer than seven times between 1881 and 1888 (see Figures 4.1 and 4.2). It was there that the inspiration for his idea of the eternal return of the same first came to him. Hausdorff ironically plays on Nietzsche as "our nearly unique one," probably because he and Lauterbach knew that Max Stirner claimed to speak with "unique" authority (for Hausdorff's dismissal of Stirner's views, see 4.6).

Figure 4.2: Nietzsche often resided here during trips to Sils Maria
www.nietzschehaus.ch/das-nietzsche-haus-i/geschichte-des-hauses

On July 8, 1894, Hausdorff wrote to Paul Lauterbach from his vacation locale in Pontresina:

> Best friend!
>
> Wouldn't you like to send me a line here in Pontresina? Along the path from head to hand you have certainly received my booklet XXVI, about which you should speak and would like not to. Isn't it so, as in mixed company, too much female? in the end even females; too little discreetness, too freshly plucked; too many prognoses when the pointer is still pendulating. [$\cdots$]
>
> These last days I gathered together my full concentration and went through Sils-Maria, but I still had a rough throat and unmelodious sensations. Dream some time against necessity: try to find an expression for what Nietzsche could still become if he were to overcome the present inferno. (*HGW* IX, 424)

This letter reveals that already by mid-1894 Hausdorff had written a large number of texts, at least 26 in number. Unfortunately, none of these manuscripts has been preserved. Nietzsche, of course, played a major role here; the "inferno" refers to Nietzsche's deteriorated mental state. It is safe to assume that Hausdorff included some of the aphorisms from this time in his first literary work, published under the pseudonym Paul Mongré.

In Chapter 1 on Hausdorff's early family life, we described the dominant role played by his father, Louis Hausdorff, a conservative Jew of firm religious beliefs. It is easy to imagine that Felix Hausdorff's interest in Nietzsche's philosophy would have made him most unhappy had he known about it or even have read any of his son's manuscripts. In any case, if any kind of ideological controversy ever developed between Louis and Felix Hausdorff, no trace of such has survived. Fate also granted Louis Hausdorff the gift of never having to encounter the work of Paul Mongré, whose ideas were in so many ways diametrically opposed to his own beliefs. His unexpected death in May 1896 at the early age of 53 may well have eased his son's conscience in pursuing his personal philosophical interests.

4.3 Sant' Ilario – Thoughts from Zarathustra's Landscape

In the summer semester of 1896, Hausdorff began teaching courses in Leipzig in the new field of financial and insurance mathematics, following requests made by the ministry and faculty. He also announced that another similar course on "Selected Topics in Financial and Insurance Mathematics" would be offered in the winter semester of 1896/97. That plan, however, never came to fruition, as Hausdorff had to take a break from teaching in order to restore his health. In fact, this would be the only semester throughout his four decades of teaching during which he took a leave of absence. Even before the winter semester began, he was suffering from a chronic problem with his respiratory tract. Perhaps the strain after the unexpected death of his father contributed to this, but in any case his physician recommended that he take a long vacation in a mild southern climate. He followed this advice by spending the winter on the Ligurian coast along the Gulf of Genoa between Nervi and Portofino, the same region in which Friedrich Nietzsche wrote the fourth book of *The Gay Science* and the first parts of *Also sprach Zarathustra*.

Hausdorff probably had not chosen this area for his stay accidentally, because here he completed his first philosophical work, later published under the pseudonym Paul Mongré. This was his volume of aphorisms entitled *Sant' Ilario. Thoughts from the Landscape of Zarathustra* (*Sant' Ilario. Gedanken aus der Landschaft Zarathustras* [H 1897b, VII: 87–473]). Although Hausdorff took advantage of these beautiful scenic surroundings to complete his aphorism volume, he later mentioned to Köselitz that its "actual origin" took place years earlier (*HGW* IX, 397). In any event, Mongré's volume appeared in mid-1897 with the Leipzig publisher C.G. Naumann, who produced it in the very same format he had used with Nietzsche's writings.

Hausdorff's chosen pseudonym Mongré stems from the French à mon gré, meaning to my taste, after my fancy, or according to my wish and will. Werner Stegmaier, the editor of Hausdorff's philosophical works (*HGW* VII), connected this with Nietzsche's manner of thinking, as the latter had a strong proclivity for everything he identified as French. Taste, as a philosophical term, played a major role in French literary and art theory of the late seventeenth century. It was discussed in detail, particularly by La Rochefoucauld, whom Nietzsche deeply appreciated. Thus, for Nietzsche, taste was a significant philosophical concept. He even placed the notion of individual taste at the very center of critical thinking, while dismissing any adherence to a priori ideas.

Nietzsche's Zarathustra thus proclaims:

> In many ways I came to my truth; not on a ladder I climbed to the height where my eye wanders into my distance.
>
> And always only reluctantly I asked for ways, – that always went against my taste! I preferred to ask and try the ways myself.
>
> All my walking was trying and asking: – and truly, one must also learn to answer such questions! But this – is my taste:
>
> – not good, not bad, but my taste, of which I have neither shame nor concealment.
>
> This – is now my way, – where is yours?' so I answered those who asked me about 'the way'. *The* way, after all, does not exist!
>
> Thus spoke Zarathustra. (Zarathustra III, [Nietzsche KSA, Band 4, 245])

The pseudonym Hausdorff chose clearly reflected a program. Its aims exalted individuality and intellectual independence, while rejecting all forms of prejudice and constraint, whether of a political, social, religious, or any other kind. Initially, the pseudonym also served as a form of protection; in a letter from 8 October, 1897 to Köselitz, Hausdorff pleaded for discretion regarding his authorship, since "as a private lecturer for mathematics and astronomy, I undoubtedly have to hide behind a pseudonym" (*HGW* IX, 394). By 1901 at the latest, however, his Leipzig colleagues knew that he was Paul Mongré. The first name Paul could have been an allusion to Paul Rée, a close friend of Nietzsche during the period from 1876 to 1882.

What about the title *Sant' Ilario – Thoughts from the Landscape of Zarathustra*? The meaning of the second part would seem obvious: thoughts from Nietzsche's environment of ideas – not Nietzsche's thoughts, of course, but those of Mongré. In the letter to Köselitz mentioned above, Hausdorff commented on his relationship with Nietzsche:

> Your musical ear will not miss my close affiliation to Nietzsche just as little as my careful restraint from Nietzsche, although I don't drum out

Sant' Ilario

Gedanken aus der Landschaft Zarathustras

Von

Paul Mongré

LEIPZIG

Druck und Verlag von C. G. Naumann

1897.

Figure 4.3: Title Page of Hausdorff's
Sant' Ilario. Thoughts from the Landscape of Zarathustra.

> either one with a stubborn expressiveness; and if you may perhaps take this latent behavior as a form of opposition, for those who are supporters it should be seen as precisely appropriate to the nature of the matter. (*HGW* IX, 393)

One can also take the title much more literally, as referring to the locality where Nietzsche's *Zarathustra* was born. In *Ecce homo*, Nietzsche reported on his stay during the winter of 1882/83 "in that graceful silent bay of Rapallo not far from Genoa, which divides Chiavari from the foothills of Portofino." He then describes two paths along which he often walked: "Along these two paths the entire first part of Zarathustra came to me, especially Zarathustra himself as a type: more accurately, *he attacked me* ..." [Nietzsche KSA, Band 6, 336]. It was in the midst of precisely this same landscape that Hausdorff now completed his first work, thus called "Thoughts from the Landscape of Zarathustra." In his own synopsis of the book, which appeared in the magazine *Die Zukunft* on November 20, 1897, he not only expounded on its relationship with Nietzsche's ideas but also expressed, in his inimitable way, the hopes and fears he had regarding the reactions of its readership:

> My book, which appears as a collection of aphorisms and would like to make a virtue of this stylistic necessity, arose from a steady surplus of good moods, good air, and bright skies: its immediate home, from which its name derives, would be found on the Ligurian Sea, halfway between impressive Genoa and the elegant foothills of Portofino. On this blessed shore, which has the advantage over the actual Italia diis Sacra of mild winters and famous palm trees, I followed in the footsteps of the creator of Zarathustra – along cliffs and wonderfully narrow coastal paths that make no room for walking abreast. Whoever wants to count me a disciple of Nietzsche, may freely refer to my own confession here. Other admirers of Nietzsche will maintain that my book pays too little homage to him. [· · ·] I have to expect it will offend fanatics and partisans of all kinds; those who tend to simplistic boasting, to world-improving pathos, or to play the "moral big mouth," [such readers] will regard my Holy Hilarius as a visitor from hell. About him only the worst will be said: he won't be boring enough for science, nor bohemian enough for literature, he will either be found lacking in system by some or in idealism by others. Perhaps I can hope, though, to attract certain sensitive aficionados with a taste for the cool, sour skepticism of my book and its courageous, disrespectful, half-hermit-like, half-sophisticated philosophy, although this has nothing to do with whether or not one believes in my "views" about culture, religion, education, women, love or metaphysics. [H 1897c, VII: 477]
>
> Paul Mongré.

It would certainly be senseless to analyze systematically a collection of aphorisms, especially when they cover such a wide range of different topics. Under number

352 alone one finds an entire chapter called "Splitters and Spines" with no fewer than 377 short sayings. In all, the book contains 411 numbered aphorisms, supplemented by a collection of elegant sonnets and rondels at its end. Any attempt to report on its contents in their entirety would lead to an even longer book and, in contrast to Hausdorff's, a very boring one. The excerpts and selections offered here represent merely a small sampling.

Just as Nietzsche emphasized the search for *his* way in Zarathustra, so do we find passages in *Sant' Ilario* where Hausdorff reflects on his way – *the way to one's own self.* In Aphorism 65, he approached this with these words:

> If in earlier times it perhaps cost real effort to breed a few social instincts into the individual, today it takes even more to produce not social, but individualistic sensibilities. It is of course a hundred times easier to renounce these and to be satisfied with the momentary advantages of a good conscience rather than to pursue one's own way forward along an uncharted path toward an uncertain goal, while parting with one's entire world as well as one's own past. [H 1897b, 66; VII: 160]

A similar viewpoint comes to the fore very nicely in Aphorism 35, which reflects Mongré's high esteem for creative individuality:

> Anyone is fertile who calls something his own, whether creatively or through enjoyment, in language or gesture, in longing or possession, in science or morality. Everything that occurs less than twice is fertile, every tree that grows from *its* earth to *its* sky, every smile that belongs to only one face, every thought that is only right once, every experience that exudes the heart-strengthening smell of the individual! [H 1897b, 37; VII: 131]

It seems likely that as a child Hausdorff was most unhappy, not only due to the strict religiosity he experienced at home but also because of the traditional educational methods he encountered in school. In Aphorism 64, he offers a promise to do this better should he have children in the future. Hausdorff was apparently also unhappy as a young man, perhaps during his adolescent years. In a number of aphorisms relating to own development, the not yet thirty-year-old author describes youth as a less than gratifying condition, as in this excerpt from Aphorism 250:

> What a miserable condition is youth! This "urgent pressure" of the developing and inhibiting organs, this chaos without polyphony, the pain of growth with its uncomfortable strains and dislocations of the soul, a spiritual mutation, when one cannot master their own voice, which only brings forth false tones! [$\cdots$] Barbaric insensitivity, on the one hand, with irritable hyperesthesia, on the other: how tortured one becomes by the sight of the sky or a girl's face – a true storm rages through the nerves and prevents the intellectual bell from ringing! [$\cdots$]

> – No, youth may have virtue, as much as it claims, but it doesn't have taste, style, reason. Youth is a martyrdom, like the beheading of St. Bartholomew or the roasting of St. Lawrence, in any case nothing that has a good smell. [H 1897b, 155–156; VII: 249–250].

In view of these lines, it is not surprising that the young Hausdorff felt attracted to Schopenhauer's pessimistic metaphysics.

One can only speculate about the experiences that led to this feeling for life. Presumably certain things came together at an early age that caused him to speak retrospectively about the "martyrdom" of youth. Certainly his gradual abandonment of his family's strong religious tradition was anything but painless and conflict-free. Yet, at the same time came another inner conflict. On the one hand, he felt a need to pursue a successful scientific career, though this initially led more and more toward a dead end. On the other, his early passion for creative activity, whether in music, poetry, literature or philosophy, grew ever stronger, as is evident, for example, in his letters to Paul Lauterbach. Beyond this, his ambivalent relationship with members of the opposite sex added to this sense of inner conflict.

Gradually, however, Hausdorff reached a kind of equilibrium by pursuing a double life. He thus found a way to balance various competing spheres of interest by keeping "separate accounts." In one, he booked his increasingly successful work in the field of mathematics with its sharp concepts and precise conclusions based on logical arguments. In another, he could record his subjective experiences of the world in all its facets as a poet, writer and philosopher. Naturally, there were many threads connecting these intellectual spheres, particularly in his epistemological works. Aphorism 252 describes the relationship between different spheres in an individual's creative life:

> One could pursue happiness in many disparate spheres of life were it not for the network of threads that connect these spheres causing them to interact with one other. In this way, they tend to align themselves in forming a joint collective sphere, a space in which they must compete. The boundaries between dreams and waking, heights and valleys, ecstasy and everyday experience are by no means mathematically sharp, however. They are rather empirically cloudy and mutually passable, and this causes conflict since they belong to the same person. They cannot be divided mathematically but only split into pieces mechanically, naturally causing pain. This disease often lessens over the years, but it can be especially acute for the young, who have not yet understood how to keep separate accounts and instead try to squeeze everything – religion, love, art, science – into a highly swollen unity of sensations [H 1897b, 156–157; VII: 250–251].

Just as with Nietzsche, Hausdorff's path to his own philosophy required overcoming all forms of metaphysics. Two aphorisms, 363 and an excerpt from 364, allude to his mature position:

> Metaphysics stands on the principle that it never remains without words when standing helplessly before a mystery, it incessantly describes the eternally indescribable by solemn demonstration, and never and nowhere keeps its mouth shut.
>
> When one is young and has not yet overcome metaphysics, one feels satisfaction in knowing the hidden hand that steers and determines the course of life. Reality appears as inexorable and at the same time as just one among millions of possibilities, each as unlikely as one to infinity, almost impossible, and yet reality, essential reality, the absence of which amounts to annihilation. [···] Later, a cooler spirit no longer believes in this transcendental determinant guiding the fate of the individual, nor in the danger of crises and the unique way out. We have come to grips with a reality and learned to love it [···] [H 1897b, 301–302; VII: 395–396].

Mongré explicitly dismisses Schopenhauer's metaphysics in a series of aphorisms, as in this excerpt from No. 395:

> Whatever else it might be, Schopenhauer's omnipotence of the will represents a departure from the present-day understanding of knowledge. Perhaps in this century we will prove the impotence of the will and reject this as the most misunderstood of all vacuous metaphysical notions [H 1897b, 334–335; VII: 428–429].

In his own principal philosophical work, *Das Chaos in kosmischer Auslese* (*Chaos in Cosmic Selection*) [H 1898a], Hausdorff attempted to pass beyond Kant, making his own contribution while declaring the "end of metaphysics." Nevertheless, he appreciated the tendencies of youth who give in to Schopenhauer's fallacies and "metaphysical needs," as he regarded this as an important life experience. He expressed this view in two sayings from "Splitters and Spines":

> The errors we later come to recognize are not for that reason to be devalued. Do we regret taking a journey because we at some time need to return home? [H 1897b, 271; VII: 365].
>
> Be not a metaphysician, except for having been one [H 1897b, 278; VII: 372].

In numerous aphorisms in his *Sant' Ilario*, Hausdorff followed Nietzsche's world of thought: more than eighty contain allusions to Nietzsche and twenty of these even contain direct quotations from his writings. Yet Hausdorff was hardly a mere partisan of Nietzsche's philosophy or an acolyte who tried to copy or even surpass the master. Rather he stood, so to speak, next to Nietzsche in an endeavor to release individual thinking and to take upon himself the freedom to question traditional norms. As Werner Stegmaier aptly put it, "what drew *him* to Nietzsche was the unity of life, thinking and writing according to one's own 'taste' and own responsibility" (*HGW* VII, 33). Indeed, after rejecting Schopenhauer's pessimism, Nietzsche made the center of his philosophy a radical affirmation of life. Like him, Hausdorff passed through Schopenhauer's pessimism to uphold a vision of life drawn from a firework of beautiful surprises, which keep life's flame burning, though these do not derive from anything monumental:

> But I am well disposed toward life, because in addition to the Cosmos there is Chaos, and this Chaos holds the most beautiful chance events, passing moments, and resonances – I look into a witch's pot, in which a devilish delicacy is always brewing. [· · ·] I am thankful for life, but what for? For those flashing moments between two parts of the world and two centuries, for high, turning, and inflection points, for mountain passes and ridges, for borders, bridges and transitions, for bright and disappearing rainbows, for everything that occurs once and never again – I thank life, mythologically speaking, that one hears for only a few seconds and not more the chords of the *pre-stabilized harmony*! "A little makes the best kind of happiness" [From *Also sprach Zarathustra* IV. [Nietzsche KSA, Band 4; 344].] If we search for the invaluable in our memory, for the traces of remembrances of enjoyment, we find it in something small, fleeting, pointless, the mosquito in the amber: maybe a glass of asti spumante [a sweet sparkling white wine], a cloud in the sky, a flower's fragrance, a look or gesture, the tone of a phrase or a tact of garden music, any more-than-nothing but less-than-something that gives the soul lust and longing for life and excites desire for it more than it leaves us saturated: a pianissimo of happiness that rewards living but not dying ... [H 1897b, 152–153; VII: 246–247].

In "Splitters and Spines," Mongré affirms, with a twinkle in his eye: "There are a great many poor places in the theater of life. But the play itself is good" [H 1897b, 261; VII: 355].

Sant' Ilario is filled with passages that attest to Hausdorff's skepticism regarding any metaphysically founded teleology and his opposition to all ideologies and schemes for improving the world, especially when promoted by those who claim to possess the true meaning and goal of the human race. Thus, he wrote in Aphorism 3:

> If not the truth itself, then the belief in found truth is to a dangerous extent hostile to life and murderous for the future. Yet none of those who deluded themselves as being blessed with the truth ever hesitated for a moment in proclaiming the grand finale or the great moment or some endpoint, turning point, high point for mankind, placing his image, his stamp, his limitation on this for all time [H 1897b, 6; VII: 100].

With regard to the ideologically driven murderous dictatorships of the coming century, the foregoing lines are truly clairvoyant.

In the first aphorism of *Sant' Ilario*, Mongré describes how our world's development depended on countless chance occurrences in the past. Had there been a few more explosions in the glowing primeval matter out of which the solar system arose or a slightly different distribution of the masses in the evolution of the planets everything could have come out completely differently. One wonders then: "Must Christianity have prevailed over other religions, did religion have to dominate over the practical cultural powers in Europe, and was the demise of Greek art necessary." The aphorism closes: "In the world there is so much outrageous nonsense, discontinuity, fragmentation, chaos, 'freedom of will'; I envy those with good synthetic eyes, who see all this as the evolution of an 'idea', only *one* idea" ([H 1897b, 4; VII: 98]). This final passage could also be seen as a strong swipe at Hegel's "absolute idea" and the Hegelian "world process." In aphorism 379, Hausdorff describes Hegel as a "genealogical metaphysicist" whose successor, Eduard von Hartmann, even managed to give the "world process" a final goal [H 1897b, 317; VII: 411]. Here he might have recalled the Young Hegelian Karl Marx, who posited that communism was the goal and ideal final state of human history.

A number of Hausdorff's aphorisms are devoted to his criticisms of religion. In aphorism 66, he first quotes – of course in Latin – the stanzas concerning the last judgment from the Catholic requiem mass. In English this reads:

The written book will be brought forth,
in which all is contained,
from which the world shall be judged.
When therefore the Judge will sit,
whatever lies hidden, will appear:
nothing will remain unpunished.

Now for Mongré's commentary:

> Oh, the modesty of these prophets! Thus, the entire earthly show and state drama that one has the temerity to call "world" history, the whole little spectacle and childish noise, shall *once again* be shown to us in extenso at the end of our days as the basis for a grotesque trial? What Peter did and Paul left undone, Hänschen's lust of the eye and Gretchen's carnal lust, evil Esau and pious Jacob, and whatever all these *Struwwelpeter* stories might be called – not enough that this commonplace, superfluous, inessential was present, even a hundred and a thousand times: now at the end it shall again be processed and recapitulated, in pleas and judicial pronouncements, in awkward cumbersome pros and cons, without any cuts, missing lines, or reductions! And we should have to endure that? and remain serious throughout?

Yet Hausdorff dealt critically not only with the Christian perspective on God but also with the Judaic variant; he thus continues:

> In order to be fair, it should be remembered that this tactlessness in divine things, in which God appears as a watchman and police offi-

cer with a fabulous memory or notebook, cannot actually be charged to Christianity, as it inherited the entire possibility of such petty anthropomorphisms from Judaism. Without the Jewish fusion of the concept of God as creator of heaven and earth and the limited notion of a national god, who designs a food menu for his chosen people, one cannot comprehend the Christian blending, which combines the eternal world order with the commonplace talk and doings of everyday persons [H 1897b, 67–68; VII: 161–162].

A second example of this can be found in Aphorism 367:

The postulates of practical reason are: *no* god, *unfree* will, and *mortality* of the soul. Theism destroys the moral responsibility of man, freedom of the will contradicts the moral evaluation of the deed, and immortality prevents balancing punishment and sin. He who God created is merely the observer of godly acts in him; the free act is only part of a random game and therefore not to be attributed to the actor; an eternity of salvation or damnation, granted on the basis of the span of a human lifetime, is decidedly too long [H 1897b, 305; VII: 399].

In some of his aphorisms, Hausdorff dealt with fashionable themes in cultural life, though mostly by way of ironic distance. We find an example of this in his biting satire of the Bacon-Shakespeare controversy. Although Hausdorff was quite unaware of this in 1897, the mathematician Georg Cantor was an active and passionate supporter of this speculative theory. This is particularly noteworthy, since Cantor was the founder of the theory of transfinite numbers, the field in which Hausdorff would later devote his principal work.

Already during the eighteenth and the first half of the nineteenth century, individual voices had raised doubts about Shakespeare's authorship of the works known by his name, particularly the dramas. The advent of a kind of movement regarding the Shakespeare-Bacon question came about in the late 1850s following publication of two works by the American Delia Bacon, who argued that Shakespeare was a mediocre and not very educated actor [Bacon 1857], [Bacon 1858]. In addition, theater people had no access to royalty, so that Shakespeare supposedly lacked the kind of knowledge of court life displayed in his tragedies. Thus, the true author must have had firsthand knowledge of English royalty, sharp eyes, and a keen philosophical head, which he safeguarded by disguising his authorship. Sir Francis Bacon fit this description very nicely, but Delia Bacon also considered Sir Walter Raleigh as another likely candidate. Bacon then emerged as the sole "true author" of the Shakespearian dramas a decade later, after publication of a book of over 600 pages by Nathaniel Holmes [Holmes 1866].

The Bacon-Shakespeare controversy only reached Germany in the 1880s, but it quickly became extremely popular. By the end of that decade more than 500 publications devoted to this theory had appeared. Numerous complaints in the yearbook of the German Shakespeare Society reveal further how this controversy played a major role in public discussions. Roughly speaking, there were two camps among the Baconians. A smaller group, which included Georg Cantor, tried to find historical sources supporting Bacon's authorship. Cantor published a long-forgotten collection of poems mourning the deceased Bacon; in one of these, he claimed to have found irrefutable evidence of Bacon's authorship ([Purkert/Ilgauds 1987, 81–92]). The vast majority of the Baconians, however, used highly dubious methods. These tried to discover hidden secrets and ciphers in Shakespeare's writings, while also investigating other sources for cryptograms that allegedly pointed to Bacon's role. Some even tried to construct anagrams from letters and texts as additional proof of Bacon's authorship. This type of mystery hunt became popular after the appearance of books by Donnelly and Owen [Donnelly 1888], [Owen 1893–1895].

In Hausdorff's Aphorism 306, *Die Shakespeare-Bacon-Frage*, he ridicules the search for hidden ciphers by constructing an anagram out of a stanza from Goethe's *Faust*, which reveals Kant to have been its actual author. Hausdorff's final verdict regarding the Bacon-Shakespeare question was as follows:

> It is wrong to view the unreasonable summarily as the opposite of reasonable; there are gradual differences, and the Shakespeare-Bacon question, like spiritism, represents perhaps a summit in the high mountains of absurdity. You decide to concede someone one or two or three assumptions that contradict bon sens, you force your brain to vibrate parallel to the vibrations of an odd-headed brain – but that doesn't help; out of the nonsensical comes the most nonsensical, the idiocy grows exponentially. Some systems are at least correctly structured on a false premise, you enter at the price of a sacrifizio dell' intelletto, but afterward you are allowed to use your reason again. And there are others, which at every turn never rid themselves of nonsense, systems that can never be fathomed by a one-off contortion of the brain; reason is shattered from below – its limbs and fingers are left mutilated! [H 1897b, 204–206; VII: 298–300].

Cantor counted himself among the serious researchers of historical literature. He totally rejected the peddling of secret codes and hidden writing; he certainly would have been amused by Hausdorff's experiment with the Goethe text. Still, he would just as surely have found this aphorism insulting. Presumably he never read *Sant' Ilario* and probably did not know who Mongré was. Hausdorff later had a warm personal relationship with Cantor, and he surely was careful not to engage in philosophical, much less "Shakespearean" debates with him. Cantor was also a determined opponent of Nietzsche, and in a letter to the Protestant church historian Friedrich Loofs he wrote the following about his work:

> Because of its stylistic charms, it finds uncritical *recognition* in our country, which seems to me *extremely dubious* in view of the perverse content and the caustic anti-Christian motifs. The desire for novelty and to fit a philosophical-historical scheme makes our philosophers *morally blind*

> and *quick to accept* anyone claiming a new system that accords with their historical account. Thus the ambitious innovator always achieves his purpose; he becomes a famous philosopher and the corruption of youth takes place on a grand scale [Meschkowski 1967, Letter Nr. 18].

Closing out theme, we cite Hausdorff's humorous approach to the composition of a drama based on mathematical principles:

> The ideal drama for n characters must contain: n persons, $\binom{n}{2}$ dialog relationships (subdramas for two actors), $\binom{n}{3}$ triangular relationships (subdramas for three), and so on, $[\cdots]$. Thus, inner completeness and exhaustion of all conceivable combinations; none of the players should be presented one-sidedly as only active or passive figures – between all existing points, one should draw all lines, planes, etc. In addition, the drama must reveal an overall plot between all n people, possibly as part of a larger one involving more than n people, i.e. it must have a background (a milieu) of a historical or other general nature. From this, one sees that the serious playwright will try hard to *reduce* the his n! [H 1897b, 201–202; VII: 295–296].

A number of Mongré's other aphorisms will be cited later in connection with his other works; here we comment on the contemporary reception of his *Sant' Ilario*. Among its readers, none approached it with more understanding than Franz Overbeck, emeritus professor of theology and a well-known church historian. Overbeck taught in Basel during Nietzsche's time there, and he became his best and most loyal friend. After Nietzsche's mental breakdown in Turin, he hastened there to help him and to bring his manuscripts to safety. He then brought Nietzsche into the care of his mother Franziska, who watched over her son up until her death. Overbeck's reading of *Sant' Ilario* is documented by sources in his extensive, now largely published estate at the University Library in Basel as well as through his correspondence with Heinrich Köselitz, and from his personal copy of Mongré's book (see *HGW* IB, 277–278).

Hausdorff must have sent a copy of his *Sant' Ilario* to Heinrich Köselitz in the early autumn of 1897. Köselitz apparently read it right away and on October 7, 1897 he wrote to Overbeck:

> I received one of the best books from the realm of Nietzsche's nutritive thoughts. It is entitled *Sant' Ilario* by Paul Mongré. The author is the astronomer Dr. Hausdorff. One imagines really hearing Nietzsche, as the writing is so close to that tone. It's a shame that this similarity in language will awaken a prejudice against its content. What interested me most was the refutation of the theory of eternal return from the standpoint of the infinitesimal calculus [Hoffmann et al. 1998, 439].

Overbeck replied to Köselitz in a letter dated March 8, 1898, after having read *Sant' Ilario* very carefully:

> For me, it also threw the right light on a review of Nietzsche's, which had been made known to me earlier by the same author, H. Gallwitz, [· · ·] Many thanks, then, for the reference to Mongré, who, by the way, as I learned by a strange coincidence the day before it happened, recently flew through here once again to Italy.
>
> Mongré has long been in my possession. Since I did not receive [the book] from you, I offered it to myself as a Christmas present [· · ·]. However, anyone who is as addicted to the daemon of individualism as I am, can in these days of "contestation" in this spirit least of all completely misjudge the friend. Thus I have not only taken an Epicurean pleasure in his sparkle, he has at times *edified* me, depending on the circumstances, and so I recently read, not without indignation, which I otherwise like to spare, a review of the book in the last issue of the Prussian Yearbooks. This threw proper light on another review of Nietzsche's work by the same author, H. Gallwitz, which I knew [· · ·] [Hoffmann et al. 1998, 447].

In another letter from July 10, 1898, Overbeck distanced himself from Hans Gallwitz and other religiously inspired atheists, contrasting such thinkers with Mongré:

> From the landscape of Zarathustra, I praise Mongré's Sant' Ilario, whatever wishes it might otherwise leave unfulfilled [· · ·]. In the meantime, I also heard, not that this increased my confidence in the book, that its author is very young. So again a precocious Jew! [Hoffmann et al. 1998, 451]

Overbeck read the Christmas present he had given himself in the spirit Mongré had wished, unlike others who constantly compared the author with Nietzsche. Thus, he took pleasure in Mongré's probing views, accepting some, reflecting on others from his own personal position, while sometimes critically rejecting his ideas. According to Overbeck's former student, Carl Albrecht Bernoulli, Overbeck regarded Mongré's book as "by far the best he had read since Nietzsche" [Bernoulli 1908, 396]. In his autobiographical reflections, Bernoulli recalled the advice Overbeck offered him when he was thinking about returning to theology. Overbeck counseled him not to do so before acquainting himself thoroughly with the modern proponents of unbelief, for which he recommended Nietzsche's writings and Mongré's *Sant' Ilario* [Reibnitz/Stauffacher 1995, 60f.]. This "unbelief" was apparently that which Overbeck found so edifying in Mongré's book, as evinced by his own personal notes:

> Dec. 31st 1897. Work on the beginnings of church history. What I read about polemics in Paul Mongré Sant' Ilario [· · ·] reveals to me why I find this self-ironic book so rewarding. One finds such reading only among 'unbelievers' – at least today – and it gives me a warm sense of the feeling for body and soul [Tetz 1962, 96].

Overbeck was particularly interested in those passages in *Sant' Ilario* that he found relevant for his own work as a church historian. Hausdorff's aphorism, cited above, about the Bacon-Shakespeare controversy was one such example. This reminded Overbeck of certain trends in source criticism that he considered similarly absurd:

> In order to understand the aberrations that have led to the strange popularity enjoyed by the new historical source criticism, one has to keep in mind that the entire method, like hardly any other, is by its very nature fixated on a "method," which finds Shakespeare in Bacon and which P. Mongré [· · ·] so appropriately characterizes as a "summit on the high mountains of absurdity" [Reibnitz/Stauffacher 1995, 262].

The review of *Sant' Ilario* that Overbeck mentioned having read, though "not without indignation", was written by the Lutheran pastor Hans Gallwitz [Gallwitz 1898]. One year earlier, Gallwitz published the article "Friedrich Nietzsche as educator of Christianity," the text which had irritated Overbeck, at least until he read the same writer's review of *Sant' Ilario.* Gallwitz really studied Mongré and his review is very detailed and full of quotations from it. His goal was to compare the work of the "pupil" Mongré with his "master" Nietzsche in order to take a close look at "the agreement as well as deep contrast between the two":

> Here stands the large gap between Mongré and Nietzsche. The latter put his philosophizing in the service of personal education; despite the fact that he calls himself an immoralist, he is quite ethically directed. [· · ·] The same cannot be said of Mongré. [· · ·] If he [Mongré] no longer recognizes in naive belief and human forms of apperception and judgment as the standard measure of things, his worldview will never find a clear horizon nor will the individual things have form and boundaries. He will no longer be able to distinguish reality and imagination and will become lost in tremendous fantasies. Nor will he recognize a certain life's task as his duty, but will instead long for ever new sensations and modes of perception, seeking more sophisticated delights from nature that she is capable of giving [Gallwitz 1898, 557, 559, 560f.].

Thus spoke the pastor, whose prophecy proved altogether mistaken. For the author he criticized opened a "second account," one in which he registered his "life's task as a duty." His achievements in that new mathematical realm were of such importance that one still speaks of them today, whereas his critic's name has long since sunk into oblivion.

Another review appeared ex officio, so to speak. Its author was the professor of philosophy in Halle, Hans Vaihinger, remembered today as the founder of the *Kant-Studien.* He kept his remarks very brief, focusing on Mongré's occasional remarks on Kant's philosophy:

> The author (private lecturer Dr. Felix Hausdorff in Leipzig) adopts a pseudonym and is, as the title already suggests, a student of Nietzsche

> and his "Gay Science." Like his master, he finds it necessary in his aphorism collection to criticize Kant's teachings with regard to the "moral world order" [Vaihinger 1898].

There follow some quotes from Mongré on Kant and finally a slight concession, namely that Mongré's "parody of the Shakespeare-Bacon theory is amusing."

A short critical review by Lou Andreas-Salomé appeared in the Viennese newspaper *Die Zeit* [Andreas-Salomé 1898]. At the very beginning, she confesses that she can only tolerate collections of aphorisms under two circumstances. They must either reflect "the creative originality of a master of imposing personality or present the scattered remarks of a writer whose life works have long attracted our strong interest." All this applies to Nietzsche, but not to Mongré. The latter's aphorisms appear like "a second outpouring" of Nietzsche's, and while "his book is very intelligent," it leaves the impression of a "chatty mind."

> In terms of form and content, his thoughts have grown, in fact, from the landscape of Zarathustra, fruits from trees that Nietzsche planted. If the author had allowed his sweet fruits to ripen in the sun in order to consume them himself, or had he wandered up and down his flowering garden to gain strength and happiness and to thrive from it – that would have been far nicer than reading all these thoughts strung out together like dried fruit on a long cord. For others already have all that he can offer them firsthand in Nietzsche's works; what purpose is there in essentially repeating this with varied twists? Many gifted thinkers appear from time to time with ideas about many varied things, but it would be sad if fashion turned to producing aphorism collections based on these.

A particularly negative review stemmed from the philosopher Erich Adickes, who later made a name for himself as a Kant scholar. For Adikes, Mongré was an unbearable imitator of Nietzsche's thoughts, one who lacked his literary gifts, "not even the dazzling quality of his thoughts: that is tasteless and inconsiderate" (*HGW* VII, 71). A similarly caustic review was penned by Michael Georg Conrad, founder and publisher of the Munich magazine *Die Gesellschaft*, who wrote:

> Nietzsche imitated to the point of deception [⋯]. As dazzling as the illusion and the nearly complete deception are in all the nuances of Nietzsche's style (from the period of *Beyond Good and Evil*), the inauthenticity breaks through everywhere. Everything is skewed in this book, everything only half-true. [⋯] I understand that Paul Mongré is a pseudonym. Nothing is real in this book, not even the author's name. [Conrad 1898]

Following these negative reactions to Mongré's *Sant' Ilario*, we close with these lines of high praise from an anonymous reviewer writing for the *Neue Deutsche Rundschau*, the leading literary magazine of the time:

> Perhaps the most intelligent book that has appeared since the Zarathustra books. A strikingly mature mind moving on the highest plane of irony speaks out about all the questions of life in aphorisms. Some come as a short paradox, others in longer essays. Not a word too much; to reproduce Mongré's thoughts one would have to repeat them. There is not a trace of Nietzsche imitation. Only the freedom of the spirit, the serene mood of Sant' Ilario is comparable to Nietzsche. There is no system in this book, except for its lack of a system. A short period of exposure to this work does not suffice to conquer it suddenly. One pores over it like the Bible. Only after some time can one engage with it in detail. (*HGW* VII, 71)

4.4 Chaos in Cosmic Selection

The autumn of 1898 saw the appearance of Paul Mongré's second book, *Das Chaos in kosmischer Auslese – Ein erkenntniskritischer Versuch* (*Chaos in Cosmic Selection – an Essay in Critical Epistemology*) [H 1898a]. Although its theme is connected with the last chapter in *Sant' Ilario*, entitled "On the Critique of Knowledge," this book has an entirely different character. For in contrast to the colorful variety of ideas provided by the aphorisms in Hausdorff's first book, he had this to say about this second work: "I find myself in the same situation as Schopenhauer, who had but a single thought to communicate but could find no shorter way to express it than by writing an entire book" [H 1898a, V; VII: 593]. Nevertheless, it is by no means easy to discover what Hausdorff meant by this single idea.

In the foreword to his book, Mongré apologizes for not being a specialist in philosophy, which prevented him from approaching his central problem in the usual way, namely by relating his own solution to previously attempted solutions. Nevertheless he wanted his readers to know that he would be the last to advocate an approach that involved abandonment of this path in favor of improvising a new one without any connection to earlier efforts. He then went on to say:

> But let's consider for once the opposite case: I don't approach the problem, but rather the problem comes to me! A flash of thought pops up that seems to carry tremendous implications as related thoughts crystallize: an entire vast philosophical context arises before a person, who by profession and personal interest is in no way equipped to deal with it! [H 1898a, III-IV; VII: 591–592].

What was this crystallizing germ, this original flash of thought? Did that moment come about during Paulsen's exercises on Kant's *Critique of Pure Reason* in Berlin? That was when the student Felix Hausdorff noted the following, writing the first words in unusually large letters, which he thickly underlined:

> One can imagine the unknown transcendental structure of things interrupting the spontaneous synthetic unity of apperception; from this event nothing at all would remain in our consciousness; thus, what we understand by the world would remain unchanged. The same holds for breaks in time millions of years long.

But perhaps he formulated this notion while reading Lotze's *Metaphysik*, which expresses a similar idea toward the end of the text?

In the very first sentence of the preface, Hausdorff reveals the goal of his book as nothing less than to attack once again "the ancient problem of the transcendental core of the world." Naturally, a massive obstacle along the way was Kant's *Critique of Pure Reason*, which posited an "unconditional dualism between appearance and the thing in itself (*Ding an sich*)." Accordingly, Hausdorff places together the concepts he wishes to employ in two columns. Those in the first column relate to the "real world," whereas those in the second stem from the world of our consciousness, the "world of appearances" [H 1898a, 3; VII: 597]:

world in itself	world in consciousness
thing in itself	appearance
noumena	phenomena
intelligible	sensible
transcendent	immanent
trans-subjective	subjective
objective	subjective
absolute, real	empirical
true	apparent

Hausdorff treats the terms within each column as synonyms, between which he freely alternates "without distinction other than respective stylistic nuance." However, he decides to avoid completely the Kantian term *transcendental*, noting that "ever since Kant's varying usages of it [⋯] this notion has taken on an impossible middle position, whereby instead of marking a sharp boundary between both worlds it occupies a transitional strip through which one world wanders into the other and vice-versa" [H 1898a, 3; VII: 597].

Kant had spoken of the "transcendental ideality" of time and space, because in his opinion time and space were a priori forms through which all conceivable empirical content are created in us. On the other hand, we experience time and space, which have an empirical reality, even though we do know whether or what they are in themselves objectively. Mongré's goal is to prove the "transcendent ideality" of time and space anew, but in *his* own way, thereby going beyond Kant:

> In the course of our considerations, it will often be emphasized that there are no such mediating areas and that no bridge leads from the empirical to the absolute and back. The "thing in itself" in appearances makes them neither more nor less clear, and our mental picture is no

Das Chaos

in kosmischer Auslese

Ein erkenntnisskritischer Versuch

von

Paul Mongré

LEIPZIG

Druck und Verlag von C. G. Naumann

1898.

Figure 4.4: Title Page of Hausdorff's
Chaos in Cosmic Selection – an Essay in Critical Epistemology.

> better nor no worse adapted to the objective factual state. We will have to show the complete diversity of both worlds and the impossibility of drawing any conclusions about transcendental foundations (in the broadest sense) from empirical findings, doing so with a comprehensive generality that takes us beyond the Kantian result in praxis and, apart from the rejection of any metaphysical positivism, motivates a new standpoint toward natural science. [H 1898a, 3–4; VII: 597–598]

The task of showing the "complete diversity of both worlds," the world as given to us empirically and the hypothetical absolutely real world, will therefore come down to "separating the one lying within the boundaries of possible experience from the one lying outside them" [H 1898a, 4; VII: 598]. Fortunately, this task can be narrowed down to the treatment of a single problem:

> This task does not require a material analysis of the entire vastly diverse manifestations in the world; it will rather already be accomplished if this separation can be determined for the *common forms* in which all empirical givens appear [$\cdots$]. In fact, such a general form embracing all empirical phenomena does exist, and this *forma formalissima* is given by *time* [$\cdots$] [H 1898a, 4; VII: 598].

Another such universal form for the phenomena of the world is space, and in the chapter "Vom Raume " Hausdorff also presents a solution to the space problem, i.e. he attempts to show that the notion of absolute space is indeterminate. We omit this topic, however, since as Moritz Epple writes: "this whole chapter [$\cdots$] is in any case not essential for the train of thought and [Hausdorff] only included it because of the beautiful parallelism that exists here between time and space" (see Epple's historical introduction to *HGW* VI). Hausdorff describes his fundamental idea for proving the transcendental ideality of time and thus the indefiniteness of the absolutely real as follows:

> We may now describe the typical form of our considerations. We want to defeat realism on its own battlefield by showing that its principles lead to absurdities. For that purpose, we take any particular property or relationship in our world of consciousness and *transfer this unchanged to the absolutely real and then try to reshape it as strongly as possible while maintaining the empirical effect as required.* If we succeed, then we have found what we need: a *negative* conceptual determination of an illegitimate world that remains correct within certain limits, an exemplar of a series of properties that we are *not* entitled to attribute to the real thing itself merely because of their agreement with its appearance. This "method," if the proud word may be allowed, is entirely sufficient to justify transcendental critical idealism, whose goal is reached with the recognition that the intelligible world is unknowable – always assuming that the transformation of the trans-subjective qualities of appearances can to some degree succeed. [H 1898a, 8; VII: 602]

Following his plan to speak the language of realism, Hausdorff must first provide a transcendental apparatus for the notion of time; he thus introduces the concepts of absolute time, timeline, and a present point of time. How he intends to apply these concepts will gradually become clear from the quotations that follow.

> The examination of *time* [· · ·] can be split into two main parts from the outset. In our conception of time two different individual ideas, which I briefly call *temporal content* (*Zeitinhalt*) and *temporal duration* (*Zeitablauf*), are easily linked to each other. On the one hand, we have a continuous series of world states, a material substrate of time, on the other, a mysterious formal process through which each world state experiences the transformation sequence of future, present, and past. Although these two special parts are always connected in our experience of time, we want to separate them conceptually and then try to determine how much each can be varied without endangering the empirical effect.
>
> By the "world state" just introduced, I understand the following precise meaning. A *world state* is either a *filled* time segment of length zero or else a *moment*, an *empty* time segment of null length. A world state has the same relation to a filled time segment as a moment to an empty one, or like a point to a line. Filled time is a simply extended continuum of world states, just as empty time is a simply extended continuum of moments and a line a simply extended continuum of points. What a filled time segment is depends on our temporal awareness, which is concerned, both receptively and productively, with nothing other than filling this temporal form. [· · ·] We cannot describe the world state itself from the point of view of temporally experienced consciousness. as it forms an unreachable empty boundary [· · ·]
>
> One conceives of the simply extended totality of all world states, whose constant succession forms the temporal world of our experience, under the image of an unlimited line, the *timeline*. Each world state forms one of its points, the process of temporal realization given by a point, the *present point* that moves along the timeline. The time in which this movement takes place is absolute time. [H 1898a, 11–12, 14; VII: 605–606, 608]

Here Hausdorff makes decisive use of the terms "essence" and "existence" to distinguish relations in the empirical and transcendent world: the *essence* refers to *temporal content*, whereas *existence* concerns *temporal duration*. The temporal content, as the collection of all world states, has the quantitative structure of the linear continuum; therefore it is referred to a *timeline* and its points are the world states. The concept of a "world state" is difficult to grasp; in fact, Hausdorff does not even attempt to describe what he imagines by this:

> Only with the help of analytical hypotheses regarding actual filled time can one analyze the world state; thus, if materialism is correct, a world state would be a resting position of material points that at a certain moment in time belong to a system of point positions in space. [H 1898a, 13; VII: 607]

Even this is problematic, however, since in classical mechanics positions alone do not define physical states. One would instead have to consider a system of points in the phase space of the overall system. Hausdorff elsewhere speaks more generally about the totality of occurrences as "the moving overall system of mechanical, organic, and psychological processes" [H 1898a, 23; VII: 617]. Here he alludes to the huge complexity that would need to be taken into account in describing the content of such a concept. As for "absolute time", the argument in later chapters shows that he conceived this as a collection with structure, once again as a line with the structure of the one-dimensional real continuum. This is never explicitly stated, but without such an assumption his references to the "movement of the present point" would make no sense. Such a movement is described by Hausdorff – to use modern terminology – as a function $f : T \to L$; here T is the absolute time and L the timeline. This function f therefore assigns to each absolute time (moment) τ a world state $l \in L$. The whole "transcendental apparatus" is therefore described using mathematical metaphors.

The use of mathematical metaphors, based on the new theory of sets created at the time, was characteristic for the epistemic-critical discourse in *Chaos in kosmischer Auslese.* In the foreword, Hausdorff wrote that his "unusual approach [$\cdots$] has not remained uninfluenced by mathematics." Only corresponding metaphors were used, however, whereas the philosophical content did not require mathematics. As Mongré put this: "When I think of mathematics, which was indispensable for clarifying my views, I also request belief in my assurances that *my representation is independent of mathematical assumptions* and accessible to every reader capable of abstract thought [$\cdots$]." Yet ultimately, Hausdorff hopes that the "fundamental mathematical questions" he touches on will also be important for philosophers:

> I would welcome it as a happy outcome, if I should succeed with this script in attracting mathematicians to this theoretical problem and, conversely, once again stimulating lively interest among philosophers in fundamental mathematical issues. Here we enter boundary regions, where an encounter between the two sciences is inevitable and the removal of the previously mutual distrust an unconditional necessity. [H 1898a, IV–V; VII: 592–593]

In his second chapter on temporal succession, Hausdorff demonstrates that the notion of absolute time is altogether vague, beginning with these remarks:

> The movement of the present point on the timeline generates the flow of empirical time for us. What can then be said about this movement? The

> naive realist can imagine nothing except for a point passing along the timeline in the same direction and with constant speed; that amounts to a direct translation of our ordinary notion of uniform time flow into the realm of the transcendent. As far as I know, the only generalization that has been proposed to this is one that would allow for a variable speed. [H 1898a, 15; VII: 609]

Here it remains unclear to whom Hausdorff was referring; probably he had in mind *Zur Analysis der Wirklichkeit*, the principal work of the neo-Kantian philosopher Otto Liebmann, who, among other things, tried to prove that the quantitative aspect of time had only subjective but no objective meaning. Liebmann relied on findings of the natural scientist Karl Ernst von Baer, who concluded from different pulse frequencies among various animals that "different subjective basic dimensions of time" prevail among living creatures. Liebmann sought to overcome the idea of an absolute time in the Newtonian sense, which in Hausdorff's terminology meant the uniform motion of the "present point" along the timeline. Liebmann summarized the quintessence of his considerations about the possible structure of an absolute time as follows:

> *Time size* is relative and subjective; in humans it differs from short-lived creatures just as it does from the toad found alive in broken, ancient rock, etc. [···]. If you therefore take away the quantitative aspect from the experience of time there remains left an objective residuum, a possibly independent reality underlying the specific limitations of our and any other intelligence: the *time order*, the ordered sequence, the *series conditionalis* in the causal chain of real causes and effects. [Liebmann 1876, 336]

Formulated in terms of Hausdorff's metaphors, the present point on Liebmann's timeline would only be allowed to move in one direction; about its speed and acceleration nothing can be said. Hausdorff now moves to eliminate this order of absolute time. After citing the last sentence in Liebmann's quintessential statement, he writes:

> Once this residuum is destroyed and the reservations have been *possibly* rejected, we may say: if you also take away the temporal succession from the conception of time, then what remains as a possible absolute reality is the temporal content, the inner constitution and linking of the world states. [H 1898a, 30; VII: 624]

In order to obtain this result, Hausdorff attempts to prove the following assertion, which he underscores as the "fundamental theorem":

> *The present point moves along the timeline in a completely arbitrary, continuous or discontinuous, manner. The transcendental succession of world states is thus arbitrary and not within our consciousness.* [H 1898a, 16; VII: 610]

He next tries to illustrate the kind of "completely arbitrary" movement he has in mind:

> A movable point can exhibit different types of behavior in relation to a fixed line during a finite time segment: it can lie outside the line, can momentarily pass through one of its points, or remain at rest in one of its points, it can finally run through any limited piece of the line in one direction or the opposite. From these elementary processes, the movement of the present point on the timeline can be composed in any arbitrary way [$\cdots$] [H 1898a, 16; VII: 610].

For the above-mentioned function $f : T \to L$ on a finite time segment $J \subset T$, these four "elementary processes" correspond in the order given above to: 1. f is not defined on J. 2. f is only defined for one point $\tau \in J$; $f(\tau)$ is the point on the timeline that the present point momentarily passes through. 3. f is constant on J. The value of f assumed on J is the point on the timeline where the present point rests. 4. f maps the interval J bijectively and strictly monotonically on an interval $f(J)$ of the timeline L; there are two cases distinguished by whether f preserves or reverses the orientation. The "composition" from those elementary processes would then give rise to a function $f : T \to L$ pieced together from these four types.

Hausdorff describes the effect of an arbitrary movement of the present point composed from these elementary processes on a hypothetical "transcendental world viewer":

> [$\cdots$] any transcendent world viewer, who would experience this form of absolute time, would live through the tissue of simultaneous changes that constitutes our empirical world in a completely arbitrary arrangement, in any order, direction, speed, and with arbitrary interruptions, resting points, repetitions, jumps, and reversals – without disturbing our awareness of the well-known and familiar *fluxus temporis*. [H 1898a, 16; VII: 610].

Of course, this is hardly a proof, but immediately after this the author presents a proof of his fundamental theorem in less than one page:

> The previously promised proof of our fundamental theorem takes the clear form of a mere syllogism. Let us assume that a certain selected succession A of world states generates our empirical world phenomenon. If another succession B were to be chosen, and if this change registered in our consciousness, it would have to be felt at some point, say, during the empirical time segment t. However, since we ourselves have incorporated with our consciousness the quintessence of all world states, the timeline, this means that by swapping A with B something would have entered the time segment t which was not present before. Accordingly, this exchange would not only have changed the order but also, contrary

> to assumption, the inner structure of the world states. Our separation of the temporal content from the temporal duration thus carries the strange implication that the latter behaves with complete indifference toward the former, as indifferent as the shape and nature of a line with regard to the possibility of a point movement. [H 1898a, 16–17; VII: 610–611]

If we examine the syllogistic form of this proof, we may note that it refers to the "inner structure of the world states," about which Hausdorff earlier asserted that this could not be analyzed without additional hypotheses. Such hypotheses might perhaps describe more precisely the relations between consciousness and temporal content, but one finds in the final pages of the book only the claim that "the slightest change in the world of consciousness also means a change in consciousness" [H 1898a, 205; VII: 799]. In discussing the relationship between consciousness and the world of consciousness, Hausdorff at times used the term "incorporated" (*eingegliedert*) but more often "intertwined" (*Verflochtensein*) to refer to this inseparable connection. Using the example of time reversal, he illustrated what this metaphor of woven interdependence meant. Thus, for example, in the act of seeing the processes of time experience are already so complicated that "the inversion from the positive to the negative temporal image of the world must completely fail" [H 1898a, 27; VII: 621]. Or, more generally, "as a subjective carrier of time, our inner structure is woven into the structure of the positively directed timeline and excluded from the negative one – or, in other words, we call the direction to which we belong positive" [H 1898a, 28–29; VII: 622–623].

Here the terminology "transcendental apparatus" expresses less than is meant by the metaphor "interwoven," since mathematically the transcendental apparatus consists of absolute time, timeless temporal content, and movement of the present point. Thus, the reversal of time "only" amounts to a mathematical change in sign. A world state, with its entire tissue of infinitely many contexts and details, is only a single point from this standpoint. As an empirical effect, however, the flow of time that the movement of the present point generates in us is a "primal phenomenon," one that Hausdorff does not attempt to derive or explain:

> The timelessness of the material substrate that we have underscored several times and continue to emphasize will elicit a desire for more detailed information, just as will be the case with the stable continuum of world states that only manifests itself by unfolding successively through time. I confess not to know an answer, and I agree with Lotze in saying that it is not our task to *make* things like Being, Becoming, Act; these can be left to bring themselves about. We are faced here with matters whose simple structure does not allow for a breakdown into simpler elements, a primeval phenomenon that can be recognized and conceptually exposed but not derived or explained. The appearances of our consciousness are neither persistent individual states of the world nor a multiplicity of persistent world states but rather only that flowing

> or sliding continuum of world states, the peculiar behavior of which we can only imagine symbolically as the image of a point movement. And on the basis of this fundamental property of our temporal existence, whose origins remain unfathomable, must arise that rigid unchangeable system (the timeline), which as soon as it appears from its imaginary foundation shall *be*, passing by the interwoven subjects of knowledge as flowing and inexorable *Becoming*. [H 1898a, 21–22; VII: 616–617]

Hausdorff then took account of what followed from his proof of the fundamental theorem:

> Summarizing the previous developments, there arise for temporal successions that fix the empirical world image such an inexhaustible abundance of metaphysical possibilities that already at this level there can no longer be any question of a transcendent reality of time. Any relation introduced in the absolute chain of temporal elements in order to secure the empirical ones represents an unnecessary limitation. The temporal sequence in general – as the fundamental form of the world as a changing, flowing, transient series of states – is metaphysically indefinite and indifferent. A definite reality can at most be ascribed only to the material substrate of events, to the rigid, unchangeable, [$\cdots$] timeline as the indestructible, inexhaustible, always complete and ever present original source of reality. [H 1898a, 22–23; VII: 618–619]

Thus, following the separation of temporal *essence* and *existence*, by the second chapter of his book Hausdorff had answered the question of existence by showing that the temporal sequence is metaphysically indefinite. There thus remains the task of debunking the essence of time, a problem taken up in the final chapter *Transcendental Nihilism* as well as in the previous chapter *The Temporal Plane*. The intervening chapters prepare this task by extending the previous approach to the timeline, which is no longer restricted to a linearly ordered set of points. Hausdorff thus introduced "a point continuum of any number of dimensions" [H 1898a, 190; VII: 784], thus an abstract set "of higher power than the linear continuum" [H 1898a, 190; VII: 788], namely "das Chaos," that something $= X$, a term already used by Nietzsche (see *HGW* VII, 49–61).

At the outset of the way to be taken, Hausdorff introduced a multitude of present points. "The next step in ascending generality," he notes, "leads us to multiply the number of present points, which in the absolute time and independently from one another undergo *at the same time* arbitrary movements on the timeline" [H 1898a, VII: 57–58; 651–652]. He then demonstrates that this is allowed without disturbing the empirical effect. The multiplicity of present points means "the simultaneous existence of arbitrarily many world states in *a single* moment of absolute time" [H 1898a, 59; VII: 653]. With the earlier assumptions for the transcendental apparatus this implies that the "arbitrary functions are many-valued," [H 1898a, 59; VII: 653], i.e. $f : T \to L$ is no longer a function but rather a multi-valued mapping.

Liebmann's idea that different individuals or different living organisms could experience time spans differently would correspond to a plurality of present points in Hausdorff's model. Whether inspired by Liebmann or not, Hausdorff assumed that among his objects of consciousness there are "those (people, animals, perhaps organisms)" whose subjective existence "is such that beyond the actual substrate of their world of consciousness (timeline) they also take account of temporal duration (movement of the present point)" [H 1898a, 68; VII: 662]. Nothing, however, justifies viewing this process as identical for all these subjects, as Hausdorff made clear:

> If it is intuitively most convenient to treat the timeline [$\cdots$] as an objective common reality valid for all conscious beings with access to it, we must take account of the diversity of individuals and assign to each one their own present point. These individual present points move about arbitrarily on the timeline independently of one another, and it would be a most unlikely accident if they were to collapse to a single point, whether forever or for a period of time [$\cdots$] The 'simultaneity of time in different minds,' that most astonishing miracle, probably does not exist, just like the ubiquity of time in space. [H 1898a, 68–69; VII: 662–663]

This last remark might be described as a remarkable premonition, as only with the acceptance of Einstein's special theory of relativity did physicists come to realize that the "ubiquity of time in space" has no objective justification.

We now want to turn briefly to Hausdorff's self-assigned task of dismantling the notion of temporal essence in his chapter "Das Princip der indirecten Auslese" ("The Principle of Indirect Selection"). In his investigation of temporal succession in the second chapter, it turned out that only the fourth elementary case, involving a positively oriented interval e of the timeline, gave rise to empirically perceptible experience. The other types of movement of the present point and their combinations produced either the absence of consciousness or, in the case of movement in the reverse direction, a world of consciousness inaccessible to us. By considering a plurality of present points, other empirical courses $e_1, \cdots$ take place in the absolute time if our present point undergoes movements that are empirically discernible. "For us, since we are condemned not to observe them, we can remain indifferent as to whether the transcendent, empirically unperceived gaps in our experience are filled with nothing or something other" [H 1898a, 125; VII: 719]. Hausdorff here connects this with his principle of indirect selection, which is inspired by biology, along with mathematical metaphors from the theory of probability:

> Thus, we will be constantly filled with e the empirical time, whereas in the absolute time one can transform e arbitrarily with not-e, *for example* e with the transformed processes $e_1, e_2, e_3, \ldots$. In all of these cases, only e is an object of our experience, whereas the cases not-e are not. In other words, the case e has a subjective *certainty* and an arbitrarily small objective *probability*. I would like to call this principle,

> which has important consequences for the theory of knowledge developed here, the *principle of indirect selection.* The word indirect, which might be replaced or sharpened by *automatically* or the like, should be understood, as in biology, not as a case selected from the outset as the only real one among all conceivable cases, but rather as arising from actual competition with them, whereby its inner peculiarities enabled it to stand out as unique. In this struggle [· · ·] those survive on account of their capability for life; out of the entanglement of chaos and cosmos, only the cosmic enters our sphere of view due to its relationship with our consciousness. [H 1898a, 125–126; VII: 719–720]

Out of the transcendental chaos our consciousness selects a world, our world of consciousness, our cosmos due to its own external and internal possibilities. However, Hausdorff does not regard this selection as a psychological process controlled by the will; it happens instead automatically. "For us, the connex of consciousness with its world of consciousness is defined purely conceptually; a continuum c of world states and its appearance e belong inseparably to an intertwined intellect i" [H 1898a, 137; VII: 731]. In the course of further discussion, the transcendental chaos will be seen as a set of immense cardinality, lying far beyond the timeline considered at the outset. The principle of indirect selection provides

> a subjective, unbreakable unity and lawfulness that turns out to be a mere excretion from a vast overarching diversity and randomness, to which it stands like a finite entity to one that is infinite to an infinitely high degree – I do not know how high one must climb in the series of "transfinite numbers" in order to count the set of all transcendental possibilities. Our usual symbols for infinities of the first and second power are completely inadequate here, as the human brain itself fails in abstracto to oversee the arbitrariness of the transcendent world to its full extent. [H 1898a, 145–146; VII: 739–740]

The realm of transcendental possibilities leads Hausdorff to considerations that he takes up in the chapter on "The Time Plane." Here, in analogy with the introduction of multiplicities of present points, he discusses the idea of increasing the number of timelines.

> If more than one present point, then more than one timeline! This means there does not need to be only one continuum of world states in which we are entangled; alongside it lie arbitrarily many other conceivable worlds of varied content represented by timelines along which present points move. Thus, we are in a position to recognize a *pluralité des mondes* precisely because we *cannot* convince ourselves of them through experience, as we always remain tied to our single timeline.
>
> And what about the proof for the transcendental admissibility of a plurality of timelines? We know the scheme for this proof already,

> for if A is the timeline belonging to our empirical world, then the existence of other timelines $B, C, \ldots$ causes no discernible reaction in A and thus escapes empirical detection, thereby showing their transcendental conceivability. [H 1898a, 152; VII: 746]

Here it is interesting to observe how Hausdorff introduces spatial mathematical metaphors, whose vivid pictorial character evidently guided his thinking:

> When we analyzed the arbitrary movement of the present point, we also considered the case where the point no longer falls on the line. We might then ask, where else could it be? In this way, we imagine it moving in some outside spatial structure, i.e. we relate its timeline to some overarching spatial manifold, such as a surface or solid structure. Nothing prevents us from mentally applying this graphic illustration, though a plane surface will suffice provisionally as a symbol for such higher-dimensional temporal structures [$\cdots$] [H 1898a, 153; VII: 747].
>
> Our *time plane*, which can be expanded immediately to a *time space* or any multiply extended continuum of world states, is of course only a visual symbol and in this regard alone constitutes a new element in our considerations. [H 1898a, 155; VII: 749]

One might doubt, however, that this "time space" is merely a "visual symbol" for the timelines, as opposed to representing a "new element." For as soon as this space is imagined, its points all become world states, since everything points to the understanding that "*each point in the plane* [or in time space] *signifies a world state and every linear structure a worldline*" [H 1898a, 153; VII: 747]. So time space has a structure, otherwise one could not observe linear structures. In suggesting this conception for the set of world states, Hausdorff neglects the more general view presented in his chapter "The Principle of Indirect Selection." As he readily admitted:

> In one respect, we remain behind the state of generality obtained so far – we restrict ourselves to viewing the ensemble of world states as a point continuum, thus giving it only "second power" – in another, we go far beyond it by filling the time plane [resp., time space] with world states of *arbitrary content.* [H 1898a, 157–158; VII: 751–752]

Moreover, the time plane (resp., time space) should not be regarded as a representation of the "real structure of the ensemble of world states":

> It is only a symbolic expression arising from our efforts to dispose of the definite, qualitatively delimited *temporal content* as done earlier with the specific, clearly prescribed *temporal duration.* We want to assign something arbitrary that lacks properties and qualities as a transcendent substrate to our qualitative, content-determined empirical world; we want to recognize irrevocably the "thing in itself" as something $= X$ [$\cdots$]. [H 1898a, 158; VII: 752].

Hausdorff initially notes that the time plane, "as an expression of the fully arbitrary character of all world states," leaves open the problem of extracting timelines from them. "We have so far only embedded our timeline in the time plane, but have not yet dissolved and decomposed its qualitative structures" [H 1898a, 158–159; VII: 752–753]. He thus poses the question whether it is restrictive to assume that the time plane containing the set of all world states also contains our timeline, our cosmic world, as one of its curves. In the chapter *The Temporal Plane*, Hausdorff appeals to various analogies which suggest that this question can be answered in the negative.

In the concluding chapter on *Transcendental Nihilism*, however, he admits these arguments have only a provisional character:

> In the end, generality is a matter of interpretation, and whoever wishes to could consider it a kind of restriction if, for example, on the basis of a complicated and wonderfully purposeful structure of the empirical world, its integration into the overall transcendental existence were no longer to be seen as completely unbound and free of attributes. Thus, in order to save the general arbitrary character of our transcendental world in the face of this logic, we must turn to the second main question: *If* the empirical world constitutes a restriction of the transcendental, is such a restriction *necessary*? [H 1898a, 80; VII: 774]

A negative answer to this question would lead to the main theorem Hausdorff hopes to prove as the final touchstone for his apagogical argument refuting the "world in itself." He formulated this as follows: "Empirical existence proves nothing pertaining to transcendental existence (leaving aside the uniqueness of transcendental existence). *Empirical reality and transcendental unreality are compatible*" [H 1898a, 180; VII: 774].

In order to prove this theorem, Hausdorff reverts back to his separation of temporal content from temporal duration as well as the notion that the form of empirical reality for the flow of time, the form of absolute existence in the momentary present, is the time point. This absolute existence, however, is completely inaccessible to our consciousness:

> Only the current present as an extensionless world state can be grasped and fixed as absolute Being, though it escapes perception: every conscious action, however small its amplitude and however slow its creeping pace, always implies Becoming and Changing, a succession of infinitely many different, steadily following world states. [H 1898a, 182; VII: 776]

From here, Hausdorff maintains that it only requires "a trace of logical impulse to draw the short, obvious, and so unmistakable conclusion *from the incomprehensibility of the present to the unprovability of existence*!" [H 1898a, 183; VII: 777]. Not that Hausdorff chooses to draw such a short conclusion; instead, over the course of several pages, he uses "the incomprehensibility of the present" in attempting to show that our empirical cosmos does not require that the transcendental chaos contain even a single line element of "cosmic texture," however short in duration; it thus persists independent of our empirical cosmic world.

About this, he tries to erase all doubt:

> A person may insist, however, that the concept of transcendental arbitrariness can only be taken as completely valid when not even sporadic lawfulness obtains, for whoever finds too much purpose in nature when it leaves one germ among millions alive, that too we must give up in pursuing to the limit the unprovability of transcendental existence [· · ·], revealing that our empirical being never and nowhere requires the correlate of an absolute being [· · ·]. Why shy away from the name? Our idealism, when followed to its last consequences, takes us to the sharp and dangerous edge of a transcendental *nihilism* [· · ·] [H 1898a, 187–188; VII: 781–782].

In fact, this is a type of hairsplitting logical argumentation that makes this part the most problematic in the whole book. As we will see below, Rudolf Steiner dealt with it critically in his review.

Having reached his goal, Hausdorff now gave a brief summary of his findings:

> Up to now, we have followed the analytical path of successive generalization; step by step absolute Being was deprived of the properties ascribed to it by Realism on the basis of a premature translation of findings from the empirical sphere into the transcendental. Instead of a smooth and uniform flow course of time, we obtained an arbitrary flow of the present point, whether continuous or with jumps, whether moving forward or backward; instead of one present point, arbitrarily many; instead of a timeline, an arbitrary manifold of world states; instead of a qualitatively limited cosmos, a chaos without boundaries or qualities. [H 1898a, 188; VII: 782]

Any attempt to follow the path backward in order to reduce the empirical cosmos from this transcendental chaos, so to speak, is doomed, of course, to fail from the outset. Hausdorff's critique leads to a variety of cosmic worlds, a "pluralité des mondes," each of which amounts to a special case with respect to the transcendental chaos. They possess reality only in the consciousness of the observer, not in the world itself, as the origin of this "pluralité des mondes" resides in the principle of indirect selection:

> Every arbitrarily chosen selection principle, if it is sufficiently strict and otherwise corresponds to our concepts of an empirical world, separates a cosmos out of the chaos, i.e. out of the totality of all world states, it determines a (linearly extended) totality of *definite* world states, which represents itself approximately to a consciousness contained therein as we represent our empirical world. [H 1898a, 207–208; VII: 801–802]

At the end of the book, Hausdorff offers these summary remarks concerning his critical theory of knowledge:

> So we may abandon cosmocentric superstition, just as before with geocentrism and anthropocentrism. Let us recognize that innumerable sets of cosmic worlds are spun into the chaos, each appearing to its possessors as the unique and exclusively real world, thereby tempting them to attribute their qualitative characteristics and peculiarities to the transcendental world core. And yet this world core escapes all chains, however loose, to maintain the freedom to represent cosmic phenomena in an infinitely varied way. It allows for the juxtaposition of all of these phenomena, which are contained as specific possibilities within a conceptually delimited part of its universality – indeed, it is nothing more than this juxtaposition and thus transcendent for individual appearances, which possess in themselves their own closed immanent domain. [...] When passing to the transcendental realm, the whole wonderful and richly integrated structure of our cosmos flattens into sheer chaotic indeterminacy. Accordingly, the attempt to construct even the most elementary forms of consciousness as necessary incarnations of the appearances must fail. With that all the bridges that metaphysicians have fantasized connecting chaos to cosmos have collapsed, and thereby *the end of metaphysics* is declared [···]. [H 1898a, 208–209; VII: 802–803].

This ends our sketchy presentation of the content of Hausdorff's *Das Chaos in kosmischer Auslese.* As mentioned earlier, we omitted any discussion of the spatial problem because its treatment is unnecessary for understanding Hausdorff's overall concerns in the book. We will, however, briefly discuss this chapter later in connection with Hausdorff's examination of the space problem in general as well as in his inaugural lecture from 1903. Otherwise, we can only recommend study of the book itself to anyone who wishes to gain a deeper impression of Hausdorff's philosophical world in its full diversity as well the relationship between his ideas and those of Kant, Schopenhauer, Nietzsche, Lotze, and many other thinkers. Whether such a reader will find Hausdorff's arguments and conclusions really convincing is an entirely different matter that surely depends on a person's own experiences and philosophical positions.

Hausdorff also wrote an announcement for *Das Chaos in kosmischer Auslese*, which like the one he wrote earlier for *Sant' Ilario* appeared in the periodical *Die Zukunft* [H 1899c, VII: 811–813]. He chose to write it in the form of a poem in 10 stanzas that conveys a humorous dialogue between a philosopher (Mongré) and the Kantian "thing in itself." The stanzas follow the same strict scheme: the eight lines in each stanza always contain 11, 10, 11, 10, 11, 10, 11, 11 syllables with the rhyme scheme *AB AB AB CC*. The result is a remarkable and truly artful introduction to the central theme of the book, the author's philosophical relationship to his great predecessor, his principal concerns and hopes for a responsive audience, all poured into this strict lyrical form and with a wink of self-irony. He wrote this poem in Binz, a resort village on the Baltic Sea, perhaps just the right environment to create something special like this.

4.5 The Reception of Chaos in Cosmic Selection

As so often happens with original creative work, the reception of Hausdorff's booklet proved disappointing for its author. As he wrote in the foreword, he wanted "to stimulate interest among mathematicians in the epistemological problem and conversely promote awareness among philosophers of the fundamental mathematical issues" [H 1898a, V; VII: 593]. But hardly anyone showed real understanding for this balancing act, which required moving across the boundary between two fields. When Fritz Mauthner not long afterward attempted a similar border crossing between disciplines in his three-volume study *Beiträge zu einer Kritik der Sprache* (1901–1902), Hausdorff expressed skepticism about the prospects for success. Writing on 1 September 1901, he alluded to the failure of his own *Chaos*:

> We serious authors (I have to explain the immodest We some other time, since you will not guess it from the signature of my name), cannot count on a polyphonic echo, and that is all the more hellishly difficult for those who work in some "boundary area," i.e. outside the conventional divisions of fenced-in special fields who engage in tearing down and building anew. (*HGW* IX, 453)

Hausdorff was surely correct in assuming that Mauthner would not have connected his name with that of the well-known author Paul Mongré. In view of the reactions to Hausdorff's book up until 1901, one can easily appreciate that he experienced a "hellishly difficult" time.

His friend Köselitz had, of course, received a copy immediately after the book came out. On October 12, 1898, Hausdorff wrote the following lines, which clearly reveal the years of tedious work he invested in it:

> My new book, which you will receive in the coming days, will hopefully not give me the reputation of being an oligographer [occasional writer]; its origin, like that of S. Ilario, actually dates back some years ago. The printed version is, roughly calculated, the third one, and if I hadn't finally turned up the heat to finish the manuscript, it might have again ended up going through revision. (*HGW* IX, 397)

Since Hausdorff knew that Köselitz was particularly interested in the possibility of eternal return of the same, he informed him that this notion meant the "identical return of any arbitrary empirical segment of time in absolute time." This interpretation, according to Hausdorff, had "nothing to fear from any empirical science." In any case, a letter from Köselitz to Overbeck, dated November 14, 1898, shows that he did not hold Hausdorff's new work in high esteem:

> Dr. Hausdorff (Paul Mongré) has published his promised book "Chaos in Cosmic Selection." For many readers of Sant' Ilario it will be a disappointment because it loses itself entirely in the abstract and mathematical realm. I have to admit that not everything in it is clear to me. How he understands *"time"* is probably to be blamed. [···]
>
> To speak of a time*line*, as Mongré does, goes entirely against my way of thinking. That is the most primitive physicist's conception: the line as a moving point! the line as one-dimensional being! – admittedly, I am not yet done reading the book: hopefully things will come to the fore that I have not thought of before. (*HGW* IB, 356)

Overbeck replied on December 31, 1898, presumably before he had read the book himself:

> Thanks first of all for the interesting report on Hausdorff's newest work. I can no longer fully venture onto these frosty heights, but as far as I follow your hints about H's speculations on time, there is nothing for me to gain from this, the only corner from which the concept of time is still accessible and interesting for me. [Hoffmann et al. 1998], Brief Nr. 237.

Thus, Hausdorff could not expect understanding and recognition from his friends in the Nietzsche circle. Nor did his book fare much better in the few reviews it received (see *HGW* VII, 73–74). Already in 1899, *Kant-Studien* published a 1-page review penned by Fritz Medicus, a young philosophy professor at the ETH in Zurich. The first half summarized the author's intention, emphasizing that the "epistemological radicalism of the book, which is compensated with mathematical ingenuity" often reverts back to Kantian ideas. The reviewer then focuses his criticism directly on the problematic point noted above, namely Hausdorff's argument for *transcendental nihilism*:

> The conclusion that M[ongré] draws from it is the complete independence of the empirical cosmos from the transcendental world, which, without our noticing it, can continuously transform like a Proteus, indeed, does not even need to be present at all and yet still presents "a sufficient equivalent" (!) of the empirical cosmos. The author describes this "final consequence" of his idealism as "transcendental nihilism." It is not surprising that the path leading to this goal is rich in paradoxes. [···]
>
> [Mongré has presented a book] that employs a strong, but nevertheless fine sense of exaggeration that starts from a thoroughly unorthodox position to develop over the course of 200 pages its strange consequences. Lovers of witty absurdities will read this work with pleasure. (*Kant-Studien* 4(1899), p. 339)

In discussing the reception of *Sant Ilario* we already quoted from Michael Georg Conrad's thoroughly damning review. This same author also wrote an article entitled "The Fight for Nietzsche" that appeared in the Viennese journal *Die Waage*, a weekly devoted to politics, economics, literature and art. There he dealt with *Chaos in Cosmic Selection* in a similarly hateful manner.

An anonymous reviewer writing for the *Literarischen Centralblatt für Deutschland* noted that the author of *Chaos* had only a minimal familiarity with the philosophical literature but was mathematically educated and displayed an "independent mind." As to the book's intentions, nothing at all was said, except to cite a line at the end of the chapter "Against Metaphysics," where Mongré says: "When the crimson falls, so must the duke follow: the philosophical delusion of other worldliness may follow where the religious one preceded it." The first part – "When the crimson falls, so must the duke also follow" – comes from the 5th act of Friedrich Schiller's play *The Conspiracy of Fiesco at Genoa.*

Max Nath, later known as a pedagogue and school director, pointed out in his brief review that the author was not a professional philosopher. He found the book very interesting and stimulating, but offered no substantive comments about its contents.

The only detailed discussion of *Chaos in Cosmic Selection* appeared in a review by Rudolf Steiner that appeared in *Magazin für Literatur*, which he was editing himself at that time. Steiner had already emerged as an authority on Goethe and Nietzsche, and by 1900 he was involved in public discussions with Elisabeth Förster-Nietzsche and the Nietzsche archive, as was Hausdorff as well. Steiner later became particularly well known as an esotericist and founder of anthroposophy. He began his discussion of Mongré's book with the following comments:

> Some time ago an extremely remarkable book was published which, like similar literary publications of its genre, shares the present fate of receiving far too little attention: *Chaos in Cosmic Selection* by Paul Mongré. Indeed, it deserves a different fate. [Steiner 1900, Column 569]

Steiner found Mongré's book unusually "stimulating, yes, even exciting for those who are intensely interested in the highest questions of existence." He discussed in detail the problems that arise from his point of view when mathematical thinking is applied to "Reality." Mathematical thinking "as such has nothing directly to do with reality":

> When making decisions, the mathematician is accustomed to consider only the necessities of his thought. And he is also accustomed to finding his truths unconditionally valid in reality. With such feelings he enters basically every sphere in which life leads him.
>
> And with such feelings, Paul Mongré treads the ground of the great question of existence. That is his danger. [Steiner 1900, Column 571]

Steiner illustrated this by pointing to the same problematic passage that Medicus had noted in his critique. He only quotes the last sentence from Hausdorff's text, but underscores "the sharp and dangerous edge" toward which his idealism ultimately leads and which he calls "transcendental *nihilism*":

> Paul Mongré's mathematical thinking has now seduced him to such an extravagant conception. In his thought, the mathematician separates time as well as space from the other content of the world and then handles these as abstract structures. He can then speak of the course of time as existing *alongside* the sequence: father – son – grandson. But in reality this sense of time is not at all present as such. It is not separate from the *contentual* sequence father – son – grandson. [Steiner 1900, Column 574]

Steiner therefore cast doubt on Hausdorff's argument because of this separation of time from temporal content or the separation of existence and essence. He then compared his own position with that of Mongré:

> I agree with Mongré – as one can see from my *Philosophie der Freiheit*, which appeared several years ago – insofar as I, too, view all reflections on the world as limited by the experiences given to us, but also in rejecting any conception of another (transcendental) world. [Steiner 1900, Column 575]

However, Steiner was decidedly against Mongré's "principle of indirect selection," just as he rejected the possibility of alternative cosmic worlds that exist alongside ours as a "pluralité des mondes" (*HGW* IB, 356–359).

The neglected interest in Mongré's *Chaos in Cosmic Selection*, already bemoaned by Steiner, would continue for a very long time. Not until 22 years after its publication did Hausdorff have the pleasure of reading appreciative words from an important scholar, namely the philosopher Moritz Schlick. By that time, Hausdorff had established himself as a full professor in Greifswald, having long since bidden adieu to his philosophical alter ego Mongré.

Schlick had studied mathematics, natural sciences, and philosophy; he took his doctorate in 1904 under Max Planck with a study on optics. In 1911 he habilitated in Rostock with his book *Das Wesen der Wahrheit nach der modernen Logik* (*The Essence of Truth according to Modern Logic*). Seven years later, Schlick published his principal work, *Allgemeine Erkenntnislehre* (*General Theory of Knowledge*) [Schlick 1918]. In October 1921 he was appointed to a full professorship at the University of Kiel, but already one year later, as a result of efforts by the Viennese mathematician Hans Hahn, he received a call to the Chair of Natural Philosophy at the University of Vienna. This position was created in 1895 for Ernst Mach, who taught empiricist philosophy, in particular history and theory of the inductive sciences. After Mach's retirement, he was succeeded from 1902 to 1906 by Ludwig Boltzmann. In Vienna, Schlick found a favorable environment and like-minded colleagues with whom he founded the so-called Vienna Circle, a philosophical direction usually characterized as logical empiricism. Moritz Schlick was murdered on June 22, 1936 by a former student, who acted on ideological and political as well as private motives. This crime spelled the end of the Vienna Circle, which had been a thorn in the side of anti-Semites at Vienna University. Not surprisingly, once National Socialist Germany annexed Austria, the government released Schlick's murderer after he had spent only two years in prison.

Schlick was one of the first philosophers who dealt seriously with Einstein's theory of relativity and fully appreciated its importance for philosophy. An article from 1915 about this appeared in the *Zeitschrift für Philosophie und philosophische Kritik* and found recognition among like-minded philosophers. After Einstein presented his general theory of relativity in *Annalen der Physik*, Arnold Berliner, the founder and co-editor of *Die Naturwissenschaften*, asked Schlick to write about general relativity theory for his magazine. This led to Schlick's essay "Raum und Zeit in der gegenwärtigen Physik" ("Space and Time in Current Physics") [Schlick 1917a]. One year later, Springer brought it out as a booklet [Schlick 1917b], expanded by a final chapter on the relationship of general relativity to philosophy. In a letter to Schlick, Einstein praised its "supremely clear explanations," including those in its final chapter. A friendly relationship soon developed between these two thinkers, as Einstein was pleased to promote Schlick's booklet, which enjoyed great success. Already in 1919 a second expanded edition came out, and one year later a still larger third edition appeared; a final fourth edition was published in 1922.

Schlick's essay contains passages reminiscent of Hausdorff's argument in *Chaos* and in his inaugural lecture "Das Raumproblem" [H 1903a], which aimed to prove the transcendental ideality of space. By 1918, at the latest, Hausdorff knew Schlick's essay from *Die Naturwissenschaften*. Moreover, he had dealt with Einstein's theory of relativity in Greifswald, where he probably gave a lecture about it based on a 36-page manuscript in his estate (*HGW* VI, 451–470). This text gives a clearly understandable presentation of special relativity with some brief remarks toward the end about the general theory of relativity.

In 1919 Hausdorff and Schlick exchanged letters, only two of which, both from Hausdorff, have been preserved. The first reveals that this contact came about through Franz Selety, a Viennese private scholar. Selety and Hausdorff were involved in a controversial discussion about the concept of reality, and in connection with this Selety asked Schlick to send his article "Appearance and Essence," which had appeared in *Kant-Studien*, to Hausdorff. Schlick immediately complied with this request, and on February 23, 1919, after reading this text, Hausdorff wrote back to him:

> I was very pleased to receive your "Erscheinung und Wesen" ("Appearance and Essence") along with your kindly accompanying letter. I wanted to answer only after first carefully studying your article to the extent needed in order for me to be able to convey my views about it. This mental state, however, has not yet occurred, even though I have

> read your essay twice. This is obviously due to my no longer being accustomed to philosophical thinking: in corresponding with Dr. Selety I also realized that I can engage with philosophical arguments, but only very slowly. [...]

Hausdorff then briefly reported on the nature of his differences with Selety and then concluded by remarking:

> Some time ago I read your "Space and Time in Current Physics" and found it so good that I assumed you were a physicist! [···]
>
> It would interest me greatly to learn what you would say about my philosophical book (P. Mongré, Chaos in Cosmic Selection), which, of course, today after 21 years I would write differently and, above all, much shorter. (*HGW* IX, 583)

Selety also referred to Mongré's book in a letter to Schlick, dated April 15, 1919. There he wrote that its basic idea involves "an epistemological-metaphysical principle of great boldness and originality." To this he added: "[Mongré's] work has become very little known and I would like to recommend it highly to your attention."

These references to *Chaos in Cosmic Selection* led Schlick to study the booklet and in the third edition of his *Space and Time in Current Physics* he added a footnote praising Hausdorff's account. This footnote must be seen in the context of Schlick's text, which stressed how epistemological criticism had long striven to free the basic concepts of the sciences from hidden untested hypotheses:

> Thereby it has already developed thoughts about the relativity of all spatial relations, as the consequent development and application of which we can regard the space-time conception of Einstein's theory. From those thoughts leads a continuous path, along which emerges increasingly clearly the meaning of the question of the "reality" of space and time, and which we want to follow here to access these new ideas. [Schlick 1920, 24]

Schlick then brings up the old thought experiment that Poincaré popularized in the French debates over the space problem, namely that if all the bodies of the world expanded by a certain factor overnight, we could not take notice of this change because our body and measuring instruments would undergo the very same transformation. Schlick also considered more general transformations of space that would produce the same effect, to which he added this footnote:

> Unfortunately, only after publication of the second edition of this text did I get to know the extremely ingenious and fascinating book: *Chaos in Cosmic Selection*, an epistemological essay by Paul Mongré, Leipzig 1898. The fifth chapter of this work gives a very complete presentation of the above discussions in the text. Not only Poincaré's argument but

> also the others discussed above have already been anticipated there. [Schlick 1920, 24]

Schlick sent Hausdorff a copy of the third edition, which made him very happy. In a letter, dated July 17, 1920, he thanked Schlick somewhat belatedly, giving the same reason as before. Having first read Schlick's booklet, he wanted "to collect his thoughts, now somewhat weaned from philosophy, to reply to the epistemological content of your letter." But he then conceded:

> This last has not occurred, and so I have to postpone this to a later opportunity, hopefully on vacation; but I can no longer postpone, finally – and requesting forgiveness for this delay – expressing my heartfelt thanks for your courteous dedication and in particular my great joy over the recognition you have bestowed on *Chaos in Cosmic Selection* in your letter as well as in your booklet. Since you were able to determine that I was not spoiled in this regard by specialists in philosophy, you will be able to imagine my joy over this belated, yet ultimately noticeable impact of my essay from that time.

In the meantime, Hausdorff acknowledged – now that he himself had become a specialist in another field – he had come to understand why his book made such an amateurish impression on experts, thereby frightening them away from getting to know it closely.

> And so I find it all the more delightful that a scholarly philosopher has nevertheless pushed beyond the inadequacies and penetrated to the durable core of my thoughts, and the extremely honorable and flattering form in which you have testified to this is the maximum that my ambition ever dreamed of attaining. (*HGW* IX, 587)

In fact, it must have appeared "extremely honorable" for Hausdorff to see his critical study accorded a place in the prehistory of Einstein's theory of relativity. Schlick also mentioned Hausdorff's *Chaos* in the second final edition of his *General Theory of Knowledge*. In Section 28 on the subjectivity of time, he wrote:

> Still another train of thought can be pointed out here, as this is likely to make the subjectivity of the temporal in the sense explained particularly clear, and which we find developed sharply by P. Mongré (*Chaos in Cosmic Selection*, Leipzig 1898). [· · ·] If we imagine the current of our consciousness broken down into successive sections, and these individual sections then exchanged with one another in any arbitrary way so that the order of our experiences is completely mixed up, we can ask ourselves what difference this transformation would make for our experience; then we must answer: none at all! [Schlick 1925, 229]

Kurt Gödel, probably the most important logician of the twentieth century, also read and referred to Hausdorff's essay. Gödel was a colleague and friend of

Einstein at the Institute for Advanced Study in Princeton, where toward the end of the 1940s he worked intensively on relativistic cosmology. In 1949 he published "An Example of a New Type of Cosmological Solution of Einstein's Field Equations of Gravitation" [Gödel 1986–1995, Vol. I, 190–198] in *Reviews of Modern Physics.* In this work, he demonstrated the existence of solutions that exhibit rotation of matter relative to a local inertial system (existence of "rotating universes"). One year later, at the International Congress of Mathematicians held in Cambridge, USA, Gödel presented a lecture entitled "Rotating Universes in General Relativity Theory." In his paper, he examined a special cosmological model with a number of peculiar properties. Two of these nine properties are particularly interesting in connection with Hausdorff's cosmological reflections:

> (5) It is not possible to assign a time coordinate t to each space-time point in such a way that t always increases, if one moves in a positive time-like direction; and this holds both for an open and a closed time coordinate.
> (6) Every world line of matter occurring in the solution is an open line of infinite length, which never approaches any of its preceding points again; but there also exist closed time-like lines. In particular, if P, Q are any two points on a world line of matter, and P precedes Q on this line, there exists a time-like line connecting P and Q on which Q precedes P; i.e., it is theoretically possible in these worlds to travel into the past, or otherwise influence the past. [Gödel 1986–1995, Vol. I, 191].

Shortly before the appearance of this work in 1949, Gödel published a short essay for the well-known collection *Albert Einstein: Philosopher-Scientist* edited by Paul Arthur Schilpp. Gödel's contribution was entitled "A Remark about the Relationship Between Relativity Theory and Idealistic Philosophy" [Gödel 1986–1995, Vol. II, 202–207]. Several longer preparatory drafts for this article, which dates from 1946, were later printed in [Gödel 1986–1995, Vol. III, 202–259]. According to his own testimony, Gödel was not a Kantian. Nevertheless, in this essay as well as in his preparatory work, he argues that the results of the theory of relativity, including the properties of his solution of the Einstein field equations as described in 1949, support Kant's position regarding the transcendental ideality of time. In footnote 4 of this article, Gödel makes a direct reference to the views of Mongré as well:

> One may take the standpoint that the idea of an objective lapse of time (whose essence is that only the present really exists) is meaningless. But this is no way out of the dilemma: for by this very opinion one would take the idealistic viewpoint as to the idea of change, exactly as those philosophers who consider it as self-contradictory. For in both views one denies that an objective lapse of time is a possible state of affairs, *a fortiori* that it exists in reality, and it makes very little difference in

> this context, whether our idea of it is regarded as meaningless or as self-contradictory. Of course, for those who take either one of these viewpoints the argument from relativity theory given below is unnecessary, but even for them it should be of interest that perhaps there exists a second proof for the unreality of change based on entirely different grounds, especially in view of the fact that the assertion to be proved runs so completely counter to common sense. A particularly clear discussion of the subject independent of relativity theory is to be found in *Mongré 1898.* [Gödel 1986–1995, Vol. II, 202f.].

One might well wonder, how did Gödel manage to learn about Hausdorff's book? It should be noted, though, that Gödel regularly attended meetings of the Vienna Circle as a young student, and so he knew all the personalities in that group (Hans Hahn and Karl Menger were his teachers in mathematics). It thus seems very likely that Moritz Schlick made him aware of Mongré's book.

Finally, we turn to a late response to Mongré's idea of a cosmic world selected from chaos, a cosmic world that arises as an unlikely special case from the infinity of possible worlds. Hausdorff no longer lived to experience the later reception and transformation of his ideas – as in the case of Gödel – but he nevertheless knew one of those who adapted these ideas in his own creative work, namely Max Bense. In 1930, Bense was a student in Bonn with wide-ranging interests in physics, chemistry, mathematics, geology and philosophy. He also attended Hausdorff's lectures and, as Brieskorn learned from his widow, he often spoke of the great impression Hausdorff made on him. He was twice invited to Hausdorff's home, where they engaged in controversial discussions over the importance of mathematical logic. In 1937, Bense took his doctorate under the philosopher Oskar Becker with a dissertation on quantum mechanics and the relativity of existence. After working as a physicist and serving in the military, he habilitated in Jena in 1946 with a postdoctoral study on history of mathematics. In 1950 he was appointed associate professor for philosophy and scientific theory at the Technical University of Stuttgart, where he taught as a full professor from 1963 up until his retirement in 1978.

Bense published numerous works on scientific theory, logic, cybernetics, semiotics, natural philosophy, history of philosophy, and many other topics, including an approach to aesthetics based on information theory. He also wrote experimental novels, poetry, as well as essays and plays for the radio. His theory of aesthetics abandoned the usual categories of content and form in order to represent aesthetic processes in a technically scientific way by making use of mathematical language. This conception of aesthetics was presented to the public between 1954 and 1960 in four booklets *Aesthetica I-IV*, which Bense then compiled into a book in 1965: *Aesthetica. Introduction to the New Aesthetics* [Bense 1965]. Its fourth chapter deals with "historical, social, scientific, and aesthetic consciousness" as integrated into a single reality grounded in the technical consciousness of our modern civilization, in a consciousness of the *fundamental feasibility* of the world [Bense 1965,

257]. Scientific and artistic productivity are thus seen as essentially related rational processes. This reductive view is concerned with "a principle that is interested in expelling everything on the *horizon of making* of a civilization that does *not* result from its theoretical structures" [Bense 1965, 261]. Seen from this perspective, the "physical as well as the aesthetic process appears as *more selective* and affected by chance." This allows for the use of notions such as *entropy* in a generalized form, and one can also apply the terminology of mathematical information theory to treat artworks as aesthetic information and to analyze processes of aesthetic communication [Bense 1965, 264f.]. Bense introduced the concept of *realization* as a "constituent category of aesthetic information and communication" [Bense 1965, 278].

> Signs form arrangements, distributions that realize information, whether documentary, semantic or aesthetic. As documentary and semantic information, they can have a scientific importance, in aesthetics they can involve syntactic (symbolic), semantic (iconic), or pragmatic (indicator) information. Realization is an expression of a selection function. This is the classical concept of realization, as in Leibniz and also Whitehead: creation as realization, as realization of the selection. [$\cdots$] If a (finite) number of possible worlds (cases) exists, from which one is to be realized, this number of possible worlds denotes the contribution (*Realisationsbeitrag*) to the realized world. These worlds can also be possible texts or possible arrangements of signs from which one is realized. [Bense 1965, 274]

Here one begins to see how and why Bense's aesthetics was connected with Mongré's critical theory of knowledge. This connection appears explicitly in the section on cosmological aesthetics in *Aesthetica IV*. It was only here, thus in 1960, that Bense referred to the names Paul Mongré and Felix Hausdorff for the first time. In this section, Bense discussed the pre-history of the sign- and information-aesthetics created by him, Helmar Frank, and Abraham Moles.

> Yet pure sign-aesthetics, as can be built from [the work of] Peirce and Morris, is not the only starting point for recent statistical information-aesthetics. Rather, the prehistory of this very complex theory also involves a set of considerations found in various authors between Nietzsche and Whitehead, on the one hand, and between Boltzmann and N. Wiener, on the other. These do not directly concern aesthetic problems, but they have influenced their solution. [Bense 1965, 284]

Among the authors Bense named in the first group was Hausdorff-Mongré, "as a mathematician world-famous, as a philosopher, quite unknown."[5] In three

[5]Bense himself tried to remedy the latter circumstance somewhat. Hausdorff's *Chaos in kosmischer Auslese* was no longer easily accessible, so in 1976 Bense reissued it in his Agis-Verlag, though with new title (while noting the original): Felix Hausdorff (Paul Mongré), *Zwischen Chaos und Kosmos oder Vom Ende der Metaphysik*, prefaced by a long introduction (see the

places, Bense cited passages from Mongré that related directly to his own position. The first presents the aesthetic problem in the following terms:

> The *existentia potentialis* immortalizes impartially all fragments of becoming, regardless of whether the world reads empirically like a meaningful text or like a mere brew of letters .... In fact, if one wants to insist on the diverse nature of both worlds by forming polar opposite aesthetic values, there remains no other choice: the transcendent world appears, measured by its own standards, as the most nonsensical, unbearable, most unreasonable of all world forms, as the deepest devaluation of human values, as the cruelest mockery itself of the basic prerequisites for appreciating and creating the values of life! [Bense 1965, 285]; the full text is [H 1898a, 50–51; VII: 644–645].

The second place where Bense refers to Hausdorff's *Chaos* relates to "cosmic structure," whereas the third concerns the improbability of the cosmic world. About these two passages, Bense writes:

> Finally, the sentence appears, which even if it refers to the concept of the "cosmic structure of our empirical world" can almost be seen as a definition of cosmological aesthetics: "Hereby I want to understand ... everything that becomes visible in the world from ordering, relating, decorating, shaping principles, all the relations, connections, regularities, which we bring together in our sense of 'cosmos' in opposition to 'chaos' ..." (p.159). In connection with this, there is thirdly the "principle of the exceptional case," formulated as the "cosmic type." This makes the "qualitative certainty of the empirical world" into an "excerpt from an indefinite random realm." I here break off this discussion of what I call the cosmological aesthetics of Mongré. It is evident that it is accompanied by statistical ideas as well as cosmological conceptions and that the comparison of (transcendental) chaos and (immanent) cosmos anticipates the relationship between repertoire and information. [Bense 1965, 285f.]

Bense also referred to Mongré/Hausdorff in many places in his other works, especially in connection with his own semiotic studies. His final work was published in 1992 by his widow, Elisabeth Walther-Bense, from his estate. Entitled *Die Eigenrealität der Zeichen* (*The Proper Reality of Signs*), it ends with a quote from *Sant' Ilario*: "man is a semiotic animal." This aphorism continues:

> Man is a semiotic animal; his humanity does not manifest itself through natural expressions of his needs and satisfaction but rather these are made only indirectly understandable through his acquisition of a conventional, symbolic sign language. [H 1897b, 7–8; VII: 101–102]

remarks of Werner Stegmaier in *HGW* VII, 80).

4.6 Mongré on Stirner

Hausdorff's essay "Stirner" [H 1898d, VIII: 381–390] was occasioned by the appearance of John Henry Mackay's biography of this long forgotten figure. Max Stirner was the pseudonym of the philosopher, writer and translator Johann Caspar Schmidt, whose principal work, *Der Einzige und sein Eigenthum* (*The Ego and Its Own*), was first published in Leipzig in the autumn of 1844. Since then, this book has been translated into 19 languages. Stirner's philosophy represents absolute egoism: the unique self as owner rejects everything beyond himself; nothing is holy, not God, church or society, and especially not the state and its institutions, not even if they are socialist or communist in design. General social norms of morality, law, marriage, or property are of no concern to the unique individual, who alone matters. These notions only hinder him from forming his own ideas as his property and means of livelihood. The preface to his book ends with the following passage:

> The divine is God's concern, the human that of "humanity." My concern is neither with the divine nor human, nor with the true, good, just, free, etc., but rather only with *mine*, and this is no general matter, but rather – it is *unique*, just as I am.
>
> For me nothing lies above me! [Stirner 1893, 14].

After some initially violent counter-reactions among the so-called Young Hegelians – especially from Ludwig Feuerbach, Bruno Bauer, Arnold Ruge, and Karl Marx – Stirner's book, like the author himself, was forgotten for almost half a century. Nor did many take notice when a new edition appeared in 1882, but by the 1890s this situation had changed. Mongré's friend Paul Lauterbach convinced the Leipzig publisher Philipp Reclam jr. to put out a new edition of *The Ego and Its Own* in 1893, to which Lauterbach added an eight-page introduction. There he pointed to Nietzsche as a follower of Stirner, in a certain sense, although Nietzsche in all likelihood never knew this work. Still, Lauterbach made this connection: "We are finally given an echo of Stirner's problem in the works of his great successor, the man who expanded and recreated the ego theory – Friedrich Nietzsche" [Stirner 1893, 8]. Lauterbach then quotes from two of Nietzsche's aphorisms, "The Immoralist Speaks" from *Twilight of the Idols* and "Aboard Ships" from *The Gay Science*. By drawing this connection, interest in Stirner was certain to grow along with Nietzsche's fame, especially since Lauterbach attributed a political role to Stirner's work. In a letter from 24 April 1893 to Köselitz, he underscored the "political role Stirner and Nietzsche have to play!"

We can assume that Hausdorff spoke to his friend Lauterbach about Stirner and his work. At a young age, Hausdorff appears to have been highly impressed by Stirner's radical questioning of all societal norms and institutions. Thus, in Aphorism 278 of *Sant' Ilario* he wrote retrospectively:

> Those were light, youthful times, when the pure Ego was enjoyed like a juicy fruit [· · ·]. One could only say Me! Me!, stammering, craving, but with this hollow Stirnerian Ego one could play ball; it was a soap bubble, but in paradisiacal colors. Out of this nothingness you spanned your whole world, from this flattest and grayest of all abstractions emerged high mountains and purple sunsets: with it you experienced dramas in which every person was called Me. – Today one has become older, more realistic, more prone to complaint, and the pure Ego loses its good taste [· · ·] [H 1897b, 178–179; VII: 272–273]

Hausdorff probably also met Mackay after he came to Leipzig for several months in 1897. During his stay, Mackay established contacts in the literary society to which Hausdorff belonged, and he often brought up Stirner and his work. In his essay, Hausdorff initially informed the reader about Stirner's biography by drawing on data taken from Mackay's book, though he often interpreted this in an ironic and distanced manner. Mackay had spent several years carefully reconstructing the details of Stirner's life, an accomplishment Hausdorff fully recognized, while rejecting his "monstrous overestimation" of Stirner's importance. Although Stirner's views were reputed to be a "philosophy of egotism," Hausdorff considered this merely a "philosophy of indifference." Moreover, he predicted "there will be no Age of the Unique One" since:

> the normative impression of the Unique One is exactly null. [· · ·]
>
> Stirner reveals the more or less conscious drive to self-assertion and self-promotion in our behavior and then gives an excellent *atomistic* account of social life: how it builds up from the Ego, which never fails, like the body from molecules, the chalk cliffs from infusoria. But who can overlook the countless *levels of transformation* that lead from the mere ego to the social forms and again, at a higher level, to the fully developed individual, to a personality in Goethe's sense? [H 1898d, 70b–71a; VIII: 385–387].

Hausdorff's conclusion regarding Stirner's importance is short and to the point: "Stirner, the *devaluer* of all values, will not lead to a *transvaluation* of values" [H 1898d, 71b; VIII: 388].

Hausdorff was particularly concerned with clarifying the relationship between Stirner and Nietzsche, a topic of lively controversy at the time:

> One often hears Stirner praised or condemned as a spiritual ancestor of Nietzsche; I can hardly imagine a more misleading attribution in genealogy. Nietzsche is the very opposite of a nihilist or indifferent thinker: his superman (*Uebermensch*) is the ghost of all ghosts, as compared with the Unique One who can be easily realized at anytime, and the Stirnerian anarchists are right when they count the inventor of this new "fuss" among the moralists, romantics and religious founders. As an "ethicist,"

> Stirner is irrefutable because he is contentless and never runs the risk of bumping into realities; [· · ·] Nietzsche himself was a passionately desiring idealist, miles apart from Stirner's comfortable ego-word-games. That his Zarathustra spoke only for "healthy, sound self-discipline" and not simply arbitrary [behavior] might be shown by citing a few random passages from both men. [H 1898d, 71b; VIII: 388–389]

Following these remarks, Hausdorff presented some quotes from *The Ego and Its Own*, comparing these with others taken from Nietzsche's writings. Incidentally, no one has fully clarified whether Nietzsche actually had read Stirner, and the relationship between their works still continues to inflame controversial discussions.

4.7 Hausdorff's Relations with the Nietzsche Archive

Hausdorff was also involved in an unrelated controversy involving Nietzsche's legacy, a matter connected with Mongré's short essay under the title "Nietzsches Wiederkunft des Gleichen" ("Nietzsche's Return of the Same") [H 1900c, VII: 889–893]. This essay arose out of a public dispute over the manner in which the Nietzsche Archive oversaw the ongoing edition of Nietzsche's works. In order to understand Hausdorff's concerns in this article, we must first briefly describe his relations with leading figures associated with the Nietzsche Archive.

For a number of years, Friedrich Nietzsche enjoyed a good relationship with his sister Elisabeth, who was two years younger than he [Hayman 1980]. At times, particularly when he was in bad health, she even managed his household in Basel. Their relations soured, however, in August 1882, when Nietzsche and his sister were vacationing in the spa town of Tautenburg. They were staying there with Lou Andreas-Salomé, with whom they conversed at length every day. Elisabeth soon became embittered over her brother's friend and began intriguing against her as well as against a second friend, Paul Rée. In mid-September, Nietzsche wrote to Rée that his sister had turned against him, despising his philosophy and showering him with mockery and ridicule. Further alienating the two siblings was the circumstance that Elisabeth decided to marry Bernhard Förster, a fanatical anti-Semite, in May 1885. Nietzsche himself loathed antisemitism, and he probably felt a sense of relief when in 1888 the Förster couple, together with like-minded others, moved to Paraguay, where they founded a colony called "Nueva Germania."

Not long afterward, on January 3, 1889, Nietzsche suffered a major mental collapse in Turin. His friend Franz Overbeck helped him initially, after which Nietzsche stayed in various psychiatric clinics before finally coming into the care of his mother in Naumburg; she also became his legal guardian [Hayman 1980, 333–342]. In the meantime, the colony project in Paraguay had fallen into financial difficulties that led to Bernhard Förster's suicide in June 1889. Eighteen months later, Elisabeth returned to Germany in December 1890. She then entered into negotiations with Nietzsche's last publisher, a company headed by Constantin Georg Naumann and his nephew Gustav Naumann, as well as with the former publisher Fritsch and Schmeitzner. As a result, the firm of C.G. Naumann received the rights to publish all of Nietzsche's writings in the future. After another stay in Paraguay, she moved back to Germany in September 1893 and settled there permanently. Once the colony project had failed, she turned to the next great goal for her life's work. She described this in the *Bayreuther Blättern* in 1894: "Another great task in life: from now on, taking care of my only dear brother, the philosopher Nietzsche, caring for his works and describing his life and thinking, that takes all my time and energy." [Hoffmann 1991, 12].

Indeed, Elisabeth Förster pursued these goals with great energy, but also without scruples when it came to editorial methods. She strove to maintain near complete control over the interpretation of Nietzsche's life and thought by means of the editing and evaluation of his work. To accomplish this, she secured the assistance of a wide network of Nietzsche enthusiasts, friends, and supporters, through whom she later attempted to create a Nietzsche cult in Weimar modeled on the Wagner cult in Bayreuth. Her aspirations to hold interpretative sovereignty over Nietzsche's work went so far that in some cases she even engaged in forging documents.

Still, the accomplishments of Elisabeth Förster – since 1895 she used the name Förster-Nietzsche – are not to be denied. She spared no effort or expense in collecting and acquiring Nietzsche's manuscripts and letters, and in 1894 she founded the Nietzsche Archive. Initially in Naumburg, it moved to Weimar in 1896 and was housed there from 1897 in the Villa Silberblick, where Elisabeth also lived and cared for her brother (their mother died that year). Her goal in founding the archive was to bring Nietzsche's manuscripts, correspondence, and personal papers to one place in order to order all this vast material and prepare it for publication. This would take the form of a complete edition of Nietzsche's writings, the only problem being that such an edition had already begun.

Heinrich Köselitz (Peter Gast), Nietzsche's faithful friend and assistant, had taken over responsibility for Nietzsche's estate together with Franz Overbeck after Nietzsche's collapse. Köselitz immediately began work on a Nietzsche edition, and between 1892 and 1894 he was able to publish five volumes with the firm of C.G. Naumann. After returning from Paraguay in September 1893, Elisabeth made clear that she did not want Köselitz as editor of her complete edition. She further demanded that Köselitz return all the manuscripts he had received from Overbeck and that he terminate any further work on his Nietzsche edition; the printed volumes were thereafter withdrawn and pulped. Köselitz then withdrew to his hometown of Annaberg in the Ore Mountains, from where he remained in touch with Nietzsche connoisseurs such as Franz Overbeck, Josef Hofmiller, Paul Lauterbach, Gustav Naumann, Kurt Hezel and Felix Hausdorff.

For her new complete edition, Elisabeth Förster won over Fritz Kögel, a Nietzsche admirer whom she had met in Berlin in 1891 through the mediation of Cosima Wagner. The son of a Protestant pastor, Kögel had studied philosophy, German and history from 1878 to 1881 in Munich, Halle/Saale and Göttingen. In 1883 he received his doctorate with a work on forms of poetry based on the aesthetics of Hermann Lotze. Even before completing his dissertation, however, Kögel began writing cultural criticism for well-known newspapers and magazines, including the Berlin *Tägliche Rundschau.* During the mid-1880s, he briefly edited an encyclopedia until his cousins Reinhard and Max Mannesmann persuaded him to work for them in Remscheid, where they had begun setting up tube-rolling mills. With the founding of the German-Austrian Mannesmann Company, Fritz Kögel became its first administrative director with headquarters in Berlin. There he moved in the highest circles, maintaining friendly relations with the chairman of the company's Supervisory Board, Werner von Siemens. In addition to this activity, he continued to follow his literary inclinations in Berlin. In 1892 he published anonymously a volume of aphorisms entitled *Vox humana, a Confessional Book* and one year later, this time under his own name, he brought out *Hospitality. Sayings of a Wanderer.* In *Vox humana* he revealed himself to be a disciple of Nietzsche. Kögel also lectured on Nietzsche at the Free Literary Society in Berlin, and he generally enjoyed a distinguished reputation in Berlin's Nietzschian circles. In 1893 he left Mannesmann when the company's headquarters was relocated from Berlin to Düsseldorf.

In the fall of 1893, Elisabeth Förster approached Kögel to inquire whether he would agree to assume the editorship of a complete edition of Nietzsche's works. He accepted her offer, and beginning in April 1894 quickly got down to work at the Nietzsche Archive. Already by 1895, eight volumes of the complete edition had appeared in quick succession, all but one of them edited by Fritz Kögel. These first eight volumes, which contained Nietzsche's writings printed before his collapse, proved to be unproblematic, unlike the still unpublished works, which led to major problems. In 1896 and 1897, Kögel brought out Volumes IX to XII, both devoted to Nietzsche's previously unpublished writings. The last volume in particular, Volume XII, which contained material in the estate from the years 1881 to 1885, soon came under criticism. Kögel's edition of *Ewigen Wiederkehr* in this volume was a posthumous compilation of a "non-aphoristic work" with a "systematic quality," a work, however, that in truth never existed [Hoffmann 1991, 51]. What did exist were numerous more or less elaborated notes by Nietzsche from different times, occasions and contexts, which in their entirety were far from constituting a work on the topic of "Eternal Return." On the other hand, some of Nietzsche's notes on "eternal return" were not included in the compilation because Kögel thought they did not belong to Nietzsche's coherent train of thought on the subject.

During Kögel's editorial work in the archive from 1894 to 1897, he naturally came into contact with various Nietzsche connoisseurs and admirers, such as Josef Hofmiller, Gustav Naumann, Kurt Hezel, and – last but not least – Felix Hausdorff. In all probability, Hausdorff's earliest contact with Kögel came about during his first visit to the Nietzsche archive, still in Naumburg, in June 1896. This invitation from the archive's manager, Elisabeth Förster-Nietzsche, was arranged by Gustav Naumann. On this occasion, Hausdorff presumably was allowed to see the ailing Nietzsche, as the archive director routinely granted this favor. In a letter, dated July 12, 1896, Hausdorff warmly thanked her for the hospitality she had extended to him. He also praised the archive as a way to "get away from the horrors of the downfall [Nietzsche's mental derangement] and to celebrate the triumph of what remains: You have taken this path, which will lead no less to fame because it was the only way of saving it" (*HGW* IX, 267). Hausdorff's appeal to the director in the final portion of his letter was particularly clairvoyant:

> In your archive I have silently admitted to myself how little the "From Him" has come to signify next to everything "About Him," and how meanwhile the deepest Nietzsche connoisseur would do wisely and honestly to allow the master himself to speak. Apostles and evangelists will be superfluous for a long time, since the fading sounds of the great Sermon on the Mount can still be heard. (*HGW* IX, 267)

With the hope of "not having entered the Nietzsche archive for the last time," the letter closes with greetings and the request to extend thanks to Dr. Kögel (*HGW* IX, 268).

In the years that followed, friendly relations developed between Hausdorff and Kögel. Unfortunately, no correspondence between the two has survived, but various postcards and letters from the years 1896 to 1898 clearly reveal that the quartet of Nietzsche admirers – Fritz Kögel, Gustav Naumann, Felix Hausdorff, and Kurt Hezel – formed a small circle of friends. A nice example is a postcard from the quartet to Heinrich Köselitz, dated October 30, 1896, in which they conveyed warm greetings to Nietzsche's friend and "master" Peter Gast from their cheerful group, written after hearing a lecture by Kögel.

By the end of 1896, however, the relationship between Elisabeth Förster-Nietzsche and Fritz Kögel had deteriorated. References in their letters suggest that Kögel's engagement to 19-year-old Emily Gelzer played a role in this falling out. Still, the statements about the crisis made by Förster-Nietzsche as director of the archives were contradictory and in some cases demonstrably untrue, so the motives behind her intrigues remain difficult to discern. Another key factor was surely her developing relationship with Rudolf Steiner, who was then working in the Goethe-Schiller Archive in Weimar. In the spring of 1895, while Kögel was staying in Italy, Förster-Nietzsche engaged Steiner to take up some urgent work in the Nietzsche Archive, and in January 1896 he also organized the materials in Nietzsche's former library. After the archive moved to Weimar, Steiner offered Elisabeth, at her request, private lessons in philosophy twice a week, in particular "on the relationship between her brother's philosophy and other philosophical currents" [Hoffmann 1991, 197]. In doing so, the mistress of the archive, who enjoyed these philosophy lessons very much, drew Steiner closer to her ambitious project. At the same time, she managed to alienate Kögel, who did not wish to have anyone next to him as editor. On December 3, 1896, Steiner wrote to his wife: "Kögel and Frau Förster are moving more and more apart. She is now clearly working towards putting the chair in front of the door for him. Terrible scenes have occurred" [Hoffmann 1991, 203].

Förster-Nietzsche wanted to exploit the materials in her brother's estate to construct his supposed principal work, *Transvaluation of All Values*, and she hoped to win over Steiner as editor for that project. They held a confidential conversation about this matter, but he declined her offer. Nevertheless, on the day following this conversation, she informed Gustav Nauman that Steiner "can and will do the *Transvaluation*" [Hoffmann 1991, 545]. Naumann then informed Kögel about this, which resulted in a hostile confrontation between Kögel and Steiner, though this ended when it became clear that Förster-Nietzsche had lied. By this time, Naumann had had enough of her scheming. In December 1896, he summarized his views in a pamphlet entitled *The Case of Elisabeth Förster-Nietzsche*, which Kögel and Steiner both received. As a result of this polemic, Naumann left the publishing house of C.G. Naumann in 1897; he resigned because he was unwilling to comply with his uncle's desire to refrain from taking independent action in matters involving the publication of Nietzsche's works, preferring instead to show loyalty toward the archive and its director. Steiner also broke off contact with the archive in mid-1898.

On April 1, 1897, Frau Förster-Nietzsche notified Kögel of his termination as editor, effective on July 1. In the years following, she then confiscated and pulped the posthumous volumes of the Kögel edition. In 1898 Kögel married his fiancée and resumed work in industry. He and his wife also wrote a successful children's book, but in 1904 he died tragically in Jena as a result of a bicycle accident. Hausdorff's friendship with Kögel survived through the turbulent years. In a letter to Fritz Mauthner, dated November 2, 1904, Hausdorff apologized for not being able to keep an appointment, explaining that on that day "unfortunately I had to be absent to bury my friend Fritz Kögel" in Jena (*HGW* IX, 473). Kögel's widow suffered from depression and committed suicide in August 1906. In a letter to Mauthner, whom he had met while on vacation in Zermatt, Hausdorff wrote: "On August 23, one day before we met in Zermatt, an excellent friend of mine and my wife threw herself out of a window in Berlin: Frau Emily Kögel" (*HGW* IX, 482).

Hausdorff's attitude towards Elisabeth Förster-Nietzsche and the archive also changed as a result of the events described above. This passage from a letter to Köselitz, dated October 7, 1897, reveals his clearly critical view of the situation:

> We, who first learned from Nietzsche that sovereign skepticism, disgust with moral trumpeters and opinion shouters and Bayreuth-blinking partisanship, for us it would be a bad thing [· · ·] to speak of our "master," to let him be systematized and us catechized. Today we see around Nietzsche a Bayreuth, a Delphi, whose Pythia you know; the same stiff, pedantic, solemnly narrow-minded party atmosphere that drove Nietzsche from Wagner wants to wrap itself in a cloud of incense around the holy image of Zarathustra. Thus it would seem not untimely to provide material evidence showing that there will never be and should never be a species of Nietzscheans in the same sense as Wagnerians, that *we* at

Figure 4.5: Portrait of Nietzsche by Edvard Munch, 1906
Source: Wikimedia Commons

> least won't allow ourselves to be intimidated into suppressed, kneeling, anxious worshipers who squint before his Mistress and tradition. (*HGW* IX, 393–394)

After Kögel's departure, the archive remained without an editor until the autumn of 1898. Then the music critic Arthur Seidl took over as editor for one year, supported by Hans von Müller, a student. After Seidl stepped down, the young philologist Ernst Horneffer, who had since 1896 been lecturing in various German cities about Nietzsche, became the archive's new editor, joined by his brother August Horneffer. Ernst Horneffer soon began an analysis of Kögel's withdrawn Volume XII, undertaken while he worked on an "Essay against Kögel," which aimed to justify the confiscation and pulping of the posthumous volumes Kögel had published. On October 8, 1899, the editors Horneffer and von Müller met with the publisher C.G.Naumann to discuss further plans. This meeting, to which Felix Hausdorff was also invited, took place at the archive. Presumably Förster-Nietzsche hoped to interest him in preparing a new edition of the "Eternal Return." Hausdorff recalled this visit in a later letter to her, in which he sought to clarify the miseries associated with past editions of Nietzsche's works:

> May I remind you, madam, of the two times I had the honor of being a guest of the Nietzsche Archive. The first time, in the summer of 1896, the philologist and unauthorized editor of Nietzsche's works, Mr. Köselitz, was terminated by Dr. Kögel as his executor; the second time, in the autumn of 1899, the same scene played out again with truly ridiculous exactitude, only with a different casting – Herr Dr. Horneffer acted as judge and Kögel was then executed. (Letter from 3 August 1900, *HGW* IX, 280.)

Horneffer's essay was published by C.G. Naumann in early December 1899 in the form of a brochure with the title *Nietzsche's Doctrine of Eternal Return and its Previous Publication.* This presented a scathing criticism of the scrapped Volume XII from Kögel's edition. Horneffer did not stop at personal attacks and defamation; he even went so far as to call Kögel a "scientific charlatan." This triggered a months-long public dispute, sometimes accompanied by violent polemics, over the publications of the Nietzsche Archive and the role of its manager. A first reaction to Horneffer's brochure came from Gustav Naumann, who sharply criticized it in the preface to part II of his Zarathustra commentary. Far more sensational, though, was Rudolf Steiner's unusually sharp attack on Horneffer's pamphlet, which appeared under the title *The Nietzsche Archive and its Accusations against the Previous Editor, a Revelation.* Horneffer replied to this with an article entitled "A Defense of the so-called 'Return of the Same' by Nietzsche," whereupon Steiner followed up with a counter-reply. Finally, in April 1900, Elisabeth Förster-Nietzsche intervened in the dispute.

Immediately after Horneffer's pamphlet appeared, he sent a copy to Hausdorff, who responsed on a printed business card (*HGW* IX, 270). Already in

January 1900, thus before the publication of Steiner's polemic, Hausdorff wrote two articles about Nietzsche's doctrine of the return of the same. The first was a polemical discussion of Horneffer's pamphlet and the work of the Nietzsche Archive, whereas the second gave a compact presentation of Nietzsche's idea of eternal recurrence, its history, and Nietzsche's attempt to prove it based on scientific knowledge. Hausdorff's articles only appeared some months later in the weekly periodical *Die Zeit*, the first on May 5, the second on June 9, 1900.

The starting point for Hausdorff's first article was the observation that, in addition to Nietzsche's enormous published work, there was a "second treasure chest not much smaller in scope in the estate of the living deceased, and this contained an immense wealth of valuable material: preparatory work, variants, addenda to finished works, fragments, drafts, thoughts, book plans, nearly complete writings alongside tiny workshop items and germs!" [H 1900c, VII: 889]. When planning the publication of writings from an estate, Hausdorff contended, the question always arises whether a selection should be made and arranged according to subject matter or whether consideration should be given to publishing *every* item in the estate, or even doing this so that *every* note is printed in the exact order found when it was left behind. One can raise many arguments against the latter method, but for personalities such as Goethe or Hebbel "every scrap is dear and important just because it stems from them." As for Nietzsche, "we are too close to this powerful one and his sphere of activity" to reach agreement on whether he belongs to this circle. And yet

> the experiences made thus far with publications from the estate speak *for* the paper scrap method, if for no other reason than this: that one would be released from the arbitrariness of independently "thinking" editors [$\cdots$]. The Nietzsche archive has paid an unusually high price by adopting the [other] method of sifting through and grouping fragments from the estate to form clear structures [$\cdots$] [H 1900c, VII: 890].

This passage is followed by a brief critical discussion of Horneffer's brochure. Hausdorff noted that, despite its sober-sounding title, this was basically nothing more than an indictment of the preceding editor, Fritz Kögel. He then castigated the "bad manners" of the attacker by citing a series of examples. While conceding the weaknesses of Kögel's volume XII, Hausdorff vehemently disputed Horneffer's assertion that "one was completely misled by Kögel's twelfth volume about Nietzsche's conception, justification and utilization of the idea of the eternal recurrence of the same." On the contrary, he contended that through Kögel's volume a discerning reader could "obtain a clear picture of Nietzsche's theory of the return" [H 1900c, VII: 890, 892]. Finally, Hausdorff formulated – so to speak as advice to the archive – his suggestions for the edition of Nietzsche's estate. One should

> completely renounce one's own arrangement and interpretation, printing the manuscript books as Nietzsche wrote them, perhaps in chronological order and eliminating those aphorisms that have appeared in

> the finished works unchanged. Everything else – assigning individual thoughts to larger groups of thoughts, connections to the finished works, structuring of the notes according to the presumed intention of the author, omission of the insignificant and immature – all this could be avoided in the future, and this would again grant the editor that freedom and trust, whose discharge can later be denied. In short, we repeat our advice to undertake paperwork and publish works of the estate in their entirety, thereby staying on the safe side in this respect. [H 1900c, VII: 891–892]

Both Horneffer's response to Hausdorff's article [Horneffer 1900] and Förster-Nietzsche's detailed written response (*HGW* IX, 271–273) reveal that the archive paid no heed to Hausdorff's advice or had no interest in changing its practices. On the contrary, compared with Kögel's edition, the archive published in 1901 a far more problematic posthumous compilation, volume XV *Der Wille zur Macht*. We shall return to this in connection with Hausdorff's last essay on Nietzsche. It should also be noted that the editorial principles of the now generally recognized critical edition of Nietzsche's works by Giorgio Colli and Mazzino Montinari largely correspond to those proposed by Hausdorff. The fact that they made no mention of this indicates that they were probably unaware of Hausdorff's essay.

4.8 Essays on Nietzsche

4.8.1 Nietzsche's Theory of the Return of the Same

Hausdorff's second essay was entitled "Nietzsche's Doctrine of the Return of the Same" [H 1900d, VII: 897–902]. It begins with comments about the history of the idea of a cyclical time, the "idea of always returning periodically to the same states in renewing the course of the world." Nietzsche himself had already noticed in 1881 that this idea already occurred among the Pythagoreans. Hausdorff quotes from Eudemos and Porphyry, thereby confirming Nietzsche's remarks. He also gives further quotations showing that the Stoics grasped the idea of cyclic return even more sharply; in fact, Diogenes even gave the length of the cycle, the "great cosmic year," as 365 times 18,000 solar years. Still, Hausdorff did not want to pursue these ideas historically, noting however "their most recent reappearance in Nietzsche, Blanqui and Le Bon." To this he added that the idea of eternal recurrence belongs "to the earliest childhood of philosophy," though this hardly entitles contemporary philosophers to glorify their ideas in the light of earlier metaphysics. Today, Hausdorff asserted, "a lot of new metaphysics is still produced, which stands in relation to the older like a Greek temple to a garrison church in Brandenburg." [H 1900d, VII: 897–898].

Following this, Hausdorff took up Nietzsche's attempt to prove eternal return mathematically and scientifically. Although Nietzsche never published these findings, relevant quotations can be found in his estate (*HGW* VII, 44–45). According to Hausdorff, Nietzsche had "nothing less than the mathematical *necessity* of the eternal recurrence in mind." Moreover, the basis for his proof was easy to grasp: "the number of possible world states is definite and thus finite, even if practically immeasurable, so the infinity of time can only be filled by an eternal repetition of the same thing" [H 1900d, VII: 899]. Hausdorff noted, however, that this conclusion was untenable. Indeed, "an analysis carried out by me shows that the exact opposite is the case: the content of the world is not too poor to fill time, but rather time is too poor to exhaust the possible contents of the world." Here he neglected to add that the reader could find this analysis carried out in his *Chaos in kosmischer Auslese.*

In this context, Hausdorff took the opportunity to level a strong criticism of the Nietzsche Archive. Ernst Horneffer had tried to deny that Nietzsche ever attempted a proof of this kind. In countering this claim, Hausdorff asserted that the only motivation for this was Horneffer's desire to refute what Lou von Salomé had asserted in her book [Andreas-Salomé 1894], since "Lou Salomé must have lied at all costs" [H 1900d, VII: 899]. Hausdorff's remark was surely aimed at Förster-Nietzsche, who was bitterly hostile towards Salomé and her book. As for its scientific basis, he emphasized that the second law of thermodynamics undermined the hypothesis of an eternal recurrence of the same.

At the end of this essay, Hausdorff discussed the role of "return as an inspiration and religious idea." When the thought of the eternal return overcame Nietzsche on the lake of Silvaplana, he felt deeply troubled by that "which must disturb any *sufferer*: the eternal perpetuation of a painful life, rigid and unchangeable, devoid of salvation and hope, the collapse of any prospect of 'redemption' as the body returns to the past with its torments and martyrdom" [H 1900d, VII: 901]. Nietzsche carried these conclusions around with him "like a terrible secret" for years,

> since he felt neither himself nor mankind strong enough to proclaim and incorporate it. A heroic, desperate struggle of the soul of the thinker with his thoughts! [⋯] A struggle that ended with victory for the thinker, with his dithyrambic "Yes and Amen" [H 1900d, VII: 901–902].

At this point, Hausdorff quotes from Nietzsche's aphorism "The Greatest Weight," in which the concluding sentence reads: "Or how should you become good to yourself and life, to ask for nothing *more* than this last eternal confirmation and seal?" [Nietzsche KSA, Band 3, 571]. According to Hausdorff, Nietzsche saw in the ethic of eternal recurrence the hope

> of replacing the age of religions with its teaching: no longer looking to the hereafter, life on earth would now gain the heaviest accent; all human insight, love, sensitivity must turn away from transcendent goals and toward life in order to create a work of art from it, one whose eternal repetition may be embraced. [H 1900d, VII: 902]

While readers of this essay were confronted with the weak scientific foundations supporting Nietzsche's theory of eternal return, Hausdorff did offer a way to salvage this principle:

> Those who still want to hold on to this as a religious emotion and normative focus will have no choice but to adopt the new concept of return, which arises as the first stage of generalization in my idealistic decomposition of our *idea of time*: this leads to the possibility of reproducing each individual time segment, a possibility which can neither be confirmed nor disproved by experience and in no way depends on whether the time line as a whole follows a cyclic or linear course. [H 1900d, VII: 902].

Here, too, Hausdorff gave no indication of where he had developed this "new concept of return," namely in *Chaos in Cosmic Selection*.

4.8.2 The Will to Power

The point of departure for Hausdorff's last essay on Nietzsche was the publication in 1901 of *The Will to Power* with the subtitle *Attempt at a Transvaluation of all Values*. This work, a compilation produced by the Nietzsche Archive, appeared as volume XV in the complete edition published by C.G. Naumann in Leipzig. The archive alleged that this represented Nietzsche's principal work, and indeed its impact far surpassed that of Kögel's earlier and since discarded Volume XII. Regarding its origins, the following should be noted. Elisabeth Förster-Nietzsche succeeded in April 1899 in obtaining the services of Heinrich Köselitz, who was employed by the archive up until 1909. His participation in the publication of Nietzsche's posthumous works was indispensable, since he alone was able to decipher the handwriting in Nietzsche's late manuscripts. In his letter of August 3, 1900 to Frau Förster-Nietzsche, cited above, Hausdorff expressed the hope that "the cooperation of Peter Gast" offered the guarantee that earlier mistakes would not be repeated (*HGW* IX, 280–281). This proved to be fundamentally mistaken, however, because unfortunately Köselitz completely subordinated himself to the mistress of the archive. In writing about this period, David Marc Hoffmann offered this view of the situation: "While he was working in the archive (1900–1909), Köselitz, as a helpless tool, made himself complicit in EFN's suppressions, corrections and falsifications in the biography, in the letter edition and in the posthumous compilation 'Der Wille zur Macht"' [Hoffmann 1991, 45].

Volume XV, *Der Wille zur Macht* (*The Will to Power*), was the work of Köselitz and Ernst and August Horneffer; the introduction was written by Förster-Nietzsche. Among Nietzsche's many dispositions on the subject of the *Will to Power*, the editors selected one and declared this his "definitive work," which they accordingly divided into four books. These books were divided into chapters and then further into sections, producing a largely constructed framework into which they arranged 483 aphorisms and fragments. Förster-Nietzsche's introduction to

Volume XV, describing the development of Nietzsche's plans for a work entitled *Wille zur Macht*, also failed to give an account of the much more complex history of these plans.

In 1906 – by which time Ernst Horneffer had resigned from the archive – a second version of this work was published, produced by Frau Förster-Nietzsche and Köselitz. This concoction, containing no fewer than 1067 aphorisms and notes, was long considered canonical and tainted the image of the late Nietzsche in a disastrous way. It played an important role in the efforts of the National Socialist philosopher and educator Alfred Baeumler to appropriate Nietzsche for the National Socialist ideology [Baeumler 1931]. The aged Elisabeth Förster-Nietzsche took a benevolent view of such political appropriation. For her part, she established friendly relationships with Benito Mussolini already in the 1920s, when he supported the archive financially. She was also on good terms with Adolf Hitler; when she died in November 1935, the dictator insisted on personally attending her funeral.

In the first year after the publication of *The Will to Power* there were about a dozen reviews of this "posthumous work." Hausdorff's essay *The Will to Power* [H 1902b, VII: 905–909] came out in December 1902. Oddly enough, Hausdorff did not explicitly mention the problematic character of the compilation alleged to be a major work by Nietzsche, though he must have been aware of these problems from the discussions about Kögel's compilation. Nevertheless, he simply spoke of "fragments from Nietzsche's main work" [H 1902b, VII: 905]. Hausdorff had no interest in discussing this work, as would usually be done in a review, by analyzing its overall conception and content. Toward the end, he even writes: "We leafed through a little: one does not discuss such a work." What was important for him was not the book itself but rather the weaknesses of the late Nietzsche. He wanted to measure these against the early Nietzsche, the Nietzsche who had so decisively accompanied him along his path to becoming the philosopher and writer Paul Mongré. Asked why he picked out "precisely those points [that] provoke contradiction," he replied:

> Because Nietzsche is popular precisely due to his weak points, just as, according to his own judgment, Wagner and Schopenhauer fared; because these crackling neon signs and spotlights – "Antichrist," "Will to Power," "Master's Morality," – filter out the milder ones that glow in the starry sky. [H 1902b, VII: 905].

A prominent point for Hausdorff in the late Nietzsche is his fanatical hatred of Christianity. Hausdorff not only quotes from *The Will to Power* but also from *The Antichrist* in order to characterize this change from the younger Nietzsche. In the latter work, Nietzsche wrote:

> I call Christianity the one great curse, the one great inmost depravity, the one great instinct for revenge, for which there is no means poisonous enough, secret, subterranean, or *small* enough – I call it the one immortal disgrace of mankind ⋯ [Nietzsche KSA, Volume 6, 253].

Mongré comments:

> [$\cdots$] It was philosophical when he marveled at the homeopathic dosage of Christianity in the alleged modern Christians and spoke with a smile about the gentle death in old age and the euthanasia of Christianity. Why kill a dying man? Ten years later, he considers this necessary; the kind, knowing rejection is kindled into burning hatred and the once overcome stage of development turns into a new dungeon and nightmare, the "immortal disgrace of mankind" [H 1902b, VII: 906].

A second, particularly provocative point for Hausdorff concerned Nietzsche's views regarding human breeding. His notes on this topic were later classified under headings such as eugenics and "racial hygiene":

> A fanatic glows in Nietzsche. His morality of breeding, built on our present-day foundations of biological and physiological knowledge: that could become a world-historical scandal compared to which the Inquisition and witch trials would fade into harmless aberrations. "Life itself recognizes no solidarity, no equal rights between healthy and degenerating parts of an organism: the latter must be *cut out* – or the whole thing perishes." The papal judges of heretics and also Dominican friars say roughly the same thing, not even with slightly different words; cutting out diseased limbs was a popular formula for Sprenger and Institoris, Arbuez and Torquemada. Shall we once again experience how to philosophize with the witch's hammer? [H 1902b, VII: 907].

A third point in Hausdorff's critique also stems from his aversion to fanaticism, but he then expands this into a comprehensive criticism of Nietzsche's thinking in its final phase:

> Nietzsche is a fanatic like all founders of religions. [$\cdots$] The old moral fanaticism has been replaced by a new one that systematizes and dogmatizes its reality no less violently. Hasn't one noticed how "generous" and primitive Nietzsche's final worldview is? "The poor and the weak impoverish life; the rich and the strong enrich it. The former are its parasites, the latter its patrons. How is any mix-up possible?" It all sounds extremely simple, suspiciously simple. History, art, morals, philosophy are all incorporated into this twofold schema: affirmation of life or renunciation of life. But we distrust these arithmetical examples that work out so smoothly, and we distrust those discoverers of a world-dividing dualism who are blinded by a preconceived category of polarity: love and hate, soul and body, extension and thought, force and matter, master morality and slave morality, ascending and descending life – it is too improbable that the sonorous play of reality should dissolve so completely into theme and counter-theme, thesis and antithesis. [H 1902b, VII: 908]

The following passage, which expresses Hausdorff's closeness as well as his distance to Nietzsche very aptly and beautifully, closes the essay:

> Only this should still be said, that precisely where we lament the unrest and injustice of this very last Nietzsche, we do so by taking our critical standard from none other than Nietzsche himself: from that kind, measured, understanding free spirit Nietzsche, from the cool, dogma-free spirit and system-free skeptic Nietzsche, and from his triumphant life-embracing song of the world-blessing, all-affirming ecstatic Zarathustra. [H 1902b, VII: 909].

4.9 Foundations of Geometry and the Space Problem

In his letter to Franz Meyer, quoted at the beginning of this chapter, Hausdorff described his personal journey with the words "From Wagner to Schopenhauer, from there back to Kant and forward to Nietzsche, while from mathematics some touching with the philosophy of the exact sciences – that's all!" We have already traced his path to Nietzsche in detail, but what did he mean by "mathematics [···] touching with the philosophy of the exact sciences" in this description of his personal journey? What he surely had in mind were two topics that occupied his attention since his student days and that led up to his interest in the general theory of relativity, namely the foundations of geometry and the space problem.

Geometry took a tremendous upswing in breadth and depth during the nineteenth century, occasionally called the great century of geometry. Among the many important developments were the discovery of non-Euclidean geometries, the expansion of projective and synthetic geometry, the theory of geometric constructions and their algebraization, the emergence of differential geometry and the theory of n-dimensional vector spaces, alongside the investigation and classification of transformation groups. This upsurge also led in the last third of the century to an increasing interest in an overarching unifying view of the various new branches of geometry together with probing studies of the foundations of classical Euclidean geometry, a discipline par excellence for over two thousand years. The culmination of these latter efforts took place in 1899 with the appearance of David Hilbert's treatise "Grundlagen der Geometrie" ("Foundations of Geometry"), published in a *Festschrift* celebrating the unveiling of the Gauss-Weber Statue in Göttingen [Hilbert 1899]. This work proved to be of fundamental importance not only for geometrical research but for the whole development of modern mathematics [Gray 2000, 48–55, 104–108], [Volkert 2015, 195–238].

Moritz Pasch's *Lectures on Modern Geometry* [Pasch 1882] was probably, prior to Hilbert's work, the single most important contribution to this field. Pasch presented a deductive construction of projective geometry independent of the axiom of parallels. From this, he then derived Euclidean and non-Euclidean geometries by means of the so-called Cayley-Klein metric. Pasch was thus the first to design and carry out a program for a strictly axiomatic construction of geometry, though he did not speak of axioms but rather of principles. In his *Foundations of Geometry*, Hilbert then gave the axioms a new and more abstract sense quite different from the previous conception. By so doing, he promoted a tendency toward dissolution of the relations between the objects and statements of an axiomatic theory and any facts of experience or extra-mathematical meanings describable by them. Hilbert's position on the fundamental underpinnings of mathematical knowledge came to be called "formalism." Pasch's work thus occupied a middle position within this long-range process [Tamari 2007].

A comparison of the approaches of Pasch and Hilbert shows that both employed a modern system of axioms to deduce geometrical theorems rigorously. This, therefore, was *not* the special feature of Hilbert's treatise, as Pasch had already accomplished this in 1882. To gain a sense of the truly novel feature of Hilbert's work, one need only cite the final sentence of his *Foundations of Geometry*:

> The present investigation is a new attempt to establish for geometry a *simple* and *complete* system of mutually *independent* axioms and to derive from the same the most important geometrical theorems in such a way that thereby the meaning of the various groups of axioms and the scope of the conclusions to be drawn from the individual axioms become as clear as possible.

What was new, then, with Hilbert's axiomatic method was not the rigorously deductive approach but rather his astute mathematical-logical analysis of a system of axioms. In particular, Hilbert posed the question of proving the consistency of such a system as well as the problem of proving the independence of individual axioms from other axioms or axiom groups. Since the fundamental work of Kurt Gödel in 1930, it has been known that one cannot prove consistency in an absolute sense. This means that it is impossible to prove that one could never derive a contradiction from the axioms in a sufficiently strong system, as Hilbert had originally assumed. Nevertheless, relative consistency proofs are not only possible but also highly fruitful. Thus, Hilbert was able to show that his axiom system for Euclidean geometry was necessarily consistent so long as the same was true for the axioms of arithmetic. His proof merely depended on the construction of a suitable model, just as he used cleverly constructed models to prove the independence of individual axioms. Thus he proved the independence of the Archimedean axiom by constructing a *non-Archimedean* geometry which satisfies all the other axioms. As another well-known example, the Kleinian model of hyperbolic geometry proves the independence of the parallel axiom.

From Hausdorff's chapter "On Space" in *Chaos in Cosmic Selection* it is evident that he was up to date on all the great breakthroughs in geometry of the past century. This concerns, in particular, the relevance of the non-Euclidean geometries for the space problem, including the fundamental contributions of Riemann and the debates initiated by Helmholtz from the standpoint of physics [Riemann 1868], [Helmholtz 1868], [Helmholtz 1870]. Hausdorff acquired this deep geometrical knowledge largely on his own. A strong motivation for doing so surely came from realizing the relevance of these mathematical questions for epistemology, i.e., for understanding the relationship between spatial perception and consciousness, a problem that occupied him for a long time.

It should come as no surprise that someone with Hausdorff's particular mathematical and philosophical background would have been a great admirer of Hilbert's *Foundations of Geometry*. Already on October 12, 1900, he wrote to its author, his earliest surviving purely mathematical letter, which began:

> Dear Professor,
> Allow me to make some remarks about your Festschrift treatise on the "Foundations of Geometry," among whose ardent admirers I allow myself to be counted. Reading your text with its extremely high critical level leaves the reader in just such a philosophical mood, and you must ascribe to that effect that a small thing does not appear completely correct, even in your extremely perceptive and careful formulations. (*HGW* IX, 327)

Hausdorff then actually goes on to criticize three "small things." What is more, before the second edition appeared in 1903, [Hilbert 1903] Hilbert improved all three passages criticized by Hausdorff. Considering the manner in which Hausdorff had previously learned to pursue mathematics, Hilbert's *Foundations of Geometry* acquainted him with an entirely new approach. This pertained not only to the conceptual objects of mathematics but also to the way in which mathematics actually constitutes its object. Hilbert's views concerned both the way in which foundations should be laid out for various fields of pure mathematics as well as the relationship of mathematics to its applications and the larger epistemological role of mathematics as an exact science.

In Hilbert's famous lecture "Mathematical Problems," presented in August 1900 at the International Congress of Mathematicians in Paris, axiomatization played a major role. His sixth problem posed, among other things, the task of placing probability thoery on an axiomatic basis [Hilbert 1900, 272]. Probably inspired by this, Hausdorff addressed this problem in a lecture course on probability offered during the winter semester of 1900/1901. His approach was based on four conditions, though it is easy to see that these were not all independent of each other. Moreover, the only possible model for this axiom system turns out to be classical Laplacian probability. That is, it leads to the definition: "the probability of an event is the ratio of the number of valid to the number of possible events" (Hausdorff estate, Capsule 2, Fascicle 10, p. 4). In his article [H 1901a] (see 3.2.4), Hausdorff included everything of importance from this lecture course except for his attempt at axiomatization; apparently, he himself was not convinced by it.

Hausdorff's estate contains two drafts for articles that were nearly ready to be printed. These provide insights into his fundamental views on the position of mathematics in relation to the natural sciences and to philosophy. They reflect, on the one hand, considerations relating to the role of non-Euclidean geometries in this context and, on the other, aspects concerned with the autonomy of mathematics in connection with logical formalization. The first is a 111-page manuscript entitled "Nichteuklidische Geometrie," not quite finished and probably written during the period 1902 to 1904. It was clearly intended for publication in a natural science journal, as it begins with the question "How does non-Euclidean geometry happen to appear in a general science journal?" (Hausdorff estate, Capsule 48, Fascicle 994, reprinted in *HGW* VI, 347–389). Which journal Hausdorff had in mind is unclear, nor do we know why his manuscript was never published. Part I begins with an introduction, before he develops hyperbolic, spherical and elliptic geometry in Parts II and III. He then takes up arguments against the prejudice of many philosophers, who continued to regard Euclidean geometry as conceptually necessary:

> But that one definite system is taken as a basis, so to speak, as the normal system, against which already by name all other systems [non-Euclidean geometries] are opposed as mere negations, abnormalities, varieties [··· yet Euclidean geometry] is logically considered not one iota more justified, more natural, more coherent than any of the others. In this respect, mathematics cannot contradict sharply enough the prejudice, popular among many philosophers, that seeks to regard Euclidean space with all its special peculiarities as a "product of logical positing" a priori. (Hausdorff estate, Capsule 48, Fascicle 994, p. 5, *HGW* VI, 350.)

On the relationship between mathematics and philosophy, Hausdorff writes in the introduction:

> Mathematics, which in this respect is independent of any philosophical or epistemological standpoint, is satisfied with the *conceivability* of the various geometries, in order to accord them equal rights and legitimacy as objects of mathematical research. Whatever holds a definite place in the context of our thinking as a well-defined concept is granted the right of mathematical "existence," whether or not it is represented in experiential reality.

Philosophy concedes to mathematics the right to subject "certain entities to its investigation," but denies it the right and competence "to form a summarizing descriptive idea of a system of certain things and their interconnections and to call this system a space" (Hausdorff estate, Capsule 48, Fascicle 994, pp. 8–9, *HGW* VI, 351).

The second manuscript stems from the beginning of a book project that Hausdorff had envisioned beginning around 1902/1903. In fascicle 1067 of the estate one finds a kind of table of contents for a book about space and time, which was to have consisted of three parts:

> I. The formal or mathematical problem: space and time as logical constructions.

> II. The objective or epistemological problem: space and time as realities.
> III. The subjective or psychological problem: space and time as contents of consciousness.
> (Hausdorff estate, Capsule 49, Fascicle 1067, p. 13.)

The titles of the first five chapters in Part I are also given on sheet 13: "1. Formalism, 2. Axioms of Time, 3. Axioms of Euclidean Space, 4. Non-Euclidean Geometry, 5. Other Geometric Systems." The first chapter, "Formalism," consists of 12 pages in print-ready form (*HGW* VI, 473–478); here are two characteristic passages:

> Everything real permits doubt, all knowledge of the real is by its nature incomplete and capable of various interpretations. The only kind of certainty that we know is that of *formal logic*. [· · ·]
> Thus, geometry, as a doctrine of real space, remains in a state of dependence awaiting foreign decisions and is thereby driven into a state of *heteronomy*. The philosophical dispute over its bases, with all its fluctuations and shocks, would be felt up to the highest floors of the mathematical building. The historical facts show conversely that mathematics has always been considered as a type of strict, independent, autonomous science and has asserted itself as such the more purely, the less it found the patience to wait for the results of philosophical efforts. There is only one explanation for it: geometry does *not* speak about real space, it speaks about a product of freely creative thinking. The certainty of geometry is the certainty of formal logic. (Hausdorff estate, Capsule 49, Fascicle 1067, p. 2, *HGW* VI, 473–474)

Hausdorff's lecture course "Zeit und Raum," which he held in the winter semester 1903/04 for a wide circle of listeners (*HGW* VI, 391–450), might also have served as preparation for his book project. The manuscript begins with the following lines:

> time and space
>
> Old problem, thousand opinions. Not pure lecture, but discussion, searching together. My passion for this problem. (Hausdorff estate, Capsule 24, Fascicle 71. p. 1, *HGW* VI, 391)

In fact, this was a *very* "old problem;" already ancient philosophers thought about space (and, of course, about time) [Jammer 1960]. Hausdorff did not publish his own mathematical contributions to the problem of space, but he made his philosophical considerations public for the first time in the chapter "Vom Raume" in *Das Chaos in kosmischer Auslese*, which we passed over in our discussion above.

In 1903 Hausdorff summarized the results of his philosophical reflections on the problem of space in "Ostwalds Annalen der Naturphilosophie" [H 1903a, VI: 281–303]. This publication was occasioned by Hausdorff's appointment as irregular associate professor in Leipzig, a title the Leipzig Faculty of Philosophy proposed to the Saxon Ministry on 5 November 1901. In support of this, Hausdorff's former teacher Heinrich Bruns wrote a very positive report which ended:

> In summary, it may be said that Dr. H., through his literary activity to date and through the services rendered as a teacher within the academic teaching organism, thoroughly deserves the recognition applied for. (The complete text appears in [Beckert/Purkert 1987, 231–234].)

Although only the title "associate professor" was at stake (irregular meant that no salary of any kind was attached to this appointment), the dean felt compelled to append the following addendum to the faculty vote:

> The faculty, however, considers itself obliged to report to the Royal Ministry that the foregoing motion was adopted at the faculty meeting, held on 2 November of that year, not by all members, but by a vote of 22 to 7. The minority voted against the proposal because Dr. Hausdorff is of Mosaic faith. [Beckert/Purkert 1987, 233]

"Mosaic faith" referred to Hausdorff's membership in the Jewish community, not, of course, his "beliefs" regarding the religion of his ancestors. This remark by the dean illuminates in a glaring way a quite general attitude of undisguised antisemitism – no attempt was even made to seek another pretext for rejecting the application. Leipzig, at the time, was a center of the anti-Semitic movement that flared up particularly strongly after the Panic Crisis of 1873 *Gründerkrach*, especially among the student body. The Ministry in Dresden, however, was not to be deterred and informed the Leipzig faculty on December 6, 1901 that Hausdorff had been so appointed; the faculty was thus to notify him and urge him to deliver the prescribed inaugural lecture.

Hausdorff, however, took an unusually long time to do so; he delayed giving his inaugural lecture until July 4, 1903. On that same day, he received the ministerial letter of appointment (*HGW* IX, 692). A possible cause for this delay may have been his mother's illness; she died on December 5, 1902. In a letter to Fritz Mauthner, written shortly afterward on December 16, 1902, Hausdorff wrote: "For the last few months I have lived (due to my mother's illness and death) in an exceptional state and have allowed relations with the outside world to loosen somewhat" (*HGW* IX, 461). On the other hand, it may have been that this "honorific title" was not particularly important to him.

The topic of his inaugural lecture was "The Space Problem" (*Das Raumproblem*); his publication in Ostwald's Annalen represents the text of this lecture supplemented by notes and references. The title *Das Raumproblem* with the definite article is some sense problematic in that it raises the question of just what is meant by "**das** Raumproblem." The opening sentence of Hausdorff's lecture points to the diverse character of the problem by referring to its interdisciplinary nature: "No less than five sciences are involved and interested in the solution of the space problem: mathematics, physics, physiology, psychology and epistemology" [H 1903a,

1; VI: 281]. He then distinguishes between three aspects of the concept of space that he wants to consider, namely:

> First, a certain free creation of our thought, not subject to any other constraint than that of logic, a system of arbitrarily chosen presuppositions, so-called axioms, together with conclusions derived deductively from them: this is the space of thought, the space of geometry, *mathematical space.* Secondly, a system of real experiences, the phenomenon of the space-filling external world present in our consciousness: this we want to call the subjective-psychological space, the space of consciousness or experience, *empirical space.* Third, finally, a certain structure of things independent of our consciousness, which is presupposed to account for our view of space: this would be objective-naturalistic space, the "intelligible" or *absolute space.* [H 1903a, 1–2; VI: 281 – 282].

Among these three aspects, Hausdorff treats *mathematical space* most extensively. About half of his text is devoted to this subject, whereas the remainder deals in roughly equal parts with *empirical space* and *absolute space.* With regard to mathematical space, one has the freedom to choose between a vast number of hypotheses based on three aspects:

> [· · ·] turning first to mathematical space, the majority of hypotheses and the freedom of choice between them is guaranteed to us in three ways: by the *scope of thought*, the *scope of contemplation*, and the *scope of experience.* [H 1903a, 3; VI: 283].

The scope of thought was expanded through mathematical developments of the nineteenth century, ranging from the discovery of non-Euclidean geometries to the logical analysis undertaken in Hilbert's Festschrift. But since geometry "protrudes at the same time into pure and applied mathematics," and since as applied mathematics it is "the most perfect natural science" – Hausdorff here echoes an expression of Hilbert – certain geometries possess, beyond their logical right to exist, an "empirical purpose for existence." Taking this into account, the scope of thought is "limited, but not completely annihilated." Experiential conditions still leave us scope, since empirical validity "is not an exclusive prerogative of Euclidean geometry but must also be granted to those non-Euclidean geometries whose deviation from the Euclidean remains below our observational threshold" [H 1903a, 3; VI: 283]. Concerning the scope of imagination, Hausdorff refers to Helmholtz's successful attempt "to picture non-Euclidean spatial perceptions," and declares the question of imaginability to be a "psychological personal question."

Following this basic discussion of our available latitude in choosing spatial concepts, Hausdorff introduces "the most interesting spatial forms" to his audience. He does this so as to highlight four characteristic aspects of these spatial concepts, about which he inquired:

> with what justification or in which sense do we consider our space to be a space *with vanishing curvature*, a space *with free mobility*, a space *with*

> *simple coherence*, or a *three-dimensional continuous* space. [H 1903a, 7; VI: 287].

In the interest of "saving space," we pass over his discussion and enumeration of conceivable spatial forms; Hausdorff himself emphasizes that his remarks by no means exhaust "the entire realm of possibilities" (see *HGW* IB, 521–528).

After this, Hausdorff attempts to show that "*absolute* space, which supposedly explains the objective spatial nature of things underlying our sensory perceptions, suffers from complete indeterminacy and indeterminability" [H 1903a, 13–14, VI: 293–294]. Kant had already rightly asserted this, but his argument was invalid. Hausdorff therefore wants to sketch here "his own method of proof," namely the one given in his chapter "On Space" in *Chaos in Cosmic Selection*; this follows the pattern of proof for the ideality of time presented earlier in the section "Temporal Succession." He thus begins: "One first of all thinks, like the realist, of a faithful transcendent and real archetype of empirically given space and then one seeks to vary it as much as possible without destroying the empirical effect" [H 1898a, 73; VII: 667]. To explain his method of proof, Hausdorff offers an analogy by comparing empirical space with a map. The shape of the original (meaning absolute space) cannot be inferred from the map alone: "we must also know the mapping procedure, as with our world maps, where the scale and projection are noted in a corner." Thus, he suggests:

> our empirical space is just such a corporeal map, an image of the absolute space; but we lack the corner notation, we do not know the projection, and consequently we do not know the archetype. Between both spaces there is an unknown arbitrary relation or correspondence, a completely arbitrary point transformation. But the orientation value of the empirical space does not suffer from it; we find our way on our map and communicate with other map owners; the distortion does not fall into our consciousness, because not only the objects, but also we ourselves and our measuring instruments are equally affected by it. [$\cdots$]
>
> If this view is correct, it must be possible to subject the original image to an arbitrary transformation without the image changing: just as it is impossible to tell from a map whether it was drawn from the original or from another map. [H 1903a, 15; VI: 295].

It follows from this that absolute or transcendent space is in no way accessible to us.

At the end of his inaugural address, Hausdorff turns to empirical space, about which he remarks: "Even if we have to abandon the notion of absolute space as a determinable concept, this does not mean giving up the regularity of empirical space" [H 1903a, 17; VI: 297]. Thus, even though Hausdorff admits that the mental processing of empirical elements enters into our geometry, he defends – against, for instance, a radical empiricist like John Stuart Mill – the justification of mathematical theories, especially geometrical structures created by our minds without recourse to experience. He then illustrates this by means of a thought experiment:

> Let us imagine a world in which there would be no rigid bodies, i.e. not even nearly rigid bodies, only liquids and gases; we ourselves would be, say, ethereally dissolved but mentally endowed inhabitants of the Orion Nebula. Then we would not be able to make any measurements and could do nothing at all with our arithmetically derived trigonometric formulas, which would apply to the outside world just as little as for instance to our dreams. We would not even be able to decide whether our space admits free mobility, let alone whether it is Euclidean or non-Euclidean. Given this objective behavior of things, we would not have the slightest psychological incentive or clue for developing a spatial conception, thus for the application of geometry to reality. A rash empiricist like J.St. Mill would then add: we would have no geometry at all. That, however, confuses psychology with logic: for we would be no more hindered by this from constructing a logical or arithmetical geometry than we are in our speculations about non-Euclidean spaces by the fact that we live in Euclidean space. [H 1903a, 18; VI: 298].

Hausdorff thus pleads in conclusion for a *prudent empiricism*:

> Thus, in the end, the two paths we have taken lead to a *besonnenen Empirismus*; from the indeterminacy of mathematical as well as of absolute space we are referred back to the only one given, empirical space. I do not want to claim that the scientific elaboration of this prudent empiricism will be altogether easy; in any case, the present forms of empiricism remain below its level. [H 1903a, 18; VI: 298].

As a matter of fact, Hausdorff never developed this *prudent empiricism*; he barely even hinted at what he meant by it (see [Epple 2006] and *HGW* VI, 189–207).

Hausdorff's intensive occupation with the space problem, as evidenced by his inaugural lecture, reflects an important phase of conceptual development in mathematics, a trend that would continue with even greater intensity in the decades that followed. If one wanted to characterize this process with a single word, then this was a trend toward *spatialization*. This does not refer to the spatialization of non-mathematical objects, however, about which some philosophers have complained. It concerns rather the spatialization of mathematical objects that were not originally conceived in spatial terms, for example certain sets of functions which came to be seen as constituting *function spaces*. These new conceptions facilitated the access to and the treatment of the objects investigated, not least because one could employ well elaborated general theories, such as those developed for topological or metric spaces. Today this process has advanced very far, as can be seen by looking up the keyword "space" in the index volume of the *Encyclopaedia of Mathematics*: about 100 different types of spaces are listed. Felix

Hausdorff knew that the ability to conceive of collections of objects as mathematical spaces can facilitate productive thinking, sometimes in quite unexpected ways. Moreover, he also believed that transforming a problem into something spatial can be of help in other spheres of life. He expressed this viewpoint in a discussion of Gustav Landauer's paper *Die Welt als Zeit*, which appeared in the journal "Die Zukunft" on May 17, 1902. Hausdorff replied under his pseudonym in an open letter that we briefly describe here [H 1902d, VIII: 529–533].

Landauer knew Fritz Mauthner's *Sprachkritik* very well; he had supported Mauthner in the final version of this work and had himself tried to further deepen its skepticism towards language and conceptual thinking. Furthermore, he was inspired by the writings of Meister Eckhart to develop his own new, mystically tinted vision of the relationship between the self and the world. Drawing on Mauthner's critique of language, Landauer formulates the following central thesis in *Die Welt als Zeit*:

> We can no longer seek absolute truth, since we have recognized that the world cannot be conquered by words and abstractions. Still, we feel an urge, so strong as humanly possible, to form out of the manifold images that our senses feed us a unified picture of the world, one whose symbolic significance compels us to believe. [· · ·] In science one finds everywhere the scattered fragments of such symbolism, which one day will take the place of the allegedly positive part of our abstract knowledge. Yet before it comes to this, before it appears possible to form a world picture from the results of scientific research, a great reorientation seems necessary: renouncing an ancient metaphor and replacing it by another. Space must be transformed into time. [Landauer 1903, 107–108].

To understand what Landauer meant by this in his essay "The World as Time," we need to consider his preoccupation with mysticism. According to Landauer, truth is "a thoroughly negative word, the negation in itself," and he identifies this with the thing-in-itself. Thus: "What is behind our reality? Something else! How is the world in itself? Different!" [Landauer 1903, 99–100]. The basic question for Landauer is thus: How should one relate to this unbridgeable gap between the inner experience of the I, which is a purely temporal phenomenon, and the "thingness" outside of me, the otherness, which can only arise through the conception of "space"? Here he finds an ally in Meister Eckhart : "It was certain for him [Eckhart] that what we find in ourselves as mental experience is closer to the true essence of the world than the world perceived outside it" [Landauer 1903, 103]. Thus, in order to be able to approach this essence of the world at all, space must be transformed into time.

In his open letter in reply to Landauer, Hausdorff took a different view, writing:

> Don't we all suffer today from internalization or, as you say, from temporalization? And late-coming prophets like Maeterlinck promise an

> "awakening of the soul": I think we have decidedly overproduced soul and should strive to get rid of this easily perishable commodity in the open air as soon as possible. The temporal arts, music and poetry, unpack so much soul that doesn't want to stay together; [···] you want even more soul, even more form of the inner view, even more "time"? [H 1902d, 442; VIII: 529].

This polemic reveals a notable inner transformation in Hausdorff. In the mid-nineties, when he was writing the poems that appear in the "Falterflüge" section of his *Ecstasies*, he also celebrated the ecstatic moment (as we know from his letters to Paul Lauterbach). Back then, when he identified with the proper time of the sacrificial artist and the temporal flow of inner experience, he might have agreed with Landauer when he wrote: "[···] the extensity of outer things must be an image for us of the intensity of our ego feelings" [Landauer 1903, 120]. But now, a different view prevailed in Hausdorff's work, the striving for sure scientific knowledge in thinking about the mathematical concept of space and transfinite sets; this sphere is timeless and void of inner feelings. Of course, for Hausdorff there were still vast areas that remained inaccessible to scientific study as well as some that should remain so. Nevertheless, he could in no way accept the rigorous rejection of Landauer, who in his struggle against the reductionism of materialistic science wanted to come close to the "true essence of the world" by reducing it to inner experience, while dismissing science as the "knowledge of what is not."

On several occasions, Hausdorff dealt with feeling or "mood" (*Stimmung*), a term that had become very fashionable around this time. He also addressed this in his open letter to Landauer, where he expressed himself very clearly by saying:

> [···] much that is temporal (not all!) can still be transformed into space and many a soul can be crystallized into a thing, so we would do well, for a change, after the eternal inwardness and mollusc-like "moods" of the last decades to seek out once again the objective side, in clear figures and sharply drawn pictures, quite spatially and substantially. [H 1902d, 445; VIII: 533]

Chapter 5

Paul Mongré: Author, Essayist and Critic

5.1 The Author Mongré and his Circle

The first traces of Hausdorff's talent as a poet date back to his last two years at the Gymnasium. Unfortunately, nothing remains from the poems he composed at that time, but we know that they must have been of remarkable quality. As described earlier, he enjoyed the distinction of being asked to read his poetry, including poems written in Latin, on ceremonial occasions held at the school. His further literary development was intertwined with his path in philosophy, as discussed in the first section of the previous chapter. He thereby gained a circle of friends among Nietzsche's admirers, including Kurt Hezel, Fritz Kögel, Heinrich Köselitz, Paul Lauterbach and Gustav Naumann, all of whom had artistic inclinations as writers or musicians. They worshiped Nietzsche, not only as a philosopher but also as a creative linguist and original poet. It is therefore not surprising that Hausdorff's debut in the field of philosophy, the aphorism volume *Sant' Ilario*, was also a remarkable literary work in the style of Nietzsche. This can be seen from the short quotations taken from poetic works that Hausdorff placed at the beginning of the chapters in his book, for example two sayings from the *Aegineten* of his friend Paul Lauterbach. The final chapter consists of Hausdorff's own lyrics in the form of sonnets and rondels.

Already during his years as a student, Hausdorff got to know the jurist Kurt Hezel as a fellow member of the *Akademisch-philosophischer Verein zu Leipzig*; he soon became one of his good friends. The personality of Hezel is reflected in an obituary published in *Juristische Wochenschrift*:

> Hezel was a lawyer, – an excellent lawyer, a brilliant jurist with wide-ranging knowledge, great erudition, and an unparalleled, almost linguistically creative eloquence. But he was not only a lawyer and not

E. Brieskorn, W. Purkert, *Felix Hausdorff*, Vita Mathematica 21,
https://doi.org/10.1007/978-3-031-52135-5_5

Figure 5.1: Exlibris Felix Hausdorff
Design from 1902 by Hans Jacob Ferdinand Zarth. Source: Kirsten Büsing, Anne Büsing: *Alumnen und ihre Exlibris – 600 Jahre Universität* Leipzig: Vieweg und Teubner, Wiesbaden 2009, p. 75.

> even primarily a lawyer and jurist; the strength of his being rested on his altogether *artistic* personality, in the charm which this personality of his radiated on all who approached him. What those who were united with him in an inner circle knew, can now be expressed in public: Hezel stood at the center, at least among jurists, of artistic life in Germany. From a young age he was one of the confidants of [Wagner's] Haus Wahnfried, and he was the intimate friend of Otto Erich Hartleben, Richard Dehmel, and not least Frank Wedekind. The latter dedicated his last work to him, the tragedy *Heracles*, and earlier he had already memorialized him in one of his dramatic figures. To list the names of Hezel's friends and admirers would be to summarize the new literature of Germany and far beyond to other circles, including philosophers, philologists, artists, and especially musicians. [$\cdots$] He, who had the ability to peer sensitively into the souls of artists, because he himself was an artist of gifted talent, became the legal friend of numerous artists and poets, among them the best of their time and, as the course of events and politics often required, not infrequently as their criminal defense attorney. (*Juristische Wochenschrift*, 7, 1922, pp. 418)

Kurt Hezel was an ardent admirer of Wagner as well as Nietzsche. During his student days, probably in the summer semester of 1886, he established a reading circle in Leipzig whose participants studied Wagner's 1850 article "The Artwork of the Future" (*Das Kunstwerk der Zukunft*) and Nietzsche's 1872 treatise *The Birth of Tragedy from the Spirit of Music* (*Die Geburt der Tragödie aus dem Geiste der Musik*). As mentioned in Chapter 4, it was Hezel who initiated Hausdorff's contact with Heinrich Köselitz and thus also to a wider circle of Nietzsche's followers.

Not surprisingly, Hausdorff, as a young man of letters, was pleased to meet interesting and for him valuable acquaintances through Hezel and his circle of friends. During Hausdorff's student days and to some extent later, this circle included Oskar Bie, Hermann Conradi, Otto Erich Hartleben, Hans Heilmann, Ferdinand Pfohl and Sebald Rudolf Steinmetz. (Another Leipzig circle of intellectuals and artists founded by Hezel was the *Bungonen*. Hausdorff later frequented this circle when he himself was already a well-known literary and philosophical writer). Circumstantial evidence and passages in the literature indicate that Hausdorff had personal relations with Bie, Conradi, Hartleben and Heilmann, though more detailed autobiographical evidence for this is lacking. For Paul Mongré the relations to Bie and Hartleben were of particular importance.

Oscar Bie studied philosophy as well as art and music history in Breslau, Leipzig and finally Berlin, where he took his doctorate in 1887. From 1894 to 1922, he was editor-in-chief of the cultural monthly *Neue Deutsche Rundschau* (from 1904 *Die neue Rundschau*). With a sure sense for quality, Bie turned this magazine into a leading organ for modern literature and critical essays. Works by writers of the first rank were featured in it, leading to the popular saying:

> Bedenk, o Mensch, daß Du vergehst,
> selbst wenn Du in der Neuen Rundschau stehst.
> (Remember, oh man, that you will pass away,
> even were you in the Neue Rundschau one day.)

Bie must have known and appreciated Hausdorff, for of the fourteen longer essays Hausdorff wrote under the pseudonym Paul Mongré, ten appeared in Brie's journal, just as did his play *The Physician of His Honor* (*Der Arzt seiner Ehre*).

Otto Erich Hartleben arrived in Leipzig in the fall of 1886 to continue his legal studies, which he had begun in Berlin and Tübingen. In 1889 he became a court clerk in Stolberg (Harz) and Magdeburg, but soon gave up his legal career and began living as a freelance writer in Berlin. After a few contributions to anthologies, his first independent work *Studenten-Tagebuch* appeared in early 1887 when he was still a student in Leipzig [Hartleben 1887]. Some of the poems in this diary, for example "Studentisches or *Die Apotheose des Duells*, satirize the lifestyle of the "German student" [Hartleben 1887, 28, 29]. Others, like *Berliner Christenthum* or *Ein christlich Gebet für Berlin*, pillory the antisemitism of the court preacher Stöcker [Hartleben 1887, 69, 71f.], while *Gottvertrauen zum Bayonette* mocks Prussian militarism and the Berlin police [Hartleben 1887, 6–8]. But the author also sings about his love of life, with drunken nights and romantic experiences. One can well imagine that this author impressed the young student Hausdorff, and Hartleben surely appreciated Hausdorff as well.

Hartleben was a very popular author for the theater during his time: he wrote eleven plays, six of which were comedies. Particularly successful was his play *Rosenmontag. An Officer's Tragedy in Five Acts* from 1900. In addition, he composed three volumes of poetry, a number of stories, including novellas as well as a parody. All of his works are all but forgotten today with a single exception, which continues to attract attention to this day: his translation (or better poetization) of a collection of poems by the Belgian symbolist Albert Giraud (actually Emile Albert Kayenbergh).

Hartleben's interest in these poems came through his contact with the Dutchman Sebald Rudolf Steinmetz, who also frequented the Hezel circle. Steinmetz came to Leipzig in 1886 to complete his studies under the experimental psychologist Wilhelm Wundt, but he was also interested in literature and particularly enthusiastic about Giraud. Later he taught at Utrecht and Leiden, where he is remembered for founding the discipline of sociology in Holland. Steinmetz's acquaintance with Hartleben led to an interesting episode in intercultural transfer, the beginning of which Hartleben described as follows:

> One fine evening, with a mischievous grin, [Steinmetz] pulled a small narrow booklet out of his breast pocket [· · ·] and said, "You see, gentlemen, this is not nothing. This is the kind of thing you would like to do: something new." [Tack 2004, 88]

The booklet was entitled *Pierrot lunaire, Rondels bergamasques* and contained fifty rondels penned by Kayenbergh and published under the pseudonym Albert Giraud in Paris in 1884. The figure of Pierrot has a long history, beginning with the Italian *commedia dell'arte* in the second half of the sixteenth century and extending to the literature of the décadence, such as the Pierrot of Paul Verlaine in 1881. The Pierrot of the Symbolists is an eccentric and melancholy figure, a self-dramatizing aesthete and dandy, occasionally a lecherous coward and liar with the feelings of a block of ice. He appears in the bare moonlight with his face made up like a mask, often accompanied by a longing for death, delusional inspirations and visionary dreams.

The form of the poems from Giraud's *Pierrot lunaire* also stems from a very old "forme fixe," appearing as early as the fourteenth century in the rondels of Guillaume de Machaut. A rondel consists of three verses, two quartets followed by a five-line verse; lines one and two of the first quartet are repeated as lines seven and eighth in the second. The first line of the poem is – sometimes slightly modified – also the last line. All lines have the same number of syllables (eight in the original French) with stress on the last or penultimate syllable. An additional difficulty arises in the rhymed form of a rondel, where only two rhymed endings occur; in this form, the first and third stanzas have embracing rhymes, the second alternating ones.

One can well imagine how extraordinarily difficult it would be to translate a rondel into another language. This involves not only reproducing the content but also preserving the verse and syllable scheme, and in addition finding two suitable rhyme endings in the other language. Nevertheless, Hartleben accepted Steinmetz's challenge and began translating *Pierrot lunaire* into German. He succeeded in producing 50 poems in the verse form of a rondel that reflect the content, mood, and spirit of the French originals, though he had to do without rhymes. In his diary he noted:

> It takes subtle chisel work to render such poems, and I have sat on it for many hours. And yet – it does not remotely achieve the charm of the French "rondels." Just think that these continuous gracious refrains in French are merged with the rest by the sounds of rhymes. What a pleasant sound, and how barren next to them stand the rhymeless translation! But it is – for me at least – impossible to bring out the rhymes: the frame is just too narrow! [Hartleben 1906, 38].

Hausdorff attempted this nearly impossible task, and he succeeded in creating rhymed translations for four of Giraud's rondels: *Poussière rose* (Morning Mood), *Parfums de Bergame* (Home Scent), *Déconvenue* (A Misery), and *Pierrot Dandy* (Pierrot Dandy). He published "Morgenstimmung" ("Morning Mood") in 1897 in his volume of aphorisms, and the other three appeared in 1900 in his volume of poems entitled *Ekstasen*. We can only speculate about when he produced these; nor do we know whether he attempted to work on other rondels from *Pierrot lunaire*. One circumstance suggests, however, that he may have written these translated and adapted poems during the early 1890s, as Hartleben had finished transcribing all 50 rondels in 1892. At that time, he published them under the title *Albert Giraud. Rondels. Pierrot Lunaire. German by Otto Erich Hartleben* with 60 autographed prints produced at his own expense. Two dozen of his rondels bore dedications to various poets, writers, and artists, as well as to good acquaintances and friends. Hartleben dedicated the rondel *Rote Messe (Messe Rouge)* to Felix Hausdorff, which suggested that he may have known and also appreciated Hausdorff's rhymed translations.

Various composers set the rondels from Hartleben's *Pierrot Lunaire* to music between 1905 and 1912, including Otto Vrieslander, Ferdinand Pfohl, Max Kowalski and Joseph Marx. The latter also set Hausdorff's reworking of *Pierrot Dandy* as well as some of his other sonnets and rondels. All of these musical works, however, have been forgotten, whereas Arnold Schoenberg's setting of *Pierrot Lunaire* marks a turning point in music history as one of the major works of early modernism. This was Schoenberg's opus 21 from 1912 with the full title "Dreimal sieben Gedichte aus Albert Giraud's 'Pierrot Lunaire"' ("Three times Seven Poems from Albert Giraud's 'Pierrot lunaire'. German by Otto Erich Hartleben. For one speaking voice, piano, flute (also piccolo), clarinet (also bass clarinet), violin (also viola) and violoncello."). It was commissioned by the singer and actress Albertine Zehme, who, like Schönberg, came from Vienna. Earlier she had worked at the court theater in Oldenburg, before accepting an appointment at the municipal theater in Leipzig. There she married the lawyer Felix Zehme – the same Dr. Zehme whose law firm employed Kurt Hezel.

In 1911, the year before she approached Schönberg, Albertine Zehme performed 22 songs from Otto Vrieslander's setting of *Pierrot Lunaire* in Berlin, so she was already familiar with Hartleben's rewriting. She and Schönberg shared an enthusiasm the striking expressivity inherent in Hartleben's texts. Schoenberg thus began composing, noting in his diary:

> Yesterday, March 12, I wrote the first of the "Pierrot lunaire" – melodramas. I think it turned out very well. It gives a lot of ideas. And I am absolutely, I feel, moving towards a new expression. The sounds here become an almost animal-like direct expression of sensual and spiritual movements. [Brinkmann 1995, 228]

The work premiered in Berlin on 16 October 1912 with Albertine Zehme as the speaking voice. Felix Hausdorff must have known her personally, since he wrote the following in a dedication copy of his aphorism volume *Sant' Ilario*:

> To the friend of Italy, Frau Albertine Zehme, this book of southern heavens is honorably dedicated, the author.

Probably already in 1898, Hausdorff joined an informal group of Leipzig intellectuals, literati and artists called the *Bungonen*. Its members usually met once a week, from which developed – beyond the regulars' table – a number of personal and family friendships. What the name means and where it came from remains a mystery, but it likely stemmed from the founder of the Bungonen circle, Kurt Hezel. An early mention of the Bungons can be found in a letter, dated June 15,1898, written by Hausdorff's friend Gustav Naumann to Heinrich Köselitz, in which Naumann reports on his first work *Antimoralisches Bilderbuch* [Naumann 1898]. At the end of this book, Naumann presents a small gallery of three nineteenth-century writers, for which he attached to each of their "images" a quotation from one of their works. These three anti-moralists were Stirner, Nietzsche, and – Mongré. To Köselitz he writes:

> Hausdorff has a new book ready: "Das Chaos in kosmischer Auslese," which will appear with Naumann. [···] I met Kühn down in Rapallo; Kögel also made their honeymoon there, and Hausdorff was down there with me; so the Bungons all luxuriated there, only Hezel, involved in a dangerous and adverse love affair, stayed in Leipzig; happily he survived it. (Gustav Naumann to Heinrich Köselitz. Goethe and Schiller Archive Weimar, Gast Collection, call number GSA 102/462.)

The writer Gustav Herrmann wrote retrospectively in the magazine *Das neue Leipzig* about the debates at the Bungonen regulars' table: "[···] the spirits of Wagner and Nietzsche hovered over the battles of words; the only pass required to join was originality of worldview and high striving freedom in thought and word" [Lange 2016, 75]. Hausdorff had acquired such a "pass" at the latest with the appearance of *Sant' Ilario*. A written source confirming Hausdorff's membership comes to light in a congratulatory letter Hausdorff sent to the physician Wilhelm

His (jun.) and his wife Lili His-Astor on the birth of their first child on January 8,1901. The letter consists of a witty poem related to the joyous event; the concluding greeting read:

> Warmest greetings and best wishes
> for the well-being of mother and child!
> Your Bungo Felix Hausdorff
> (*HGW* IX, 347)

Wilhelm His (jun.), son of the well-known Basel anatomist and physiologist Wilhelm His (sen.), had habilitated in 1891 in Leipzig, where he was appointed associate professor before taking a position as senior physician at the Dresden-Friedrichstadt Hospital. He later held professors in Basel and Göttingen, before assuming a directorship at the Charité in Berlin, where he was employed from 1907 until his retirement in 1932. As a leading researcher, his most significant discovery was the atrioventricular bundle as part of the heart system (today known as the His bundle). Lili His-Astor was the daughter of Edmund Astor, managing director and, from 1884, sole proprietor of the Leipzig music publishing house Rieter-Biedermann.

Two members of the Bungonen left autobiographical accounts that mention the group. One of these, *Erzähltes aus sieben Jahrzehnten (1863–1933)*, was written in the 1930s by the Germanist and literary scholar Georg Witkowski before he emigrated to Holland [Witkowski 2010]. The other, *Vom Jungsein und Altern* [Eggebrecht 1935], was begun by the Leipzig physician Ernst Eggebrecht shortly before his 70th birthday. Witkowski recalled these details:

> I do not know where the name came from. The circle gathered every Thursday after the Gewandhaus concert in a separate room of the Ratskeller. Here they drank heavily into the morning, and their spirits fired each other up until sometimes the champagne bottles shattered against the walls with a tremendous sound. The speeches from all sides flew just as violently over the round table, loaded with bold thoughts and often pouring out in German, but even more often in Latin and Greek verse. [Witkowski 2010, 261]

There then follows a longer paragraph about the man at the center of the circle, Kurt Hezel:

> He mastered the ancient languages as well as German, preferring to write his letters and postcards in Greek, and with his knowledge and understanding of the ancient philosophers he could rival any expert. If he found an able partner, he would discuss for hours individual propositions of Heraclitus and the other pre-Socratics, passionately even here, but when the subject turned to Wagner and his art, most excitedly to the point of thunderous roars. [Witkowski 2010, 264]

Georg Witkowski, son of a Jewish merchant, first attended a Gymnasium in Berlin. After the family moved to Leipzig, he entered the Nicolaigymnasium, the same humanistic school Hausdorff later attended. Witkowski then studied in Leipzig and Munich, where he received his doctorate in 1886 on the life and work of Diederich von dem Werder. In 1889 he habilitated in Leipzig and became a private lecturer in German language and literature, but he only received a salaried associate professorship in 1919 and finally a full professorship in 1930. Although he had been awarded the Goethe Medal for Art and Science by Reich's President Paul von Hindenburg, Witkowski was forced to stop teaching in April 1933. In 1937, the Gestapo took him into custody for two weeks after they had intercepted letters addressed to him which contained disparaging remarks about the Nazi state. In early May 1939, he emigrated to Holland, but he died there only four months later.

Ernst Eggebrecht's remarks about the Bungonen also focus on Hezel:

> He presided over a Thursday evening table (*Stammtisch*) that met almost regularly. It wasn't at all civil there, you were always a bit on the edge of your seat, as the conversations and their subject matter really weren't for anyone else. You heard a lot of swear words expressed loudly and very exuberantly. The center of attention was without question Dr.Hezel. [$\cdots$]
>
> On good evenings there was a coming and going, at times with out-of-town men who came by to visit us if they happened to be in Leipzig on a Thursday. A number of poets have erected monuments to our friend Hezel in their works; he was a widely known man, who during his university years interacted with several up and coming poets. No one attached any importance to formalities; it was simply a social group of men with its advantages and conveniences [$\cdots$] [Eggebrecht 1935, 68–69]

Among those who attended the Bungonen sessions during their visits to Leipzig were Richard Dehmel and Frank Wedekind, both eminent writers and poets. Although evidence relating to their relationship with Hausdorff is limited, these were more than just passing acquaintances. One indication of this comes from a letter Hausdorff wrote to Dehmel on December 3, 1907, which relates to an important episode in the latter's life. In 1896, Dehmel published *Weib und Welt. Poems and Fairy Tales*, which contained his controversial poem "Venus consolatrix." Because of it, he was convicted of "injury to religious and moral feelings," with the consequence that the poem was blackened out in all extant copies of *Weib und Welt.* A decade later, Dehmel included "Venus consolatrix" in a reworked form in his *The Metamorphoses of Venus.* Around the same time, though, he ordered some copies with the original version for distribution to trusted friends. Hausdorff's letter refers precisely to these circumstances and thus begins:

> Dear Sir,
>
> It is with joyful surprise that I learn I am one of the few in the elect, whom you deem worthy of an unmutilated copy of your "Metamorphoses of Venus," and it is my wish to tell you at once how strongly I feel about this distinction. I would wish to tell you much more, for example, that you are for me one of the very few who really force the elements out of a final state of dissociation into new forms,[$\cdots$]. (*HGW* IX, 181)

According to Hausdorff, he lacked the time at that moment to discuss this in more detail; however, he hoped for a verbal conversation soon: "Our mutual friend Hezel gave me hopes of seeing you in Leipzig one day soon [$\cdots$]" (*HGW* IX, 181). From a much later letter that Hausdorff wrote to Ida Dehmel, it appears the two families stood in good relations (*HGW* IX, 183).

In 1912, the first Wedekind cycle at the Deutsches Theater in Berlin proved to be an enormous success. In this context, a banquet in honor of Frank Wedekind was held on June 18, 1912, at the Esplanade Hotel in Berlin. His wife Tilly could not attend as she was ill, but Wedekind sent her a letter the next day to tell her about the evening [Wedekind 1924, 269]. He noted that Gerhart Hauptmann and Richard Dehmel had "stayed away" and Max Liebermann, likewise, because of illness, but that the latter had sent "a very dear telegram." From these few lines it is clear that personal invitations had been sent to those who attended banquet. In reporting on the course of the evening, Wedekind mentions some participants by name, such as Alfred Kerr, Maximilian Harden, Paul Cassirer and three journalists from the "Berliner Tageblatt;" the letter then says "Reinhardt came with Felix Holländer, Kahane and Hausdorff." Max Reinhardt, Felix Holländer and Arthur Kahane were the leading lights of the Deutsches Theater and, so to speak, the *crème de lá crème* of the Berlin theater scene; thus Hausdorff truly found himself in good company. The fact that he was one of the twelve participants mentioned by name in the letter (out of around 70 who attended) indicates that he must have been quite well known to the Wedekind family. Toward the end of his *Bungonenabende*, Eggebrecht wrote: "After Hezel's death [in 1921], this altogether unbourgeois Stammtisch immediately disintegrated." [Eggebrecht 1935, 68].

Felix Hausdorff also befriended the physician Ernst Eggebrecht, as the latter noted in his autobiography [Eggebrecht 1935, 80]. Eggebrecht came to Leipzig in 1889 and soon became an assistant to the professor of internal medicine Heinrich Curschmann. After working for several years as an assistant physician, Eggebrecht opened a highly successful practice as a resident physician in 1898; he worked there until 1934, in the final years on a reduced scale. Hausdorff always paid careful attention to his health, which gave him good reason to value this friendship. Still, there was another reason, as Ernst Eggebrecht shared Hausdorff's passion for music. Since his childhood he played the cello, and also through his professional practice he came into contact with many musicians and music-loving people. In Eggebrecht's autobiography there is an evocative passage about an evening that he, Hausdorff, and Max Reger spent at the home of the sculptor Max Klinger:

> Klinger's house and garden was situated on a quiet headland, like a peninsula, surrounded by the Pleisse River. Leaning against the house were marble blocks along with abandoned or provisional attempts in gypsum and clay. From the riverside one could see far across the then empty fields to the forests north of Leipzig. It was a beautiful spot in this scenically poor, flat terrain. I was invited to a company of about 30 gentlemen. In a festive manner, we sat around the large table as Master Klengel played beautifully on the cello. Gradually, the crowd diminished, daylight quickly beckoned (it was July), and finally only Klinger, Reger, the mathematics professor Hausdorff, and I sat together with the rising sun. Now, one must not imagine that the gods' conversation between these great ones as all too ambrosial. They were walking in Arcadia, but bartering with each other as if they were on the Brühl [Leipzig's marketplace]. After the two Maxs had drunk brotherhood, Reger went on the attack. "Max, why don't you give me the model of Liszt's bust!" After some back and forth, Klinger conceded. Then Reger asked further for Brahms' death mask, promising to dedicate his next opus to Klinger. Finally, we were ready to leave at eight o'clock. We road in a horse-drawn carriage to Reger's apartment – an illustrious company: two dead men and three only barely alive. In front of Reger's house, we exchanged roles. I took on the task of guarding the cab horse, although this felt like a lively hold. Liszt was shouldered by the coachman, Brahms found his resting place under Reger's arm, and Hausdorff went up with him as the guide. It dawned on me that I had assumed the task of being Reger's guardian. When I was again in the possession of my friend Hausdorff, we left. The night was gone, and I drank a mocha somewhere. As a colleague used to say to his patients in such cases, we had been ordered to attend to Prince Solms! [Eggebrecht 1935, 80], published in: [Huschke 1938, 863].

It should be noted here that Eggebrecht was Max Reger's physician during the latter's Leipzig period. We return to Hausdorff's friendship with Max Klinger later in this chapter.

Georg Witkowski met his future wife Petronella in Paris in 1897. She was the Dutch daughter of a museum director from Leiden, known only as Pietje or Pietchen to all her closer acquaintances. The couple's first child, Käthe, was the same age as the Hausdorff's daughter Nora. In part through their children, a friendly relationship developed between the Witkowski and Hausdorff families. About Käthe and Nora, Georg Witkowski reported:

> We thought it was unproblematic for Käthe to start her schooling earlier than usual. The first year she was taught at home by an elementary school teacher along with little Nora Hausdorff and two other ABC students. [Witkowski 2010, 203].

Witkowski's work as a researcher and academic teacher focused on the history of German literature and drama from the seventeenth to the nineteenth centuries. He published a number of books on these subjects and was known as an important Goethe scholar. His biographies of Goethe, Lessing, and Goethe's sister Cornelia were widely read. Witkowski also prepared historical-critical editions of the works of Schiller and Lessing, while elaborating his views on such editorial work. In addition to his academic profession, Witkowski was co-founder of the "Society of Bibliophiles" and for many years editor of its *Zeitschrift für Bücherfreunde.* His autobiography, which breaks off suddenly in 1933, has a melancholy flavor as he looks back on a life full of work, but rich in encounters with artists and musicians, theater people, writers, book lovers and publishers. At one point, he writes that "certainly among the best gifts of my life were the many valuable people I have met" [Witkowski 2010, 233]. Among the valuable people to whom Witkowski was particularly attached as a book lover were two of Hausdorff's good friends: the publisher Gustav Kirstein and the book illustrator and typographer Walter Tiemann. Nora Hausdorff recalled from her childhood in Leipzig how Kirstein, Tiemann and her father all seemed to belong together.

Walter Tiemann studied painting and drawing, starting in 1894, at the Royal Art Academy in Leipzig and later at the already venerable Art Academy in Dresden. After a period of study in Paris, he returned to Leipzig and began his career as a book designer. By the early 1900s, he began teaching at the institution where he had begun his studies, now called the Royal Academy for Graphic Arts and the Book Trade. There he taught a master class that covered the "overall field of book design, illustration, and applied graphics." When Max Seliger died in 1920, Tiemann took over as director of the academy, after which time the institution flourished. He always worked closely with others involved in the book trade: writers, publishers, and printers. Of great importance was his friendship and close collaboration with Carl Ernst Poeschel, whose father was co-owner of the Poeschel & Trepte printing company. In 1907, Poeschel and Tiemann founded *Janus-Presse* in Leipzig, which became a trend-setter in the next decade for artistic books based on high-quality typographic designs and printing. Tiemann and Poeschel worked together for decades for the *Insel-Verlag*, founded in 1901, creating works that became famous in the German book trade, though Tiemann also worked with other well-known publishers, such as Rowohlt, Rütten & Loenig, and S. Fischer. He created over twenty new typefaces over the course of his life, including *Tiemann-Mediäval, Tiemann-Fraktur, Tiemann-Antiqua, Orpheus* and *Daphnis.*

Of these, *Tiemann-Mediäval* had a special relevance for Hausdorff, that is for the poet Paul Mongré, because that typeface was used for a special publication that featured one of his works. In 1904, Tiemann's friend Georg Witkowski had initiated an organization for local book lovers, the Bibliophile Evening. Once a year its 99 members published their own literary anthology as an art book, a special limited edition produced in 99 numbered copies. A page was inserted into each copy with the name of the designated member. It was, of course, a high honor for

Figure 5.2: Felix Hausdorff as poeta laureatus, 1910
Woodcut by Hanns Alexander Müller after a Graphic by Walter Tiemann

a writer to have one of their works chosen for this publication. Besides Witkowski, other good friends of Hausdorff were active in the Bibliophile Evening: Tiemann, Eggebrecht and the publisher Gustav Kirstein. So it would seem hardly surprising that Hausdorff, too, was so honored in 1910 for his play *Der Arzt seiner Ehre*, which appeared in the *Fifth Annual Publication of the Leipzig Bibliophile Evening*. This edition was adorned with woodcuts made by Hans Alexander Müller, based on drawings made by Tiemann. One of the woodcuts shows Hausdorff as *poeta laureatus*, while others show characters in the play as Tiemann imagined them in their respective roles. In her conversations with Egbert Brieskorn, Nora Hausdorff noted that her father held Walter Tiemann in high esteem and regarded him as an amusing and clever man. She herself continued to correspond with Tiemann after the had ended.

Beginning in 1912, Gustav Kirstein took over chairing the Leipzig Bibliophile Evening at Witkowski's request. Their two families enjoyed a warm friendship, which profited from the coincidence that Kirstein's birthday coincided with that of Witkowski's wife Pietje. The families thus took up the custom of celebrating this occasion at each others' homes in alternating years. After Hausdorff departed from Leipzig, his relationships with several of his former friends naturally loosened, though not with Gustav Kirstein, his wife Cläre and their daughter Marianne. In fact, the couple had two daughters, Marianne and Gabriele, the first born in Leipzig in 1905, the second in 1907. Marianne later married the well-known mathematician Reinhold Baer.

Gustav Kirstein was born in Berlin in 1870 as the son of a Jewish doctor. After attending school, he completed an apprenticeship as a pharmacist and then as a bookseller. For several years he worked as a volunteer and assistant for the firm E.A. Seemann in Leipzig, a time during which he traveled to Paris, Berlin and Vienna. This publishing house was particularly known for its high-quality art books and the periodical *Zeitschrift für bildende Kunst*, among other publications. Founded in 1858 by Ernst Arthur Seemann in Essen, the company soon thereafter moved to Leipzig, which was then the center of the German book industry. In 1899, Seemann handed over the publishing house to his son Arthur, who later won over Kirstein as a partner. These two joint owners enjoyed success by using halftone printing to reproduce color paintings. The first Seemann art prints appeared in 1900 in their series "Old Masters" (*Alte Meister*). Later, other series were added – "Contemporary Masters" (*Meister der Gegenwart*) and "Masters of Color" (*Meister der Farbe*) – as well as albums and artists' portfolios along with the first books with colored illustrations. By 1909, Arthur Seemann largely withdrew from the business and left the management to Kirstein, who was keen to make the works of famous artists accessible to broad sections of the interested public through high-quality color reproductions.

Hausdorff must have met Kirstein around 1902, shortly after the latter returned to Leipzig after briefly working in Berlin. Among Hausdorff's papers, there is a note in Kirstein's hand inviting him to take a walk together. Their first acquaintance soon turned into a friendship between the families that lasted for several decades. Kirstein's daughters called Hausdorff "Uncle Felix" – a term of endearment that Marianne Baer still used in 1938 when she wrote from Urbana (Illinois) in 1938 on the occasion of Hausdorff's 70th birthday:

> Dear Uncle Felix,
>
> this letter has a long way to travel before it can bring my congratulations to you. I wish we lived a little closer to our friends – or better yet, our friends lived closer to us – then all three of us would come to you on your birthday and help you celebrate. (*HGW* IX, 138)

Hausdorff's daughter Nora recalled visits over Whitsuntide to the Kirstein's villa, located in the health resort of Lindhardt about 20 kilometers southeast of Leipzig. Several Leipzig entrepreneurs had built similar villas there, and the Kirsteins also owned a nearby country house, where they liked to stay during the summer months. Nora also remembered that they always invited many artists, especially visual artists. Kirstein owned an impressive collection of paintings, drawings, sketches, prints and sculptures, including works by famous artists such as Lovis Corinth, Max Klinger, Käthe Kollwitz, Max Liebermann, Éduard Manet, Adolph von Menzel, Edvard Munch, Emil Nolde, Max Slevogt, Carl Spitzweg, and Heinrich Zille, to name but a few [Knopf 2003, 304f.].

A special place in this collection was occupied by the works of Max Klinger, since Kirstein and Klinger were good friends. In 1912, the very first year that he chaired the Bibliophile Evening, Kirstein managed to have its annual publication decorated with works by Klinger. The cover of the volume, set in Tiemann medieval and bound in vellum, reads: "Twelve poems by Richard Dehmel with three drawings on wood by Max Klinger." Later, Kirstein also edited a book entitled *Die Welt Max Klingers* ("The World of Max Klinger"). It seems likely that Kirstein was also a member of the Bungons. When Richard Dehmel announced a forthcoming visit to Leipzig, Kirstein wrote back that he had admonished "Hezel, the archibungo, to arrange [···] a Bungon evening in his garden [···]" (Kirstein to Dehmel, June 22, 1918, Dehmel Archive in the Hamburg State and University Library. Signature DA:BR:K 689.)

On February 24, 1930, Gustav Kirstein celebrated his 60th birthday. His friends had prepared a very handsome gift for that day, which – like some of the other gifts for him – they prepared with affectionate mockery. The year before, the publisher E. A. Seemann had brought out a booklet with the provocative title: *The Woman of Tomorrow, as We Wish Her to Be* (*Die Frau von Morgen wie wir sie wünschen.*). This collection of essays by seventeen authors – all of whom were men, almost all known today in the history of literature from Max Brod to Stefan Zweig – addressed the phenomenon of the "new woman" in the Weimar Republic. And now, in 1930, the "publishing house" *Friends of Kirstein* issued a booklet imitating to the hair on its cover and in print the *Frau von Morgen*. Its editor was Walter Tiemann and the booklet was entitled: *The Publisher of Tomorrow, as We Wish Him to Be* (*Der Verleger von Morgen wie wir ihn wünschen*). In 31 contributions, friends and associates paid homage in an amusing way to Gustav Kirstein and his wife Cläre. Hausdorff's contribution was an "Acrostic for February 24, 1930" in the form of a sonnet (*HGW* IB, 438).

Three years later, after the National Socialists had seized power, Kirstein was forced to resign from his honorary posts in the German Publishers and Booksellers Association (*Börsenverein der Deutschen Buchhändler*), an organization to which he had rendered lasting service. He also had to give up chairing the Leipzig Bibliophile Evening, and its large contingent of Jewish members was also prohibited from participating. The "Leipzig Ninety-Nine" reacted to this forced departure of its"non-Aryan" members by no longer holding meetings after May 1933. The National Socialist takeover also ended Kirstein's involvement with the E. A. Seemann company, which he had run jointly with Arthur Seemann since 1902 and after 1923 with his son Elert (who happened to be a convinced National Socialist and member of the NSDAP). As compensation, Kirstein was given a small part of the publishing house, which he continued to operate under the name Seemann & Co., whereas Elert Seemann took over the lion's share of the original company and kept the original company name E. A. Seemann.

On February 14, 1934, Gustav Kirstein was found dead in his apartment. That same day, Cläre Kirstein telegraphed a very good friend of the family: "Gustav died of heart attack tonight." A few weeks later, on March 29,1934, she wrote a letter reporting on the first weeks after her husband's death, having since decided she would continue his business (later under the name "Masters of Color").

She reported on the assistance of their Leipzig friends: Witkowski, Tiemann, and Anton Kippenberg, and of visits, "[···] then for two days the infinitely dear and tender Hausdorffs." In December 1938, her publishing house was aryanized, after which she hoped to emigrate and join her children in the USA. The Gestapo kept delaying her departure, however, and finally they prevented it altogether. Cläre Kirstein then took her own life on June 29, 1939.

Having reported on important personalities from Hausdorff's circle of friends and acquaintances, we now turn to his literary writings.

5.2 Paul Mongré's Essays from the Years 1898–1899

We have already mentioned above that through the Hezel circle Hausdorff got to know Oskar Bie, editor-in-chief of the monthly *Neue Deutsche Rundschau*, later renamed *Die neue Rundschau (Freie Bühne)*. Under his editorship, it soon became a leading journal for modern literature and essays, making it Paul Mongré's publication organ of first choice, especially for his essays.

5.2.1 Mass Happiness and Individual Happiness

In 1898 Mongré made his debut in *Neue Deutsche Rundschau* with the essay "Massenglück und Einzelglück" ("Mass Happiness and Individual Happiness") [H 1898b, VIII: 275—288]. With a wink toward Kant, he sets the goal of undertaking a "critique of pure socialism" [H 1898b, 65; VIII: 275]. In this first essay – as a reflection, so to speak, of his unusually varied education – Hausdorff shoots off a veritable fireworks display of mathematical, scientific, historical, philosophical, sociological, theological, and literary references; one can well imagine that Bie was very much taken by it. The substantive background for a "critique of pure socialism" may well have come from the fierce attacks on Nietzsche and his followers undertaken by social democratic publicists. One of these was Franz Mehring, who after joining the SPD in 1891 regularly contributed to its theoretical journal *Die Neue Zeit. Revue des geistigen und öffentlichen Lebens*. There, in 1892, Mehring published a review of Kurt Eisner's *Psychopathia spiritualis. Friedrich Nietzsche and the Apostles of the Future*, in which he wrote that "the entire Nietzscheanism with its Byzantine cult of the superman and its rampant blasphemy against the proletariat [···] is a brutal and thoughtless crudity." As for the meaning of the doctrine of the superman, it was this: "Curse the 'crazed brothers' (*Schnapsbrüdern*) from the social democrats who see in the misery of today the hope of tomorrow, but hail the 'supermen, [those] free, very free spirits,' who harshly and pitilessly crack their whips over the herd of humanity" [Schneider 1997, 115f.].

With philosophical considerations reminiscent of the argument in *Chaos in Cosmic Selection*, Mongré tries to show that concepts such as society or collective do not correspond to anything objectively real. In view of this understanding with regard to the reality of life, one need "no longer avoid [···] saying a few nasty things to the face of the masses, society, or the collective animal" [H 1898b, 69; VIII: 281]. By reflecting on the theoretical problem of "many individuals," the author of this essay wishes "to embarrass modern social ethicists and cast some doubt on their presuppositions." He takes the view that one should examine an axiom "which seeks to give the basis of a *morality*" with the "same rigor of epistemological reexamination" accorded the axioms of geometry, all the more so for a morality that imagines it can effect "a substantial and influential change in the total picture of existence" [H 1898b].

At the outset, Mongré offers a reflection on the ambiguous problem of individuality, whereby the emergence of "I-feeling" is seen as a "mysterious process of realization,"

> as though an absolute I-point wanders in the world of subjects, and whenever it meets with an individual point its individual existence and I-feeling are awakened – just as an absolute present point gliding along the series of world states awakens and forces them to become real.

Indeed, "every subject says I, every present calls itself Now" [H 1898b, 65; VIII: 276]. The "theoretical ambiguity of the ego problem" leads to various philosophies, and these "ideologies of the philosophers [can lead to a] world-historical shift of values," an *ethical injustice* of the first order. Because of such ideologies people can "beat their heads bloody" [H 1898b, 65–66; VIII: 276–277]. How can such ideologies be countered?

> Well, with a stupid humanity one must behave stupidly, and in order to dissuade a ruminating head, one must ruminate oneself. With unspoiled people of a certain *intégrité de l'instrument mental* I would not trust myself to make an impression with speculative apparatus; but here, where it concerns first of all a question of returning to nature, a nature that has been distorted and bent by theory, here something can perhaps be hoped for by way of a counter-theory. Idea against idea, abstraction against abstraction, perhaps nonsense against nonsense! [H 1898b, 66; VIII: 277].

Nature with its "struggle for existence," i.e., standing apart from human civilization, "is basically aristocratic and promotes the emergence, though not always the existence, of the higher organized entities over the lower ones" [H 1898b, 66; VIII: 277]. The matter stands differently with human nature and the turmoil and deficiencies of human cultural enterprises. Here the suspicion arises that the ideals of historical humanity are neither desirable nor tenable.

> Or, if we confine ourselves to modern, democratic, herd-animal mankind, as depicted by those who know it, and its specifically modern ideal, which consolidates itself ever more firmly and depresses other ideals: does the concept of society give a tenable ideal, a just mode of compensation in the struggle of interests? – Certainly not, as soon as it

> is proven that with it not even a reality, an epistemologically tangible something is encompassed. [H 1898b, 67; VIII: 279].

The section following, which is supposed to prove that nothing real corresponds to the term "society," pursues a line of philosophical speculation of the kind Hausdorff himself characterized by the words "maybe nonsense against nonsense." Namely, he applies the reasoning in *Chaos in Cosmic Selection* to aggregates of human individuals. Perhaps it was this encroachment of considerations about space and time on the realm of social phenomena that led him to later call "Mass Happiness and Individual Happiness" a "philosophical oddity." As the end of this philosophical speculation, he comes to the following conclusion:

> If a hundred normal men, each in the dull isolation cell of his ego, enjoy one and the same dull normal man's happiness, how high, calculated realistically and critically, does the centurion's happiness stand above his unit, above the individual discomfort of the individual normal man? Relations, whose *stringent* fulfillment would melt those hundred realities into one, are – sadly! sadly! – simply not stringently but rather only approximately fulfilled; the congruence is not an absolute one, but rather still allows room for individual difference; that is the entire factual basis of "society," which to me does not weigh heavily enough to serve as support and anchor for a value system that prioritizes the masses over the individual [H 1898b, 70; VIII: 282].

Finally, in accordance with the title, Hausdorff sets the individual happiness of the "higher" man against the happiness of the mass of "normal" men, from which he concludes:

> I repeat: not society, but the individual is a suitable object of hedonistic activity, above all the superior, distinguished individual, whereby as a measure of superiority should serve sharper individuality, finer organization, and more developed sensibilities, at least so long as no more objective criteria of valuation are available. [H 1898b, 72, 74–75; VIII: 284, 287].

Hausdorff's "nastiness" was directed at "opponents of vaccination and alcohol, protectors of animals, homeopathic doctors, pacifists, apostles of chastity, 'some' Christians, Bayreuthers, and assorted other do-gooders," with whom it would be possible to talk about their cause, but with whom it is impossible to talk because of their "solemn mindlessness" [H 1898b, 71; VIII: 283]. The following passage, however, suggests that his sharpest critique was probably directed against Karl Marx, whose ideas were originally influenced by Hegel, and also against social democracy:

> Those who, in fact, place the value of the community above that of the individual, who fall in line with the agitating cries of the economic materialists and their preaching, as if this constructive falsification of

> history marked the beginning of science rather than just the end of Hegelianism, and who hypostatizes the representative presence of the masses in the individual, which nobody denies, into an independent reality of the masses outside the individual – he may build his world view on altruism [· · ·]. We ourselves build it on egoism, on the individual, on the only provable and controllable I [· · ·]. [H 1898b, 74; VIII: 287].

5.2.2 The Impure Century

In 1898, the *Neue Deutsche Rundschau* also published a second essay by Hausdorff, "Das unreinliche Jahrhundert" ("The Impure Century") [H 1898c, VIII: 341–352]. Though not taking himself entirely seriously, he here uses his linguistic gifts and fine sense of irony to gloss ambiguities in everyday life as well as in philosophy, morality, religion, and in the role of social institutions. His theme, in short, was the "impurity" in language, thought, and action of the past century as it came to an end. What was needed, in Hausdorff's view, were clean distinctions, since these alone enable one to make clear decisions that can lead to a definite orientation in life.

He begins quite harmlessly by remarking on how as a traveler one not infrequently heard sweeping judgments such as "I find the French so nice and amiable" or "those obnoxious Englishmen" or "Switzerland is decidedly more beautiful than Tyrol." He then went on:

> And indeed these unprepossessing lump-sum, survey judgments fall not only from the lips of women but also from the discerning man, even the exact-thinking university professor, who in his field is obsessed with divisions and subdivisions and razor-sharp distinctions, but allows himself to be caught employing rough, blunt, or dull generalities while speaking of *the* Englishman or *the* Swiss, somewhat like the Belgian statistician Quételet, who spoke of *the* typical man. The difference, though, is that the statistical normal man, *l'homme moyen*, was at least numerically averaged from the deviations of individuals, while that normal Englishman, whom one finds between his fish and roast to be obnoxious, serves only as a fictive carrier for scattered and incoherent casual observations. Supposing there really were what one calls *the* Englishman, supposing two or three characteristics could be found in the face, gesture, and behavior that would point to the member of a certain people [· · ·], well, the foreigner, hurried by the travel demon with his focus on the momentary and conspicuous, will certainly be the last to discover those hidden traits. [H 1898c, 443; VIII: 341].

Hausdorff located one source of the problem at the beginning of the "impure century." He found this in

> the most astonishing *dualism* that ever found word and language: the *Kantian philosophy*, the consciously tolerated, even demanded contra-

> diction between theory and practice, the canonized "lie of life," to put it in modern terms. [H 1898c, 444–445; VIII: 343]

He especially criticizes the author of the *Critique of Practical Reason.* Kant had by no means wanted to undermine speculative proofs for the existence of God. On the contrary, he wanted

> to have God completely outside the sphere of provability and refutability; it was hard on him to realize that this fixed idea of his was exposed to the changing fates of philosophical discussion. [H 1898c, 445; VIII: 343]

Kant brought God to safety in transcendental territory,

> where he could say: prove or refute me, in the end I exist after all! The formula for this is the Christian credo *quia absurdum* with a modern twist: the ideas of reason must be considered true for practical purposes, even if their truth is not provable theoretically – in short: one cannot know, therefore one must believe. [H 1898c, 445; VIII: 343]

The nineteenth century did not need to think twice before putting Kant's duality to practical use. Thus arose a kind of division of consciousness, also in moral terms, where one says:

> On the one hand, science is right, and on the other, catechism; under the door we chat with the doctor, while through the garden gate the brother Capuchin sneaks into the nursery.
>
> When the government clerk puts his hymnal into his Sunday jacket and goes to church, [· · ·] he thinks nothing of it, because in the right pocket of his jacket he carries an edifying booklet with gold edges, in the left pocket the sports newspaper; he harasses horses, rabbits and social democrats as "appearances," whereas he feels the "thing in itself" for the weary and burdened. Schopenhauer hurls one of his most powerful curses against American slaveholders, who on weekdays profit from twitching human bodies, but on Sabbaths follow the pious litany: well, that's just dualism! [H 1898c, 445–446; VIII: 344–345]

Topping it all off, certain manifestations of this duality have been "legally nailed down." The blasphemer is punished, meaning "the 'appearance' is imprisoned if it fails to express itself in parliamentary language about the 'thing in itself'." The situation is similar with *lèse majesté*, an offense based on medieval convictions regarding a "nimbus of divine grace" and a "sphere of mysticism and transcendence that surrounds the monarch" [H 1898c, 447; VIII: 346]. Hausdorff also shoots sharp arrows against dueling, as a residual concept of honor inherited from the Prussian nobility and officer corps, but which was becoming increasingly anachronistic in the developing bourgeois society of the late nineteenth century.

This critique became the centerpiece for his satirical play *Der Arzt seiner Ehre* (The Physician of His Honor).

However, the end of the century, the *fin de siècle*, suffered from a special "dualism": on the one hand, a highly developed scientific culture with enthusiasm for technological progress, on the other, obscure esoteric beliefs and spiritualism:

> We have tremendous research and a deep and sharp recognition of the interconnections of nature – and alongside it spiritualism: the dull superstition of our nurses and great-grandmothers draped in an impudent sham science. Next to the exaggerated specialization in science we have the most fundamental obscurity about the limits and competence of knowledge, next to the luminous exactness the most nebulous haunting of occultists, obscurantists and mystics; think of the great magician Schopenhauer and his success, which he himself ascribes to a "metaphysical need" of contemporary mankind not satisfied by the church. [H 1898c, 451; VIII: 350–351]

In the end, it is left to the individual to decide what to think and to believe, not to society with its moral claims and legislation:

> Whether the Parisian goes to the Buddhist Club or attends an electrotechnical conference, whether one pulls Helmholtz or Jacob Böhm out of the lexicon for fashionable table talk, is ultimately a matter of indifference; it is less indifferent that fragments of the law and morals from the Thirty Years' War still lie all around us. One can stumble over them – I hinted at a few individual cases in order to conceal very many others. If one clears them away, it makes life easier for some exposed people and creates a smooth path by performing an act of housecleaning with which every departing century should commend itself to its successor. [H 1898c, 452; VIII: 352].

Hausdorff's criticism of Kant's "dualism," which comes across quite loosely, was bound to set the Neo-Kantians on edge. Hans Vaihinger, professor of philosophy in Halle, published a short review [Vaihinger 1899] of Mongré's essay in *Kant-Studien*, a journal he had founded. Its author, says Vaihinger, "makes fun of our century in the style of his master Nietzsche" by asserting that beyond all the modern advances it "has left medieval cobwebs hanging." The passing century was thus "enlightened on one side, mystical on the other," which is why superstition rules "alongside natural science." Then Vaihinger continues: "And who is to blame for this "modern halving and doubling"? Can you guess, patient reader? None other than Kant, of course!" Vaihinger then quotes from the relevant passages on Kant and then concludes:

> In this wittily quipping tone, the Nietzschean continues to speak, blaming Kant's "idealism" for all sorts of moral excesses of the time, such as dueling and the like. Indeed, one cannot imagine a more frivolous

> distortion than this. Never has Kantian idealism offered a hand to a moral duality, as the author imputes to it. (*HGW* VIII, 377–378)

5.2.3 Death and Return

The essay "Tod und Wiederkunft" ("Death and Return") [H 1899b, VIII: 415–429], also published in *Neue Deutsche Rundschau*, takes the form of a letter from a fictitious female friend of the author, who seeks guidance regarding Schopenhauer's views on the subject of death and suicide. This literary framework has famous models, such as Fontenelle's *Entretiens sur la pluralité des mondes* from 1686. Mongré begins with a hefty dig at Schopenhauer:

> My dear friend, you have read Schopenhauer's chapter on suicide and the indestructibility of our true being through death, and you feel a little troubled in your urge for clarity: little wonder with this twilight thinker, who has learned his trade by watching bats fly on the evening divide between day and night, between luminous prudence and darkest mysticism. [H 1899b, 1277; VIII: 415]

So expressed, Mongré's friend should not let herself be painted over by "metaphysical tendrils," as Schopenhauer has done when intoning that death relieves us of the burdens of life that would have come afterward. The demagogues of mysticism all point to suicide as "the natural way out" for those who find life no longer bearable.

This discussion of Schopenhauer is actually only a first point of entry; above all, the essay leads to the main ideas in *Chaos in Cosmic Selection*, though in a more popular way, thus "without grinding the philosophers' conceptual mill." Hausdorff begins by stating that nothing absolves us from the course of life that begins with birth and extends to death. Sensing his counterpart's doubts, he then asks:

> That which once was remains having been, and you mean, precisely because of that it does not concern us any more? Yes, what other form of being do you actually know besides having been? The present, the momentary instant? [H 1899b, 1279; VIII: 417]

Here the language virtuoso Mongré is called upon. With impressive pictures he tries to make clear to his friend that the "puzzling being known variously as moment, point in time, or state of the world" is not accessible to us, because "only continua of fulfilled moments produce a conscious process" [H 1899b, 1280; VIII: 418]. From this it follows that:

> [· · ·] for our experience, no being other than past being exists; because the *present* in the strict sense, as an extensionless boundary between past and future, would be a *point in time*, and thus incapable of entering consciousness. [H 1899b, 1280; VIII: 418]

The conceivable movements of the present point on the time line, which Hausdorff considered in *Chaos in Cosmic Selection*, permit among other things pauses, repetitions, jumps, and reversals that do not disturb our consciousness of time, the familiar empirical flow of time. In particular, any number of identical repetitions of a time stretch would be possible. This thought, which leads to the eternal return of the same, Hausdorff tries to enliven with the following image:

> We cling to the nutshell of the present until death snatches it from our hands, and now we see that actually countless such nutshells float on the waves, from which we only need to grab one by chance again. What will then happen? We will cling to that one again, as if it were the only one; then we say Now again, as if no other time would say Now; then we are interwoven again into the fabric of the course of the world and know nothing more of the fact that we had already once slipped out. ... Do you understand this, my friend? We live a span of life for the second time, but exactly as if we lived it for the first time! [H 1899b, 1281; VIII: 419]

At this point, Mongré arrives at his main theme, the "*Eternal Return*, that solemnly brilliant formula of Zarathustra." He is first concerned, though, with *his* conception of the eternal return, which differs from Nietzsche's, and rests on the principle that every period of time is capable of unlimited *identical* repetitions. We would not notice these, however, because each repetition is identical and because there is no counting mark for the different reproductions. Therein lies precisely the *possibility* of this whole structure, "because that which is accessible to us alone, our world of consciousness, is not affected by it." As an "explanation of the world phenomenon present to us" it accomplishes exactly the same thing as the "naively realistic hypothesis of the unique and irrevocable course of things." [H 1899b, 1282; VIII: 420–421]. After this, he continues:

> More I cannot convey to you here; should these strange and offbeat speculations rob you of your slumber, then as a narcotic you can consult a book I published under the title "Chaos in Cosmic Selection." There you will enter a spherical-cyclic magic cabinet from which you can observe from the outside the circle-, ring- and ball game played by us fools of consciousness, with peace of mind. Time, space, order, natural law – I grab onto all these illusions as if they were the beautiful, regular, but not very durable bodies of jellyfish. [H 1899b, 1282; VIII: 420–421]

Hausdorff now comes to Nietzsche's view of the eternal return as compared with his own. His own hypothesis has no bearing on the world of consciousness and consequently can "expect neither confirmation nor refutation from the side of experience; whereas Nietzsche's doctrine attempts to say something positive about the structure and connection of reality." In this respect, it does not belong to philosophy, but ultimately to natural science. Nietzsche also intended to pursue scientific studies in order to prove the eternal return of the same as an exact truth. He would have failed in this effort, though he did not live to experience this failure. In the meantime, theoretical physics has adopted the law of constantly increasing entropy, thereby accepting the "final solidification of the universe" for an absolute finale as "against a purely periodic course of the world." Of course, besides our cosmos, which presumably does not run cyclically, there could be other worlds with other laws of nature that lead to closed time lines, i.e. physical states that return to themselves, but we could never learn about them and they would know nothing about us.

Hausdorff had already refuted Nietzsche's "materialistic-atomistic proof" in aphorism 406 of *Sant' Ilario*, which ends with the following remark:

> Nietzsche's materialistic proof for the necessity of the eternal return may thus be considered refuted – something completely different than judging the *conceivability* of this hypothesis or else, even again, its poetic, ethical, and speculative value. The eternal return is a powerful conception, a mystery, which already as a possibility excites, shakes, and opens tremendous consequences. We do not tread at all closely to this "abysmal thought" simply by rejecting its superficial proof. [H 1897b, 354; VII: 448].

In the remaining pages of "Death and Return," Hausdorff demonstrates to his fictional friend, with imaginative, impressive, and yes – even harrowing – images, some of the "unheard-of inferences" that arise from the "mystery" of eternal return.

5.3 Relations with Women, Ideal and Real

When he married, Felix Hausdorff was thirty years old. What experiences he may have had with women until then, we do not know. Only a few hints have survived in his letters and postcards to good friends, one of whom was Paul Lauterbach. In a letter to him, dated February 2, 1894, Hausdorff writes of an experience from his previous summer vacation:

> So that you can see that even in bad times I also have had good feelings about "the" woman, I am sharing a portrait from my private gallery; it is from the Alps, summer 1893.
>
> That I come from a naufragium in eroticis [ship-wrecked in love matters], you probably finally guessed yourself; insofar I feel a little ashamed not to have kept my mouth shut better. In the aftermath of such things, one does not have good manners, one does not know how to keep silent, for example. (*HGW* IX, 419)

The "Portrait" from the "Private Gallery" is a poem enclosed with the letter, entitled "Miss Ellen"; Hausdorff later included it, slightly modified, under the title "Herbstwunsch" ("Autumnal Wish") in his volume of poems *Ekstasen* (*HGW*

VIII, 187–188). In this letter, but much more so in numerous aphorisms and poems published later, Hausdorff reflects on his erotic relations "to the woman." The chapter "Pour Colombine" in *Sant' Ilario* is the most extensive devoted to this subject. Its aphorisms often reflect a mood of melancholic resignation and sometimes also a distant tone, quite as if one or the other "naufragium in eroticis" was having its effect. As an example, we cite these two:

> Nr. 120: The man can demand at least from the woman a special disappointment, a personal rejection, or an individually delivered crushing no; he does not want to be spurned en masse and to be dragged along in the troop of infatuated worshipers. [H 1897b, 102; VII: 196].
> No. 155: Whoever has once gone through in full sharpness and violence that attack on the nerves, called love, will no longer believe what sensitive poets sing about the harmony of souls. Perhaps love is the widest gulf that can be placed between two people. All understand each other when the need arises, buyer an grocer, pupil and teacher, friend and enemy – only lovers do not understand each other. [H 1897b, 113; VII: 207].

In general, the image of women in the aphorisms of *Sant' Ilario* is ambivalent. On the one hand, Mongré remains bound to the contemporary role model of women and the attendant sense of intellectual superiority of the man. On the other, all the efforts of the "intellectual man" ultimately serve only one goal, namely to impress women.

More than two dozen of the aphorisms in the chapter "Pour Colombine" deal with the subject of marriage, and in all of them Hausdorff distances himself from this institution, although he still had no personal experience with marriage at the time. These two aphorisms are in this sense representative:

> No. 171: The mode of existence of the serious-minded is the moment; repetition, duration, habit are already parodistic. Marriage, too, is a parody – of love. [H 1897b, 118; VII: 212].
>
> Nr. 223: That two can get along with each other, for that love is hardly a help. Marriage means inventing a thousand clever tricks to perpetuate one folly. [H 1897b, 135; VII: 229].

Perhaps one should not take too seriously Hausdorff's aphorisms on "the woman" in general, or "marriage" in particular. One should, in any case, remain mindful of his mocking motto in the chapter "Splitters and Spines" of *Sant' Ilario*: "What young blood wrote down, gives gray heads something to think about" [H 1897b, 277; VII: 371]. Perhaps we gray heads should not think too much about what Hausdorff wrote regarding marriage as a twenty-nine-year-old bachelor. But then, what was life like in the small Hausdorff family once he did marry?

In *Sant' Ilario*, Felix Hausdorff made his aversion to moralists of all sorts plain, placing himself in an entirely different category, "as a creator [he upheld] the highest ethos attainable on earth" [H 1897b, 60; VII: 154]. This ethos determined his life, but also, in a very different way, the life of the woman who shared her life with him. In conversations with Egbert Brieskorn, Nora Hausdorff reported that her mother was only there for her father. That had been natural, of course, also for her father. Their motto was: "Dienen lerne das Weib beizeiten" (Let the woman learn to serve in time), a quotation from Goethe's *Hermann und Dorothea* that continues:

> Let the woman learn to serve in time according to her purpose.
> For through service alone she attains finally to rule,
> To the deserved power, which belongs to her in the house.

The motif of serving and ruling appears clearly in Nora Hausdorff's memoirs and also in those of Else Pappenheim. According to Nora, her father absolutely dominated over his family. He was demanding, but modest in personal matters such as his clothing, since he was indifferent when it came to such externals. Because Felix always impressed himself on others, Charlotte appeared outwardly very withdrawn, perhaps even repressed. Yet the "power in their house" belonged to her in the sense that she organized the household and daily life in the broadest sense, thereby keeping her husband's back free. She provided understanding and loving support with regard to his peculiarities, especially his pessimism, which occasionally fed periods of depression and hypochondria. Charlotte embodied what was expected of a good wife in the upper middle classes of that time, and Felix knew what he had in her.

Charlotte Hausdorff also ran a hospitable home. When guests were present, her husband's depressive moods were never noticeable, since he was always able to come out of his shell in a cheerful way when in a pleasant social setting. He liked to eat well and loved good wines; his wine cellar was richly stocked and a bottle was always on the table in the evenings with the family. In appropriate company, Hausdorff loved witty conversation and puns. He could be very charming, but he sometimes kept an ironic distance. He maintained some older friendships from his bachelor days in Leipzig, especially with those close to the Bungon circle.

For Hausdorff, work and creative activity always remained the central focus; everything else played a subordinate role. He was extremely sensitive to noise, a factor that played a decisive role when choosing an apartment and the houses he later purchased in Greifswald and Bonn. At home, the study became his "sanctuary," and he loved to work at night when all was very quiet, often until around 3 o'clock in the morning. He smoked fine imported cigars, often one after the other, while drinking strong tea. Then he went to bed and slept until around 11 o'clock. During the morning, the word in the house was: "Quietly, quietly, father is sleeping." In the evenings, before his nightly routine, he usually set time aside for family conversation and music making. Hausdorff played the piano very well and especially loved Beethoven and Mahler. He also liked to accompany singers, including his daughter, who had a very beautiful voice. He also wrote and composed songs for her; unfortunately, these compositions have not survived.

Figure 5.3: Nora Hausdorff (second from left) with friends from Greifswald
Source: Gift from Dr. Armin Danco.

Hausdorff was happy to leave the education of Nora entirely to his wife. Already three years before the birth of his daughter, he had announced the following in an aphorism:

> I will probably become a poor educator, in the sense of our present-day, education-addicted civilization, and if I love my children I should warn them about me, which I herewith want to do in advance. This much or this little I believe I can say in my own favor: that every outburst of self-will, wantonness, or freely wild naturalness would please me aesthetically too much for me to decide, with anxious calculation of the consequences, to play the authority-lusting exercise master and pedagogical sergeant. [H 1897b, 62; VII: 156]

According to all we know, Felix Hausdorff kept his promise. Charlotte's upbringing of their daughter, however, was very strict – too strict, as Nora later found in retrospect.

5.4 Mongré's Poetry Volume *Ekstasen*

In 1900, the publisher H. Seemann brought out the third book by Paul Mongré, his poetry volume *Ekstasen* (Ecstacies) [H 1900a, VIII: 39–190]. The title was certainly chosen with due care, for the term "ecstasy" was then highly fashionable and was often used to characterize modern art and literature of the *fin-de-siècle*. One usually associated it with an individual's search for dissolution of boundaries or the artist's stepping outside his social existence to surrender himself to moods and the boundless self-reflections of the ego. These often took form as quasi-religious states in the realm of the irrational and mystical. Hausdorff, for his part, always viewed these latter tendencies very critically. In aphorisms from *Sant' Ilario*, his skeptical view of corresponding manifestations in the modern art and literature of his time is evident:

> We have discovered the inner life of the soul and, as true artists, Germans and obscurantists, we rejoice that once again there is a quality that cannot be proven, but only believed. Intimate! a new gospel [···]. The linguistically gifted, the form-perfect, the creator of works – these are poor superficial [artists]; but he who is silent, or stammers, or acts deliriously, he is intimate, he leads an inner life. [···] Thought and contemplation are antiquated; we want "mood" [···] [H 1897b, 186–187; VII: 280–281].

> What is not haunted today? The symbolism. The neo-idealism. The spiritism. The Japanism. Satanism. The sexual, no longer in forty but in four hundred varieties. The mysticism, no longer of Catholic churches, but of the giant temple of elephants. [···] In this vortex we moderns live and tumble. [H 1897b, 243–244; VII: 337–338]

In fact, what the typical reader around 1900 associated with the title "Ekstasen" and what Hausdorff himself intended by this were likely very far apart. One might well say that the author undertook a risky tightrope walk with the title he chose for this volume of poetry.

He divided it into four parts: "Falterflüge" (Moth Flights), "Sonnets," "Rondels," and "Vermischte Gedichte" (Mixed Poems). One must imagine these were the results of many years of studying poetry and personal experiments writing such works, since we know that Hausdorff's interests in poetry date back to his grammar school days. From his letter of June 3, 1894, to Paul Lauterbach, we know that by that date he had already composed some "two dozen rondels" as well as some poems under the motto "Falterflüge," thus years before their publication in the *Ekstasen* (*HGW* IX, 422). Hausdorff's preoccupation with the rondels from *Pierrot Lunaire* and his own contributions to this poetry also predate the *Ecstasies* by several years. Hausdorff produced a particularly impressive rondel, entitled "15. October 1894," on the occasion of the 50th birthday of Nietzsche. He sent this to Heinrich Köselitz, but it also appeared, slightly modified, under the title "Katastrophe" in the *Ekstasen*.

The "Falterflüge" present a cycle of thirty poems, dithyrambic songs in the spirit of Nietzsche's "Dionysos-Dithyramben." Hausdorff, however, breaks away from Nietzsche's thematic attachment to Dionysos. As Friedrich Vollhardt emphasized, for Hausdorff "the modern dithyramb is a form of lyric poetry in monologue form capable of expressing ecstatic-hymnal enthusiasm and offering the speaking I space for thought experiments" (*HGW* VIII, 23). The poems of "Falterflüge" also reflect Hausdorff's engagement with Nietzsche's philosophy and, to a lesser extent, with Schopenhauer's. The numerous key words from the philosophical debate occurring in it – being, world, will, space, time, eternity, infinity, past and future life – all bear witness to this. Titles such as "Dem Geist der Schwere" ("To the Spirit of Gravity") or "Wiederkunft" ("Return") allude directly to Nietzsche's *Also sprach Zarathustra.*

The poems of this cycle are perhaps also a narrative of Hausdorff's own journey through deep skepticism and pessimism to the affirmation of life. This is also suggested by Aphorism 353 in *Sant' Ilario*, which celebrates an optimism that has "crawled out of the wildernesses and cave passages of a pessimistic youth" [H 1897b, 285; VII: 379]. At the end of a "long, protracted critique of the 'lies of life'," stands

> a new, skeptical, almost scientific love of life: the pessimistic storm has cleared the air and swept away the clouds from the nearby mountain tops! [· · ·] Yes, it is something thoroughly wonderful, this newly created, no, only newly discovered *empirical* world, which is now able to take into itself everything human without exception, also the purest and most spiritual – while that other "higher" world, in which one once thought to administer a part of human affairs, now surges as a chaotic abyss at our feet. Will we continue to sacrifice our best to this transcendent fiend, called "Something = X?" by throwing into this gurgling gap our desires and values, our golden jewels and ornaments, in the fantastic hope that it will keep them for us and one day give them back in greater plenty? No – we, who have hung on the edge of the abyss and have rescued ourselves, shuddering, to solid ground, know how to treasure the solid, accessible ground of reality. [H 1897b, 285–286; VII: 379–380]

Felix Hausdorff dealt extensively with the poetic form of the sonnet. He published 86 of them in all, 70 in the section "Sonnets" of *Ecstasies* and 16 in the final chapter of *Sant' Ilario.* The sonnet form can be traced back to thirteenth-century Italian poets, but it became especially famous through Francesco Petrarch's *Canzioniere* (first edition 1470). Celebrated writers of sonnets in the German-speaking world included Johann Wolfgang Goethe, Heinrich Heine, Friedrich Hebbel, Hugo von Hofmannsthal and Rainer Maria Rilke. The sonnet is an elaborate poem with a very strict form consisting of four stanzas. Two four-line stanzas (quartets) are followed by two with three lines (tercets). The individual lines of verse are arranged metrically with either eleven syllables, where the stress falls on the penultimate syllable, or ten syllables, where the final one is stressed. The rhyme scheme is also very strict with two basic options: either ABBA – BAAB – CDC – DCD or ABBA – ABBA – CDC –DCD. For the quartets, especially the tercets, poets sometimes deviated from this strict rhyme scheme; Hausdorff's sonnets show a great variability in this respect. From these few remarks, one can see that it takes considerable creativity to mold the content of a poetic idea into this form. Hausdorff's Aphorism 331 from *Sant' Ilario* makes evident why he devoted so much attention to this form of poetry, in particular.

> Strict poetic forms, such as the sonnet, are certainly detrimental to factual subtlety; in the majority of cases, rhyme and rhythm force one to say, either somewhat more or somewhat less, but much differently than imagined in the original conception. [· · ·] But precisely in this danger of the strict and absolute forms lies their disciplinary value. Above all, it must be prevented that the poet finds it easy to write verse, that it can be written "without any trouble;" this is precisely what a demanding formalism serves, which, as soon as it is learned and no longer fits tightly enough, should be exchanged for a more stringent one. In this manner of creating, much will fall by the wayside as a provisional effort and, as said, will be less successful than would be the case were the factual elaboration alone to be considered; but precisely this selection is necessary. Up to now, it was carried out by the reader, the editor, best of all by the poet himself; now it will be carried out already in the poetic method, and what leaves the workshop will offer quite different guarantees for authenticity and durability than those that were once before demanded. [H 1897b, 230; VII: 324].

Another strictly regulated form of poetry is the rondel. This consists of three verses: two quartets, followed by a five-line verse; lines 1 and 2 of the first quartet are always repeated as lines 7 and 8 of the second. Furthermore, the first line of the poem is – sometimes slightly modified – repeated in the last line. The rhymed form of the rondel poses the additional difficulty that it uses only two rhyming endings. Hausdorff published a total of 42 rhymed rondels, 32 appear in the "Rondels" section of the *Ecstasies* and 10 in the final chapter of *Sant' Ilario*. His creative exploration of the rondel form was undoubtedly stimulated by Otto Erich Hartleben's German adaptation of Albert Giraud's *Pierrot Lunaire*, mentioned above in 5.1. The composer Joseph Marx wrote songs based on Hartleben's version of *Pierrot Lunaire*, but he also adapted at least three poems by Hausdorff to music: the sonnet "Dein Blick" and the rondels "Pierrot Dandy" and "Dem Genius des Augenblicks."

As an example, here the already mentioned rondel "Catastrophe," in which Hausdorff deals with Nietzsche's terrible fate. At the same time, he hints at the gap between the late and the middle Nietzsche as well as the controversies surrounding his legacy and the Nietzsche Archive.

Katastrophe

Held, dein letzter Wille ist vollstreckt:
Rissest selber ihn in Fetzen!
Hat ein Irrlicht dich zu Schätzen
In des Grauens tiefsten Grund geneckt?

Nacht, daraus kein Sonnenaufgang weckt,
Fing dich ein in schwarzen Netzen.
Held, dein letzter Wille ist vollstreckt:
Rissest selber ihn in Fetzen!

Jäger einst, dem sie die Hand geleckt,
Wild nun, das die Hunde hetzen!
Um dein Ende schwebt Entsetzen;
Die dir folgen, hast du heimgeschreckt.
Held, dein letzter Wille ist vollstreckt.
[H 1900a, 158; VIII: 147]

Catastrophe

Hero, thy last will is done:
Tear it to shreds yourself!
Has a madman teased you
To treasure in the deepest pit of horror?

Night, from which no sunrise wakes
Caught you in its black nets.
Hero, thy last will is done:
Tear it to shreds yourself!

Hunter, whose hand they once licked,
Now Hunted, whom the hounds chase!
Around your End hovers horror;
Your followers, you have scared home.
Hero, thy last will is done.

The final section of the *Ecstasies*, "Mixed Poems," contains 25 poems; some are composed in free verse, others are rhymed. A number of these poems relate to relationships with women, mostly expressing disappointment, resignation, and distance; they suggest that his early *naufragium in eroticis* confessed to his friend Paul Lauterbach, had recurring effects. The last poem of this section, "Die drei Altäre," probably has more to do with Hausdorff's development as a philosopher and artist than any of the others. There he erects three altars, giving thanks to "Dir heil'ges Mißlingen," ("To You, Holy Failure"), and pays homage to "holy falsehoods" and "my holy pride" in "Dir, heil'ge Lüge," and "Dir, mein heil'ger Stolz."

The reception of *Ecstasies* was very quiet; in fact, despite a thorough search, only three reviews could be found. One reason for this dearth of interest was surely the enormous competition in the poetry market during the second half of the nineteenth century. In addition, Hausdorff's poetry aimed to reach a special circle of readers, just as he had before with his *Sant' Ilario*, which was written only for "that species [· · ·] of free, hedonistic, good-humored people, who have outgrown all the solemn narrow-mindedness and blustering self-righteousness of the lower levels of culture" [H 1897c, 361; VII: 477]; that restriction certainly applied all the more to his *Ekstasen*. This was poetry for an elite, a sense captured by Max Messer in his succinct review from 1901:

> The author, who has the reputation of being a follower of Nietzsche, surprises us with these strange, deep and, from a purely artistic point of view, extraordinary poems. Here intone songs of a truly lonely spiritual man, who speaks not out of arrogance or impotence but by virtue of an all too refined organization that stands outside the life of the "all too many." The poet speaks to himself. He renounces any echo from the outset. It suffices that his own soul develops through new creations, always richer, finer, rarer. Here and there is one who will be seized by these sounds that proclaim the same suffering. The banner of the loneliest marches ahead of these verses. But it makes these songs sweet for us, when Zarathustra's beloved voice resounds from some words. No one at all capable of enjoying this poetry will doubt that this is no mere copyist, but a blood relative speaking, perhaps a grandson or nephew, if not even a son.

Leonhard Adelt, in his review in the Viennese journal *Die Zeit*, places Paul Mongré in the ranks of "l'art pour l'art devotees and neo-Romanticists." He considered the title *Ecstasies* as only partially appropriate, applying to the majority of the poems in "Falterflüge," several of the "Vermischten Gedichte," and also some rondels. "Happily," he added, "it does not cover the most distinguished section of the book, the sonnets, at all." Adelt's criticism was directed mainly against the "Falterflüge," calling these "for the general public [· · ·] indigestible pastry." He then concluded by saying:

> On the whole, however, the Rondels are more detached, more controlled than the "Falterflüge" and the related part of the "Mixed Poems." At times taking a tone of bloody humor and the grotesque, at others glowing with a sensuality reminiscent of Dehmel, some even evince serious beauty of thought or symbolic-lyricism, leading over to the most valuable gift of this volume: the sonnets.

Finally, as part of a collective review of poetry volumes under the title *Neue Gedichtbücher*, Wilhelm von Scholz devoted a few lines to *Ekstasen* in *Das literarische Echo.* He criticizes the free rhythms in this work, claiming they seduced the beginner "irretrievably to breadth and garrulity." This is all the more to be regretted, he says, because Mongré "deals courageously with great questions," even though these poems were "not poetry in the proper sense." Nevertheless,

> his sonnets, among which some beautiful ones can be found, show how well he handles a taut form. On the whole, I believe that one should place hopes in his development; this book has sensuality, life, color.

5.5 The Veil of the Maya

Paul Mongré's *Der Schleier der Maja* (The Veil of the Maya), published in 1902 in *Neue Deutsche Rundschau* [H 1902a, VIII: 453–466], has a very unique character that in no way resembles a classical essay with strictly constrained subject matter. Here he instead presents a collection of thirteen short stories, parables, and allegories that play through various philosophical themes, such as eternal recurrence, chaos and cosmos, plurality of possible worlds, time and space, individual vs. society, and his critique of word realism and word fetishism. According to Werner Stegmaier, these stories are, in the best sense of the word, "thought-poetry that shows Felix Hausdorff's literary possibilities" (*HGW* VII, 62).

The title harkens back to ancient Indian philosophy. Hausdorff's preoccupation with Indian philosophy presumably began with the study of Schopenhauer, who saw in Indian philosophy as expressed by Adi Shankara (late 7th/early 8th century CE) a confirmation of his own monistic idealism. Hausdorff employed the image of the "veil of Maya" already in 1898 in *Massenglück und Einzelglück*, where he wrote that Schopenhauer's "metaphysical thinking" was dominated by a "highly picturesque, but also highly unscientific fundamental viewpoint":

> To him the forms of cognition are deceptive juggling arts, which fragment metaphysical unity and changelessness into an empirical multitude of changes. [···] Thus, consciousness would be a kind of alcoholic intoxication, a double and multiple vision of the simple, a device producing illusions, like faceted glass and kaleidoscopes, and it would remain the philosopher's prerogative to lift this "veil of Maya" and to reveal true being, the All-One behind the magic trick of the restless many. [H 1898b, 68; VIII: 279–280]

The title "The Veil of Maya," when seen against this background, was probably meant allegorically, as an invitation to the reader to seek and find what Hausdorff really wants to say through his stories.

Hausdorff had earlier addressed the relationship between the individual and the masses in several aphorisms from *Sant' Ilario* and also in *Massenglück und Einzelglück.* For him, as for Nietzsche, the individual is not a mere figure in a historical process that forces him to subordinate his individuality to a "higher destiny." On the contrary, widespread freedom of the individual serves as the guarantor of a prosperous society. To see how Hausdorff treats such a serious subject in *Der Schleier der Maja*, here is one of its thirteen stories:

The Balance of the Whole

> Two ships ran side by side on the sea.
>
> On one of them there were jolly thoughtless people, who had no cares about the ship or about one another. They sat, stood, walked, danced, and did whatever came to mind; one squatted in the sun, the other in the shade under a tent roof, some climbed up into the rigging, and many lay seasick down below in the cabins. And since everything whirled around in disorder, from the right to the left and from above and below, these several displacements balanced out on the average and the ship glided through the waves, gently and inconspicuously, without particularly rocking.
>
> Over on the other ship, however, things were serious and purposeful, guided by people who "kept an eye on the well-being of the whole." They had set up very fine precious apparatuses, above all a level, which indicated whether the quiet fluctuations of the ship were stable rather than tilting over to the right or to the left. Whenever the mercury in the level fluctuated, the captain commanded ten or twenty or a hundred men to move from one side to the other to restore the ship's balance. Thus those aboard ship had to watch diligently and had no time for sitting, dancing, or walking. And then came a somewhat stronger wave that caused the mercury to sway alarmingly. The captain cried out to all these purposeful people, who were keeping the balance of the whole, commanding them to rush toward the windward side, so that the ship would not tip over leeward. But then the ship tilted to windward, and those wise, serious people who were concerned about the welfare of the the whole drowned. [H 1902a, 994; VIII: 464]

5.6 Max Klinger's Beethoven

In the section "Thinking, Speaking, Forming" in his *Sant' Ilario*, Hausdorff dealt in detail with questions of art [H 1897b, 181-246; VII: 275–340]. A true work of art – as he puts it so expressively in Aphorism 328 – must polarize:

> Let us look at the truly soul-stirring artists and their works: all of them kindle furiously flaming quarrels over pros and cons, not only in the individual's sensibilities when they encounter them on "some chance occasion," but even more so in the sensibilities of many contemporary as well as posterior observers [···]. The boundlessly enchanting effect on me is met by the boundlessly repulsive one on you, while only the mediocre remain unnoticed and undisputed. [H 1897b, 225–226; VII: 319–320].

In the field of fine art, Hausdorff had experienced just such a dispute in 1894 over Max Klinger's polychrome sculpture *Neue Salome*, which he discussed with his friend Paul Lauterbach. Then, in the spring of 1902, he himself intervened in the renowned *Zeitschrift für bildende Kunst* concerning the controversy over an even more famous work of art with his essay *Max Klinger's Beethoven* [H 1902c, VIII: 489–500]. First, however, a few remarks about the prehistory are required.

On April 3, 1897, the "Vereinigung bildender Künstler Österreichs - Secession" – known as the "Vienna Secession" for short – was constituted. This organization, supported by young artists around Gustav Klimt who were dissatisfied with the older established art business, split off from the "Gesellschaft bildender Künstler Österreichs - Künstlerhaus." These Secession artists developed a particular variant of Art Nouveau called "Secession Style" or "Viennese Art Nouveau." They financed their own exhibition building, which was completed in 1898 and served as their answer to the Künstlerhaus, and the Secession's exhibition activities played a significant role in Vienna's artistic life from the very beginning.

An outstanding event in this context was the 14th exhibition of the Secession, which opened on April 15, 1902. It was dedicated to the memory of Ludwig van Beethoven and was staged shortly after the 75th anniversary of his death. The centerpiece of the exhibition, soon the subject of heated discussions throughout Viennese artistic circles and beyond, was Max Klinger's sculpture *Beethoven*. The art critic and journalist, Ludwig Hevesi, an enthusiastic supporter of the Secession, greeted the Leipzig sculptor's conception for his *Beethoven* by writing: "Klinger is in Vienna and Vienna is full of him" [Hevesi 1906, 382].

Klinger stood in friendly contact with members of the Vienna Secession from the beginning and he soon became a corresponding member as well. He and the artists of the Secession forged a united effort to bring painting, sculpture and architecture together; they aimed to create a unified spatial effect that would lead to a new type of "spatial art." The 14th exhibition represented such a "total work of art," combining all branches of the fine arts with architecture. Klinger's *Beethoven* stood in a specially constructed hall, decorated in the left aisle by Klimt's now famous Beethoven frieze. In Hausdorff's article for the *Zeitschrift für bildende Kunst* one finds a photograph of that room (*HGW* VIII, 500).

Klinger's *Beethoven*, however, was not a work commissioned by the Secession. In fact, he already had the idea for this work in Paris during the mid-1880s, thus long before the founding of the Vienna Secession. By the end of 1886, after making

Figure 5.4: Max Klinger: Beethoven
The sculpture stands today in the Museum der bildenden Künste zu Leipzig.

sketches in plasticine, he completed a plaster model that he painted with vivid colors. This composition already came quite close to what one sees in the final execution of the sculpture. Klinger regarded the color scheme of the sculpture to be a constituent element in his concept of spatial art. In the finished sculpture, though, he realized this not by painting, but by using colored materials: various kinds of marble, Tyrolean onyx, bronze, gold cut, ivory, amber, millefiori glass and opal.

Klinger's *Beethoven* sits with a naked, bent-over upper body on a massive bronze throne chair, his right hand clenched into a fist, his gaze directed outward into an indefinite distance. The face was modeled on the mask taken by the sculptor Franz Klein in 1812. The crossed legs are covered by a flowing robe made of yellowish onyx. At his feet sits a life-size eagle of veined black marble, about to soar, its amber eyes fixed on the sitter. At the upper edge of the backrest are five ivory angel heads surrounding Beethoven, whose throne rests on a pedestal of dark uncut marble with a kind of ledge in front on which the eagle clings. The back and outer sides of the throne are decorated with reliefs. The right side portrays the fall of man: Adam and Eve under the tree of knowledge. The left side shows guilt-ridden figures from the ancient world: Tantalus and a woman suffer the torments of Hades. Various figures are depicted on the back side: below the naked Aphrodite, with raised arms as she emerges on a shell from the sea, accompanied on her left by a Nereid; above the crucifixion of Christ on Golgotha. In the center, John of the Apocalypse, pointing toward the sea with a threatening gesture at the ascending Aphrodite. On the upper edge of the backrest we also see some small naked figures, who are crawling, sitting, or lying about.

Klinger's *Beethoven* was – quite in the sense of Hausdorff's aphorism quoted above – a "soul-stirring work of art," inducing euphoric enthusiasm among some and severe criticism, even mockery, among others, and not only among contemporaries but also later observers and critics. The above mentioned Ludwig Hevesi, wrote in *Max Klinger in Vienna*:

> Most common adjectives fail, if one wants to attach them to [Klinger's] Beethoven [$\cdots$ which is] a thing for itself, as it is made once, at the end of an earlier time and the beginning of a new one. Maybe sculptors will take this as marking a new era. Who can know that today? [Hevesi 1906, 384].

A few days later, in his article *Max Klinger's Beethoven*, he tried to interpret the meaning of the different materials used:

> This is the work of a sculptor who is a painter, just as was Phidias. A painter-sculptor who is completely saturated in the highest music. Beethoven is Max Klinger's Ninth Symphony. [Hevesi 1906, 388].

Numerous other articles celebrated Klinger as a genius of his time.

Criticism from the ranks of experts focused on two aspects: the lack of inner cohesion in the composition, but even more, the fact that this image of Beethoven in no way corresponded to what they imagined as an appropriate monument to the composer. Moreover, as with Klinger's Salome, the coloration of his sculpture became the subject of lively discussion and criticism. The archaeologist Heinrich Bulle, a recognized specialist in Greek sculpture, claimed, for example, that Klinger had not been able to control the effect of the colored materials and had become, in effect, a slave to them. The well-known art historian Julius Meier-Graefe regarded Klinger's work as closer to craft art than to to artistic sculpture. For the Leipzig art historian August Schmarsow, the Vienna exhibition and especially Klinger's *Beethoven* was one big disappointment. His opinion of the work bordered on caricature.

Although it had nothing to do with the artistic value of Klinger's work itself, some criticism came from influential anti-Semitic circles in Vienna. Karl Kraus reported on this in his magazine *Die Fackel* on June 16, 1902: "Since, moreover, the 'Deutsche Volksblatt' soon reported that Max Klinger was of Jewish descent, the general indignation of anti-Semitic Vienna broke out over the Secession. [···] Liberal Vienna did not behave much better." Even the prudish moralizers of the K.u.K. Monarchy spoke out in large numbers. Hermann Bahr published an entire collection of attacks on Klimt and the Secession taken from articles in newspapers and magazines. The following quotation appears in one of these: "The naked has never celebrated such orgies at any Viennese exhibition. [···] Apparently, the gentlemen of the Secession believed they needed to surpass the half-nakedness of Klinger's Beethoven by modernizing nudity into the pathological-allegorical."

A particularly infamous critique appeared in the *Deutsches Volksblatt*; it insinuated that Klinger's Beethoven meant to represent him as an "Urning," i.e. a homosexual man. Klinger's work was also mocked in a series of caricatures, one of which shows the enthroned Beethoven with his face distorted in pain, his mouth uttering a scream. He stretches his naked right foot toward the eagle, which pecks his big toe with its pointed beak; its caption reads: "The eagle as corn-eye surgeon."

Among contemporary accounts of Klinger's *Beethoven*, Mongré's essay from the *Zeitschrift für bildende Kunst* occupies a special place. Having maintained friendly relations with the artist, Hausdorff was very familiar with his ideas and intentions, including those that led to the creation of his *Beethoven*. He knew, for example, its colorfully designed predecessor, which had never been shown in public. Only in a few places does he address the criticisms heaped on Klinger's sculpture, and even then he writes in a very subtle and indirect way without naming anyone. Instead, he mainly celebrates the work enthusiastically, and his opening remarks set the tone for the entire essay: "The Easter days of this year saw an artistic celebration of secular rarity and significance. Klinger's Beethoven revealed itself to the gaze of the public, almost exactly around the 75th anniversary of Beethoven's death" [H 1902c, 183; VIII: 489].

Hausdorff first briefly reports on the six-day public exhibition of the work in Klinger's Leipzig studio and how it was disassembled subsequently and then transported to Vienna, where this "reassembled giant work forms the center and main event of the Vienna Secession." Hausdorff even expressed the hope that the city of Leipzig, in competition with Vienna, would succeed in purchasing the work once the exhibition had concluded and, indeed, this actually transpired. In the summer of 1902, a committee of Leipzig art lovers raised the enormous sum of 250,000 gold marks for Klinger's work, which was then donated to the Leipzig Museum of Fine Arts in 1912. Following a period with much confusion, it can again be seen there today in a specially designed room in the new museum building, together with two other polychrome sculptures by Klinger, *Neue Salome* and *Kassandra*.

Before describing the genesis of Klinger's Beethoven and its place within his artistic oeuvre, Hausdorff offers a striking general appreciation of the work:

> For the present generation, which is otherwise so liberal in speaking of "milestones" and new epochs, well, here something truly, not merely proverbially monumental has been created: a new bodily wonder of the world, which to behold gives us an indelible feeling of happiness and gratitude. [H 1902c, 183; VIII: 489]

For Hausdorff, Klinger's *Beethoven* was one of those works of art that break the bonds of their time,

> which go beyond the framework of formal and historical and artistic art, which speak to us of great things of humanity and are "eternal" in that sense in which humanity may be allowed to utter this word; works one cannot overlook if one hopes to have lived and not merely vegetated as a dreary guest on the dark earth; works in which the soul loses and renews itself, from which a power of transformation flames forth like a purifying fire; works in which not just the individual but humankind manifests itself, and with which the telluric brotherhood virtually marks itself out against other planetary peoples by sending out its slogan. [H 1902c, 183; VIII: 489–490]

At the center of Hausdorff's description and interpretation stands, of course, the figure of Beethoven. Here he takes his place as the new ruler of Olympus: "Zeus-Beethoven enthroned with bent forward naked upper body; the right leg crossed over the left, on the raised thigh rest the clenched fists, the right in front of the left." [H 1902c, 185; VIII: 491]. A passage from Homer, suggesting a reference to the statue of Zeus by Phidias, leads to a contrasting view of Klinger's intentions; Hausdorff points to the latent tension and energy in his composition, as the eagle awaits the command from its god to fly forth. A brief side remark contains a subtle Nietzschian criticism of how "educated" Germans fail to understand the dynamism underlying classical Greek art.

Hausdorff then turns to Beethoven's countenance in Klinger's original model, which he compares with previous representations but also with the finished work:

> What an ascent of spiritual liberation and technical mastery of expression from this model of Beethoven's head to the one in marble! Here,

> too, gloom and the defiance of fate still glows: but no longer with the dull, helpless personal bias, the predatory gaze behind bars, but rather as universal depth of suffering, a crushing and uplifting realization of necessity and the tragic innermost foundations of the world. Beethoven no longer suffers personally, he suffers as humanity; he has extended his person to them; he has created the Ninth Symphony. Zeus is, after all, not omnipotent – the Greeks were too healthy for such an incomprehensible brain-twisting conceptualization: Moira sits enthroned above the gods. The Olympian stares into the unchangeable ... [H 1902c, 186; VIII: 493].

Beyond the main figure, Hausdorff also analyzed the reliefs Klinger used to decorate the throne, comparing these with his monumental painting "Christ in Olympus": "On the mighty back surface of the throne, however, Klinger, as in 'Christus in Olympus,' once again lets the world-historical contrasts collide: Nazarene flight from the world and pagan joy of the senses" [H 1902c, 186; VIII: 494]. He then offers this résumé:

> A world history with its opposite driving forces is banished into these bronze reliefs: lost paradise and unfulfilled longing for happiness, self-indulgence and self-sacrifice, sensualism and spiritualism, Cyprus and Golgotha – out of this interplay of intersecting contrasts, from its warp and weft, the tapestry of life weaves itself. Klinger's spatial art reaches here for the highest: it transforms Beethoven's musical hieroglyphics into visible symbolism, it speaks of the fate of mankind. [H 1902c, 187; VIII: 495]

In the final section of his essay, Hausdorff raises a question about Klinger's underlying conception in associating Beethoven with Zeus – "may one place him on Olympian clouds without evoking trepidation at his likeness to God?" To this, he answers that what matters is not whether such an association may be inferred from Beethoven's life and work or from our knowledge of the Greeks' conception of the figure of Zeus. For him, the legitimacy of Klinger's conception depends solely on the profound effect his work produces on the viewer:

> Might one think of Beethoven differently, then let it come down to a test of strength comparing your inner picture with that of Klinger's work – which of the two will annihilate the other? Klinger's is the psychically stronger one; like those Indian magicians, he can force you to see what he wants, not what you want to see or what realistic reality dictates. [H 1902c, 188–189; VIII: 497].

The essay concludes by praising Klinger as a bold creator and designer. Hausdorff emphasizes the contrast between Klinger's art and conventional contemporary tastes by citing the same fashionable word in the art world that he had earlier ridiculed in Aphorism 282 from *Sant' Ilario* (*HGW* VII, 280f.):

> A favorite word today is "mood" (*Stimmung*), an avowal of passivity that moves around with soft fish fins, kneading amorphous symbols of its own inadequacy in hopeless songs that fail to mark a stone. [···] Yet "shaping" (*Gestaltung*) is a superior [word], inwardly Seeing with rough grasping hands molded in hard material: man's will in sufficient amount (*quantum satis*) to wrest a cosmos from chaos. Klinger's Beethoven is such a shaped thing, brought out of stone and metal like the ordered world out of the "boundless" realm of Anaximander, like the rule of Zeus out of the battles with the Titans, like Beethoven's sharply composed tone characters out of labyrinthine emotional turmoil, like every animated work of art arising out of the undivided whirling, senselessly roaring elements of reality. [H 1902c, 189; VIII: 498]

Here and elsewhere, Hausdorff's essay verges on the pathetic, leaving no doubt that he was deeply impressed by and enthusiastic about Klinger's art. Clearly, he tried to convey this to the reader by means of powerful language. He might well have shown more restraint in his formulations under other circumstances, but given the quite crude criticism of Klinger's work he surely wanted to make an emphatic case in support of his friend. In fact, they shared a wide range of common interests. Klinger, like Hausdorff, was a great music lover and a talented pianist. He was on friendly terms with local musicians and often organized house concerts in his studio. He was also exceptionally well read in literature and history.

The earliest evidence of Hausdorff's friendship with Klinger comes from a letter Klinger wrote to his partner Elsa Asenijeff, probably in 1901. In it, he informed her that he had invited the Hausdorffs and another friend to hear a local clarinetist perform a piece by Brahms in his studio. A later postcard from Klinger to Hausdorff, probably from 1908, depicts one of the four genii surrounding and looking up to the figure of Brahms, as seen on Klinger's monument in Hamburg. The short text, addressed to Hausdorff and his wife, reads:

> Dearest! your thought is much too beautiful to come true! At least this year! I'm thickly engaged with what stands here [the Brahms monument] and Frau Asenijeff with moving (1.Oct.). Oh God would that be nice. Do not forget Ischia.
> Best regards from both of us your M.Klinger

What Hausdorff proposed to Klinger and his companion is not known, perhaps a joint vacation on the volcanic island of Ischia in the Bay of Naples – or perhaps they had already been together there some time earlier. We already quoted extensively from Ernst Eggebrecht's account of the festive evening at Klinger's with Hausdorff and Max Reger, a gathering that carried on into the next morning.

We know nothing about the circumstances that prompted Hausdorff to publish his essay in *Zeitschrift für bildende Kunst*, though it seems likely that his good friend Gustav Kirstein, the managing director of E.A. Seemann in Leipzig, approached him to write about Klinger's *Beethoven.* Whether or not this was the case, though, we can be sure that Kirstein liked the essay very much, because he had it published separately in the same year. In 1904, when he founded the new periodical *Meister der Farbe*, Kirstein began by publishing an illustration of Klinger's Beethoven together with a short accompanying text by Hausdorff [H 1904e, VIII: 685–686].

5.7 Language Criticism

Fritz Mauthner's three-volume *Beiträge zu einer Kritik der Sprache* (*Contributions to Language Criticism*) was published in 1901 and 1902. Like Schopenhauer and Nietzsche, Mauthner was not an academic philosopher. In Prague, he had studied law, but without obtaining a degree. There he also attended lectures on philosophy, archeology, music history, medicine, and theology. After working briefly as a feature writer for the *Prager Tagesbote*, he moved to Berlin in 1876 and became a permanent staff member and later feature editor with the *Berliner Tageblatt.* He also contributed to other newspapers and magazines, whether as a writer, editor, or publisher, eventually becoming well known as a literary and theatrical critic. Beginning in 1905, he worked as a freelance writer, first in Freiburg and then from 1909 onward in Meersburg on Lake Constance. Throughout these years, Mauthner produced an extremely extensive and varied literary and philosophical oeuvre. He achieved his breakthrough with the public early on with parodies of well-known authors, published collectively under the title *Nach berühmten Mustern* (*After Famous Patterns*) (30 editions between 1879 and 1902).

Following his *Contributions to Language Criticism*, he also published a two-volume *Wörterbuch der Philosophie - Neue Beiträge zu einer Kritik der Sprache* (1910, 1911). Mauthner's importance as a philosopher emerged from these critical language studies. Indeed, his critique of language had a lasting influence on a number of important twentieth-century thinkers, including Georg Simmel, Ernst Cassirer, Ludwig Wittgenstein and Walter Benjamin. His ideas also made themselves felt among literary figures, such as Christian Morgenstern, August Stramm, Gustav Sack and Hugo Ball.

Already as a student, Felix Hausdorff showed an interest in the philosophy of language, especially the relationship between thought and language. Early hints of his own critical reflections on language appear in the chapter "Thinking, Speaking, Forming" in *Sant' Ilario*, for example in Aphorism 283, which addresses incongruences between language and reality in art. In Aphorism 327, he warns against the "injustice" of language, which often tends to overhear and exaggerate, pretending through "cutting ability" (*Schneidigkeit*) to command authority:

> Language, the ancient tool of injustice, has heaped around us modern men a ballast of arbitrary and presumptuous *façons de parler*, from raging partisanship and saucy irony to categorical loudmouths and impertinent superlatives: it teems in all languages, from decisive and un-

> conditional to *absolument* and *senza dubbio* – nothing goes more against the grain of modern linguistic tone than a little calmness, modesty, skepticism. [H 1897b, 224; VII: 318]

In view of such thoughts, it comes as no surprise that Hausdorff took a lively interest in Mauthner's *Sprachkritik.* In a letter from the resort of Cortina d'Ampezzo, dated September 12, 1901, he sought to establish personal contact with him.

> Dear Sir,
>
> I have taken your linguistic criticism to these summer climes, and already along the way, still a few steps before page 657 [volume 1 has 657 pages], I feel an unscientific need to say thank you, as one might say to someone who has kept a person good company a few hours before parting.

He then predicted that Mauthner could "not count on a many-voiced echo," since all authors "have a hell of a time when they work in a 'boundary area" (*HGW* IX, 453). Hausdorff spoke here from his own experience as the author of *Chaos in kosmischer Auslese*, which caused little "reverberation" after its publication. He then apologized for not yet having many cogent things to say:

> If I must thus refrain from saying anything in depth to you today, I will at least justify the existence of this letter with the general confession that your basic tendency brilliantly suits me ("in den Kram passt"). [···] You have applied what might be called "iterative" skepticism, criticism of oneself or the critique of critique, especially in the highly refined exploitation of the idea that the critique of language must, after all, be undertaken by means of language [···]. (*HGW* IX, 453–454)

After getting hold of the third volume of Mauthner's *Beiträge zu einer Kritik der Sprache*, Hausdorff wrote to him again on December 16, 1902:

> I have not yet read [Volume 3 of the *Critique of Language*], but I will probably undertake a review of the entire work for the Neue Deutsche Rundschau – a promise that in any other case would be unforgivable frivolity. But be altogether assured that, even without this inducement, I would have read it with the seriousness and conscientiousness that your work deserves. (*HGW* IX, 462)

Whether Hausdorff offered to review Mauthner's work for *Neue Deutsche Rundschau* or instead had been approached by the editor, Oskar Bie, remains unclear. In any case, it took him several months to complete this task, after which he wrote to Mauthner on August 25, 1903:

> Dear Mr. Mauthner,
>
> My wife and I are celebrating with a bottle of sparkling wine that I finally, finally have finished my essay on your linguistic criticism. It is

> not free from contradiction and, as time went on, it became somewhat more sober than it should be. I wish nothing more than that my review may woo even more friends to your valiant work!
> With gratitude your Felix Hausdorff. (*HGW* IX, 465–466)

The essay appeared in the December 1903 issue and was by far the most extensive of Mongré's contributions to *Neue Deutsche Rundschau.*

The first section, a kind of prologue to the essay proper, begins with a narrative that reads like a satire on the beginning of John's Gospel and the Book of Genesis. Instead of "In the beginning was the Word," the text reads: "And in the beginning was dreamless depth." Out of this turmoil new things emerge and take form, finally producing life. Hausdorff's description of these transformations, covering only a single page, expresses magical poetic thoughts for which a divine creator is not needed. An equally metaphorical narrative then deals with the emergence of language, where a horde of "human animals" produce phonetic signs for things and processes. Eventually, they invent visible signs to represent these phonetic symbols, the first stage in the development of written language. But then,

> [· · ·] a strange thing happens; the sign detaches itself from the signified thing and takes on its own life.
>
> Mankind made God in the image of himself, but God sets up altars and calls on mankind to kneel before them and to sacrifice. [H 1903b, 1234; VIII: 552]

And man created word ghosts, word fetishes, even dangerous and murderous words:

> From words sprout poison and fire and slow agony. [· · ·] Words without reality shall heal the sick, speak to spirits, and cause springs to gush. Words command how the world is and how it must be, and in the halls of learning people argue about the right way to connect words with words. [H 1903b, 1235; VIII: 553]

Finally, after more than two millenia, a young heroic generation appears on the scene, "that throws all idolatry, ghostly delusion and word superstition behind." The earth is no longer a disk, the orbits of heavenly bodies are no longer circles, many "dare now to see what is, and renounce the fear of words." Yet "flying word ghosts and clouds throw shadows over sunny land, confusing us [with] old word delusions," which need finally to be shaken off: "The day of realities has come!" [H 1903b, 1235; VIII: 553].

In the section that follows, Hausdorff extols the truly philosophical spirit of Mauthner's work:

> A chilling shower plunges forth in this critique of language like that in the *Critique of Pure Reason*, though this mixes with a gulf stream bringing tenderness and distant floral fragrance. How quiet, kind, resigned

> is the voice of this critic of language, despite the great seduction of speaking harshly from a higher vantage point about what one sees from above. [· · ·] A philosopher must have a free view, a wide panorama: he should go beyond what pertains to the here and now or what circulates as the true, the good and the beautiful within a few hundred square miles of a country or a few decades in the lives of its people. But Mauthner's mission is to lift barriers, to open the barriers of space and time! [H 1903b, 1236; VIII: 554–555]

Mauthner had, indeed, opened the barriers of space by directing attention to numerous peoples and their languages; he also broke with the conventional limitations of time by taking prehistorical developments into account and thus vast periods of language formation and development.

Hausdorff's admiration for Mauthner's "iterative" skepticism, as expressed in his first letter to him, takes a somewhat different form in his essay, where he writes:

> There is still much in this book that is philosophical, though not in the narrower sense of the term. Mauthner found himself in a position similar to Kant's and all the sharper critics of knowledge since: one has to think about thinking, to mark out the limits of reason with the help of reason, to undertake linguistic criticism by means of language. [H 1903b, 1237; VIII: 555]

Hausdorff employs witty remarks and poetic imagery to explain this paradox and its consequences, both for Mauthner as author as well as for his readers.

As a reviewer, Hausdorff recognized that his first task was to describe and critically analyze various aspects of Mauthner's work. Before doing so, however, he wanted to confess his great sympathy for the author's endeavor: "Mauthner's book gave me extraordinary pleasure, and [· · ·] I would like to entice a good number of readers to grant themselves the same pleasure" ([H 1903b, 1238; VIII: 556]). Hausdorff was well aware that Mauthner's critique of language faced two main criticisms, which were repeatedly brought forth even by friendly reviewers. First, he was charged with being a dilettante in the many fields of knowledge, human history and culture that he touched upon in his work. Second, his studies were regarded as unsystematic, unstructured, and in some parts redundant but also difficult to read. Hausdorff attempted to address these criticisms and finally concluded:

> Mauthner's book has the faults of its merits; it is a life's work that grew out of research notes written over decades, nourished by a thousand associations, fed by daily sources of experience. [· · ·] Though it may take us on many detours, we see by the end that we have been in captivating company. [H 1903b, 1239; VIII: 557–558]

In Hausdorff's view, Mauthner, as a literary critic, journalist, and writer was much better suited than an academic philosopher to expose "the daily misuse of language, the prostitution of language," and to expose "language as an expression and means of general disorientation, as a kind of hypnosis promoting crime and artificial madness" [H 1903b, 1239–1240; VIII: 558–559]. These conditions reflect the intellectual situation that led to the "language crisis" at the turn of the century.

Hausdorff turns next to Mauthner's thesis asserting that language prevents people from understanding each other. Mauthner's position was based on two further claims:

> Language, however, never exists for itself alone, but always only between individuals.
> There is only individual thinking and only individual languages. Everything else is abstraction.

These circumstances, according to Mauthner, lie at the root of constant misunderstandings:

> We do not know, even in the case of the simplest concepts, whether we share the same idea when we use the same word. [· · ·] When I say tree, I imagine – I personally – something like a twenty-year-old linden tree, whereas the listener perhaps thinks of a fir or an oak tree several hundred years old. And these are the simplest notions. Words referring to inner mental processes depend, of course, on many values or concepts and are therefore never the same for any two people. [· · ·] (F. Mauthner, *Beiträge zu einer Kritik der Sprache*, Vol. I, p. 54).

Hausdorff managed to bring Mauthner's ideas to the reader in a concise and masterful manner, as illustrated by this striking passage:

> The real life of language would have to be studied microscopically, based only on temporary individual languages; even dialects are abstractions, whereas standardized or written languages represent the very highest abstraction. [· · ·] One speaks the same words and means different things, one speaks opposite words and means the same things; a shared situation brings understanding without words, or even in opposition to words, but words apart from a shared situation do not bring understanding. The culmination of misunderstanding through language is self-misunderstanding, which can happen to even the keenest thinkers. [H 1903b, 1241; VIII: 560].

Hausdorff noted that similar thoughts, though "perhaps less pointed," had been formulated by others. Moreover, he also saw a certain danger, especially in Mauthner's tendency to exaggerate, which might then lead him to fall into speechless mysticism, much like Gustav Landauer. Regarding this suspicion, Hausdorff turned out to be right. If already in Volume 3 of Mauthner's work such tendencies were clearly visible, they intensified in his *Wörterbuch der Philosophie*, and by the end of his life, Mauthner confessed:

> Language criticism was my first and is my last word. Looking backward, language criticism is a skepticism that crushes everything; looking forward, playing with illusions, it is a longing for unity, it is mysticism. (Fritz Mauthner: *Der Atheismus und seine Geschichte im Abendlande.* Volume IV, p. 447.)

Mauthner based his epistemology on the conviction that "there is no thinking apart from speaking." Language, in turn, uses words as "memory signs" for the ideas our senses have given to us. But our senses, which have evolved over the course of time, are "accidental senses;" they make accessible to us only a narrowly limited part of the world. Language is therefore "worthless for our knowledge of the world of reality" (Mauthner, Volume 1, p. 37):

> Language can never become a photograph of the world because the human brain is not an honest camera obscura, because purposes dwell in the brain and these have shaped language in accordance with utilitarian principles. (Mauthner, Volume 1, p. 48.)

Here the phrases "knowledge of the world of reality" or "photograph of the world" should be understood as knowledge of the "world in itself" in the metaphysical sense. Mauthner, like Hausdorff, rejects all metaphysics, but whereas for Hausdorff the "world in itself" is nothing more than structureless chaos, Mauthner feels stricken by the bankruptcy of metaphysics; he is driven by a longing for metaphysical knowledge.

In his essay, Hausdorff characterizes Mauthner's epistemological convictions in a few sentences. The concluding passage reveals the fundamental differences between their viewpoints with respect to human cognition:

> With regard to epistemology, Mauthner is, to express it briefly in the usual jargon, a sensualist without being a materialist. Locke's thesis, according to which nothing resides in the intellect that was not given before by the senses, reads with him: nothing resides in the *memory of the senses* that was not before in the senses. Thus the intellect is memory of the senses, which for their part are merely accidental senses and control only occasional sections of reality; he thus rejects independent thinking as a factor of cognition, including the mental forming and shaping of sense perceptions into ordered experience. [H 1903b, 1242; VIII: 561]

Mauthner's rejection of thought as an independent factor in cognition implies a fundamental critique of science, which Hausdorff deals with in detail at the conclusion of his essay.

Another section of Hausdorff's essay takes up Mauthner's critique directed at the science of linguistics. At the center of his attacks stood the linguistic theory pertaining to the existence of an Indo-Germanic language, in particular, a universal language whose laws had fallen from heaven, so to speak. Some had supposed such a language was once spoken by an ancient Aryan people and that its roots could be traced through studies of Sanskrit. Hausdorff was absolutely correct in stating that "Mauthner's attack in many cases comes too late and thus storms already abandoned positions" [H 1903b, 1243; VIII: 562]. Even for an outsider like him, it was apparent that the current trend in linguistics had turned against speculative hypotheses in favor of empirically based research. Even so, Hausdorff was convinced that Mauthner's studies had a place in modern linguistics:

> As a thoughtful contributor to the work of modern science, Mauthner may perhaps be more welcome to linguists than he himself thinks and than they will at first admit. Only he should put aside the dynamite box and join the ranks as a peaceful reformer. [H 1903b, 1243; VIII: 563]

In suggesting this, Hausdorff refers to some of Mauthner's ideas that he perhaps felt would be worthy of "joining the ranks" of linguistics. He thus points to Mauthner's rejection of laws that constrain the development of languages. In opposing this view, Maunther thought languages had an "accidental history" dependent on and determined by culture, education, religion, custom and tradition, but also affected by factors such as war, trade, and migratory movements, including even the whims and misunderstandings of individual people. Hausdorff underscored how Mauthner emphasized the role of metaphors and their transformations over time, including his conception of language as a "dictionary of discolored metaphors." Finally, he noted his extended investigations of word compounds as mixtures of images (*Bildermischungen*) that have nothing to do with each other in the original literal sense; Mauthner called them seesaws ("Wippchen"). Hausdorff points further to Mauthner's understanding of grammar, its "indeterminacy and lack of relation to the real world," as highly interesting and rich "in fine suggestions," [H 1903b, 1245; VIII: 564], a number of which he discusses in detail. In the long run, Mauthner's work did exert an influence on linguistics, but Hausdorff was quite correct in assessing its weakness. Contemporaries reacted without sympathy to its sweeping attacks and wholesale condemnations, which limited the effectiveness of his arguments; Mauthner should have set his "dynamite box" aside.

Hausdorff also dealt with Mauthner's critique of classical logic, which focused on Aristotelian ideas. He here criticized Mauthner for satisfying himself with demolishing this ancient approach. Rather than rehashing familiar eighteenth-century ridicule, Hausdorff thought he should have taken aim at the "formalisms offered especially by English logicians and logical algebra" [H 1903b, 1246; VIII: 566]. These included especially the contributions of Augustus de Morgan and George Boole, but also the logical piano invented by Stanley Jevons, topics which Hausdorff found lacking in Mauthner's discussion of logic.

Jevons, a student of de Morgan, constructed this mechanical computer in 1869. It performed various logical operations based on the premises one entered, producing outputs in the form of logical conclusions drawn from these. By using fictitious premises, Hausdorff gave an example illustrating how this device functioned. He called it a "thinking machine," even though he regarded formal logic as unsuitable for describing human cognition. According to Hausdorff, "logic is condemned to barrenness because [· · ·] its exclusive arrogance is indifferent to real thinking activity" [H 1903b, 1247; VIII: 567]. Nevertheless, he saw progress in efforts like Jevons' logical piano:

> [· · ·] to be just, it must be conceded that only in this clean mechanical concept-calculus does deductive logic really take on a complete and precise form, next to which the profundity of the Stagirite's [Aristotle's] much admired Organon appears as a provisional attempt at orientation, afflicted with impurities and contingencies. We now at least master the chaos of logical processes from a unified point of view, and the intricate system of judgments, conclusions, and chains of conclusions with their figures and modes is revealed as a simple game of logical equations in which we substitute like for like at will. It is no accident that logic has thus become mechanizable and ripe for automata [· · ·] . [H 1903b, 1247; VIII: 567]

Seen from today's vantage point, with our knowledge of what has emerged from such seemingly primitive beginnings, these comments are remarkably clear-sighted.

Hausdorff took up the subject of logic in a later letter to Mauthner, dated June 14, 1904, where he argued:

> I cannot concede as a mathematician – the last decades will appear in a future history of the exact sciences as a period in which mathematics underwent complete logical formalization – that thinking is speaking, and that the power of logic extends no further than to proclaim that Cheshire is a cheese [· · ·]. My anti-logical period is behind me now, at least my first one, which does not exclude the possibility that I will spiral back to that again. (*HGW* IX, pp. 466–467)

In fact, throughout his life Hausdorff consistently underestimated the importance of logic for basic mathematical research, as described later in Part III.

Hausdorff particularly appreciated the way Mauthner fought against word realism (*Wortrealismus*), meaning the belief that abstract concepts designate real things. This mistaken viewpoint leads to sinister "word fetishes" that can be used to control the actions of people or even to unleash terrible excesses:

> Priests and popular orators know the fascinating power of words, which can hypnotize humanity into wars, witch hunts, and mass delirium. After people have played with words long enough, their words become insolent and play with people, e.g. the word "honor," or the words state, monarchy, fatherland, or hereditary enemy. How often have the names Jesus, Mohammed, or the three words of the French Revolution played with humanity! [H 1903b, 1249; VIII: 569]

Hausdorff also shared Mauthner's critical view of what he called category realism, or "seduction by grammar." In Hausdorff's view this was "perhaps the seduction by language taken to the highest level: a scholastic realism no longer confined to individual words still half alive, but rather embracing the pale floating categories and forms of relations" [H 1903b, 1250; VIII: 571]. Thus the noun merges with the subject concept, and inanimate objects become acting subjects through language. If, as Hausdorff imagined, we happened to live in a world without approximately rigid bodies, "a language with a more *verbal* character might have developed and, as a corresponding natural philosophy, a kind of functionalism that did not stress the flowing boundaries of natural things but rather modes of action and mutual relatedness" [· · ·] [H 1903b, 1251; VIII: 572]. Mauthner had "pursued word and category realism far into modern philosophy and researches in natural science," [H 1903b, 1251; VIII: 572] and Hausdorff gave a number of instructive examples of this interest. At the same time, he criticized those places where Mauthner, in his opinion, makes exaggerated claims and unwarranted attacks against science. He addressed this concern in the final section of his essay, which aimed to salvage the positive aspects of Mauthner's work:

> Linguistic criticism is an act; in order that it may also become an event, its proponents must not leave one thing unattended, namely, the need for *attenuation*. Without this, all extreme expression becomes lifeless thought-play, producing no further effect, no possibility of fruitful exchanges or absorption into organic contexts. [· · ·] At the risk of seemingly throwing away the deepest and most unique aspects of language criticism, I must confess that I can make nothing from Mauthner's final world-dissolving consequences [· · ·]. Nothing from the degradation of science to disputes over words and Talmudic arguments, nothing from the despair over thinking as worthless tautology, nothing from the suicide of language and its redemption in speechless mysticism. [H 1903b, 1253; VIII: 574]

Hausdorff motivated these remarks by way of his own personal experiences, namely the disinterest accorded to *Chaos in Cosmic Selection*, though he does not mention this work by name:

> It is out of personal resignation that I assume this mitigating and mediating or parliamentary function, so to speak; I too have ventured to the limits of humanity with a work of destruction yielding a void that found no echo. [H 1903b, 1253; VIII: 574]

Hausdorff's "parliamentary function" was mainly concerned with defending science against Mauthner's "world-dissolving consequences." Every science must necessarily make use of terms from a general language without any possibility of clarifying their complexity. A science must first be built over a long period of time before the clarification of its foundations can begin. The uncritical collection and ordering of facts always precedes strict systematization. If mathematicians had begun by developing a rigorous theory of numbers and functions, we would probably still be without logarithm tables today. If physiology had not progressed before strictly defining the basic concepts of life, cell, and organ, we would probably still know nothing about the circulation of blood. If "no concept is allowed without its complete definition, one cuts off all further conversation" [H 1903b, 1253; VIII: 575].

Hausdorff thus rejected Mauthner's attempt to identify thinking with speaking, just as he saw no merit in the latter's death sentence on language. In doing so, Hausdorff defended the interests of science with particular vigor, while openly confessing his hopes that "despite Mauthner's death sentence the days of human thinking will not yet be numbered" [H 1903b, 1255; VIII: 576]. Moreover, he pointed to mathematics as "a science of independent, though not obviously self-evident products of human thought," whose propositions "are not tautological, they bring novelty" [H 1903b, VIII: 1255; 577]. As an example, Hausdorff mentioned the squaring of the circle, a famous geometrical problem dating back to the ancient Greeks that was finally solved at the end of the nineteenth century. Moreover, the solution was found by purely arithmetical means, thus with no recourse to experience, which for Mauthner was the only source of knowledge. Mauthner's position on mathematics remained simply unclear to Hausdorff; it seemed to him

> if one equates *thinking and speaking*, then one renounces a clear relation to this science from the outset. If speaking and thinking are the same thing, then one must either reckon numbers, symbols, and formulas to language or withhold from the intellectual activity of the mathematician the title of thinking, which in Mauthner's case is not even an honorary title. [H 1903b, VIII: 1256; 577]

Hausdorff then turns to examples from physics that further undermine Mauthner's radical critique of science. About this he writes: "No, the highest and most general syntheses of our natural science are indeed something more than empty shells and deaf nuts" [H 1903b, 1257; VIII: 579].

In the concluding passage of his essay, Hausdorff eloquently summarized his general consensus with Mauthner, while making clear that his dissent was due in no small part to concerns about the impact of his work on linguistic criticism:

> We expect great and lasting effects from linguistic criticism. It will sharpen our intellectual conscience and our sense of the factual, while sensitizing our mistrust of words. It offers us lightning flashes to smote down the wretched cult of catchwords, the semi-educated party bickering of our days. It once again calls to account the most stubborn forms of word fetishism, religion and morality. It takes the side of living language over paper language, favoring development over rigidity, intuition over abstraction. It has a say in revising our scientific concepts by eliminating the last residues of word realism and scholasticism. It will have to be consulted if the present dispute over logic and psychology, epistemology and metaphysics is ever to end. All this is the function [of

> linguistic criticism]; but may it renounce in due time the herostratic ambition to burn down the temple of human language and reason. The enemies of knowledge will also exploit this new skepticism like any other that suits their sense, and ecclesiastical as well as occultist clergymen, who always rush to act as provisional executors of not yet legally binding death sentences, will raise a frightful clamor and blood cry in judging the witch of science. For the sake of its historical task, linguistic criticism will not be spared, just as the Reformation was not to be spared, the need to shake off vigorously and ruthlessly the swarming mob and murderers who will constitute its next following. [H 1903b, 1258; VIII: 580]

From 1904 onward, Hausdorff and Mauthner were on friendly terms. They corresponded, met occasionally, and exchanged publications. Thus, on November 2, 1904, Hausdorff wrote to Mauthner:

> I gratefully greeted your Aristotle as a commemorative token of our first meeting and sipped it immediately upon receipt as a Sunday treat. [$\cdots$] In the absence of recent noteworthy achievements, I therefore take the liberty to send the "Chaos" or to bring it to you on occasion. I would like to create something great again! very naive, isn't it? (*HGW* IX, 473–474)

Yet Hausdorff did create something truly great soon afterward, though not in philosophy or literature, but rather in the field of mathematics: his theory of order types (*Untersuchungen über Ordnungstypen*). He even sent Mauthner the first three parts of these investigations, published in 1906. Some of Hausdorff's letters also touch on family matters; others allude to meetings with Mauthner in Berlin or on vacation in Switzerland. The last surviving letter from Hausdorff to Mauthner was dated February 25, 1911. Presumably, the relationship broke off not long afterward, possibly as a result of Mauthner's growing mysticism, as reflected in his *Wörterbuch der Philosophie.* Political attitudes might also have played a role, as once World War I broke out, Mauthner threw himself into the wartime propaganda and behaved like a nationalistic "Alldeutscher." On the occasion of Mauthner's death on 29 June, 1923, the publicist Paul Fechter asked Hausdorff if he would agree to write an article about Mauthner, without mentioning the purpose, probably because he could not imagine Hausdorff did not know it already. Hausdorff refused on the grounds that he was now completely addicted to mathematics; he then added: "By the way, since I only read a local newspaper, I do not even know the occasion: has Mauthner died or is he turning 80?" (*HGW* IX, 248).

5.8 God's Shadow

Only a month after Hausdorff's *Sprachkritik* appeared in *Die neue Rundschau*, the January 1904 issue published his short essay "Gottes Schatten" ("God's Shadow") [H 1904c, VIII: 663–666]. In it, using very drastic words, he turns against the irrational tendencies of the *fin de siècle*, including mysticism and metaphysical speculations of any kind. Following Mauthner's practical language criticism, he lashes out against popular word fetishes, carried forward like banners for new and dangerous ideologies. God had already been proclaimed "very dead" around the middle of the past century, "at the time of Hegel, Feuerbach, Schopenhauer, Darwin, Marx, Helmholtz" [H 1904c, 122; VIII: 663]. Now, however, at the turn of the century, Hausdorff took note of the renaissance of religion: "God is more alive than ever, it smells alarmingly of theoform in the European hospital for the soul!" [H 1904c, 122; VIII: 663].

In this short essay, Hausdorff was not concerned with outmoded or conventional religions and their institutions but rather with the manifestations of numerous new substitute religions. These took on many diverse forms and operated with very diverse intentions at the turn of the century:

> [$\cdots$] new religions and restored old religions are offered everywhere. That many of these new God-seekers hate the old ecclesiastical God and call themselves antichrists or atheists is quite trivial and must not deceive us. After all, they cannot be blamed for preferring to worship everything possible and impossible – humanity, freedom, nature, evolution, violence, blond beast, Second Coming, chance, chaos, nirvana – than that "gas-shaped vertebrate" [Ernst Haeckel's satirical notion of God], discredited by two millennia of worshipers, who have poured out the nonsense of their heads and the filth of their hearts. [H 1904c, 122–123; VIII: 663–664]

The new believers "no longer allow themselves to be offered God, but approve of God's shadows, masks, and disguises; they have exchanged for theism, if the word is permitted, a theomorphism" [H 1904c, 123; VIII: 664]. In particular, Hausdorff attacks new forms of mysticism that fall back on mystics of earlier times or foreign cultures, replacing God with all sorts of "shadows," with something like "all-soul, world-breath, central-fire," or with the "unconscious," "instinct," "animalist" or "vitalist." In this group also belongs the mysticism of pure *blood, race, heredity*; [$\cdots$] not to forget, *home feeling, earth smell* [$\cdots$][H 1904c, 123–124; VIII: 664–665].

Just how prescient Hausdorff was in warning against this kind of "God's shadow" became apparent some 30 years later. A particularly infamous example was the book *Der Mythus des 20. Jahrhunderts* written by the chief ideologist of the National Socialist Party, Alfred Rosenberg. In the chapter "Mystik und Tat" ("Mysticism and Deed"), Rosenberg "interprets" the writings of the medieval Christian mystic Meister Eckhart, claiming these embodied "Nordic being, [$\cdots$]

the innermost, most tender and yet strongest essence of our race and culture." The effect of Eckhart's teaching, Rosenberg asserted, was this:

> In place of the Jewish-Roman world-view, the Nordic-Occidental confession of soul now takes its place as the inner side of the German-Germanic man, the Nordic race." [···] Next to the myth of the eternal free soul stands the myth and the religion of blood.

There is no need to go into the vast discrepancy between the Eckhart studied by research scholars and these fantastic elaborations.

Hausdorff was also suspicious of the "declared atheism of the natural sciences," which exchanged a "primitive concept of God for a late, refined, veiled one":

> Development, natural law, cosmos, these are the last masks of God; down to the most lifeless theory of force and matter, world events are expected to perform intelligent feats that belong at worst to mythology or theology. [H 1904c, 123; VIII: 664–665]

Hausdorff also dispatches the "fashionable words in artistic discourse" such as "mood," "sensation," "personality," and finally "worldview," but also a slogan that dare not be missing: "*life*! that abyss of the irrational into which Zarathustra's ecstasy fluttered" [H 1904c, 124; VIII: 666].

Hausdorff's little essay ends with the following quotation from Nietzsche's *The Gay Science*:

> Centuries after Buddha was dead, one was still shown his shadow in a cave – a tremendous eerie shadow. God is dead: but given the way people are, there will perhaps for millennia be caves showing his shadow. And we – we must also still conquer his shadow! [Nietzsche KSA, Vol. V, 2, 145]

5.9 A Satirical Play: The Physician of His Honor

As noted earlier in Chapter 2, Felix Hausdorff completed two periods of exercises as a reserve officer trainee in 1893 and 1894. In his notes on the officer's position, presumably taken during one of these courses, he wrote: "What should the officer above all uphold? – Honor (Must write about it)." A decade later, Hausdorff realized this intention, and the August 1904 issue of *Die neue Rundschau* carried Paul Mongré's one-act play "The Physician of His Honor" (*Der Arzt seiner Ehre*) [H 1904c, VIII: 767–792]. This was a biting satire exposing the outdated culture of dueling, a practice which had spread from aristocratic and military officer circles into wide parts of the middle classes, but particularly the academic world.

The content of the play is easily summarized. The architect Adelung, having seduced the wife of the government councilor von Granitz, is quite madly in love with her. Von Granitz, on the other hand, is more than happy to be rid of his wife, but the social conventions of the honor code demand that he face his rival in a duel, if only for the sake of appearances. He thus instructs his second, Colonel Oldefähr, to meet with Adelung's second, the lawyer Dr. Wangerow, to propose a plan whereby both men would intentionally shoot past the other. Both seconds reach agreement on this; Colonel Oldefähr had, in fact, earlier been removed from active service because of his critical attitude toward dueling. Only Adelung takes the matter terribly seriously, and even the eloquent Wangerow cannot convince him at first.

Having settled these preliminary negotiations, the parties decide that the duel shall take place near a small secluded village. Since this place offers only one hotel, both parties must spend the night in the same locale, where only a single room is heated; thus they are forced to pass the evening together. Gradually, though, after emptying many glasses of champagne and Bordeaux, even Adelung becomes fickle, and in the end the rivals drink to brotherhood. It should be noted that both men have substitute seconds, though Adelung's, a certain Baron Freisleben, never appears. Von Granitz's backup second, the brewery owner Lohm, "Reserve Lieutenant in the Xth Infantry Regiment, 'Prince Y,' Number Z," was initially present. He, however, became indignant over the attitudes of the others, so he decided to leave prematurely. He also threatened to inform the "Honorary Council" about these monstrous incidents, a threat Oldefähr acknowledged by remarking that von Granitz would now probably also – like he himself before – receive the "blue letter."

Then, around dawn, a letter arrives from Baron Freisleben, who excuses his absence, while urgently advising both parties to avoid any killing. He supports this well-meaning advice by informing Wangerow that he has just left for Nice with the lady in question. Learning this amusing turn of events, the whole group joins in a mood of song and din as they move out to the "field of honor." A hotel servant, holding a black box in his hands, yells out after the departing carriage: "Gentlemen, the pistols!"

In the year of its publication, *Der Arzt seiner Ehre* was performed at two theaters: it premiered on 12 November 1904 at the Deutsches Schauspielhaus Hamburg and then played on New Year's Eve at the Lessingtheater in Berlin. As a one-act play, it was too short to fill the theatrical program for an evening, so it usually appeared together with one or two other shorter plays. On New Year's Eve 1904, for example, it was paired with Otto Erich Hartleben's *Im grünen Baum zur Nachtigall*. *Der Arzt seiner Ehre* was Hausdorff's greatest literary success with nearly 300 performances in some thirty cities between 1904 and 1912. In Budapest it was staged with a Hungarian text, a translation produced by the young Georg Lukács, later a famous Marxist philosopher.

Hausdorff thought that a successful work of art needed to polarize its viewers, and this his play certainly achieved. While it flopped in Hamburg, where the theater canceled all performances after the premiere, it celebrated triumphs elsewhere. At the "Kleines Theater" in Berlin, where Curt Götz was cast in the role of

Adelung, there were 106 performances during the 1911/12 season. Certainly, the quality of the performances varied a great deal, which consequently had a major effect on the reception of the play itself. Nevertheless, the following small selection from contemporary theatrical reviews clearly suggests that the play triggered very different reactions from audiences and critics precisely because of its content.

We begin with two excerpts from reviews of the premiere:

> [*Hamburger Nachrichten*:] The evening ended with the one-act comedy "Der Arzt seiner Ehre" by Paul Mongré. [In it] the duel, which was the only conceivable and moral solution of the conflict in the present situation, is made to look ridiculous and the sanctity of marriage is belittled in cynical speeches. The play [· · ·] brought out hefty laughter in places, but ultimately met with deserved rejection amid some applause. (*HGW* VIII, 837)

> [*Hamburg Echo*:] · · · the performance was good, and the comedy itself is exceptionally interesting in subject and not unskilful in design, but – performed on a Saturday – a theatrical audience will always fail to appreciate it.
>
> Paul Mongré presents the madness of dueling from such a startling new angle and with such unusually subtle humor that the audience in a large public theater could not follow. Perhaps this comedy would find understanding with an unprejudiced, small literary circle. On Saturday, at any rate, it found rather little understanding, even though the noise effected by the clapping outweighed that of the hissing. Still, there were many non-hissers who also did not clap. Mongré has the courage to show the duel in the light it deserves. He treats it as a comedy about which one can very well agree over a glass of wine, [· · ·]. (*HGW* VIII, 837–838)

The last cited anonymous review in the *Hamburg Echo* found no fault with the play, but rather attributed its lack of success to its unsuitability for a broad audience. The following two reviewers also assessed Hausdorff's play quite differently. The first excerpt appeared in the newspaper *Bohemia* following the Prague premiere on February 4, 1905; the second was published in *Bühne und Welt*, reporting on the New Year's Eve performance at the Lessingtheater in Berlin:

> [Emil Faktor in *Bohemia*]
>
> [· · ·] the comedy "Der Arzt seiner Ehre" by Paul Mongré [· · ·] is a grim satire on the "well-understood wife," who sets the ideally feeling lover against the husband who trampled on her soul, and who, just before the duel is to take place, elopes with the second of her adored one. The author of this play, which in its scenic elements is a bold persiflage of that widespread opinion that adultery demands a duel, is said to be a Leipzig astronomer. In any case, he is a highly educated

Figure 5.5: Advertisement for a Performance of Felix Hausdorff's
Der Arzt seiner Ehre
Düsseldorf, 24 April 1906

> man who has forged very sharp weapons in the silence of the scholars' room. I first encountered his name in the Neue Rundschau, where his ingenious review of Fritz Mauthner's giant work on linguistic criticism caught my eye, with its outstanding use of poetic imagery and expression. In Mongré's satire, whose title recalls a drama by Calderon, he mocks the stiff concepts of honor, exposed in caustic dialectics offered by the rivals, who choose to fight against the necessity of the duel. That this disintegrating mood occurs in the circle of the duelists themselves, therein lies the bold ingenuity of the play [⋯]. (*HGW* VIII, 843)
>
> [Heinrich Stümcke in *Bühne und Welt*:]
>
> [⋯] If the Leipzig professor of mathematics, who hides behind the pseudonym Mongré, wanted to show us that there are also strumpets with the marriage ring on their finger, women unworthy of causing two honest fellows to fire pistol bullets or cross rapier blades, then the story is plausible. For the concept of honor in general, however, this farce proves nothing, and it would be more than frivolous to deduce from it that every duel arising from the injured honor of a spouse is only fencing with a mirror. Even if we consider that we are dealing with a grotesque comedy, some specifics of this nocturnal pub crawl touch a sensitive spectator unpleasantly as a result of the situation. (*HGW* VIII, 842)

Such often vehement reactions, pro and con, can be found in the numerous reviews of *Der Arzt seiner Ehre*, from the first performances up to an extraordinarily successful one at the "Kleines Theater" in Berlin (*HGW* VIII, 836–857), which received high praise:

> [Arthur Eloesser in *Vossische Zeitung*]
>
> Paul Mongré is a university professor, an outstanding mathematician, a witty essayist, and a literary son or at least nephew of Nietzsche; he thus has no need at all to write plays. That he did so, however, was entirely for his own pleasure and for ours as well. This "Physician of his Honor" has already delighted us once before at the Lessingtheater, and even on a New Year's Eve, where one watched with silent and loud amazement a most witty and happily contrived satire, instead of the usual tale, which the occasion would have excused. The reader may remember that [the play is a spoof] against the notion of honor. [⋯] Paul Mongré has written a truly bizarre satire, and this requires exaggerations of reality.

Hausdorff added an epilogue to his play for the special edition of it published by the Leipzig Bibliophilen-Abend. In this short scene, von Granitz happens to meet Daisy, his divorced wife, now a Countess Nazzaro, at a Swiss hotel near the Matterhorn, two years after the memorable "duel." A somewhat stiff, but then steadily more intimate dialogue unfolds, until von Granitz orders a room for two to enjoy an "echoing of the irretrievable." This epilogue was staged 96 years later in Bonn in September 2006 in a performance of the *The Doctor of his Honor* for the annual conference of the German Mathematical Society. Recently, an English translation of the play was also published.[1]

After 1904, the author Paul Mongré disappeared from the scene for several years. His disappearance can no doubt be explained by the fact that Hausdorff had begun working intensively on mathematics; during these years he produced his major contributions to the theory of ordered sets. Around Christmastime in 1908, he sent a postcard from Rome to the librarian Franz Meyer in Jena:

> So the production is already set? Dear God, I am already to the point that the reception can start as well. Rome, Naples glides by without soulful tumult. Hibernation, without metabolism, completely snowed in with pure mathematics. (*HGW* IX, 508)

It was not until 1909 that Paul Mongré once again awoke from "hibernation" with a few short essays, as we shall see in the next chapter.

[1] Paul Mongré (Felix Hausdorff): *The Doctor of his Honor.* English Translation by Volker Vetter and William P. Hausdorff. Hausdorff Center for Mathematics, Bonn 2021.

Chapter 6

Mathematics takes First Priority

6.1 Hausdorff finds his Research Field

In Section 3.2 we saw how Hausdorff, at the beginning of his career, was still searching for orientation as a researcher. But soon he found his field, set theory, in which he quickly achieved important results.

General set theory at its core concerns the theory of transfinite ordinal and cardinal numbers, which goes back to Georg Cantor's works from the years 1872 to 1887. Cantor later published a summary of this theory in a large two-part work in the *Mathematische Annalen* [Cantor 1895/1897], later published in English translation as "Contributions to the Founding of the Theory of Transfinite Numbers" [Cantor 1915/1955]. Up until the First International Congress of Mathematicians in Zurich in 1897, very few mathematicians were concerned with Cantor's theory, as can be seen from the abstracts published in the *Jahrbuch* up until 1904, when works in set theory were placed under the rubric of Philosophy and Pedagogy.

Hausdorff was no exception in this regard. One can assume with some certainty that Hausdorff's interest in set theory was mainly aroused by his philosophical interests. In Aphorism 406 [H 1897b, 349–354; VII: 443–448] in *Sant' Ilario*, Hausdorff gives a critique of Nietzsche's alleged proof of his idea of eternal return. Nietzsche had argued that, whereas time itself is infinite, the content of each time segment consists of an enormously large, but nevertheless finite number of different combinations of atoms, from which the eternal repetition of time segments with the same content would follow. Hausdorff accepts the validity of this argument, but not the underlying assumption, because the time content is also an infinite quantity, in fact, "a much more comprehensive, extensive one, an infinity of a *higher dimension*". Then he goes on:

> Without this distinction, borrowed from mathematics, of manifolds of different dimension and infinities of different rank and extent, no one should dare to approach epistemological matters; how can one deal with

E. Brieskorn, W. Purkert, *Felix Hausdorff*, Vita Mathematica 21,
https://doi.org/10.1007/978-3-031-52135-5_6

> infinity if one merely recognizes it as opposition to the finite without distinguishing the stages and degrees of the infinite? [H 1897b, 350; VII: 444]

This perhaps sounds like set theory, but Hausdorff probably did not yet know Cantor's theory at the time of the final editing of *Sant' Ilario* in early 1897. In this aphorism, the stages and degrees of infinity refer to the different dimensions of point sets: thus there are ∞ points on a line, ∞^2 on a surface, and a solid contains ∞^3 points. Cantor had already proved in 1878 the seemingly paradoxical result that $\mathbb{R}^n$ and $\mathbb{R}^m$ have the same cardinality for $n \neq m$. Hausdorff wanted to "approach epistemological things," which explains his burning interest in a mathematical theory of "infinities of various rank and extent." Over the course of 1897, thus before he completed the final version of *Chaos in kosmischer Auslese*, Hausdorff must have come across Cantor's set theory, since he wrote in a note:

> ...the examples given by G. Cantor and G. Peano show, however, that if one dispenses with a one-to-one *continuous* mapping, then the points of a surface, a cube, or a bounded manifold of n dimensions can be made to correspond with the points of a line segment. Cantor's foundations of a general theory of manifolds. Due to this, revise my remarks about Nietzsche's eternal return. (Hausdorff estate, Capsule 49, Fassicle 1076, p. 52)

He made that revision in the text of *Chaos in kosmischer Auslese*, where his refutation of Nietzsche's "proof" shows he was completely oriented regarding set-theoretic matters. The set of world states is not finite, as Nietzsche believed, but rather is even "of a greater *power* than the linear continuum." As noted earlier, Hausdorff established a connection between philosophical argumentation and set-theoretical concepts and theorems in several places in his *Chaos*.

It is not known exactly when Cantor and Hausdorff first met in person, though it was probably in August 1897 at the International Congress of Mathematicians in Zurich. At this congress, Adolf Hurwitz gave a plenary lecture "On the Development of the General Theory of Analytic Functions in Modern Times," in which he emphasized the importance of set theory for recent research in this field. It is significant that he did not assume his audience was familiar with Cantor's ideas, and so he explained the necessary fundamentals of transfinite numbers and their application to point sets in some detail.

In the summer semester of 1901, Hausdorff taught a course on set theory in Leipzig, only the third time such a course had ever been offered. Twice earlier lectures courses had been given in Göttingen, the first by Arthur Schoenflies in 1898 and then by Ernst Zermelo during the winter semester 1900/1901. Cantor himself never gave a lecture course on set theory during his many years in Halle, and only three auditors attended Hausdorff's course, which closely followed the contents of Cantor's two papers in *Mathematische Annalen*.

In this lecture, Hausdorff also recorded his first discovery in this field, though he learned soon afterward that this was no longer a new result. He proved that the type class $T(\aleph_0)$ of all countable order types has the power $\aleph$ of the continuum. His proof was based on the Cantor-Bernstein equivalence theorem, which he applied in showing that:

> a) a subset of the class of countable types is equivalent to the continuum; and
> b) a subset of the continuum is equivalent to the class of countable types.

Hausdorff added in red pencil in the upper margin:

> a) according to an oral communication from Cantor,
> b) by me. Presented on June 27, 1901.
> Dissertation from F. Bernstein received June 29, 1901. [Felix Bernstein had also proven claim b) in his dissertation.]

Hausdorff's note a) is the first written evidence of personal contact between him and Cantor. There were regular opportunities for such contacts, though, as mathematicians from Leipzig and Halle had formed a joint colloquium, which met in sessions that alternated between the two cities. Most of those who attended were younger people, though Cantor and Otto Hölder also took part regularly.

In his lecture course, Hausdorff carefully separated provable theorems in set theory, for which he gave detailed arguments, from the propositions that Cantor had only conjectured to be true. Hausdorff called the latter postulates, which he formulated as follows:

> Postulate II: The power of the continuum is equal to that of the second number class.
> Postulate III: Every well-defined set can be placed in the form of a well-ordered set.

Postulate I. asserts that cardinal numbers are comparable, which follows if Postulate III. holds. Postulate II. is Cantor's continuum hypothesis or the continuum problem, when viewed as the question as to which place $\aleph$ occupies in the series of alephs $\aleph_\alpha$. This was the first of the 23 problems David Hilbert brought to the attention of mathematicians as challenges for the next century in his famous address at the 1900 International Congress of Mathematicians in Paris. Hilbert's lecture helped to give Cantor's continuum hypothesis special allure, and Hausdorff worked on it for decades (see 6.2 below).

Hausdorff's interest in undertaking a systematic study of ordered sets was motivated not least by the continuum problem, since he saw in the theorem on the power of $T(\aleph_0)$ a strategy for attacking this problem from a direction which Cantor had not yet attempted. This is clearly expressed in his first set-theoretical publication, "Über eine gewisse Art geordneter Mengen" (On a Certain Type of Ordered Sets) [H 1901b, IA: 5–20]. First, he notes that not only the field of ordinal numbers but also that of order types must be thoroughly investigated,

> if only because the old question of the power of the continuum might possibly in this way be brought closer to a solution. Since according to a theorem of Cantor and Bernstein[1]) the power of the second *type* class (the class of countable types) equals the cardinality $\aleph$ of the continuum[2]), whereas the power of the second *number* class is denoted by $\aleph_1$, so it may be important for comparing $\aleph$ with $\aleph_1$ to determine intermediate levels, i.e. to pick out a narrower group from among the entirety of all countable sets that contains all the countable well-ordered sets. The proof of Cantor's hypothesis that $\aleph = \aleph_1$ promises to become easier by introducing intermediate levels, whereas conversely, if instead $\aleph > \aleph_1$ this way would be rather a detour. [H 1901b, 460; IA: 5]

Footnote [1]) is a reference to Bernstein's Göttingen dissertation. In footnote [2]) Hausdorff writes that Cantor "soon plans to publish" a proof of the well-ordering theorem, another indication of an oral exchange between them, since Cantor never published such a proof.

In this work, Hausdorff considered a class of ordered sets – today one would say linear ordered sets – which he calls graded (*gestuft*). A linear ordered set M is graded when two different segments are never similar to each other, i.e. for $a, b \in M\,; a \neq b$ there is no order-preserving isomorphism mapping $\{x \in M; x < a\}$ onto $\{x \in M; x < b\}$. Every well-ordered set, in particular, is graded, and Hausdorff was able to show fairly easily that his theorem on the power of $T(\aleph_0)$ also applied for graded sets. Thus the set of all countable graded order types has the power of the continuum $2^{\aleph_0} = \aleph$. Hausdorff seemed to have thus taken a first step in the program he had formulated. However, he made no further attempts in this direction, and we know today that this strategy cannot lead to a solution of the continuum problem. The same was true of Cantor's strategy, which aimed to generalize the Cantor-Bendixson Theorem to arbitrary uncountable point sets. This idea of Cantor's turned out to be the germ for the development of descriptive set theory during the 1920s, a theory that gained a major impulse in 1916 through a result proved by Hausdorff and Paul Alexandroff.

The main result in Hausdorff's paper is the generalization of the Cantor-Bernstein Theorem to sets of arbitrary cardinality. Hausdorff formulated it as follows:

> If M is a graded set whose cardinal number $\mathfrak{m}$ equals its square, then the set of all order types of the power $\mathfrak{m}$ has the power $2^{\mathfrak{m}}$.

Hausdorff noted that this theorem applies, in particular, for all cardinals $\mathfrak{m}$ which are alephs, so that $T(\aleph_\alpha)$ has the power $2^{\aleph_\alpha}$. He formulated this result so carefully because in 1901 neither the well-ordering theorem nor $\aleph_\alpha \cdot \aleph_\alpha = \aleph_\alpha$ had been proven. Moreover, in the course of his proof, Hausdorff used results that were equivalent to the axiom of choice, which Zermelo would first formulate three years later [Ebbinghaus 2007b].

Hausdorff's first deep result in set theory, the recursion formula for aleph exponentiation that bears his name, is related to a sensational event that took place at the Third International Mathematical Congress held in Heidelberg from August 8–13, 1904. There, in discussions both during and after the congress, Cantor's continuum problem and the question as to whether the continuum can be well ordered played a special role.

On August 10, the Hungarian mathematician Julius König presented a lecture on the continuum problem. He first proved that the sum $\mathfrak{a}$ of an increasing sequence of cardinal numbers satifies the inequality $\mathfrak{a}^{\aleph_0} > \mathfrak{a}$. From this he used $\aleph_{\beta+\omega} = \aleph_\beta + \aleph_{\beta+1} + \ldots$ to derive in inequality

$$\aleph_{\beta+\omega}^{\aleph_0} > \aleph_{\beta+\omega}. \tag{6.1}$$

König then combined this with an equation for alephs that Bernstein had given in his dissertation, namely

$$\aleph_\mu^{\aleph_\alpha} = \aleph_\mu \cdot 2^{\aleph_\alpha}. \tag{6.2}$$

He then noted that if $\aleph = 2^{\aleph_0}$ were an element in the series of alephs, say $2^{\aleph_0} = \aleph_\beta$, then substituting $\aleph_\mu = \aleph_{\beta+\omega}$ in (6.2), one obtains

$$\aleph_{\beta+\omega}^{\aleph_0} = \aleph_{\beta+\omega} \cdot 2^{\aleph_0} = \aleph_{\beta+\omega}\aleph_\beta = \aleph_{\beta+\omega},$$

contradicting (6.1). This would then show that the cardinality of the continuum cannot be an aleph, meaning it did not admit a well ordering, thereby totally refuting Cantor's conjecture that $\aleph = \aleph_1$. In the discussion that took place immediately afterward, the congress protocol indicates that Cantor, Hilbert, and Schoenflies all spoke, though what they said was not recorded.

König's lecture caused a furor of excitement, as this result would have spelled the end for one of Cantor's principal convictions, namely the belief that every set could be well ordered. Beyond that, this finding would have solved Hilbert's first Paris problem, though not in the sense he and Cantor had hoped. In his autobiography, *Bestand und Wandel*, Gerhard Kowalewski gave a vivid account of the atmosphere afterward:

> Thus, with König's lecture two of Cantor's fundamental principles [well ordering and $\aleph = \aleph_1$] were refuted. Cantor was deeply moved when he spoke afterward, and he even expressed his thanks to God that he could witness this refutation of his mistakes. The newspapers reported on König's significant lecture, and the Grand Duke of Baden received a report from Felix Klein about this sensation.
>
> Fortunately, already on the next day it was determined that König's argument was untenable. He had relied on a theorem of Felix Bernstein, which on closer examination proved to be incorrect. Zermelo, an extremely ingenious and rapid thinker, made this important finding. In fact, he even found in those days a proof that an arbitrary set can be well ordered, a theorem he later proved a second time using completely

different considerations. [Kowalewski 1950, 202]

This is such a vivid description that a reader would almost certainly never imagine that Kowalewski had not even attended the Heidelberg Congress and was only reporting from hearsay (and nearly 50 years later). In any case, the fact remains that the name Kowalewski does not appear among the participants in the official directory.

What Kowalewski wrote about Cantor's reaction cannot be directly refuted, but this flatly contradicts the latter's deepest convictions and also stands in striking constrast to what Arthur Schoenflies reported in [Schoenflies 1922] (see below). Also Felix Klein's alleged report to the Grand Duke of Baden is pure fiction – in August 1904 the Grand Duke was visiting a health spa in Switzerland. However, these marginalia are barely worth mentioning.

The passage concerning what supposedly happened the day after König's lecture is also almost totally false, but this deserves a more thorough refutation because it has often been cited in the literature on the history of set theory and thus has become part of its mythology [Purkert 2015b]. In reality it was Felix Hausdorff who noticed that Bernstein's formula (6.2) can only be proved for those μ which are not limit numbers. Presumably the discussions at the congress and then afterward in Wengen (see below) stimulated him to deal intensively with aleph exponentiation, because only a few weeks later he published his famous recursion formula:

$$\aleph_\mu^{\aleph_\alpha} = \aleph_\mu \aleph_{\mu-1}^{\aleph_\alpha} \tag{6.3}$$

This formula has no meaning if μ is a limit number; when this is not the case, it leads to Bernstein's recursion formula.

At the end of [H 1904a], Hausdorff referred directly to König's lecture:

> The formula
>
> $$\aleph_\mu^{\aleph_\alpha} = \aleph_\mu \cdot 2^{\aleph_\alpha}$$
>
> obtained by Mr. F. Bernstein ("Untersuchungen aus der Mengenlehre", Diss. Halle 1901, p. 50) by means of unrestricted recursion is thus to be regarded provisionally as unproven. Its correctness appears all the more problematic, as from it, as Mr. J. König has shown, would follow the paradoxical result *that the power of the continuum is not an aleph and that there exist cardinal numbers that are greater than any aleph.* [H 1904a, 571; IA: 33]

If Zermelo had already pointed out in Heidelberg that Bernstein's formula had not been proven for limit numbers μ and furthermore that his proof was probably wrong, then Hausdorff would certainly not have written this passage or at the very least he would have mentioned Zermelo's name.

Hausdorff's role in clearing up the fallacy following König's lecture is also evident from Schoenflies' later account of the events in Heidelberg written in 1922. As a participant in the discussions surrounding the lecture, Schoenflies was a far more credible witness than was Kowalewski. For this reason, we cite these passages from his report:

> The highpoint of interest taken by the mathematical world in set-theoretical problems was reached at the Heidelberg Congress of 1904. What moved the mathematical world at that time was the question of the power of the continuum and the well-ordering theorem, in short, the statement that any set is capable of being well ordered. For Cantor these demands were a kind of dogma for his set-theoretical thinking and beliefs .... König's lecture, which culminated in the theorem that the continuum could not be an aleph (i.e. was incapable of being well ordered), therefore seemed astonishing, especially since it was supported by an extraordinarily well developed and precise argument. [Schoenflies 1922, 100–101]

Cantor had, however, thought from the beginning that König's result was incorrect. Schoenflies thus continued:

> After the Heidelberg Congress a kind of post-congress took place in Wengen. Hilbert, Hensel, Hausdorff, and I myself came together there quite by accident. Cantor, who was originally staying nearby, soon came over to join us, as he was filled with a constant urge to speak his mind. The theorem of König was over and again at the center of our conversations. [...] As is known, we have Hausdorff to thank for the first exact test of the range of validity of Bernstein's aleph relation; his result, which he already obtained in Wengen, leads to the above limitation of the theorem and consequently to the negation of the consequence drawn for the continuum. [Schoenflies 1922, 100–101]

The last statement is, in fact, not entirely accurate, as Hausdorff himself wrote to Hilbert on 29 September 1904, soon after he had clarified the mystery. There he explained:

> After the continuum problem had tormented me in Wengen nearly like a monomania, my first glance here of course turned to Bernstein's dissertation. The worm is sitting there at exactly the place I suspected, on p. 50: [...] Bernstein's consideration leads to a recursion from $\aleph_{\mu+1}$ to $\aleph_\mu$, but it fails for those $\aleph_\mu$ which do not have a predecessor, that is for precisely those alephs which Herr J. König necessarily requires. I had already written in this sense to Herr König on my trip back, so far as I could do so without having Bernstein's work, but I have received no answer and am all the more inclined to regard the proof as faulty and König's proposition as the height of improbability. On the other hand, you will scarcely have received the impression that Cantor has found in the last weeks what he has sought in vain for 30 years. And so your Problem No. 1 appears, after the Heidelberg Congress, to stand precisely where you left it at the Paris Congress.

> But perhaps, while I write this, one of the parties to the dispute is already in possession of the truth. I am very anxious to see the printed proceedings of the [Heidelberg] Congress. (*HGW* IX, 330)

In fact, in the published version of his lecture for the congress proceedings, König explicitly withdrew his earlier conclusion, noting that the error was due to Bernstein's faulty formula; he did not mention Hausdorff in this context.

It should also be added that there is no basis for Kowalewski's further assertion that Zermelo "even found in those days a proof that an arbitrary set can be well ordered," a statement which suggests this took place in direct connection with the congress. The known facts point to a later date: on 24 September 1904, Zermelo wrote a letter to Hilbert containing a proof of the well-ordering theorem; Hilbert then forwarded this for immediate publication in *Mathematische Annalen.* The proof was based on the axiom of choice, which requires "that the product of an infinite collection of sets, each of which contains at least one element, is itself not empty" [Zermelo 1904, 516]. Zermelo was present at the Heidelberg Congress and it is quite possible that the discussions there inspired him to undertake this work, though he makes no mention of this. Concerning its origins, he merely wrote that "the corresponding proof came about through conversations I had the previous week with Mr. Erhard Schmidt ..." [Zermelo 1904, 514].

Heinz-Dieter Ebbinghaus published a postcard that Zermelo sent to Max Dehn about one month later in connection with his proof of the well-ordering theorem. This card shows that Zeremlo had also found the mistake in Bernstein's dissertation independently of Hausdorff. When he actually confirmed this remains unclear, but probably both had discussed this possibility already in Heidelberg or afterward in Wengen. In any case, based on the available sources, Zermelo's discovery played no role in the published clarifications that followed. In his postcard to Dehn, he wrote:

> So you still do not know the fate of König's talk? Innocent soul! Nothing else was really discussed during the entire vacation. K[önig] solemnly informed both Hilbert and me that he revoked the Heidelberg proof. B[ernstein] did the same with his theorem on powers. K[önig] can be happy that the library in Heidelberg had closed so early; otherwise he might have made a fool of himself. Thus I had to wait for checking until my return to G[öttingen]. Then it was instantly obvious. In the meantime, however, K[önig] had found that out himself when elaborating his proof. [Ebbinghaus 2007a, 431]

Zermelo's proof of the well-ordering theorem unleashed a long and impassioned discussion concerning the legitimacy of the axiom of choice (see [Moore 1982]).

6.2 Hausdorff's Views on Foundations of Mathematics

In this section we no longer adhere closely to the chronology in order to consider Hausdorff's views on foundations over the course of his lifetime. At the beginning of his scientific career as an astronomer and as an author of works on optics and actuarial mathematics, fundamental mathematical questions had little relevance for him. But already with *Chaos in Cosmic Selection*, his philosophical interests motivated an examination of fundamental questions in geometry, especially non-Euclidean geometry and the space problem. These interests deepened after the appearance of Hilbert's *Foundations of Geometry* in 1899. After studying this work, Hausdorff began to develop his own standpoint on questions concerning the epistemological foundations of mathematical theories. Through his pioneering work on set theory and topology, he eventually became one of the foremost advocates of modern abstract mathematics during the first decades of the twentieth century.

Especially revealing in this context is the unpublished fragment "Der Formalismus" (*HGW* VI, 473–478), which was probably written around 1904. In it, Hausdorff says that formal logic is the "only certainty we know," and he praises the trend toward rigorization through logic and formalization that dominated the history of pure mathematics during the second half of the nineteenth century. By *formal logic*, Hausdorff understood the usual rules of logical reasoning employed in mathematical discourse. He thus distinguishes this from *symbolic logic*, based on the logical calculi found in artificial symbolic languages, a new research field whose main representatives at that time were Ernst Schröder, Gottlob Frege and Giuseppe Peano. Hausdorff was very critical of symbolic logic, as can be seen from his review of Mauthner's *Sprachkritik*, though he relativized this somewhat in a letter to Mauthner. After reviewing Bertrand Russell's *The Principles of Mathematics* [H 1905], he made no further public comments about symbolic logic.

The development of set theory after the turn of the century was accompanied by controversial debates, especially over the antinomies of set theory and the legitimacy of the axiom of choice [Moore 1982, 85–141], [Ferreirós 2007, 299–336]. In his review of Russell's book, Hausdorff comments for the first time on the antinomies. Russell's antinomy, he argues, "has turned out to arise from the philosophically well-known paradox of the ill-defined *allness* category." The word "all" does not always denote an attainable property, so that we may

> in some cases imagine assigning to *any* object a certain definition, but we cannot collectively imagine *all* objects "uno intellectus actu."

In particular, Hausdorff rejects Russell's definition of the cardinal number of a set as the class of all sets co-equal with it, because

> here we have again the incomprehensible notion of "all" sets, and at the summit of arithmetic would stand, as a primordial contradiction, the number one in its contradictory form as a class *of all* single things.

One would thus have to check the definitions of those sets one operates with to ensure they are free of contradictions;

> such classes as the set of all natural numbers, the set of all points of space, the set of all functions seem to present no objections, whereas the set of all infinite cardinal or ordinal numbers [...] belong to the inadmissible class concepts. [H 1905, 123–124; IA: 485–486]

Of course, Hausdorff left open the difficult question of how one might decide which infinite sets are transfinite, and thus logically unobjectionable, and which are absolutely infinite in Cantor's sense, i.e., represent classes whose elements cannot all be grasped "uno intellectus actu" [Purkert 1986]. Cantor spelled out this distinction in several letters he communicated to Hilbert, but unfortunately we do not know whether he also discussed his views on the question of antinomies with Hausdorff. It seems, in any case, that Hausdorff trusted his intuition when it came to recognizing shaky ground, and so thought one could circumvent such problems by taking appropriate caution. He expressed this same general optimism at the beginning of his *Investigations of Order Types.* As an example of how he sought to avoid the dangers of antinomian sets, he wrote:

> We assume the theory of well-ordered sets to be known. We think of a sufficiently large ordinal Δ, which exceeds in power all sets we wish to consider, and understand by W the set of all ordinals $< \Delta$; only the elements and subsets of W are considered in what follows here. [H 1908, 441; IA: 219]

Hausdorff's aspirations were always directed toward extending the mathematical substance of set theory itself, for example by creating and investigating new order structures, but also in opening up new fields of application for it, such as topology or probability theory. He was happy to leave the expansion and fortification of its foundations to others. He surely made his position clear in lectures and discussions, as confirmed by Gerhard Hessenberg in a postcard, dated September 20, 1907, sent to his friend Leonard Nelson. Hessenberg was reporting from the Dresden meeting of the German Mathematical Association, at which Hausdorff presented results from [H 1907b], about which Hessenberg wrote the following:

> Furthermore, Hausdorff pleased me excellently. He is the only productive set theorist who is as little irritated by the paradoxes as Newton etc was by eleatic jokes (Nelson estate, Correspondence H, 65–66, Folder Hessenberg 3).

Hessenberg here alludes to the fact that the founders of the infinitesimal calculus proceeded boldly, ignoring famous difficulties with the infinite already discussed in antiquity, such as the paradoxes of Zeno of Elea.

In [H 1908], Hausdorff briefly outlined his standpoint on the discussions surrounding the antinomies and the axiom of choice:

> In particular, [this paper] takes the position that the totality of "all" ordinal or cardinal numbers does not exist without contradiction, either as a set or as a subset of a set; it accepts Cantor's well-ordering theorem in the formulation proved by Zermelo and attaches no importance to the statement that part of its results could also be derived independently; it makes use of "choice postulates," even without noting them, and only meets a *perhaps* legitimate concern [...] by occasionally noting that a transfinite set of successive choices, each of which presupposes the earlier ones, can be avoided on the basis of a well-ordering or, what amounts to the same thing, replaced by a simultaneous set of independent choices. [H 1908, 436–437; IA: 214–215]

Hausdorff consistently held to the views formulated here in later years. In his principal work, *Grundzüge der Mengenlehre* [H 1914a], he introduced its central concept as follows:

> A set is a gathering (*Zusammenfassung*) of things into a whole, i.e. to a new thing. One will hardly accept this as a definition, but only as an illustrative demonstration of the concept of set [...] [H 1914a, 1; II: 101]

In speaking of mathematical objects as "things," Hausdorff probably adopted Hilbert's language. In his lecture "Über die Grundlagen der Logik und Arithmetik" at the International Congress of Mathematicians in Heidelberg, Hilbert spoke of "an object of our thought" as "a *thought thing* or briefly a *thing*" [Krazer 1905, 176].

Hausdorff considered Cantor's greatest merit his willingness to "overlook popular prejudices and authoritative philosophical pronouncements" in order to investigate infinite sets and in the process found a new science, set theory, which served as "the foundation of all mathematics," since infinitesimal calculus, analysis and geometry worked constantly with infinite sets. However, there was a caveat: "complete agreement has not yet been reached regarding the foundations of this foundation, that is, on the foundations of set theory itself" [H 1914a, 1; II: 101]. Most of the "paradoxes of the infinite" are only apparent and resolve themselves, once one observes that laws which hold for finite sets need not hold in the realm of the infinite. This had already been pointed out by Bernard Bolzano in his posthumously published work *Paradoxes of the Infinite*. The fact that Hausdorff put those words in quotation marks strongly suggests that he was referring to Bolzano's work, which he certainly knew.

Hausdorff loved such apparent paradoxes, and he tried to construct other examples himself. One of his constructions became famous as Hausdorff's sphere paradox (see 7.6.4). Not all contradictions that arise in connection with the infinite are apparent, however, and the naive notion of set, as given in the above descriptive definition, produces real problems. In his book, Hausdorff shows that for every set of cardinal numbers there is always a cardinal number larger than all the numbers of the given set, i.e., a number that cannot appear in that set. This means that the "set of all cardinal numbers" is contradictory, and the same is true for the ordinal numbers. So one cannot operate with the "set of all cardinal numbers" or the "set of all ordinal numbers," as Hausdorff had already emphasized in 1908. However, that year also saw Zermelo's axiomatization of set theory [Zermelo 1908], which Hausdorff regarded as a hopeful attempt to constrain the principles of set formation in order to avoid all potential antinomies:

> E. Zermelo has hereafter undertaken the needed attempt to restrict the process of unbounded set formation by means of suitable conditions. Since, however, these extremely delicate investigations cannot yet be regarded as closed, and since such an introduction to set theory would for novices likely cause great difficulties, we shall here permit the naive notion of sets, but while holding on to restrictions that cut off the path leading to those paradoxes. [H 1914a, 2; II: 102]

Hausdorff did not elaborate on his remark stating that Zermelo's axiomatization of set theory could not yet be considered a closed matter. It may be assumed, though, that he considered Zermelo's notion of "definite property" to be insufficiently precise. This weakness in Zermelo's system would not be remedied until 1922 when it was addressed in works by A.A. Fraenkel and Thoralf Skolem [Fraenkel 1922], [Skolem 1922].

In the summer of 1923, Hausdorff's *Grundzüge der Mengenlehre* was out of print. The Berlin publisher DeGruyter requested a new edition that would appear in the series *Göschen's Lehrbücherei*, which meant it would have to be 200 pages shorter than the *Grundzüge*. Hausdorff decided to write a completely new book that came out in 1927 under the title *Mengenlehre* [H 1927a]. In preparing this work, he had to decide whether he should maintain the naive standpoint or adopt the Zermelo-Fraenkel axiomatic foundation. This decision was made much easier for him due to the fact that Fraenkel had in the meantime published his own book [Fraenkel 1923]. In a postcard sent to Fraenkel on June 9, 1924, Hausdorff wrote regarding these circumstances:

> You have rendered me a great service for the 2nd edition of my book (which I intend to write entirely anew), inasmuch as I can refer to your excellent exposition for various important matters that are rather remote for me, e.g. for axiomatics (in which you have made a considerable advance beyond Zermelo concerning the axiom of separation) and for the treatment of the antinomies. (*HGW* IX, 293)

Accordingly, the preface of Hausdorff's new book states: "Just as before, I have decided against undertaking a discussion of antinomies and foundational critique" [H 1927a, 6; III: 46]. This unchanged standpoint, which Hausdorff expresses with pointed language at the beginning of his new book is, so to speak, the position of the "working mathematician," who – as Hessenberg already appreciated in Hausdorff back in 1907 – does not become perturbed over fundamental difficulties and remains optimistic about their ultimate resolution. Returning to Cantor's original ideas, he writes:

> A set is formed by combining individual things into a whole. A set is a multiplicity, thought of as a unity. [...] A set may or may not consist of a natural number of things; depending on this, it is called *finite* or *infinite*. [...] It is the immortal merit of Georg Cantor to have dared to take this step into infinity faced with internal as well as external fights against apparent paradoxes, popular prejudices, authoritative philosophical claims (infinitum actu non datur), but also against misgivings which were expressed even by the greatest mathematicians. He thus became the creator of a new science, set theory, which today is the foundation of all mathematics. In our opinion, nothing in this triumph of Cantor's ideas is changed by the need for complete clarification and elimination of the antinomy that arises from an all too boundless freedom in the formation of sets. [H 1927a, 11; III: 55]

The notion of the cardinal number $\mathfrak{a}$ of a set A had been introduced by Hausdorff in the *Grundzügen* as a "thing" associated with the set A in such a way that exactly the same thing corresponds to two sets A and B if they are equivalent, i.e. if a bijective mapping is possible between them. John von Neumann later, in 1922, introduced ordinal numbers as canonical representatives of isomorphism types of well-ordered sets. He defined the cardinal number of a set M as the smallest ordinal number [Neumann 1922] which is equivalent to M. Thus, cardinal and ordinal numbers became specially defined sets, which was of decisive importance for a consistent set-theoretic foundation of mathematics. In his book *Mengenlehre*, however, Hausdorff retained the definition of cardinal number from the *Grundzügen*, but added this comment:

> This formal explanation tells what the cardinal numbers are supposed to be, not what they are. [...] Relations between cardinal numbers are to us only a convenient expression for relations between sets: we must leave it to philosophy to fathom the "essence" of the cardinal number. [H 1927a, 25; III: 69]

When speaking of Hausdorff's position on fundamental questions of mathematics, one cannot overlook his sharp rejection of intuitionism, even though he never dealt with this topic publicly. Not even in his posthumous studies or notes do we find ant remarks about intuitionism, and those fascicles in his estate that contain references to L.E.J. Brouwer deal only with topological topics. Similarly with Hermann Weyl: Hausdorff's notes only contain references to mathematical fields (Riemann surfaces, uniform distributions modulo 1). We only know something about Hausdorff's opinion with regard to intuitionism from his correspondence.

Hausdorff was probably unfamiliar with Brouwer's early writings on intuitionism, which were quite technical and difficult to understand. An exception is the short article "Intuitionistische Mengenlehre," which appeared in the journal of the

German Mathematical Society (vol. 28(1919), 203–208). Hausdorff probably by 1920 had read this as well as Hermann Weyl's booklet *Das Kontinuum. Kritische Untersuchungen über die Grundlagen der Analysis.* His first personal encounter with Brouwer likely took place in September 1920, when both attended the first post-war meeting of the German Mathematical Society held in Bad Nauheim as part of the large-scale conference organized by the Society of German Naturalists and Physicians. Brouwer offered a provocative lecture entitled "Does Every Real Number Possess a Decimal Fraction?" – his answer was (not surprisingly) negative. We do not know whether Hausdorff, who lectured two days later on a topic in analysis, took part in the discussion following Brouwer's lecture, but contemporary sources indicate that the reactions were either amused or hostile.

Hausdorff's first known comment on intuitionism can be found in the above-mentioned postcard to Fraenkel of June 9, 1924. There, after the passage quoted above, he continued as follows:

> You have even succeeded in making the oracular sayings of Messrs. Brouwer and Weyl understandable – without making them seem any the less nonsensical to me now! Both you and Hilbert treat intuitionism too respectfully; one should bring out bigger guns against the senseless destructiveness of these mathematical Bolshevists! (*HGW* IX, 293)

Hausdorff himself, at any rate, never brought out such ordnance; his deep disdain, however, remained. In a postcard to Fraenkel, written February 20, 1927, he thanks him for sending his book *Zehn Vorlesungen über die Grundlegung der Mengenlehre* (Leipzig 1927). He found it "so tensely dramatic" that he had "already devoured a large part of it," after which he goes on to say:

> I still cherish the hope that you, as the best connoisseur of this literature, will for once make a vigorous and witty attack on intuitionism – if it is not perhaps more advisable to let this castrato mathematics choke on its own complicated stupor. (*HGW* IX, 294–295)

After 1927, i.e. after the publication of his book *Mengenlehre*, Hausdorff did not comment publicly on fundamental questions. However, he still occasionally touched upon such matters in letters, which show that he remained informed about foundational issues. Moreover, he never lost interest in the continuum problem, which had once "plagued him like a monomania" (*HGW* IX, 330), even in difficult times. In August and September 1938, he thoroughly studied Sierpiński's book *Hypothèse du Continu* (Warsaw 1934), taking excerpts from it and transferred these into his terminology, occasionally simplifying the proofs, and adding some critical remarks.

Although he took little interest in problems related to the axiomatization of set theory – which became a major focus of foundational research in the 1920s – Hausdorff continued to work on problems in the field of transfinite arithmetic, i.e. the theory of ordinal and cardinal numbers. Part of this research interest was connected with Julius König's ill-fated lecture at the 1904 Heidelberg ICM, described in the previous section. As a positive result, König had derived the inequality $\mathfrak{a}^{\aleph_0} > \mathfrak{a}$, for $\mathfrak{a}$ a sum of of increasing cardinal numbers. He derived this result from the following theorem: if $1 < \mathfrak{m}_1 < \mathfrak{m}_2 < \ldots$, then

$$\sum_{i=1}^{\infty} \mathfrak{m}_i < \prod_{i=1}^{\infty} \mathfrak{m}_i.$$

This important theorem due to König was later generalized by the Russian set theorist and logician Ivan I. Shegalkin. In 1907, he proved the following result, which has since become known as König's Theorem in the textbook literature: If A is an arbitrary index set and the sets $\{\mathfrak{a}_\alpha\}_{\alpha \in A}$, $\{\mathfrak{b}_\alpha\}_{\alpha \in A}$ have cardinal numbers satisfying $\mathfrak{a}_\alpha < \mathfrak{b}_\alpha$ for all α, then

$$\sum_{\alpha \in A} \mathfrak{a}_\alpha < \prod_{\alpha \in A} \mathfrak{b}_\alpha.$$

This generalization was also found independently by Zermelo [Zermelo 1908].

König's Theorem is directly relevant for the continuum problem because it rules out certain alephs, which cannot be equivalent to the power of the continuum. Thus, since $\aleph_1 < \aleph_2 < \ldots$ and $\aleph_\omega = \aleph_1 + \aleph_2 + \ldots$, it follows from König's Theorem that

$$\aleph_\omega^{\aleph_0} > \aleph_\omega.$$

On the other hand,

$$\aleph^{\aleph_0} = \left(2^{\aleph_0}\right)^{\aleph_0} = 2^{\aleph_0 \cdot \aleph_0} = 2^{\aleph_0} = \aleph;$$

so $\aleph$ cannot be equal to $\aleph_\omega$. In his short note [H 1904a], Hausdorff deals not only with aleph exponentiation, for which his recursion formula (6.3) is a fundamental result, but also with various types of powers, a topic he later pursued in his studies of ordered sets.

In the mid-1920s, Alfred Tarski found a generalization of Hausdorff's recursion formula, starting with three ordinal numbers α, β, γ with $\overline{\gamma} \leq \aleph_\beta$ ($\overline{\gamma}$ is the cardinal number of γ). Tarski then showed that

$$\aleph_{\alpha+\gamma}^{\aleph_\beta} = \aleph_\alpha^{\aleph_\beta} \cdot \aleph_{\alpha+\gamma}^{\overline{\gamma}},$$

which for $\gamma = 1$ results in Hausdorff's formula (6.3). In this work, Tarski also achieved further fundamental results on aleph exponentiation by exploiting the concept of cofinality as well as the generalized continuum hypothesis (GCH).

Hausdorff introduced the notion of cofinality in 1906; two years later he formulated GCH in the form: $2^{\aleph_\alpha} = \aleph_{\alpha+1}$. If GCH holds, then for ordinal numbers α, β Tarski showed:

$$\text{if } \aleph_\alpha \leq \aleph_\beta, \text{ then } \aleph_\alpha^{\aleph_\beta} = \aleph_{\beta+1};$$

$$\text{if } \mathrm{cf}(\aleph_\alpha) \leq \aleph_\beta \leq \aleph_\alpha, \text{ then } \aleph_\alpha^{\aleph_\beta} = \aleph_{\alpha+1};$$

$$\text{if } \aleph_\beta < \mathrm{cf}(\aleph_\alpha) \leq \aleph_\alpha, \text{ then } \aleph_\alpha^{\aleph_\beta} = \aleph_\alpha,$$

where $\mathrm{cf}(\aleph_\alpha)$ is the cofinality of $\aleph_\alpha$. Thus, Tarski's result shows that, if GCH holds, aleph exponentiation is then uniquely determined. If one does not assume the validity of GCH, on the other hand, then one can hardly assert anything with regard to the exponentiation of regular alephs; for example, $2^{\aleph_0}$ can be any $\aleph_n, n \in \mathbb{N}$ (though it cannot be $\aleph_\omega$, as we saw above). Tarski and Hausdorff corresponded in connection with aleph exponentiation, an exchange that led to Hausdorff's unpublished study "Alefsätze" (*HGW* IA, 330, 469–471) from April 1924. Their letters are no longer extant, but references to them appear in Hausdorff's study as well as in Tarski's paper from the time; see (*HGW* IA, 467).

The most significant result on the continuum problem obtained during Hausdorff's lifetime was Kurt Gödel's theorem that the generalized continuum hypothesis and the axiom of choice are consistent with the Zermelo-Fraenkel system **ZF** [Gödel 1938], [Gödel 1939], [Gödel 1940]. There is no indication that Hausdorff received these papers. Since they were published in the U.S., it is very possible, given his living and working conditions at the time, that he had no chance to see them. But it is also possible that he ignored this kind of basic research. He also did not receive Gödel's completeness theorem [Gödel 1930] and his incompleteness theorems [Gödel 1931], which he could have easily studied in the *Monatshefte für Mathematik und Physik* – in fact, the name Gödel appears nowhere in his entire estate. Mathematical logic always remained foreign to Hausdorff. Despite the assurance he once gave Fritz Mauthner that he had put his "anti-logical period" behind him, Hausdorff underestimated the mathematical importance of this field throughout his life.

6.3 Hausdorff's Research on Ordered Sets

Felix Hausdorff's first work on set theory, which was briefly described in the previous section, begins with the following remark:

> In the field of order types opened up by G. Cantor, it is actually only the special realm of ordinal numbers that we know to some extent; extraordinarily little is known about general types, the types of sets that are not well ordered. And yet a more precise knowledge and a classification of types belongs to the narrower scope of problems of set theory: if only because in this way the old question about the cardinality of the continuum can possibly be brought closer to a solution. [H 1901b, 460; IA: 5]

Hausdorff saw a great challenge here, though he could not have foreseen that this would occupy him for a number of years. From around 1905 onward, he seems to have immersed himself in this research with great intensity. From his summer vacation in Bad Elgersburg in Thuringia, he wrote to the librarian Franz Meyer in Jena on August 12, 1905:

> As you can see, I am in your homeland of Thuringia, but this time less to see wonderful landscapes than longing for fresh air and quiet *to work*; I have mathematical stuff to do and I won't let this go, not even on vacation. I am now rather resigned to whether another call will come from a higher sphere. (*HGW* IX, 507)

The last sentence suggests that he probably thought important new results would take some time in coming, but this turned out not to be the case. He was on the brink of a breakthrough.

6.3.1 Investigations into Order Types I–V

Hausdorff submitted his lengthy five-part study "Untersuchungen über Ordnungstypen" ("Investigations on Order Types") to the Saxon Society of Sciences in two stages: the first three parts in February 1906 [H 1906b, IA: 41–104] and the last two almost exactly one year later [H 1907a, IA: 107–182]. In his introductory remarks to these investigations, he gave this general assessment of their place:

> As far as I know, all the *essential* results of this work are new, since the previous investigations of others have almost exclusively related to subsets of the linear continuum or to well-ordered sets. [...] The following developments tie in with the assets of pure set theory that are inventoried in the last two papers by G. Cantor [Cantor 1895/1897] [...] or in the report by A. Schoenflies [Schoenflies 1900] [...] [H 1906b, 106–107; IA: 41–42]

What Cantor "inventoried" can be briefly summarized as follows: First, he defines the concept of the linearly ordered set and then assigns to each such set M an "order type" $\overline{M}$, by which is meant "*the general concept that results from M if we merely abstract from the nature of the elements m but retain the rank order among them*" [Cantor 1932, 297]. Two ordered sets M, N are called similar, written $M \simeq N$, if there is a bijective mapping from M to N that respects the orderings, and $\overline{M} = \overline{N}$ exactly when $M \simeq N$. To each order type α there corresponds a unique cardinal number $\overline{\overline{\alpha}}$, which means if $\alpha = \beta$, then $\overline{\overline{\alpha}} = \overline{\overline{\beta}}$. The converse, however, does not hold; indeed, for any infinite cardinal number $\mathfrak{a}$ there will be an entire class of order types α such that $\overline{\overline{\alpha}} = \mathfrak{a}$. Cantor calls these the type class of $\mathfrak{a}$, which is $[\mathfrak{a}]$. The inverse ordering of an ordered set M of type α yields the inverse type, which Cantor designates by $^*\alpha$.

With these fundamental definitions in hand, Cantor introduced the arithmetic of order types. He defined addition and multiplication as follows: if α, β are two order types and M, N are two linear ordered sets with $\overline{M} = \alpha$, $\overline{N} = \beta$, then $\alpha + \beta$ is the order type of $M \cup N$ as an ordered union. Addition of order types is associative, but not in general commutative. The product $\alpha \cdot \beta$ is defined as the order type of the Cartesian product $M \times N$ ordered lexicographically. Multiplication is associative and right distributive, i.e., $\alpha(\beta + \gamma) = \alpha\beta + \alpha\gamma$, but not in general commutative or left distributive. These two arithmetical operations are independent of the chosen representatives for α and β.

Cantor next turned to investigate the order type η of the set of rational numbers and the type θ of the real numbers, both taken in their natural orderings according to magnitude. For both order types Cantor was able to give a purely set-theoretic characterization independent of the number concept. He first showed that if an ordered set M is countable, unbounded, and everywhere dense, then $\overline{M} = \eta$. From this theorem he could then immediately prove that there are only four distinct countable everywhere dense order types, namely: η, $1+\eta$, $\eta+1$, $1+\eta+1$. To characterize θ, he introduced concepts for ordered sets – closed, perfect, and dense in itself – that were analogous to properties of point sets of the real number line. Cantor could then prove that if M was an unbounded perfect ordered set which contained a countable subset S that was dense in M, then M had the order type of θ. From here Cantor took up the theory of ordinal numbers, that is the order types of well-ordered sets, devoting special attention to the second number class.

We have summed up Hausdorff's starting point, i.e. the state of the theory reached by Cantor, in order to make clear in what follows the tremendous progress Hausdorff achieved in the field of ordered sets. In the second part of his report to the DMV on the development of set theory, Schoenflies referred to these accomplishments, although he had not yet seen Hausdorff's important works from 1908 and 1909. After briefly mentioning graded sets from [H 1901b], he added that "we also owe to Hausdorff the other advances that have been made with ordered sets" [Schoenflies 1908, 40].

From a present day point of view, Hausdorff was one of the most important researchers on set theory in the generation after Cantor. His work greatly expanded the mathematical substance of set theory and extended it into vital new areas. Hausdorff's role, however, has been largely neglected in the historiography of set theory, which has focused for the most part on foundational issues. Hausdorff followed with interest the progress made on such questions, for example in works by Zermelo and Fraenkel, but he did not wish to participate in research on the foundations of set theory.

We now come to the content of the first three parts of "Untersuchungen über Ordnungstypen" [H 1906b]. These investigations were based on the general concepts of power elucidated in the first part, in which Hausdorff succeeded in generating a formidable arsenal of new and interesting order structures. Letting A, M be ordered sets with order types α, μ having the cardinal numbers $\mathfrak{a}, \mathfrak{m}$, Hausdorff introduces his power constructions by means of the covering set M^A, which consists of all single-valued mappings f from A into M. In [H 1904a] he had already introduced powers when A is a well-ordered or inversely well-ordered set. These turn out to be special cases of the general power concepts developed here.

In general, M^A cannot be ordered lexicographically because if $f, g \in M^A$ the set of a for which $f(a) \neq g(a)$ need not have a first element, so that f, g would then be incomparable. The goal was thus to identify subsets of the covering set that can be lexicographically ordered and to study their order types. For this purpose, Hausdorff fixed an $m \in M$, which he called the *principal element*; all other elements of M he called *secondary elements.* We first consider the following subset T_m^W of M^A (later called the *maximal power* by Hausdorff):

$$T_m^W = \{f \in M^A;\ \{a \in A; f(a) \neq m\} \text{ is well ordered}\}.$$

The set $\{a \in A; f(a) \neq m\}$ is defined as the support of the mapping f with respect to the principal element m, written *supp* f. Thus T_m^W is the set of all mappings from A to M, for which the support is well ordered in relation to the given linear ordering of A. For $f, g \in T_m^W$ one can then define the following (lexicographic) ordering:

$$f < g \Leftrightarrow f(\mathrm{Min}(\{a \in A; f(a) \neq g(a)\})) < g(\mathrm{Min}(\{a \in A; f(a) \neq g(a)\})).$$

The minima introduced here exist because they are formed from a subset of the well-ordered set *supp* $f \cup$ *supp* g.

For a given ordinal number ν and principal element m, let

$$T_m^\nu = \{f \in T_m^W;\ \mathrm{Kard}\,\{a \in A; f(a) \neq m\} < \aleph_\nu\}.$$

Hausdorff introduced the order type of the set T_m^ν as $\mu_m^\nu(\alpha)$, calling this a power of the $(1+\nu)$st class. The term *power* is justified by the fact that it satisfies the law:

$$\mu_m^\nu(\alpha) \cdot \mu_m^\nu(\beta) = \mu_m^\nu(\beta + \alpha);\ \ (\mu_m^\nu(\alpha))(\beta) = \mu_m^\nu(\alpha\beta),$$

where α is the *argument* of the power and the inverse type α^* is its *exponent.* This designation is related to Cantor's choice for the product definition $\alpha \cdot \beta$, where α is inserted into β; for example, $\omega \cdot 2$ is represented by $(a_1, a_2, \ldots, b_1, b_2, \ldots)$ which is $\omega + \omega$, whereas $2 \cdot \omega$ is represented by $(a_1, b_1, a_2, b_2, \ldots)$ which is $= \omega$. Cantor originally defined products the other way around, i.e. β is inserted into α, which would have been better adapted for Hausdorff's definition of power. In that case, α would have been the exponent and the order would have remained the same, thus $\mu_m^\nu(\alpha)\, \mu_m^\nu(\beta) = \mu_m^\nu(\alpha + \beta)$. Hausdorff, however, did not want to alter the product definition again because, in his opinion, this would have caused great confusion in the set-theoretical literature.

The types of the first class are those for which $\nu = 0$, so that T_m^0 is the set of all mappings with finite support (with respect to m). This class yields interesting new types beyond those already known. For example, if A is finite with k elements, then $\mu_m^0(k)$ is identical with Cantor's power μ^k independent of m. Hausdorff furthermore proved that if A has no last element, then the set T_m^0 is dense in itself.

For example, for $M = \{0,1\}$, $A = \mathbb{N}$, i.e. $\mu = 2$, $\alpha = \omega$, the types $2^0_0(\omega)$ resp. $2^0_1(\omega)$ yield the simplest representations for $1 + \eta$ resp. $\eta + 1$, whereas for $M = \{0,1,2\}$ the type $3^0_1(\omega)$ is the simplest representation for η. If $\alpha = \omega_1$ (the initial number of the third number class), then $2^0_0(\omega_1)$, $2^0_1(\omega_1)$ are examples of everywhere dense types with cardinality $\aleph_1$. In case the argument A is inverse well-ordered, i.e. the exponent is well-ordered, then the support will be finite and the powers always in the first class. This was one of the cases that Hausdorff dealt with already in [H 1904a].

If $\nu = 1$ then the powers $\mu^1_m(\alpha)$ are in the second class. Hausdorff determined that their cardinality satisfies the relation:

$$\aleph \leq \overline{\overline{\mu^1_m(\alpha)}} \leq (\mathfrak{m}\mathfrak{a})^{\aleph_0}.$$

It follows that if neither $\mathfrak{m}$ nor $\mathfrak{a}$ have a cardinality exceeding the continuum, then all such powers have the cardinality $\aleph$. If the argument A is well-ordered (α is thus an ordinal number), then the covering set M^A can be ordered lexicographically independent of the principal element. In this case, Hausdorff designated the order type $\mu((\alpha))$ and called M^A resp. $\mu((\alpha))$ a *complete power* (*Vollpotenz*), terminology he first introduced in [H 1908, 463].

In Part II, entitled "The Higher Continua," Hausdorff began by introducing two concepts – *cofinal* und *coinitial* – which proved to be fundamental for the entire theory of ordered sets. Let M be an ordered set, then the sets $M^b = \{x \in M; x < b\}$, $M_a = \{x \in M; x > a\}$, $M^b_a = \{x \in M; a < x < b\}$ are called, respectively, the initial, end, and middle segment, whereas segments that include boundaries are called intervals. Hausdorff also referred to the corresponding concepts for initial, end, and middle pieces defined, respectively, for the subset of M consisting of elements preceding or following a given subset A, or the middle piece preceding a given B and following a given A.

If A is a subset of M, then M is cofinal (coinitial) with A if for each $m \in M$ there exists an $a \in A$ such that $a \geq m$ ($a \leq m$). If M is both cofinal and coinitial with A, then M and A are *coextensive.* For example, the set of real numbers in the natural ordering is coextensive with the set of integers. These concepts are easily applied to order types. Hausdorff had recognized that the initial numbers in Cantor's number classes play a particularly important role. He here first considered the initial numbers ω $(= \omega_0)$, ω_1, ω_2, ...; ω_ν is thus the smallest ordinal number with the cardinality $\aleph_\nu$ ($\nu \in \mathbb{N}$) (in modern set theory ω_ν is identified with $\aleph_\nu$). Following Cantor, Hausdorff called ordered sets of type ω_ν and ω^*_ν, or the corresponding ω_ν– and ω^*_ν–sequences *fundamental sequences* and their limits *fundamental limits.* He also retained Cantor's definitions for a set that is closed, dense in itself, or perfect.

In this opening paragraph, Hausdorff also proved two general results:

Theorem A: If $M'' \subset M' \subset M$ and M is cofinal (coinitial, coextensiv) with M', and likewise M' with M'' , then M is cofinal (coinitial, coextensiv) with M''.

Theorem B: A type cannot be cofinal with two different numbers of the sequence ω_0), $\omega_1, \omega_2, \ldots$.

Hausdorff later used Theorem B as a building block for a more general result, which he employed in order to define the notion of the character of an element and a gap in a linearly ordered set.

In the following paragraphs of Part II, it is assumed throughout that the argument A is well-ordered and at most countable; α is therefore an ordinal number of the first or second number class. The powers of the second class are then complete powers $\mu((\alpha))$. Hausdorff first showed that for $1 < \alpha < \omega_1$ there are two order types μ, those in which $\mu((\alpha))$ are all equal and those in which they are pairwise distinct. He then investigated the extent to which properties of the base μ – being closed, dense in itself, or perfect – are inherited by the power $\mu((\alpha))$, obtaining these results:

1. If α *is a limit number or* μ *dense in itself, then* $\mu((\alpha))$ *is also dense in itself.*
2. If μ *is a bounded closed type, then* $\mu((\alpha))$ *is also bounded and closed.*
3. If μ *is bounded and closed and* α *is a limited number or if* μ *is bounded and perfect, then* $\mu((\alpha))$ *is bounded and perfect.*

Hausdorff then turned to special types, including those that are isomeric or homogeneous. A type μ is *isomeric* if all its middle segments have the same type ρ. If all of its segments have the same type as μ itself, then μ is *homogeneous.* One can easily show that every isomeric type is dense in itself, whereas every homogeneous type is both dense in itself and unbounded. The main theorem in this section asserts:

If the base μ *is an isomeric type coextensive with* $\omega^* + \omega$ *and if* $\alpha < \omega_1$, *then* $\mu((\omega^\alpha))$ *is a homogeneous type that is also coextensive with* $\omega^* + \omega$.

Part II concludes with the treatment of continuous types, where Hausdorff showed that Dedekind's concept of continuity can be transferred directly to ordered sets. Thus, μ is said to be *continuous* if for each cut $\mu = \sigma + \rho$ either σ has a last and ρ no first element, or σ has no last and ρ a first element. Every continuous type is dense in itself, and the following theorem holds for complete powers $\mu((\alpha))$:

If μ *bounded and continuous, then so is* $\mu((\alpha))$.

After proving this general result, Hausdorff takes up the special case where the base μ has the type ϑ of the closed unit interval $[0, 1]$. The decisive point here is that ϑ belongs to those special order types, for which Hausdorff proved that the $\mu((\alpha))$ are all distinct. Furthermore, ϑ is isomeric, perfect, continuous, and after removing the boundary elements it is coextensive with $\omega^* + \omega$. As a result, the earlier theorems for $1 < \alpha < \omega_1$ lead to the following principal theorem for Part II:

> 1. *All powers* $\vartheta((\alpha))$ *are distinct from one another, perfect, and continuous.*
> 2. *All powers* $\vartheta((\omega^\alpha))$ *are isomeric and, disregarding the boundary elements, homogeneous and coextensive with* $\omega^* + \omega$.
> 3. *All powers* $\vartheta((\alpha))$ *are invertible (i.e.* $\vartheta((\alpha)) = \vartheta((\alpha))^*$*) and have the cardinality of the continuum.*

The following commentary on these results reveals a bit of Hausdorff's pride, but also his surprise in finding that such a rich world of new order structures comes to light merely by considering these types of powers:

> In particular, there are with certainty $\aleph_1$ different types $\vartheta((\omega^\alpha))$, which share almost all the same characteristic properties (perfectness, isomerism, continuity, reversibility, cardinal number) with the continuum and can justifiably be called continua of a higher order. This result appears very remarkable when compared with views expressed elsewhere and it offers a deep insight into the inexhaustible abundance of higher order types. If one considers that a countable type is already uniquely determined by the single property of being everywhere dense (up to the possibility of boundary elements), it seems quite surprising that in order types with the cardinality of the continuum that demanding special properties, such as being perfect, isomeric, and continuous, still leaves a great deal of freedom. [H 1906b, 143; IA: 78]

Part III, the final part of [H 1906b] entitled "Homogeneous Types of Second Infinite Cardinality," was mainly concerned with the classification of the homogeneous types of cardinality $\aleph_1$. In its first paragraph, Hausdorff dealt with the problem of constructing this concrete type. As mentioned, Cantor had shown that there is only one homogeneous type with cardinality $\aleph_0$, namely the type η of the rational numbers. Hausdorff thus sought to construct something like rational numbers $(\alpha \mid \beta)$ with $\alpha > \beta$ from ordinal numbers α with $1 \leq \alpha < \omega_1$. His method was based on a continued fraction representation for pairs (α, β), which he obtained from the Euclidean algorithm for ordinal numbers. Using this, he derived a remarkable linearly ordered set $\mathcal{M}$, which is universal in the type class $T[\aleph_1]$; thus, every linearly ordered set of cardinality $\aleph_1$ can be embedded in the set $\mathcal{M}$ so as to preserve the ordering relation. Such a set $\mathcal{M}$ would later be called an η_1-set.

In the following, we employ the concepts of element and gap characters later introduced by Hausdorff for homogeneous sets. This was based on Alfred Tarski's extension of Hausdorff's concepts cofinal and coinitial to the notions of cofinality and coinitiality. In 1925, Tarski defined the cofinality of an ordered set M as the smallest ordinal number of any well-ordered subset cofinal with M. If M is homogeneous with $m \in M$, then m has the *character* (κ, λ^*), if κ is the cofinality of the initial segment M^m and λ^* is the coinitiality of its end segment. In this case, m is called a (κ, λ^*)-limit. An ordered decomposition $M = A + B$ into an initial and an end piece is called a *gap*, if A has no last element and B has no first. Such a gap has the character (κ, λ^*), if κ is the cofinality of A and λ^* the coinitiality of B, in which case the decomposition is called a (κ, λ^*)-gap.

In the section following, Hausdorff gave a classification of *homogeneous types* with cardinality $\leq \aleph_1$. There are four groups of such types τ represented by sets M for which:

I. Each element is an $\omega\omega^*$-limit.
II. Each element is an $\omega_1\omega^*$-limit.
III. Each element is an $\omega\omega_1^*$-limit.
IV. Each element is an $\omega_1\omega_1^*$-limit.

For example, a type τ in group III will have ω cofinal and ω_1^* coinitial with τ. Regarding the status of ω_1- and ω_1^*-sequences in τ, there are four families of types:

A. τ contains both ω_1- as well as ω_1^*-sequences.
B. τ contains ω_1-sequences, but no ω_1^*-sequences.
C. τ contains no ω_1-sequences, but ω_1^*-sequences.
D. τ contains neither ω_1- nor ω_1^*-sequences.

Thus, for families of type B. this means that a set M representing τ can be embedded in ω_1 preserving order, but not in ω_1^*. Hausdorff considered combinations of these groups and families, designating each as a *genus.* Certain combinations cannot occur, however: the families C and D do not arise in group II, nor do B and D appear in group III, and only A occurs in group IV. This means there are nine possible genera: IA, IB, IC, ID, IIA, IIB, IIIA, IIIC und IVA.

Within each of these genera further distinctions are required to determine the possibilities for gaps in τ. Concerning the elements there are again four cases: $\omega\omega^*$-, $\omega_1\omega^*$-, $\omega\omega_1^*$-, $\omega_1\omega_1^*$-gaps, which Hausdorff abbreviated as 11-, 21-, 12-, 22-gaps. Those types which possess the same kind of gaps and belong to the same genus form what he called a *species* of types. Not every combination of defining properties (group, family, kind of gaps) can occur simultaneously, however, and there turn out to be altogether 50 different possible species. Among these, 32 species have types that possess $\omega\omega^*$-gaps, whereas the other 18 do not. Thus, for example, the noteworthy powers $\vartheta((\omega^\alpha))$ discussed in Part II are order types of species "I D, without gaps," which is one of these 18 species.

Following this classification, Hausdorff turned to consider

> the fundamental question, *which of our 50 species in the case of types of second cardinality* [$\aleph_1$] *are really represented.* A definitive answer to this cannot be given at the moment, so long as the question concerning the power of the continuum has not been clarified (on the other hand, after so many other approaches have been attempted in vain, precisely here could emerge a new way of answering the continuum question). [H 1906b, 156; IA: 91]

What could this new path look like? Every homogeneous, i.e. everywhere dense type contains the type η as a subset. If it has no $\omega\omega^*$-gaps, then it also contains the linear continuum as a subset. So if one could succeed in constructing a homogeneous type without $\omega\omega^*$-gaps (i.e., a type of the last 18 species), which also had the power $\aleph_1$, then this would mean $\aleph \leq \aleph_1$, proving Cantor's conjecture that $\aleph = \aleph_1$, thus $2^{\aleph_0} = \aleph_1$. Gregory H. Moore sees Hausdorff's "extremely insightful work on order types [... as] motivated by a desire to solve the Continuum Problem" [Moore 1989, 109]. Whether Hausdorff made an earnest effort to pursue this "new path," we do not know. If any such attempts were indeed undertaken, no traces of the necessarily fruitless results can be found among his papers.

In view of the open continuum problem, Hausdorff could only take up the above "fundamental question" only for the 32 species with $\omega\omega^*$-gaps. He gave a complete solution in the following §3 by proving the following:

> *Theorem: All of the species exist for homogeneous types of cardinality no higher than* $\aleph_1$ *and with* $\omega\omega^*$*-gaps.* [H 1906b, 156; IA: 91].

Before giving the proof, Hausdorff made the following remark: "As a construction principle, we apply the power operation developed in the first note [Part I of [H 1906b]], whose fertility and scope has proven itself here in a surprising way; without this comprehensive algorithm our project would hardly have succeeded" [H 1906b, 157; IA: 92].

For this "comprehensive algorithm" Hausdorff defined eight bases $\mu_1, \dots, \mu_8$ as follows:

> $\omega_{11} = \omega + \omega^*$, $\omega_{21} = \omega_1 + \omega^*$, $\omega_{12} = \omega + \omega_1^*$, $\omega_{22} = \omega_1 + \omega_1^*$, then $\mu_1 = \omega_{11}$, $\mu_2 = \omega_{21}$, $\mu_3 = \omega_{12}$, $\mu_4 = \omega_{22}$, $\mu_5 = \omega_{21} + \omega_{22}$, $\mu_6 = \omega_{12} + \omega_{22}$, $\mu_7 = \omega_{21} + \omega_{12}$, $\mu_8 = \omega_{21} + \omega_{22} + \omega_{12}$.

He then proved the following theorem:

> *The eight types* $\mu_1, \dots, \mu_8$*, as bases of the first class, yield all 32 species of homogeneous types with cardinality* $\leq \aleph_1$ *and with* $\omega\omega^*$*-gaps. In fact, one obtains types of group I if the argument is* ω *and the principal element is any element of the base; group II if the argument is* ω_1 *and the principal element is the last element of the base; group III if the argument is* ω_1 *and the principal element is the first element of the base; and group IV if the argument is* ω_1 *and the principal element is a middle element of the base.*

The constructed types for 31 of these species have the cardinality $\aleph_1$, whereas for the species "ID, only $\omega\omega^*$-gaps" Hausdorff was only able to construct a type with the cardinality $\aleph_0$. A few years later, however, Paul Mahlo was able to construct a type of this species with the cardinality $\aleph_1$ [Mahlo 1909].

Parts IV and V of Hausdorff's investigations into order types appeared in the early summer of 1907 [H 1907a, IA: 107–182]. Part IV is entitled "Homogeneous Types with the Cardinality of the Continuum." Here, immediately following Part III, he took up the question as to which of the 50 species of homogeneous types were represented by types with cardinality $\aleph$ of the continuum. Hausdorff the following result at the outset:

An everywhere dense type without $\omega\omega^*$-limits and without $\omega\omega^*$-gaps certainly contains ω_1- and ω_1^*-sequences; if it also has no $\omega_1\omega_1^*$-gaps, then its cardinality is $> \aleph_1$.

From this it follows that there are five species for which each type from them has a power $> \aleph_1$. Thus, were it possible to find a type from one of these species with the power $\aleph$ of the continuum, this would imply $\aleph > \aleph_1$ and Cantor's continuum hypothesis would be refuted. Hausdorff therefore excluded these five species from his investigation. For the remaining 45 species, he was able to show by means of artful power constructions that types of cardinality $\aleph$ exist for each of these species.

The fifth and final part of Hausdorff's investigations into order types is entitled "On Pantachie Types." In the introductory section, Hausdorff explains that here his starting point were Paul du Bois-Reymond's original and probing studies of real-valued functions on $(0, \infty)$. In these, Du Bois-Reymond introduced a partial ordering for the set of all monotonically increasing functions on $(0, \infty)$ based on their behavior at infinity:

$$f < g, \text{ if } \lim_{x\to\infty} \frac{f(x)}{g(x)} = 0.$$

In this case, one says f is "infinitarily smaller" than g, resp. g is "infinitarily larger" than f. Furthermore, two functions f, g are "infinitarily equal" (denoted $f \sim g$), if

$$\lim_{x\to\infty} \frac{f(x)}{g(x)} = a, \quad 0 < a < \infty.$$

The objects being compared here are, more precisely, classes of infinitarily equal functions. Du Bois-Reymond called such a class an "infinitary point" or simply an "infinite," though only a partial ordering can be defined on the set U of such "infinites" because of infinitarily incomparable elements: f, g are incomparable if for $x \to \infty$ the limit of their quotient does not exist. Du Bois-Reymond discovered that the order relation defined in this way (within an ordered subset of U) exhibits a strange behavior that is completely different from the linear continuum. He expressed this as follows: "For a given infinite $\lambda(x)$ and any sequence of functions $\varphi_p(x)$, $p = 1, 2, \ldots$ approaching it, one can always find a function $\psi(x)$ such that for arbitrarily large values of p either

$$\lambda(x) < \psi(x) < \varphi_p(x) \text{ or } \lambda(x) > \psi(x) > \varphi_p(x)$$

holds" [Du Bois-Reymond 1877, 153].

It is hardly surprising that Hausdorff became interested in such ordered structures, though he was little concerned with the lively debates over du Bois-Reymond's ideas that engaged experts in analysis. He was interested instead to investigate the theory from a set-theoretical perspective by considering the order types of maximally ordered subsets of U:

> If we take as our task to relate the infinitary rank ordering as a whole with the theory of Cantor's order types, then we are forced to investigate *the most comprehensive sets of pairwise comparable functions.* Here, by the most comprehensive possible we mean that such a set cannot be extended to include other functions comparable with all the functions of the set. [H 1907a, 110; IA: 133]

Hausdorff called such a maximally ordered subset of U, following the terminology of du Bois-Reymond, a *pantachie.*

Hausdorff next showed that the order types of the pantachies remain unchanged if the "infinitary rank ordering" is replaced by a "final rank ordering" based on number sequences. He also calls this final rank ordering a "graduation by final behavior." In the "final rank ordering" a partial ordering is defined on the set $\mathcal{M} = (\mathbb{R}^+)^\omega$ of all sequences $X = (x_1, x_2, \ldots)$ of positive real numbers as follows:

> If for $X, Y \in \mathcal{M}$ there exists an n_0 such that for all $n \geq n_0$ $x_n = y_n$ throughout, then X is finally equal to Y (written $X \sim Y$), $x_n < y_n$, throughout, then X is finally smaller than Y $(X < Y)$.

If neither $X \sim Y$ nor $X < Y$ or $X > Y$, then X, Y are incomparable, designated by the notation $X|Y$. Hausdorff calls a linearly ordered subset of $\mathcal{M}$ an *ordered domain* (*Bereich*); an ordered domain in which there is no element in $\mathcal{M}$ comparable with all its elements is then called a pantachie. The existence of ordered domains is trivial, but what about a pantachie? Hausdorff answers this by noting Zermelo's proof in 1904 of the well-ordering theorem: "The existence proof is very simple if one bases this on the well-ordering of the continuum" [H 1907a, 117; IA: 140].

Hausdorff next proves a special case of his general maximal chain theorem, which today carries his name; this latter result is equivalent to the well-ordering theorem and hence to the axiom of choice. In the present context, this theorem states:

> *Every ordered domain is contained in a pantachie, which is to say that any linearly ordered subset of $\mathcal{M}$ can be expanded to a maximal linearly ordered subset.*

In his proof, Hausdorff used the fact that $\mathcal{M}$ has the power of the continuum and thus, according to the well-ordering theorem, $\mathcal{M}$ can be written in the form $\{X_0, X_1, \ldots X_\omega, \ldots X_\alpha, \ldots\}$, where all indices are less than a suitable ordinal number κ.

Hausdorff proved the following fundamental theorems for a pantachie P:

> (I) *If $\{U_n\}$ is an at most countable set of elements from $\mathcal{P}$, then there exist $X, Y \in \mathcal{P}$ with $U_n < X$ $(Y < U_n)$ for all n.*
> (II) *If $\{U_n\}$, $\{V_n\}$ are at most countable sets of elements from $\mathcal{P}$ and if*

$U_j < V_k$ for all j, k, then there exists an $X \in \mathcal{P}$ with $U_m < X < V_n$ for all m, n.

From (II) one can derive a proposition for a pantachie $\mathcal{P}$ analogous to the result of du Bois-Reymond quoted above, namely, a pantachie has no ω- or ω^*-limits. Furthermore, theorems (I) and (II) together lead to the following results:

A pantachie is unbounded, everywhere dense, and is neither with ω cofinal nor with ω^* coinitial; moreover, it has no $\omega\omega^*$-gaps.

Hausdorff called an order type with all these properties an *Eta*-type, or in print form H-type since a capital *Eta* looks like an H.

In the following §3, "On H-Types," Hausdorff's starting point is the type η of the rational numbers in their natural ordering. Cantor already knew that η is universal for the type class $T(\aleph_0)$ of all countable order types, so that every countable ordered set can be imbedded in the rational numbers. For the unbounded everywhere dense type, like η, the following holds: if A, B are *finite* subsets of such a type with $A < B$, then there are elements x, y, z with

$$x < A, \qquad A < y < B, \qquad B < z. \tag{6.4}$$

Hausdorff calls these η-types. There is only one countable η type, namely the type of the rational numbers, denoted η_0. In what follows, condition (6.4) will be extended to apply also for A, B *at most countable subsets* of the types under consideration. If one then chooses single-element sets or ω- or ω^*-sequences for A, B, one sees that a type of this kind must have the following properties: it will be unbounded, everywhere dense, and will be neither with ω cofinal nor with ω^* coinitial, nor will it have ω- or ω^*limits or any $\omega\omega^*$-gaps. Conversely, these properties imply that the condition (6.4) applies for A, B at most countable ($A < B$). Since this is a first step beyond η_0, Hausdorff designates a type of this kind as an η_1-type. According to the results in §2, H-types and η_1-types are identical. Hausdorff then proves a series of theorems for η_1-types:

(I) *Every η_1-type is universal for the type class $T(\aleph_1)$.*
(II) *If there exists an η_1-type with cardinality $\aleph_1$, then it is unique.*

(I) is the analogue to the universality of η_0 for $T(\aleph_0)$. In his proofs for (I) and (II) Hausdorff used his newly invented "back-and-forth" method, which later found numerous applications in set theory, algebra, and other areas of mathematics.

With respect to cardinality the following holds for η_1-types:

(III) *An η_1-type has at least the power $2^{\aleph_0}$ of the continuum. If an η_1-type has no $\omega_1\omega_1^*$-gaps, then its cardinality is at least $2^{\aleph_1}$.*
(IV) *For any η_1-type there is an embedding of the power second class $3_0^1(\omega_1)$ that preserves order.*

Here the base is $\mu = 3 = \{-1, 0, 1\}$, the principal element is 0, and the argument is ω_1. This type is a homogeneous η_1-type of the species "IVA, $\omega_1\omega^*$-, $\omega\omega_1^*$-, $\omega_1\omega_1^*$-gaps."

These results can now be applied to pantachies. Since $\mathcal{M}$ has the power of the continuum, every pantachie has at most the power of the continuum. It thus follows from (III): Every pantachie has the power $2^{\aleph_0}$ of the continuum. From (I) and (IV) follows: In a pantachie, every ordered set with the power $\aleph_1$ as well as every homogeneous set $3_0^1(\omega_1)$ can be embedded isomorphically. Finally, from (II) follows that, if the continuum hypothesis $2^{\aleph_0} = \aleph_1$ were to hold, then all pantachies would be order-isomorphic, homogeneous, and of type $3_0^1(\omega_1)$; there would then be only one type of pantachie.

Hausdorff next set the goal of finding universal types η_ν for the type classes $T(\aleph_\nu)$ ($\nu \in \mathbb{N}$). He did this by stepwise extending the cardinality of A and B in relation to (6.4). An η_ν-type is one in which, for subsets A, B (with $A < B$) with cardinality $< \aleph_\nu$, there always exist x, y, z that satisfy the relations (6.4). Every $\eta_{\nu+1}$-type is thus an η_ν-type. Moreover, the ideas used to prove the theorems for η_1-types can be modified in a straightforward way to obtain these results for η_ν-types:

(I) *Every η_ν-type is universal for the type class $T(\aleph_\nu)$.*
(II) *If there exists an η_ν-type with cardinality $\aleph_\nu$, then it is unique.*
(III) *An η_ν-type has at least the power $2^{\aleph_{\nu-1}}$. If an η_ν-type has no $\omega_\nu\omega_\nu^*$-gaps, then its cardinality is at least $2^{\aleph_\nu}$.*
(IV) *For any η_ν-type there is an embedding of the power $(\nu+1)$ class $3_0^\nu(\omega_\nu)$ that preserves order.*

It was in this connection that Hausdorff for the first time brought the generalized continuum hypothesis into play: "Should Cantor's hypothesis $2^{\aleph_0} = \aleph_1$ hold in an extended sense, so that $2^{\aleph_\nu} = \aleph_{\nu+1}$ for any finite ν, then there would be exactly one η_ν-type with cardinality $\aleph_\nu$, namely η_ν itself" [H 1907a, 133; IA: 156] (by the latter is meant the type $3_0^\nu(\omega_\nu)$).

The question as to whether homogeneous pantachies exist is, according to the above, to be answered positively, if one assumes the validity of $2^{\aleph_0} = \aleph_1$. In §4 Hausdorff presented a construction of a homogeneous pantachie without assuming the continuum hypothesis, about which he wrote: "All the more remarkable is the fact that with a suitable method and without any hypothesis, one can arrive at the *construction of a homogeneous pantachie*, which brings us at least one step closer to the ideal of finding a systematic procedure for obtaining these strange types" [H 1907a, 140–141; IA: 163–164]. Hausdorff's method consisted in constructing pantachies with an additional algebraic structure (see 6.3.4).

In final sections of his paper, Hausdorff dealt with the problem of finding a closer classification for the order types of pantachies within the class of η_1-types. As we have seen, if the continuum hypothesis is valid, then this problem is already settled. If, however, one only assumes that $2^{\aleph_0} \geq \aleph_1$, then a number of open problems result:

(α) *Is there a pantachie without $\omega_1\omega_1^*$-gaps?*

If so, then according to theorem (III) on η_1-types, the continuum would have the cardinality $2^{\aleph_1} > \aleph_1$, which means the continuum hypothesis would be refuted.

> (β) *Is there more than one type of pantachie, i.e. could there be two or more pantachies that are not isomorphic as ordered sets?*

A positive answer would also imply that the continuum hypothesis is not valid.

> (γ) *What are the sequences of highest cardinality that can occur in a pantachie?*

It could happen that, beyond the ordered sets of types $\omega, \omega^*, \omega_1, \omega_1^*$, others such as $\omega_2, \omega_3, \ldots$ and their inverses can be imbedded, and that among these ω_ν there is a maximum ω_μ. In this case, one would deduce that $2^{\aleph_0} \geq \aleph_\mu$.

Finally, Hausdorff formulated the scale problem:

> *Is there a well-ordered subset in $\mathcal{M} = (\mathbb{R}^+)^\omega$ that is cofinal with $\mathcal{M}$?*

Hausdorff called such a well-ordered subset a scale. In the final pages of [H 1907a] he dealt with the scale problem and proved the following theorem:

> *If the continuum hypothesis holds, then there exist ω_1-scales.*

For the proof he uses the well-ordering theorem applied to the set $\mathcal{M}$; the proof is then relatively easy. The real challenge, in fact, lies elsewhere:

> To prove the existence of an Ω-scale [$\Omega = \omega_1$], but independently of the continuum hypothesis, would therefore be an essential step forward, as this would establish the first closer connection between the continuum and the second number class; however, the difficulty seems to harmonize with the importance of the matter. [H 1907a, 155; IA: 178]

Hausdorff described the constantly recurring characteristic problem as follows:

> The difficulties of the pantachie problem, like the analogous ones for the continuum question, can be briefly expressed as follows: one is carried beyond the realm of the countable, but one does not know how far, since the starting point lies in the countable and yet no direct relation can be recognized with the next higher level, the second power [$\aleph_1$]. [H 1907a, 158; IA: 181]

The realization that the question "how far?" cannot be decided on the basis of Zermelo-Fraenkel set theory with the axiom of choice was one of the greatest mathematical achievements of the twentieth century. We will briefly return to the scale problem at the end of the discussion on ordered quantities.

6.3.2 The Dresden Lecture

In September 1907 Felix Hausdorff attended the annual meeting of the German Mathematical Society (DMV), which took place in Dresden as part of the 79th meeting of the Society of German Natural Scientists and Physicians. On this occasion, he delivered a lecture with thc title "On dense ordered types," which was later published as [H 1907b, IA: 203–208]. According to the report from this meeting, there were 44 participants; after his talk, Arthur Schoenflies made some remarks in the discussion session and Hausdorff spoke as well. His lecture was related to ideas in Richard Dedekind's famous work *Continuity and Irrational Numbers* (1872), which introduced what today are called Dedekind cuts. These were defined as ordered decomposition $M = A + B$ $(A < B)$ of an ordered set M, though this was not an abstract set but rather consisted of the set of rational or of real numbers.

In any such ordered decomposition, there are four possibilities:

(α) A has a last and B a first element.
(β) A has a last and B has no first element.
(γ) A has no last and B has a first element.
(δ) A has no last and B has no first element.

Deviating from Dedekind's terminology, Hausdorff calls the decomposition in the case (α) a jump, in the cases (β) or (γ) a cut, and in case (δ) a gap. A set is everywhere dense precisely when it has no jumps, and [H 1907b] is devoted to the investigation of such sets, which Hausdorff here calls *dense sets.*

In the following, Hausdorff considers arbitrary initial numbers ω_α, where the index α is any ordinal number rather than being restricted to natural numbers. "As usual," he writes,

> we denote the power of a well-ordered set by $\aleph_\alpha$, where the index α is an ordinal number, namely the type of the set of all the smaller alephs; the smallest ordinal number with the cardinality $\aleph_\alpha$ is written ω_α and is called an *initial number.* [H 1907b, 542; IA: 204]

The "as usual" indicates that Hausdorff assumed the general introduction of the aleph sequence and thus also the ω_α to be known; in fact, they can be found in Gerhard Hessenberg's paper [Hessenberg 1906]. Hausdorff calls sets with the type of an initial number (or of the inverse type) sequences.

Hausdorff next defined a concept that would prove to have fundamental significance for set theory:

> *The initial number ω_α is called regular if it is not cofinal with any smaller ordinal number, otherwise it is singular.*

For example, ω $(= \omega_0)$ and every nonlimit number $\omega_{\alpha+1}$ is regular, whereas ω_ω as the limit of ω_n $(n \in \mathbb{N})$ is cofinal with ω_0 and therefore singular. Since the cardinal numbers $\aleph_\alpha$ are identified with the initial numbers ω_α in modern set theory, one can say that the distinction between regular and singular cardinal numbers goes back to Hausdorff.

The following theorem is fundamental for further developments:

Every linearly ordered set without a last element is cofinal with exactly one regular initial number. Every linearly ordered set without a first element is coinitial with the inverse of exactly one regular initial number.

This theorem allows the position of elements and gaps in dense sets to be characterized by pairs of regular initial numbers (cardinal numbers): Let M be a linearly ordered dense set without a first and last element and $m \in M$. Then in the ordered decomposition $M = A + m + B$ A has no last and B no first element, thus A is cofinal with ω_α,B coinitial with ω_β^*; $\omega_\alpha, \omega_\beta$ being regular initial numbers. In this case, one calls m an $\omega_\alpha\omega_\beta^*$-element, or simply a $c_{\alpha\beta}$-element. The symbol $c_{\alpha\beta}$ is then the *character* of m. In the same way, one concludes for a gap $M = C + D$ that C is cofinal with a regular ω_γ,D is coinitial with the inverse ω_δ^* of a regular ω_δ. The gap is then called an $\omega_\gamma\omega_\delta^*$-gap or a $c_{\gamma\delta}$-gap for short, where $c_{\gamma\delta}$ is the character of the gap. The characters of elements and gaps serve to classify ordered sets and their order types.

Now let $U(M)$ be the set of element characters, $V(M)$ the set of gap characters of an ordered set M. For example, for the set of rational numbers $U = \{c_{00}\}$, $V = \{c_{00}\}$, whereas for the set of real numbers $U = \{c_{00}\}$, $V = \emptyset$. Hausdorff calls sets with $V = \emptyset$ continuous. Two ordered sets (or their order types) belong to the same *species* if they have the same sets (U, V), so that Hausdorff used this symbol to denote a species. Letting $W = U \cup V$, he further designated two species as belonging to the same *genus* if they have the same W.

Hausdorff now proceeded much like Dedekind had when he constructed the real numbers by means of cuts in the set of rationals. In the case of a dense set M of species (U, V) and genus $W = U \cup V$, Hausdorff constructed a continuous set of the same genus, thus $W = (U \cup V, \emptyset) = (W, \emptyset)$. He did so by introducing a set of ideal elements $\overline{M}$, where each new element $\overline{m} \in \overline{M}$ results from a corresponding gap in $M = C \cup D$. The order relation on M then induces an order relation on the extended set $[M] = M \cup \overline{M}$, which no longer has any gaps and is therefore a continuous ordered set of species $(W, \emptyset)$. Hausdorff calls $\overline{M}$ the *completion* and $[M]$ the *filling* of M. As a simple example, he notes that the sets of rationals and irrationals are completions of one another, whereas the reals are their common filling.

Let M again be a dense linearly ordered set without first or last element and let ω_π be the smallest regular initial number, for which no character $c_{\pi\gamma}$ appears in W, and likewise for ω_ρ^*. Then it follows for every $c_{\alpha\beta} \in W$:

$$\alpha < \pi,\ \beta < \rho,$$

where $\omega_\alpha, \omega_\beta$ are regular. If the character sets $U \neq \emptyset$ and V are given with $U \cup V = W$, then W is called *complete* if there exists a type of species (U, V)

and genus W. One of Hausdorff's key results in [H 1907b] gives necessary and sufficient conditions for a given W to be complete.

If the bounding initial numbers ω_π, ω_ρ for W have finite indices, then the number of genera and species can be calculated. In the case $\pi = \rho = 2$ there are six genera:

$$\{c_{00}, c_{11}\}, \{c_{00}, c_{01}, c_{10}\}, \{c_{00}, c_{01}, c_{11}\},$$

$$\{c_{00}, c_{10}, c_{11}\}, \{c_{01}, c_{10}, c_{11}\}, \{c_{00}, c_{01}, c_{10}, c_{11}\}.$$

If $\pi, \rho \leq 2$, there are 9 genera and 210 species, and the number grows very fast. Thus, for $\pi, \rho \leq 3$ there are already 302 genera and 243,376 species. If π or ρ are transfinite, then the set of species has at least the power of the continuum.

Hausdorff had to make a restriction on the indices π and ρ for the "bounds" on W, a limitation from which he "has not yet been able to free himself," namely, they must not be limit numbers. He would have been exempt from making this restriction if he could assert that regular initial numbers (cardinal numbers) with a limit number as index did not exist. Regarding this, he remarked: "It is very questionable whether there are any regular initial numbers with a limit index; in any case, the smallest of these would be of such an exorbitant cardinality that it would likely exceed all the sets hitherto considered or still to be considered" [H 1907b, 546; IA: 208].

Regular initial numbers (cardinal numbers) with a limit index were later called weakly inaccessible cardinal numbers. Hausdorff's remark thus sounds like a premonition that the existence of such numbers would not be provable by means of the then usual principles for constructing sets. Indeed, the existence of weakly inaccessible cardinal numbers cannot be proved from the axioms of Zermelo-Fraenkel set theory including the axiom of choice; one must – if so desired – assert their existence by means of an additional axiom. Hausdorff thus hereby hinted at what would become an extensive research area in modern set theory, the field of large cardinal numbers.

6.3.3 Hausdorff's Comprehensive Study of Ordered Sets

Toward the end of June 1907 several mathematicians from the nearby universities of Leipzig, Halle, and Jena met in Kösen near Naumburg/Saale. Among them were Felix Hausdorff and Georg Cantor from Halle. On this occasion, Hausdorff had the opportunity to present some results from his investigations on order types, which he also discussed, in particular, with Cantor. The following letter from Hausdorff to Hilbert, dated July 15, 1907, relates to their encounter:

> Professor Cantor, with whom I spent some time together 14 days ago, suggested that I work out a brief synopsis of my "Investigations on Order Types" and offer this to you for publication in Mathematische Annalen. He thought it desirable that these matters reach a wider audience than the readership of the Leipziger Berichte [a publication of the Scientific

> Society of Saxony]; he also proceeded from the not incorrect assumption that the somewhat lengthy work would become more comprehensible in a shortened and systematic presentation restricted to the essentials. [...] I allow myself thus to ask whether you would be inclined *in principle* to accept a paper for the Annalen, entitled say "Theory of Order Types" and with a length of between 16 and 24 pages.
>
> You may find the suggestion of casting a vote on a paper that has yet to be written somewhat premature. Of course, this would be a statement in principle, which does not obligate you in any way but rather reserves your right to judge the actual work at hand. It is only that I would prefer to save myself the trouble if, for example, the editorial board of the Annalen were inclined to exclude the field of set theory, which is now so widely contested (and with such medieval weapons!). (*HGW* IX, 332)

The last passage sheds telling light on the uncertain and controversial position that set theory as a mathematical sub-discipline still held around 1907 in the eyes of many mathematicians.

Hilbert must soon have reacted positively because Hausdorff's estate contains a manuscript of "Fundamentals of a Theory of Ordered Sets" (*Grundzüge einer Theorie der geordneten Mengen*) with a marginal note in Hilbert's hand: "Nov. 1907, Accepted Hilbert." The paper, which appeared under the same title in [H 1908], had the structure of a self-standing memoir or brochure with a preface, table of contents, and subject index.

In his preface, Hausdorff noted that within the field of ordered sets only the well-ordered sets and sets of real numbers, i.e. linear point sets, had been investigated in any substantial detail. The main subject of his own previous work have been

> certain types distinguished by a particularly regular structure (homogeneity) with sequences up to the second cardinality, and as far as generalizations are sought, they are limited to the next nearest levels, which correspond to the alefs with finite indices. In the present treatise, should it not swell into a book, these special types appear only in the role of occasional illustrative examples, whereas the general methods take up the foreground. The ideal in mind was to survey the world of types in the sense of building up the complex structures from elementary ones by means of generating operations [...] [H 1908, 435; IA: 213]

These generating operations are the addition over an ordered "generator" together with operations Hausdorff introduced for products and powers. The "elementary structures," his "building material"

> are the well-ordered sets and their inverses, whose powers may no longer actually be restricted to alefs with a finite index, and especially the

> *regular initial numbers* with their inverses, which can be seen as the ultimate building blocks and primordial atoms of the world of types. [H 1908, 435–436; IA: 213–214]

The main subject in [H 1907b] was to demonstrate the possibility of classifying dense sets and to show the enormous variety of order structures that thereby emerge. This remained, of course, an important part of the present work, though in a reduced form. In its preface, Hausdorff described the corresponding results with these words:

> The dense sets can be classified on the basis of the "characters" of their elements and gaps, these characters themselves arising from the regular initial numbers. With the help of generating operations of the most general kind, one finds that all a priori conceivable "species" of dense sets actually exist, and they are represented moreover not by just one, but by uncountably many essentially different types. [H 1908, 436; IA: 214]

Hausdorff referred to his own earlier work, but he emphasized in the preface that no prior knowledge was assumed. In [H 1908] some topics from his investigations into order types are repeated or developed in greater generality than in [H 1907b]. In what follows, we will only discuss results that are completely new or substantially generalized compared with the previous work.

Paragraphs 10 and 11 of [H 1908] deal with a completely new subject. In §10 Hausdorff introduces "Type Rings" as follows:

> General addition is to be seen as a fundamental operation for generating types from other types; its vast, though not unlimited scope will be evident from the following conceptual constructions. [H 1908, 454; IA: 232]

Hausdorff calls a set $\mathcal{A}$ of order types a *type ring* if

> (α) the sum of two types in $\mathcal{A}$ is again in $\mathcal{A}$,
> (β) the sum of an infinite number of types of $\mathcal{A}$ generated by a type in $\mathcal{A}$ is also in $\mathcal{A}$.

If $\mu, \nu \in \mathcal{A}$, then so is their product, and further more if $2 \in \mathcal{A}$ then (α) follows from (β). The intersection of an arbitrary number of type rings is again a type ring, if $\neq \emptyset$.

Now let A be a set of order types, then the intersection of all type rings that contain A, denoted by $[A]$, is the smallest such type ring. Hausdorff then introduced the concept of a basis, a notion of fundamental importance for the structure of a type ring. If the type β does not belong to a type ring that contains A, then it is is called independent of A, otherwise it is called dependent on A. $[A]$ thus consists of all types that are dependent on A. If no $\alpha \in A$ is dependent on the set $A \setminus \{\alpha\}$, then A is called a *basis* of the ring $[A]$.

Hausdorff emphasized that there are numerous examples of type rings, "since almost every type set that is presented naturally is a ring" [H 1908, 455; IA: 233]. Due to the aleph relations, every type class $T(\aleph_\alpha)$ is a type ring. Each Cantorian number class is also a type ring. Moreover, if ω_α is a regular initial number, then the set of all ordinal numbers $< \omega_\alpha$ is a type ring whose basis is

$$A = \{1, \omega, \omega_1, \dots \omega_\xi, \dots\} \qquad (*)$$

with regular initial numbers $\omega_\xi < \omega_\alpha$. Thus $[1, \omega]$ is the ring consisting of the first two number classes, $[1, \omega, \omega_1]$ consists of the first three number classes, $[\omega]$ contains all limit numbers of the second number class, $[1, \omega + 1]$ all finite numbers and all non-limit numbers of the second number class. The second number class itself has no finite base.

Properties that are invariant under general addition can be used to construct type rings. Such an invariant property is for a type to be unbounded. Hausdorff was, however, well aware of the danger of inconsistent "sets," so he carefully avoided speaking of the type ring of unbounded types. He considered instead the type ring of all types of power $\leq \aleph_\alpha$, from which he determined that this contains three subrings, namely: a) the unlimited types, b) the limited types, and c) the types with a last element.

In §11 Hausdorff proves a very general structure theorem. He calls an ordered set *scattered* if it contains no dense subset, and then shows for arbitrary ordered types:

> *Theorem: Every ordered set is either scattered or an ordered sum of scattered sets over a dense generator.*

For example, to build the ring of all at most countable order types, one first forms the ring $[1, \omega, \omega^*]$ of all scattered countable types. There are only four dense countable generators: η, $1+\eta$, $\eta+1$, $1+\eta+1$. So all finite types and all types from $T(\aleph_0)$ are elements of the ring $[1, \omega, \omega^*, \eta]$, which has a finite basis. About this, Hausdorff remarked, "this strange result is probably an isolated instance because, starting from the cardinality of the continuum, there are an infinite number of independent dense types" [H 1908, 458; IA: 236]. (He proved this already in §25 of [H 1908].)

The third and final part of [H 1908] deals with dense sets. Hausdorff begins by proving in full detail the main theorem formulated in [H 1907b] for complete sets of characters W. This states:

> *Theorem: If W is a complete set of characters and $U \neq \emptyset$ and V are sets of characters with $U \cup V = W$, then there exists a type of species (U, V) and the genus W. In particular, there is an irreducible type of species $(W, \emptyset)$.*

A type is called irreducible if every middle segment of a representative set M belongs to the same species as M itself. In particular, it follows from this theorem that irreducible types of any species of the same genus exist.

After having briefly studied η_ν-sets for finite ν in [H 1907a], Hausdorff developed a general theory with arbitrary ordinal numbers as index based on the following concept:

> *A dense unbounded set M is called an η_α-set if it is not cofinal or coinitial with any set of cardinality $< \aleph_\alpha$ and does not contain a pair of adjacent subsets that both have cardinality $< \aleph_\alpha$.*

Subsets A, B of M with $A < B$ are called *adjacent* if there is no $m \in M$ that lies between A and B. In this general case, too, the following theorems hold:

> (I) *η_α is universal in the type class $T(\aleph_\alpha)$, i.e., any ordered set of cardinality $\leq \aleph_\alpha$ can be embedded order isomorphically in a η_α-set M.*
> (II) *There is at most one η_α-type of power $\aleph_\alpha$, i.e., if there are two η_α-sets of power $\aleph_\alpha$, they are order isomorphic.*

Hausdorff had to leave open the question whether η_α-sets of the smallest possible cardinality $\aleph_\alpha$ exist at all. For non-limit numbers as index, however, he formulated the following result as a consequence of what he called Cantor's Aleph Hypothesis:

> *An $\eta_{\alpha+1}$-set of cardinality $\aleph_{\alpha+1}$ exists if and only if the generalized continuum hypothesis $2^{\aleph_\alpha} = \aleph_{\alpha+1}$ holds.*

For a singular $\aleph_\alpha$-set, a η_α-set is also a $\eta_{\alpha+1}$-set.

Hausdorff called an ordered dense set M a $c_{\sigma\sigma}$-set if $c_{\sigma\sigma}$ is the lowest symmetrical character in its character set W. Such a set M has neither elements nor gaps of character $c_{00}, c_{11}, \ldots c_{\alpha\alpha} \ldots$ with $\alpha < \sigma$. The penultimate paragraph of [H 1908] is dedicated above all to cardinality estimates for such sets. Hausdorff denotes the sum of the powers

$$\aleph_\sigma + \aleph_\sigma^2 + \ldots + \aleph_\sigma^{\aleph_0} + \aleph_\sigma^{\aleph_1} + \ldots + \aleph_\sigma^{\aleph_\gamma} + \ldots = \sum_{\gamma<\sigma} \aleph_\sigma^{\aleph_\gamma}$$

with $(\aleph_\sigma)_\sigma$. Then the following result holds:

> *Theorem: A $c_{\sigma\sigma}$-set M has cardinality at least $(\aleph_\sigma)_\sigma$. If it has no $c_{\sigma\sigma}$-gaps, then its cardinality is at least $\aleph_\sigma^{\aleph_\sigma}$.*

Hausdorff now assumes that σ is not a limit number, i.e. $\aleph_\sigma$ is regular, and then shows that $(\aleph_\sigma)_\sigma = \aleph_\sigma$ precisely when the GCH holds: $2^{\aleph_{\sigma-1}} = \aleph_\sigma$. Under this condition it then follows:

$$\textit{For any } c_{\sigma\sigma}\textit{-set } M,\ \mathrm{card}(M) \geq \aleph_\sigma.$$

Hausdorff showed further that one can construct a $c_{\sigma\sigma}$-set with the least possible cardinality $\aleph_\sigma$ by writing this in the form of a special power of the $(1+\sigma)$-th class. Assuming GCH, a $c_{\sigma\sigma}$-set M without $c_{\sigma\sigma}$-gaps has $\mathrm{card}(M) \geq \aleph_{\sigma+1}$.

6.3.4 Graduation by Final Behavior

After the appearance of his comprehensive study from 1908, Hausdorff wrote two additional papers on ordered sets: "Graduation by Final Behavior" [H 1909a, IA: 297–335] and "Sums of $\aleph_1$-Sets" [H 1936b, IA: 349–363]. Although closely related in content, they were separated by 27 years. The first appeared in the *Reports of the Saxon Scientific Society*, whereas the second was published in *Fundamenta Mathematicae*, the leading journal for studies in the foundations of mathematics. This may explain in part why the second work has enjoyed a reception that continues to the present day, whereas the first was largely overlooked. Vladimir Kanovei wrote an extensive essay on both works along with comments on individual passages for Volume IA of the Hausdorff Edition [Kanovei 2013]. For the second paper, he offered the following assessment of its importance:

> Judging by its impact on the course of set theoretical investigations, Hausdorff's paper [H 1936b] is one of the most valuable research articles of the pre-forcing era of set theory. The results obtained, especially, the existence of (ω_1, ω_1^*)-gaps, the concepts and methods introduced, and the problems discussed in [H 1936b], have inspired numerous set-theoretic studies, including those based on forcing and other techniques completely unknown in Hausdorff's times. [Kanovei 2013, 367]

The study [H 1909a] follows up on Part V, "On Pantachie Types," in [H 1907a]. Hausdorff recapitulated his earlier results in the first two paragraphs before turning to the maximal principle, which he used for the first time in [H 1909a]. He now generalizes this to a maximal chain theorem, which no longer needs reference to ordered sets (the usual formulation and proof of Hausdroff's maximal chain theorem appears in *Grundzüge der Mengenlehre* [H 1914a, II: 140–141].) In §3 Hausdorff introduced componentwise operations for addition, subtraction, multiplication, and division of real number sequences in the set $\mathbb{R}^\omega$, that is, for $A = \{a_n\}$, $B = \{b_n\}$ he defines

$$\frac{A}{B} = \{\frac{a_n}{b_n}\}; b_n \neq 0, n > n_0.$$

A *rational domain* (*rational pantachie*) is closed under these operations, so for any A, B in the domain (pantachie), $A + B$, $A - B$, AB, $\frac{A}{B}$ also belong to it. Hausdorff succeeded in proving the existence of a rational pantachie. The proof is quite complicated, using a number of tools from algebra and finally the general maximal chain theorem that he proved in §1 (Plotkin notes this was the first use of a maximal principle in algebra; for a modern proof, see [Plotkin 2005, 263–265]). We saw earlier that, assuming $2^{\aleph_0} = \aleph_1$, all pantachies are of the same homogeneous type η_1. For rational pantachies, on the other hand, their homogeneity follows directly from the closure of the above operations. In this way, Hausdorff succeeded in proving the existence of homogeneous pantachies "independent of the continuum hypothesis" [H 1909a, 310; IA: 311].

This is the problem he tackled in §4, entitled "Relationships to the Second Number Class." More precisely, without assuming the validity of the continuum hypothesis, Hausdorff considered the possibility of proving the existence of a pantachie that satisfies one or more of four properties:

1. cofinal with ω_1 (or coinitial with ω_1^*),
2. contains ω_1-elements (or ω_1^*-elements),
3. contains $\omega_1\omega^*$-gaps (or $\omega\omega_1^*$-gaps),
4. contains $\omega_1\omega_1^*$-gaps.

He noted that if even one of these existence statements could be proved, this "would establish a first closer relationship between the continuum and the second number class" [H 1909a, 320; IA: 321]. Although he was unable to prove the existence of a pantachie with any of the first three properties, Hausdorff did manage to show that if a pantachie satisfies any one of them, then it will meet all three conditions.

As for the fourth case, he was able to score a resounding success by proving:

Theorem (I): Within the set $\mathbb{R}^\omega$ of sequences of real numbers, partially ordered according to the sign of the last difference, pantachies exist with at least one $\omega_1\omega_1^*$-gap.

The proof, which is more than three pages long, involves many subtle intermediate steps and a great deal of ingenuity.

The last paragraph of [H 1909a] concerns applications of the theory to questions in analysis. Let $\mathfrak{Z}$ be the set of real positive number sequences, partially ordered according to the sign of the last difference. A sequence $A = \{a_n\}$ is called convergent (divergent) if $\sum_n a_n$ converges (diverges). Hausdorff ultimately wants to answer du Bois-Reymond's question as to whether there can be a "boundary between convergence and divergence." Suppose there is a pantachie $\mathfrak{P}$ in $\mathfrak{Z}$ that may contain convergent as well as divergent sequences, and let $\mathfrak{P}_c$ be the set of convergent and $\mathfrak{P}_d$ be the set of divergent sequences. Then $\mathfrak{P}_c < \mathfrak{P}_d$, which means one has to investigate the decomposition $\mathfrak{P} = \mathfrak{P}_c + \mathfrak{P}_d$. Hausdorff established the following results:

1. *The decomposition can be a cut, i.e. $\mathfrak{P}_c$ has a last and $\mathfrak{P}_d$ no first element, or $\mathfrak{P}_c$ has no last and $\mathfrak{P}_d$ has a first element.*
2. *The decomposition can be a gap, in which case this can be any of the possible gaps in a pantachie.*

In case 1. one can speak of a boundary between convergence and divergence; this is either the last convergent or the first divergent sequence. In case 2, there is no such boundary.

Hausdorff's point of departure in [H 1936b] was representations of a separable metric space X by sums of $\aleph_1$ increasing Borel sets X^ξ:

$$X = \sum_{\xi<\omega_1} X^\xi, \quad X^0 \subset X^1 \subset \ldots \subset X^\xi \subset \ldots \tag{6.5}$$

If $T^\xi = X - X^\xi$, then

$$T^0 \supset T^1 \supset \ldots \supset T^\xi \supset \ldots \text{ and } \bigcap_{\xi<\omega_1} T^\xi = \emptyset .$$

The sum (6.5) is called *c-convergent* (convergent in category), if in the limit, that is for all $\xi \geq \alpha$, $\alpha < \omega_1$, the sets T^ξ are of first category in X.
The sum (6.5) is called *m-convergent* (convergent in measure), if for any finite measure $\mu(A)$ defined on X and at least on the Borel sets $\subset X$, $\mu(T^\xi) = 0$.

As early as 1920, Waclaw Sierpiński had raised the following two questions with regard to sums of $\aleph_1$ sets:

> Can one prove each without assuming the continuum hypothesis
> – that a sum of $\aleph_1$ sets of Lebesgue measure zero does not necessarily have Lebesgue measure zero?
> – that a sum of $\aleph_1$ sets of the first category need not necessarily be of the first category?

If these questions could be answered affirmatively, this would lead to $m-$divergent, resp. $c-$divergent representations (6.5). About this, Hausdorff remarked:

> The fact that the representations known up to now, which can be defined without the continuum hypothesis, are all $c-$convergent and $m-$convergent, illuminates the difficulty of the two Sierpiǹskischen problems. [H 1936b, 243; IA: 351]

One of the representations to which Hausdorff here referred was due to him; he obtained it by means of his result proving the existence of $\omega_1\omega_1^*$-gaps in suitable partially ordered sets of number sequences. For this reason his principal Theorem (I) in §4 of [H 1909a] played a central part in [H 1936b]. To employ it, however, he needed dyadic sequences, i.e. sequences with the elements 0 and 1, to obtain representations of spaces in the form (6.5). In this context Hausdorff introduced a relation of "final smaller" ($a \leq b$) for pairs of sequences. Two sequences $a, b \in 2^\omega$ are called *final equal*, written $a = b$, if there is a $k \in \mathbb{N}$ with $a_n = b_n$ for $n \geq k$. Hausdorff denotes the usual equality of sequences ($a_n = b_n \forall n$) by $a \equiv b$. All sequences final equal with a given sequence a form a countable class $K(a)$. The relation $a \leq b$ means: there is a $k \in \mathbb{N}$ with $a_n \leq b_n$ for $n \geq k$. If $a \leq b$ and *infinitely often* $a_n < b_n$, then a is called less than b, written $a < b$. If neither $a \leq b$ nor $a \geq b$, then a, b are incomparable. The relations $\leq$ and $<$ are transitive.

A set A of comparable dyadic sequences that contains exactly one element from each class $K(a), a \in A$ forms an ordered set of dyadic sequences. Theorem (I) in §4 of [H 1909a] can now be reformulated as follows:

> *There exist two ordered sets of dyadic sequences A, B of order types ω_1, ω_1^* such that $A < B$, but no sequence x exists with $A < x < B$.*

A und B therefore form a $\omega_1\omega_1^*$-gap. Hausdorff commented in a footnote that he had already proven a theorem of this kind for sequences of real or rational numbers in 1909. Since this work was little known, however, and because certain modifications were needed for dyadic sequences, he wanted to present the entire theory again in detail. He also emphasized that this theorem was "the only connection that can be established for the second number class with the problems of final orderings independent of the continuum hypothesis" [H 1936b, 247; IA: 355], whereas point 3. from [H 1909a] remained open and thus also points 1. and 2.

In the set C of dyadic sequences let

$$a^0 < a^1 < \ldots < a^\xi < \ldots \mid \ldots < b^\xi < \ldots < b^1 < b^0, \quad \xi < \omega_1 \tag{6.6}$$

be a $\omega_1\omega_1^*$-gap. C can be made into a topological space by choosing as a basis of open sets

$$M_{a_1,\ldots a_k} = \{x \in C;\ x = (a_1, \ldots a_k, x_{k+1}, x_{k+2}, \ldots);\ \ k \in \mathbb{N}\},$$

this means that for each of these sets a certain number of initial components are given. Moreover, the sets in this basis are also closed. One can introduce a metric in C by defining the distance $\rho(x, y) = \frac{1}{k}$, where k is the first component for which $x \neq y$. The resulting metric is compact. For $a, b \in C$ with $a < b$ the set $T = \{x \in C;\ a \leq x \leq b\}$ is a F_σ;[1] for $F_n = \{x;\ a_n \leq x_n \leq b_n\}$ is closed and

$$T = \underline{\lim}\, F_n = \sum_n \bigcap_{k=0}^{\infty} F_{n+k}.$$

The gap (6.6) thus yields with the sets $T^\xi = \{x;\ a^\xi \leq x \leq b^\xi\}$ a sequence of decreasing and distinct F_σ-sets with $\bigcap_\xi T^\xi = \emptyset$. This leads to the following central result:

> Theorem I. *Every separable complete uncountable metric space X can be represented in the form* (6.5) *with increasing and distinct G_δ-sets X^ξ.*

For the proof, one only needs to observe that C is a compact subset of X and thus closed in X. The T^ξ are F_σ also in X, so that $X^\xi = X - T^\xi$ is also G_δ in X. Hausdorff then proves beyond this that the representation so obtained is c−convergent as well as m−convergent.

Theorem I suggests the following question: Can a separable complete uncountable metric space X be represented as the sum of $\aleph_1$ increasing and distinct F_σ? Hausdorff showed that this question can be answered affirmatively exactly when the same holds for Sierpiński's second question. Since §2 of [H 1936b] concerns Lusin-Sierpiński space decompositions and is not related to Hausdorff's gap theorem, we pass over it here.

[1] A F_σ-set is a countable union of closed sets; a G_δ-set is a countable intersection of open sets.

6.3.5 On the Reception of Hausdorff's Works on Ordered Sets

In looking back on all of Hausdorff's works on ordered sets, one cannot help but be impressed by the artful display of new concepts, methods, constructions and theorems. Some parts of this work soon became an integral part of modern set theory, whereas other elements have only recently served as starting points for recent and still ongoing research. The importance of this achievement was deeply appreciated by Jacob M. Plotkin, who translated Hausdorff's articles into English and provided introductions to them that are well worth reading [Plotkin 2005]. In the synopsis on the back cover of his book, Plotkin characterizes very aptly Hausdorff's position in the post-Cantorian era as follows:

> Georg Cantor, the founder of set theory, published his last paper on sets in 1897. In 1900, David Hilbert made Cantor's Continuum Problem and the challenge of well-ordering the real numbers the first problem in his famous Paris lecture. It was time for the appearance of the second generation of Cantorians. They emerged in the decade 1900–1909, and foremost among them were Ernst Zermelo and Felix Hausdorff. Zermelo isolated the Choice Principle, proved that every set could be well-ordered, and axiomatized the concept of set. He became the father of abstract set theory. Hausdorff eschewed foundations and pursued set theory as part of the mathematical arsenal. He was recognized as the era's leading Cantorian.

The following brief remarks on the reception of Hausdorff's investigations into ordered quantities are based on comments made by specialists who contributed to relevant volumes of the Hausdorff Edition. Some of the fundamental concepts dating back to Hausdorff have become the common property of set theorists who use these without attribution and usually without awareness of their provenance. This includes the concept of cofinality, which Tarski modernized, and the distinction between regular and singular initial numbers or cardinal numbers. The investigation of partially ordered sets also began with Hausdorff. Two fundamental results that today carry his name are the Hausdorff recursion formula and Hausdorff's maximal chain theorem, also called Hausdorff's maximal principle. The recursion formula, which serves as the basis for aleph exponentiation, was discussed earlier in the context events surrounding the International Congress held in Heidelberg in 1904. Hausdorff's reflections on regular cardinals with a limit index, should these exist, underscored that these were necessarily of "exorbitant size." Today the theory of large cardinal numbers occupies a prominent position within set theory. One of the standard works in this field is [Kanamori 1994/2005], which also takes historical developments into account. In the synopsis of his book, the author briefly characterizes the current importance of this field: "The theory of large cardinals is currently a broad mainstream of modern set theory, the main area of investigation for the analysis of the relative consistency of mathematical propositions and possible new axioms for mathematics."

Elaborating on the historical background, Kanamori writes:

> While Cantor had concentrated his efforts on the rational and real order types, the second-number class, and of course, the Continuum Hypothesis, Hausdorff extended mathematical investigations into the higher transfinite. Deploring all the fuss made over foundations by his contemporaries he ventured forth with vigor, pursuing structure for its own sake. His paper [[H 1908]] contains the first statement of the Generalized Continuum Hypothesis, the construction of the η_α sets – prototypes for saturated model theory – and for the first time, the following concept formulated for $\kappa > \omega$: κ is *weakly inaccessible iff* κ is a regular limit cardinal. [Kanamori 1994/2005, 16]

The first contributions to the theory of large cardinal numbers stem from Paul Mahlo during the years 1911–1913 [Gottwald/Kreiser 1984]. Readers can find much information about subsequent developments in the field in Kanamori's book.

We have seen that by means of Hausdorff's product and power concepts an enormous arsenal of new order structures was created. Yet even such a seasoned set-theorist as Abraham A. Fraenkel underestimated the importance of these innovations. In his book *Abstract Set Theory*, Fraenkel at first explains that it is impossible in general to order lexicographically the Cartesian product of ordered sets M_α (α runs through a transfinite ordered index set). He then writes further:

> In 1904, Hausdorff began to develop a theory of what might be called *substitute-products* and *substitute-powers.*[2] Its purpose is to replace the outer product in question by a subset which is just narrow enough to allow for a suitable order of its complexes. Not only are these investigations rather complicated, but the result is meagre enough. [Fraenkel 1953, 210]

In fact, it took almost 50 years to recognize the potential usefulness of the most important order structure found by Hausdorff, the η_α-sets already mentioned several times above. In their monograph on model theory, Chen Chung Chang and Howard Jerome Keisler write about this in the historical notes to Chapter 5 on "Saturated and Special Models": "The notions of α-saturated and saturated models go back to the η_α-sets of Hausdorff. [...] Their importance for model theory was not realized and exploited until the late 1950's" [Chang/Keisler 1973, 524].

The authors refer to the work of Bjarni Jónsson, who referred to Hausdorff's theory of η_α-sets in [Jónsson 1956], [Jónsson 1960]. Jónsson's starting point was the question of existence for a $\aleph_\alpha$-universal group $\mathfrak{G}$:

> Is there a group $\mathfrak{G}$ of power $\aleph_\alpha$ such that any group of power $\aleph_\alpha$ will be a subgroup of $\mathfrak{G}$?

For Jónsson's results and subsequent research on universal homogeneous and saturated models, see [Felgner 2002]. Ulrich Felgner's essay also presents more recent

[2]Here Fraenkel referred to [H 1904a], [H 1906b], [H 1907a], [H 1907b], [H 1908] und [H 1914a].

results on the cardinality of η_α-sets and on η_α-groups and η_α-fields. In the conclusion, he writes:

> Overall, the theory of η_α-sets shows itself to be a lively field. Several applications and further developments we have left unmentioned, as for example the results on partition relations for η_α-sets. Others we have explained only in a cursory fashion, such as the set-theoretical results in the realm of gap theorems (see also the extensive review article by M. Scheepers [Scheepers 1993]). The field is by no means exhausted and will certainly lead to numerous applications and further developments in the new century. [Felgner 2002, 671]

In the final paragraph of [H 1907a], "The Pantachie Problem," Hausdorff raised a number of questions. He had to leave these open, however, as he realized they were closely linked with the continuum hypothesis, whose validity he did not wish to assume. The first of these questions read:

(α) *Is there a pantachie without* $\omega_1\omega_1^*$ *gaps?*

Hausdorff showed that the existence of such a pantachie implies $2^{\aleph_0} = 2^{\aleph_1}$, which would mean that the continuum hypothesis does not hold. Between 1963 and 1974, no less an authority than Kurt Gödel struggled with (α) again and again. In particular, he believed that $2^{\aleph_0} = \aleph_2$ and thought he could prove this means of three plausible axioms together with what he called the "Hausdorff continuity axiom" (number (4) on his list).[3] Unfortunately, Gödel never made clear what he meant by "Hausdorff's axiom." Akihiro Kanamori has studied Gödel's attempts to prove $2^{\aleph_0} = \aleph_2$ in detail, and he quotes from a letter Gödel wrote to Abraham Robinson, dated March 20, 1974. There, Gödel noted that "Hausdorff proved that the existence of a 'continuous' system of orders of growth is incompatible with Cantor's Continuum Hypothesis" [Kanamori 2007, 182].

Here, 'continuous' undoubtedly means the absence of $\omega_1\omega_1^*$-gaps, since this was the property Hausdorff utilized in his existence proof. Kanamori therefore formulates the fourth axiom on Gödel's list as follows: "(4) In addition, the pantachie has no $\omega_1\omega_1^*$-gaps." However one might judge Gödel's eefforts in this direction, Kanamori's final remark is noteworthy: "As set theory was to develop after Gödel, there would be a circling back, with deep and penetrating arguments from strong large cardinal hypotheses that, after all, lead to $2^{\aleph_0} = \aleph_2$" [Kanamori 2007, 183]. Kanovei's assessment of Hausdorff's problem (α) is also particularly noteworthy: "As for pantachies containing no ($\omega_1\omega_1^*$)-gaps, the problem of their existence remains open, and in fact it appears to be the oldest open problem in set theory explicitly stated in a suitable mathematical publication" (*HGW* IA, 341).

The equation $2^{\aleph_0} = 2^{\aleph_1}$ is usually known today as Lusin's hypothesis; we will abbreviate it as **LH**. It remains an open question whether the existence of

[3]Kurt Gödel: *Some considerations leading to the probable conclusion that the true power of the continuum is* $\aleph_2$. (Nachlaß Gödel, 1970a). *Collected Works*, Vol. III. Oxford Univ. Press 1995, 420–422.

pantachies without $\omega_1\omega_1^*$-gaps can be proved or disproved within **ZFC** + **LH**. In other words, the questions

- is the existence of such pantachies consistent with **ZFC** + **LH**?
- is their non-existence consistent with **ZFC** + **LH**?

remain to be answered.

Another question that Hausdorff had to leave open in [H 1907a] is the problem of scale:

> *Is there in $\mathcal{M} = (\mathbb{R}^+)^\omega$, partially ordered by final behavior, a scale, i.e. a well-ordered subset with which $\mathcal{M}$ is cofinal?*

Hausdorff was able to show that there are ω_1-scales if the continuum hypothesis holds. Results from Robert Solovay and Stephen Hechler show that the scale problem cannot be solved on the basis of **ZFC** alone. The special considerations with regard to $\omega_1\omega_1^*$-gaps indicate a general problem which is rooted in Hausdorff's investigations and which has been pursued in various directions in modern set theoretical research. One starts with the sets $\mathbb{R}^\omega$, $\mathbb{N}^\omega$ resp. 2^ω, endowed with Hausdorff's final partial ordering or suitable modifications of the same. Now if P be one of these partially ordered sets, one seeks to determine the spectrum of its gaps, limits, towers, scales and pantachies. Of particular importance in this spectrum are the gaps, but the literature on this topic has in the meantime grown immensely.

6.4 Two Mathematical Intermezzi

During his years in Leipzig, Hausdorff published two smaller mathematical works that stood quite apart from his main field of work at that time. Both were prompted by external events. The first, *Die symbolische Exponentialformel in der Gruppentheorie* [H 1906a, IV: 431–460], was occasioned by a course Hausdorff offered in Leipzig, whereas the second, *Zur Hilbertschen Lösung des Waringschen Problems* [H 1909b, IV: 503–507], responded to a famous paper published by David Hilbert in 1909.

6.4.1 The Baker-Campbell-Hausdorff Formula

In the winter semester 1905/06, Hausdorff taught a lecture course, "Introduction to the Theory of Transformation Groups (after Sophus Lie)." In it, he closely followed the three volumes of Lie's *Theory of Transformation Groups* (I–III, 1888, 1890, 1893), but also used the books by Lie and Scheffers, *Lectures on Differential Equations with Known Infinitesimal Transformations* (1891) and *Lectures on Continuous Groups* (1893). In Lie's theory the "symbols" for infinitesimal transformations play a fundamental role; these are 1st-order differential operators of the form

$$X = \sum_{i=1}^{n} \xi_i(x_1, \ldots x_n)\frac{\partial}{\partial x_i}.$$

By means of the formal power series for the exponential function, one can introduce for the symbol X the operator e^X, which is fundamental for the theory. In the elaboration of his lecture notes, Hausdorff formulated the following theorem:

> *From the formula* $e^W = e^U e^V$, W *is defined as a 1st-order operator arising from* U, V *by repeated bracketing.*

By the bracketing operation $[U, V]$ is meant the Lie bracket $[U, V] = UV - VU$.

Hausdorff then shows how to find the initial terms of the series development for W. One sets $e^W = 1 + T$, obtaining the initial terms of the series for T by multiplying out the series for e^U and e^V, and then substituting this into the logarithmic series

$$W = T - \frac{1}{2}T^2 + \frac{1}{3}T^3 - \ldots.$$

This yields

$$W = U + V + \frac{1}{2}[U, V] + \frac{1}{12}[[V, U], U] + \frac{1}{12}[[U, V], V] + \ldots.$$

He remarked, however, that "the continuation of this calculation is quite opaque and reveals neither the law of development or the correctness of the theorem."

So it appears obvious that Hausdorff, while preparing his lecture, realized that the essence of the exponential formula was not yet expressed clearly enough in the literature available at the time. In particular, he recognized that this problem can be detached from the meaning of exponents as symbols of infinitesimal transformations and – in modern terms – in order to be treated as a problem of algebra theory. Thus, he writes:

> But a large part of group theory is quite independent of this special meaning of the symbols and rests exclusively on the *formal* circumstance that the composition of the operators defines a kind of noncommutative multiplication on the symbols X. [...] This amounts to the problem of establishing the exponential formula for noncommutative multiplication, i.e., of studying the function z of x, y defined by $e^x e^y = e^z$; a problem of the greatest consequence for the so-called fundamental theorems of group theory can, thus in turn, be treated as a task of symbolic analysis without group-theoretic presuppositions. [H 1906a, 19; IV: 431]

In modern terminology, Hausdorff's result can be formulated as follows: Let K be a field of characteristic zero and x, y two associative but not commutative variables. Let A be the algebra of all formal power series in x, y with coefficients from K and the usual multiplication. If $\mathfrak{m}$ is the set of all elements of A with constant term 0, then for every $u \in \mathfrak{m}$

$$e^u = \sum_{k=0}^{\infty} \frac{u^k}{k!}$$

is well-defined. The problem now comes down to finding the element z in $e^z = e^x \cdot e^y$, i.e., $z = \log(e^x \cdot e^y)$, where $z \in \mathfrak{m}$. One thus introduces the Lie bracket $[u, v] = uv - vu$ into A and considers the Lie algebra L generated by x and y in A and consisting of all formal power series, where the term of degree n is a linear combination of n-fold bracket expressions of x and y. Hausdorff's main result then reads: $z \in L$. The proof involves a recursive procedure for computing the terms z_n of n-th degree in the series for z. Hausdorff computed the series up to the 5th-degree terms, but he emphasizes that "this method is also suitable for the actual computation of any number of terms of z" [H 1906a, 29; IV: 441]. He did not write down the explicit recursion formula, which with $z_0 = 0$, $z_1 = x + y$ reads:

$$z_m = \sum_{k=0}^{m-1} \sum_{m_1+\ldots+m_k=m-1} \frac{B_k}{m\,k!}[z_{m_1}, [z_{m_2}, [\ldots [z_{m_k}, x + (-1)^k y] \ldots]],$$

where B_k is the k-th Bernoulli number [Czichowski 1992].

In 1947 the Russian mathematician E.B. Dynkin [Dynkin 1947] gave an explicit formula for the term z_n in the series for z. The formula is rather complicated so that Hausdorff's recursion method is easier to apply in many cases than Dynkin's explicit formula. Dynkin emphasized that the latter can be obtained by direct calculation, thus providing a new proof of Hausdorff's theorem $z \in L$.

In Part II, "Group Theoretical Applications," of [H 1906a], Hausdorff showed how Lie's fundamental theorems can be elegantly proved by means of the exponential formula. Here questions of convergence come into play for the first time, whereas these have no role in the first algebraic part. In modern formulation, the two fundamental theorems assert the following: Given a finite-dimensional real or complex Lie algebra, there exists a finite-dimensional Lie group whose Lie algebra is the one given. In concluding his commentary on [H 1906a], Winfried Scharlau summarized Hausdorff's important achievement in the field of Lie theory as follows:

> Hausdorff recognized the formal-algebraic core of a central point of Lie theory (i.e., the connection between Lie groups and Lie algebras) and brought this into a form that is still valid today. (*HGW* IV, 463)

Not long after [H 1906a] appeared, Friedrich Engel called Hausdorff's attention to a paper by Henry Frederick Baker [Baker 1905], which bore a close relationship to Hausdorff's work. In a letter to Engel, dated March 9, 1907, Hausdorff informed him that after studying this paper he unfortunately noticed "that its content really largely coincides with that of my treatise," adding that:

> Obviously, this problem was in the air, since it has been taken up and settled with more or less skill just in the last few years by Campbell, Pascal, Poincaré and by me. But specifically the last work by Baker – the only one that has remained entirely unknown to me – has the most points of contact with mine. Incidentally, if I had known it at that time I would have anyway published my work, albeit with somewhat subdued pleasure, since I have found the simplest and most transparent proof of all, Baker included. (*HGW* IX, 216–217)

Today, the result sketched above is called the "Baker-Campbell-Hausdorff formula."

6.4.2 On Hilbert's Solution of Waring's Problem

The mathematical sensation of the year 1909 was the solution of Waring's problem by David Hilbert. Already in the seventeenth century, two mathematicians had conjectured the validity of the four-square theorem (Bachet de Méziriac and Pierre de Fermat): Any natural number can be represented as a sum of at most four squares. The first proof was given by Joseph Louis Lagrange in 1770. In that same year, Edward Waring formulated his famous conjecture generalizing the four-square theorem to any given n, claiming there exists a fixed number $g(n)$ such that any natural number can be represented as a sum of at most $g(n)$ n-th powers. By the four–square theorem, $g(2) = 4$. Prior to Hilbert's breakthrough, mathematicians had gradually been able to prove Waring's conjecture for the special cases $n = 3, 4, 5, 6, 7, 8, 10$. Hilbert succeeded by analytic means in proving the general case [Hilbert 1909]. Before his article appeared in *Mathematische Annalen*, it appeared in the *Nachrichten* of the Göttingen Scientific Society, which received it on February 2, 1909. Felix Hausdorff received an offprint of this paper, and already on 3 March 1909 he wrote to Hilbert:

> Thank you very much for sending your solution to Waring's problem. Once again, this is an event! [...]
>
> May I still take the liberty of pointing out to you a simplification of Theorem I: as far as I can see, instead of the 25-fold integral one can use the 5-fold
>
> $$(x_1^2 + \ldots + x_5^2)^m = C \int \ldots \int (t_1 x_1 + \ldots + t_5 x_5)^{2m} \, dt_1 dt_2 \ldots dt_5 \,,$$
>
> extended over the domain
>
> $$t_1^2 + t_2^2 + \ldots + t_5^2 \leq 1 \,,$$
>
> from which it is still immediately clear that it is orthogonally invariant [...]. (*HGW* IX, 334–335)

Hilbert took up this suggestion in his paper for *Mathematische Annalen*, where he states in a footnote:

> In my original publication [...] I made use here of a certain 25-fold integral; the fact that the same can be replaced for the present purpose by the above 5-fold integral was very thankfully suggested to me by various parties (F. Hausdorff, J.Kürschák, and others). [Hilbert 1909, 282]

At the core of Hilbert's proof is the following Theorem II, which he derives from the above integral formula with considerable effort (5 printed pages).
Theorem II: *For a given natural number m, the following expression holds identically in the five variables* $x_1, \ldots, x_5$

$$(x_1^2 + \ldots + x_5^2)^m = \sum_{h=1}^{M} r_h (a_{1h} x_1 + \ldots + a_{5h} x_5)^{2m}$$

with

$$M = \frac{(2m+1)(2m+2)(2m+3)(2m+4)}{4!},$$

where $r_1, \ldots, r_M$ *are positive rationals and* $a_{1h}, \ldots, a_{5h}$ *are integers, all determined by* m.
From Theorem II, via a series of auxiliary theorems, follows the proof of Waring's conjecture. Hausdorff now found a second proof for this identity, which was completely different from Hilbert's. Already on March 16, 1909 he wrote to Hilbert:

> Your wonderful work on the Waring Problem has so impressed itself on my thoughts that I could not resist trying to approach some points in it in ways differing from yours. In particular, your Theorem II tempted me to look for a proof which would at the same time give a hint about the appropriate choice of the linear forms in question. Having finally found something like that, where the zeros of the derivatives of e^{-x^2} play a role, I take the liberty of sending you a short manuscript for your consideration. If you consider it worthy of the honor of swimming along in the wake of your work, it could perhaps be included as a small appendage of 2–3 pages on the occasion of the printing of your treatise in Mathematische Annalen. (*HGW* IX, 336)

Hilbert was evidently pleased to follow Hausdorff's suggestion, and so this note appeared immediately after his paper in volume 67 of the Annalen [H 1909b, IV: 503–507].

In Hausdorff's derivation of identities like those in Hilbert's Theorem II, the Hermitian polynomials $f_m(x)$ and their zeros play a crucial role. These are defined by

$$e^{-x^2} f_m(x) = \left(-\frac{1}{2}\right)^m \frac{d^m e^{-x^2}}{dx^m}$$

and are orthogonal with respect to the weight function e^{-x^2}. In his commentary, Winfried Scharlau raised the question as to "how Hausdorff came up with the original idea of using the Hermitian orthogonal polynomials [...]." One cannot answer this question conclusively, since Hausdorff made no comment about this either in his letter to Hilbert or in his paper. However, he knew these polynomials and their properties very well, because he had used them in his paper *Contributions to Probability Calculus* [H 1901a] to obtain an elegant derivation of the Gram-Charlier series of type A (*HGW* V, 551–554).

An interesting simplification of the proof of Waring's theorem was provided by the Swedish mathematician Erik Stridsberg [Stridsberg 1912]. Stridsberg used ideas from Hausdorff's work, in particular operations with Hermitian polynomials, to simplify Hilbert's long chain of reasoning in going from his Theorem II to Waring's Theorem. Stridsberg's proof culminates with the following induction result: if Waring's theorem holds for all $k < n$, then it is also valid for n, and since it obviously holds for $n = 1$ and 2, it is therefore true for all n. Georg Frobenius later modified Stridsberg's proof by using more purely algebraic methods [Frobenius 1912], but also drawing on results of Robert Remak [Remak 1912]. He begins his paper by noting:

> The famous Hilbertian proof of Waring's theorem has been considerably simplified by Hausdorff in a most astute way (Math. Ann. vol. 67). [Frobenius 1912, 666]

These words – from the leading figure in Berlin mathematics and a highly critical authority as well – must have given Hausdorff real pleasure.

6.5 Mongré Reawakens

By 1909, Hausdorff had completed his major work on ordered sets, which during previous years had taken up a great deal of his time. Thus, it comes as no surprise that he now found the leisure to return briefly to his literary interests. In the years 1909 and 1910, he wrote three essays under the pseudonym Paul Mongré for *Die neue Rundschau.* All three were also linked with concrete events; two concerned recent books that were much discussed at the time, whereas the third was prompted by the return of Halley's comet in 1910.

6.5.1 Strindberg's Blue Book

The Swedish playwright and novelist August Strindberg, already world-famous in the 1890s, had by then turned to alchemical studies and experiments, which he conducted for several years in Germany, Austria and France. He imagined he could revolutionize chemistry and even believed he might realize the old alchemist's dream of turning base metals into gold. In 1893 he published *Antibarbarus*, a diatribe directed against contemporary chemistry. Although scathingly criticized in professional circles, this reaction did nothing to deter him from continuing his alchemical experiments with great zeal. In 1896 Strindberg suffered a mental breakdown accompanied by delusions and depression, though he gradually overcame these problems through writing, especially his autobiographical novels, *Inferno* (1897) and *Legends* (1898), as well as in his dramatic works. During this crisis and afterward, he turned increasingly to religion, mysticism, and the occult, studying the writings of Emanuel Swedenborg and more recent theosophists, such as Helena Petrovna Blavatsky. In 1907 and 1908, three volumes of his *Blue Book* were published in Stockholm; Volumes 1 and 2 were immediately translated into German and appeared in 1908. In them, Strindberg offered a fundamental critique of modern science covering all possible fields, from Assyriology and mathematics to bacteriology and Chinese linguistics. This critique questioned all the achievements of modern science and revealed the author's staggering ignorance regarding the current state of the fields in question. For the most part, this sweeping form of rejection and vehement clamor was based on Strindberg's own abstruse ideas about the subjects and theories under discussion.

Mongré's essay *Strindberg's Blue Book* [H 1909c, VIII: 691–695] was the first detailed review of the two translated Blue Book volumes in the German-speaking world. During the mid-1890s, Hausdorff regarded Strindberg's criticism of contemporary natural sciences with a certain sympathy, as can be seen from Aphorism No.8 of *Sant' Ilario*, which appeared in 1897. Twelve years later, his views on such matters had changed, as in the meantime Hausdorff had on several occasions dealt critically with new forms of religiosity and irrational argumentation, including mysticism and occultism. In particular, he had written *The Impure Century* and *God's Shadows*, as described in the previous chapter. Moreover, he had criticized his friend Fritz Mauthner's wholesale rejection of science and his apparent turn to mysticism. Many readers who recognized the name Paul Mongré were by now well aware that this author had since become a leading scientific figure who could speak about such matters with authority. Strindberg's Blue Book revealed, on the other hand, that the famous Swedish author had moved very far from any real understanding of contemporary science, whereas his polemics against it had become far more radical than before. Yet the fact that he was an important and internationally esteemed writer assured him an audience from the outset; all his works received considerable attention – regardless of their content.

We do not know whether the editors of *Die Neue Rundschau* approached Hausdorff about writing a review; perhaps he took up his pen on his own initiative in order to expose and hopefully neutralize this outlandish attack on science. Whatever the case may have been, even such a caustic work as the Blue Book could not extinguish Hausdorff's admiration for Strindberg as a writer. His sympathy for his early works, which were rooted in naturalism, remained unbroken:

> Twenty years ago, we watched in fearful wonderment the reflection of a twilight of the gods in the northern sky. *The Father*, *Miss Julie*, *By the Open Sea* ... a volcanic eruption without equal. Strindberg, most glorious of the haters! [H 1909c, 892; VIII: 691]

But now the sixty-year-old author had written a work of some nine hundred pages, where

> inside this is called "the synthesis of my life," whereas the cover claims it to be Strindberg's testament to mankind. Was this once again to be another great eruption? Plutonic fire from the deepest inner-earth, rattling and shaking of the old *seismos*, trying to push a new world up to the light but must first overthrow the existing one? Disappointment! Out of its long-known cracks and fissures springs sulfur, nothing but sulfur; powerless stones and mudslides that uselessly devastate a few orchards. A spectacle without any grandeur. Here nothing is to be learned about fear.
>
> But how can this be an insignificant spectacle, when the science of our days is destroyed by an angry prophet? How so, when the cowardice and cheapness in philosophy, the mendacity in mathematics, the corruption in astronomy, the absurdity in the natural sciences, the bankruptcy in cuneiform research, and so on and so forth, when all these rotten and outdated lies of a pagan, venal, counterfeiting, youth-defying sophistry are exposed by the genius of hatred, who here speaks for the religion of love?
>
> Yes, such words appear in this book. And since they are not proved, only repeated so often that an unbiased person might finally believe them [...], it is perhaps necessary to contradict them sometime and somewhere in the name of science, so to speak, about which every arbitrary person today, from the lowest Center Party voter up to a Strindberg, brings his beasts to market with a degree of boldness, which increases in proportion to the square of his distance from science. [H 1909c, 892–893; VIII: 691–692]

Hausdorff certainly concedes that in the Blue Book Strindberg "has not yet lost some of his old charming qualities [...], that this poet is still audible" [H 1909c, 893; VIII: 692]. Should we, then, disregard the content, treat this as a literary matter, and generously allow the author appropriate liberties? Hausdorff's answer is an unequivocal No! There needs to be a limit "beyond which tolerance of the intolerant ceases":

> No one is obligated to like violin playing. But nobody is entitled to declare: I do not like violin music, *and all violinists play wrongly*. And this is all the more the case when this critic understands as much about violin playing as Strindberg does about science. [H 1909c, 894; VIII: 693]

Only naive minds – asserts Hausdorff – can believe that a single person could so master all the sciences to speak with authority about them, and only such minds will admire the wisdom "of this uncanny old man, who knows everything and knows everything better." A review board of experts in the various disciplines would certainly pass a different judgment "on this pan-historical blue book." Hausdorff could easily imagine, for example, what a chemist would say "about this belated alchemist Strindberg, who patiently writes about forming chlorine from hydrogen and oxygen, and iron from hydrogen and carbon" [H 1909c, 894; VIII: 693].

Hausdorff could speak for himself, of course, when it came to mathematics and related sciences, such as astronomy and physics, and he voiced his criticism in the form of a witty satire. For he claimed to have succeeded, by means of Herpentil's *Instructions on Black Magic* (*Anleitung zur schwarzen Magie*), in summoning the ghost of Strindberg's deceased mathematics teacher from the world beyond:

> This old man, calcified in honor, but still somewhat choleric despite his long storage on the astral plane, after glancing at certain passages of the Blue Book, which we pointed out to him with a reproachful finger, snatched the book out of our hands, pounced on a red pencil protruding from the milieu, wielded it like a wet brush and hurled colored strokes, flaming crosses and double crosses on the edge of the paper, then painted under the whole a five, resounding like a fanfare, and burst into a shattering wail:
>
> "It's exasperating! Withdraw my facultas docendi, I deserve it! No, I don't deserve it! I have talked my lungs out of my chest, and he still has not understood, this eternal grade-school pupil! He has never understood that $\frac{1}{2} \cdot \frac{1}{2} = \frac{1}{4}$; he claimed that there was a contradiction, because multiplying could not result in a reduction; he still claims it today – ! [...]" [H 1909c, 894–895; VIII: 693–694]

And so, on and on, the ferocious rantings of this spirit from beyond takes up almost an entire printed page. At the end of his essay, Hausdorff also found something useful in Strindberg's latest work. To appreciate modern enlightened Europe, his Blue Book had this to recommended it:

> An hour among troglodytes! after which you will return with a new eagerness, a new gratitude to alluvial Europe, to the fine-branched tissues of your science, to the understanding smile of your poetry, to the good manners of your unconverted paganism. [H 1909c, 896; VIII: 695]

6.5.2 Devotion to Life

Hausdorff's essay "Devotion to Life" (*Andacht zum Leben*) [H 1910b, VIII: 743–746] was written in response to a book by the theoretical physicist Felix Auerbach entitled *Ektropismus oder die physikalische Theorie des Lebens* (1910). It begins with the statement that the existence of life represents the greatest of intellectual challenges, so that "thought has contorted itself into desperate poses" in attempting to deal with it.

> Spiritualism and materialism, vitalism and mechanism, dualism and monism all offer their solutions to the riddle, some of which are no more than a refusal to seek a solution. Only the poets, those devout souls, help themselves with relatively simple means: with adoration! They worship life, the world mystery, the inscrutable, the mystery of procreation; they are the heirs of a thousand years of piety, only changing the fetish by replacing the old gods on the altars with new symbols that are less compromised in being divine. [H 1910b, 1737; VIII: 743]

Auerbach cannot be reproached for this; he "does not prostrate himself in agnostic humility before the Sphinx, though his way is not without some poetic excitement and devotion to life" [H 1910b, 1738; VIII: 743].

Hausdorff first explains to readers of the Rundschau the physical ideas at the center of Auerbach's theory, in particular, the second law of thermodynamics. Rudolf Clausius formulated this in 1865, when he declared: *The entropy of the world is striving towards a maximum* [Clausius 1865, 400]. The notion of entropy introduced by Clausius constitutes a measure for the energy devaluation in a thermodynamic system, which means that maximum entropy of such a system will be reached when all temperature differences are balanced out. Applied to the universe as a whole, the second law predicts a final state of heat death, a hypothesis that engendered a great deal of lively contemporary debate. According to this view, world events strive toward a "state of perfect equilibrium: if all intensity differences have balanced themselves and all energy in the form of tepid warmth is dispersed evenly in space, then nothing more can happen" [H 1910b, 1738; VIII: 744].

Against this pessimistic prognosis, Auerbach brings his ectropism into play. According to biological theorists, the history of life is characterized by higher development, differentiation, "increase in value." Auerbach thus attributes to organic matter an *ectropic* quality, meaning that biological development reduces entropy. The world has created life, so to speak, as a counterweight against the devaluation of energy, so that, according to Auerbach, it will not fall prey to heat death, but rather will ultimately become ectropic.

Such a speculation, representing a new kind of "admonition to life," Hausdorff could not let stand unchallenged. Thus he asked: "Does physics also flow into a religion of life, biology into a biolatry? The pathos of the devout provokes resistance" [H 1910b, 1739; VIII: 745]. First of all, he states that the hypothesis of heat death already becomes invalid if one considers the universe not as a closed system, but as a system with an infinite supply of energy. But apart from this, he raises the objection that the entropy reduction caused by life, as for example life on Earth, is vanishingly small when compared with the overall energy radiation of the Earth, but even more so for the sun and the increase in entropy caused by it. Auerbach also believed he had come close to answering the question why there is life. Hausdorff disputed that, however, by pointing out that the interpretation of increasing entropy as a "devaluation" of energy is already an anthropomorphism. "Inorganic nature," he wrote, "could be quite indifferent at to whether molecules were ordered or disordered, condensed or dispersed in swarms that danced in space" [H 1910b, 1740; VIII: 746]. And he concluded with the following admonition:

> No, you worshipers of life: the fact remains that life is a hopeless and mysterious thing, to which no dithyrambic toast should be offered! and ectropism may be a physical theory of life, a valuable contribution to the understanding of life, but a biodicy, a justification of life, it is not.

6.5.3 The Comet

In May 1910 the return of Halley's comet was imminent. On the basis of orbital calculations, it appeared probable that on May 18, 1910 the Earth would pass through its tail. Since cyanides had been detected in Halley's comet using spectral analysis, some speculated that this could poison the Earth's atmosphere with deadly hydrogen cyanide gas; the end of the world seemed to be imminent. Articles began appearing in various newspapers around the world, spreading a doomsday message and pointing to the calamitous event due to unfold on May 18. Others saw this as nothing more than hysteria, and soon newspapers and magazines carried an abundance of satirical articles, caricatures and poems that parodied the alleged danger for our planet. Hausdorff's essay *The Comet* [H 1910a, VIII: 723–727] was a satire of a different kind, directed at the press's sensationalist handling of the event. Perhaps it was also intended to parody the criteria of the *Neue Rundschau* and its publisher, Samuel Fischer, in selecting authors and contributors. In fact, *Der Komet* takes the form of a discussion between a magazine publisher by the name of "Spleen" and his editor about what should appear in the May 1910 issue:

> *Publisher*: Have you finished your essay about the comet for the May issue? Something unprecedented, ecstatically rising and falling, circling around the highest peaks, you understand, something ultraviolet and transcendental, a projection onto the non-Euclidean plane of consciousness? Do you have it? Give it.
>
> *Editor*: I have nothing, honored master. [H 1910a, 708; VIII: 723]

In fact, however, the editor had asked a dozen authors for a contribution, and all of them had since delivered manuscripts. He decided not to bother reading them at all, however, because their content was completely predictable – just as predictable as what one would find in an essay written seven years' hence on the occasion of Theodor Storm's hundredth birthday. The publisher admitted that he shared his editor's sense of resignation, but the May issue was now pressing. So they decide they should have a look at the contributions to see if there might be at least one they could use.

Before they do, however, the editor wishes to demonstrate to his master that he can already deduce a priori more or less what these will contain. For this purpose he designs a "table of categories" and distinguishes "three main tribunes from which to view the phenomenon: I. Science, II. Philosophy, and III. Mood." Browsing through the manuscripts belonging to I., it immediately becomes clear to Spleen that these are hopeless. Still, both discover to their amusement that one author apparently wrote his essay using an outdated encyclopedia; he had nothing to say about α-particles, β-particles, γ-rays, etc. But also with category II. there was nothing of any use: "What was there to be philosophically excited about?" Even the category "mood" at first yielded nothing of interest until the publisher came across a name neither he nor the editor knew; the latter then began reading from this manuscript:

> "In Lubienitzky's History of Comets it is said that in the year 1000 of our era a burning torch with a long flashing tail fell from an opening in the sky, and when the cleft in the sky gradually disappeared, a figure with a serpent's head and blue feet was seen; this fictitious animal is even illustrated in the book, and one finds collected there many similar reports which are quite incredible."

Then, while the editor read aloud, continuing to pull out this and that from the "mood" category, the publisher sank into deep thought and no longer listened at all. Suddenly, he jumped up excitedly and shouted out:

> Lubienitzky! We are saved. What do we care about Halley, the son of the wealthy soap boiler, what do we care about perihelion and aphelion, parabolas and hyperbolas, 75 years of orbital period, nucleus, coma and tail, split tail, sixfold tail, a whole forest of tails! We need an experience with soul, a thing with a snake's head and blue feet ... Phone the book decoration department! We are doing a bibliophile reprint of Lubienitzky!!! [H 1910b, 712; VIII: 727]

In 1910 a new artistic-literary journal was founded in Munich, "Licht und Schatten." In the spring of 1913, Hausdorff's close acquaintance Hans Heilmann – known to his friends as Papa Heilmann – took over the editorship of the journal. It can be assumed that he asked Hausdorff, then already an associate professor in Bonn, for a small contribution. This took the form of a short essay entitled *Biologisches* [H 1913, VIII: 757–758] that reflects close ties with the earlier essay *Andacht zum Leben.* Here, in a few lines, Hausdorff contrasts the beautiful world of art with the horrors in the real world, revealing his by now pessimistic views on the eve of the First World War. With *Biologisches*, the philosopher and man of letters Paul Mongré finally departed from the public scene.

Part III

Hausdorff's Crowning Mathematical Works

Chapter 7

Hausdorff at the Pinnacle of his Career

7.1 Associate Professor in Bonn

In 1910, Gerhard Kowalewski left Bonn to accept a full professorship at the German Technical University in Prague. This left the associate professorship in mathematics at the University of Bonn vacant, and in January 1910 the faculty voted on its list of recommended candidates. They decided unanimously to propose first and *pari passu* Georg Landsberg and Felix Hausdorff, second Hans Hahn and third Georg Faber. This proposal originated with Eduard Study, who had been full professor (Ordinarius) in Bonn since 1904. The Ministry of Culture in Berlin decided in favor of Hausdorff, who was also apparently Study's preferred candidate. On a postcard sent to his friend Friedrich Engel from Rome, Study wrote: "I have just received the news that Hausdorff is coming to Bonn. You will probably have heard about it elsewhere, and are just as happy about this" [Ullrich 2004, 398].

In fact, Engel had already congratulated Hausdorff on his call to Bonn even before the latter had responded to the offer. Hausdorff answered Engel by telegraph on April 4, 1910, the very day he accepted the call:

> We thank you and your wife most sincerely for your kind congratulations, and from me especially to you for your share in the matter – whether it was, as you wrote, quite small or somewhat larger or even very large. Of course, I have accepted, although it is difficult for me (not for academic reasons!!) to leave Leipzig. (*HGW* IX, 218)

Hausdorff later sent Engel a letter that sheds considerable light on his chances for obtaining a full professorship in Greifswald, where Engel had been teaching since 1904. To appreciate the content of that letter, however, requires first tracing back to some earlier events. When Engel left Leipzig for Greifswald in 1904, he

E. Brieskorn, W. Purkert, *Felix Hausdorff*, Vita Mathematica 21,
https://doi.org/10.1007/978-3-031-52135-5_7

Figure 7.1: Eduard Study, Hausdorff's friend, benefactor and colleague in Bonn
Source: König Estate, Nr. 172

succeeded Eduard Study, who was on his way to Bonn. At the time of Study's call, Kowalewski held an associate professorship in Greifswald, but he now took the opportunity to exchange this position for the one in Bonn (which Hausdorff then gained in 1910). One easily sees from these circumstances that relationships between mathematicians at these two universities were closely entangled. This reflected the fact that these three – Engel, Study, and Kowalewski – all began their careers in Leipzig and, like Hausdorff, all had been connected with Sophus Lie. Eduard Study, a prominent geometer and invariant theorist, was the central figure involved in these various appointments. His own academic career pattern, in fact, essentially mirrored Hausdorff's, beginning with his initial appointment in Bonn in 1894, when he assumed the associate professorship previously held by Hermann Minkowski. Three years later, Study gained a full professorship in Greifswald, before returning to Bonn to succeed Rudolf Lipschitz in 1904; there, he would later become one of Hausdorff's closest colleagues. Taking these earlier events into account, we can now turn to Hausdorff's letter to Engel from 1911.

When Engel left Leipzig for Greifswald in 1904, his arrival coincided with that of Theodor Vahlen, who had assumed Kowalewski's former position. The senior mathematician in Greifswald was Wilhelm Thomé, who however died in 1910. Thereafter, in early 1911, Vahlen was appointed as Thomé's successor, though only because the Prussian Ministry overrode the decision of the Greifswald faculty, which – following Study's advice to Engel – had placed Hausdorff first on its list of suitable candidates. Not surprisingly, Hausdorff afterward learned about what had happened, which led him to write Engel on February 21,1911:

> Dear Professor,
>
> I had intended for some weeks to write you this letter, namely ever since I learned through Study and London that I was on the list in Greifswald, moreover *primo loco* and placed above a whole aggregate of second-order members. For this great honor, which I owe to you, I would like to express my heartfelt gratitude, which of course is not diminished in the least by the fact that the highly benevolent government in the Wilhelmstrasse has decided otherwise. I would have liked very much to come to Greifswald and the cooperation with you would have sincerely pleased me; since nothing has come of that now, I have at least had the pleasure, firstly, to know that you think something of me, secondly, after years of standing still on a dead end track in Leipzig, to finally be in circulation again and to gain consideration at all. This was doubly soothing for me in view of the disgraceful treatment I received in Leipzig, a song with several stanzas that you well know how to sing. (*HGW* IX, 218–219; about the song Engel also "knew how to sing" in view of his treatment in Leipzig, see [Purkert 1984]).

One can only speculate about the reasons why the Ministry of Culture ignored the proposal made by Greifswald's faculty and instead chose to appoint Theodor Vahlen as Thomé's successor in 1911. In any event, Hausdorff probably soon forgot about this mysterious incident, since not long afterward, in 1913, Engel left Greifswald to take a professorship in Giessen, and this time he managed to engineer Hausdorff's appointment as his successor. Study would then later bring him back to Bonn as a full professor in 1921. Having sketched the circumstances surrounding Hausdorff's various academic appointments, we now turn back to his first period in Bonn.

7.2 Hausdorff's New Surroundings in Bonn

In Leipzig, Hausdorff had lived for many years in an apartment at Lortzingstrasse 13, before moving to another at Nordplatz 5. Both buildings were located on the edge of the Waldstrassenviertel, a distinguished residential area near the Rosental. After moving to Bonn, he rented an apartment at Händelstrasse 18 in the bourgeois Musikerviertel, located on the edge of the city center near Beethovenplatz. A small park, the Baumschulwäldchen, graced the immediate vicinity. He needed no more than 15 minutes to walk from his apartment to the main university building on the Hofgartenwiese, where he held his lectures.

Hausdorff soon felt very much at home in Bonn, and he especially enjoyed the atmosphere among his colleagues. In the letter to Engel cited above, he went on to write:

> It was only here in Bonn that I really became conscious of the fatally mandarin-like unpleasantness of the Leipzig hierarchy – in contrast with here, where the Privatdozent has the status of a human being whose visits are reciprocated and who receives invitations to the rector's dinners. In Bonn, even those who are not full professors (*Nicht-Ordinarius*) feel they have a real legitimacy (*förmlich existenzberechtigt*), a sentiment I could never bring myself to feel on the Pleisse [in Leipzig]. By the way, I do not need to add that especially my mathematical colleagues in Bonn have played the most outstanding part in promoting my feelings of comfort in the local milieu. (*HGW* IX, 219)

Hausdorff was referring here not only to Study, but also to his other two "mathematical colleagues," Franz London, who was promoted to full professor in 1911, and the private lecturer Johann Oswald Müller. London, son of a Jewish merchant, had habilitated in Breslau in 1889, where he taught as a private lecturer (Privatdozent) until coming to Bonn in 1904. His research interests originally concerned topics in elementary projective geometry, but after his appointment in Bonn he took little interest in publishing and instead devoted all his energy to teaching. He also suffered from a heart ailment, which prematurely took his life in 1917, before he had reached the age of 54. In his obituary for London, Study paid tribute to him with the following words:

> Franz London [$\cdots$] was a teacher with heart and soul; for him, no effort was too great if he could thereby support his numerous pupils, who followed his lectures with enthusiasm. To them, in his dedication to duty, he devoted an unusually large part of his time. [Study 1918, 155]

On London's initiative, the seminars in Bonn were organized on three levels: 1) exercises for introductory lecture courses; 2) presentations by participants on simpler topics; 3) presentations by members on advanced topics. He himself was especially concerned with the seminars on the lowest level. Five years after his death, when the university applied for funding to support the teaching of exercises in 1922, it was noted that previously Professor London had paid for an assistant from his own funds [Schubring 1985, 155]. London also solicited funding to promote mathematical research from the wider circle of his family, arranging for his wealthy brother-in-law, Alfred Hamburger, to endow 30,000 marks to the University of Bonn. Unfortunately, he himself did not live long enough to witness the establishment of this Hamburger Foundation.

London and his wife Louise, *née* Hamburger, often welcomed guests, as Eduard Study described in his obituary article:

> Through his openness and mental vigor he enlivened his home, which exuded a comfort, cheerfulness and happiness that he was only too glad to share with others; thus the Londons' house was a center of noble sociability. For those who ever had the pleasure of being a guest there, London's witty, spirited conversation, his amiable humor, and

> his encouragement to the joy of life will remain as unforgettable as the warm feeling and understanding which he showed to others in joy and sorrow. [Study 1918, 156]

In view of these warm words, it comes as no surprise that the Hausdorffs quickly became part of this circle of friends. At the time of Franz London's death, they were living in Greifswald, but they remained in contact with his widow, Louise. Thus, when they resumed living in Bonn in 1921, the Hausdorffs were among her closest friends up until her emigration.

Hausdorff's other colleague, Johann Oswald Müller, was younger than the three Bonn professors, having received his doctorate in Göttingen in 1903 under David Hilbert. Müller's dissertation contained an improvement of Hermann Amandus Schwarz's proof of the isoperimetric property of the sphere. Subsequently, he studied in Paris for a year and then continued his mathematical education in Göttingen. In 1909 he habilitated in Bonn under Study with a postdoctoral thesis on the calculus of variations. After becoming a Privatdozent, Müller abandoned mathematical research and devoted himself exclusively to teaching. In 1917 he was awarded the title of professor, and in 1921 he gained a salaried lectureship in higher algebra, number theory, and probability. He also became a member of the examination committee for candidates seeking certification to teach higher-level courses in the schools. However, in 1937 the Nazi government withdrew his right to teach, and he died three years later. Müller was thus Hausdorff's colleague not only during the first, but also throughout the second Bonn period. One must assume they got along well, as Müller stood by Hausdorff after 1933, when both of them experienced difficulties (see 9.1.3).

Hausdorff's academic and personal situation improved dramatically once he arrived in Bonn. As he noted in his letter to Engel, it was not for academic reasons that he regretted leaving Leipzig. His departure meant leaving behind many friends, a circle of intellectuals and artists that he had enjoyed for several years. His new surroundings in Bonn, however, offered not only warm collegial relationships among the mathematicians but also the chance to engage in stimulating teaching activity. In Leipzig, Hausdorff only taught set theory once, in 1901, despite the fact that for many years this field stood at the center of his research interests. Set theory was undoubtedly not very popular in the Leipzig mathematical community. Quite possibly, some there even fought against it "with medieval weapons," as Hausdorff once complained to Hilbert, though in a different context. In any case, Günther Eisenreich noted with regard to the most important mathematician in Leipzig at that time: "[Otto] Hölder was always skeptical about operating with infinite sets" [Beckert/Schumann 1981, 163].

In Bonn, Hausdorff had the opportunity to teach a wide spectrum of courses. In a letter to the Bonn faculty, dated April 18, 1910, he already announced a 2-hour lecture course on set theory for the coming summer semester (*HGW* IX, 693). Probably he knew that his future colleagues took a very liberal view regarding set theory, and they articulated their position on such modern subjects soon after

Hausdorff's arrival. At that time, they composed a successful application to have Franz London's position upgraded to a full professorship. After underscoring that the number of students had now doubled, the faculty put forward this additional argument:

> At the same time, mathematics teaching in particular has expanded significantly in recent decades. Subjects never before treated, such as set theory, group theory, and descriptive geometry, now require special regularly recurring courses. (Geheimes Staatsarchiv Preußischer Kulturbesitz, Berlin. Rep. 76 V a, Sekt. 3, Tit. IV, Nr. 55, Bd.4, Bl. 24)

In the winter semester of 1910/11, Hausdorff taught "Introduction to Group Theory," and during the summer semester of 1912 he offered a completely revised version of his earlier course "Introduction to Set Theory." That summer he also lectured on elliptic function theory, whereas the previous winter he taught a special course on Fourier series and related infinite series. It should also be noted that Hausdorff taught special courses at the Business College (*Handelshochschule*) in Cologne, namely, "Introduction to Insurance Mathematics" and "Political Arithmetic."

7.3 Work on his opus magnum

Like his colleagues, London and Müller, Hausdorff focused primarily on teaching during this first period in Bonn. In fact, his only publication from these years was [H 1911, IV: 513–518], a review of the two volumes of Edmund Landau's *Handbook for the Theory of the Distribution of the Prime Numbers* (*Handbuch der Lehre von der Verteilung der Primzahlen*), a mammoth work of nearly one thousand pages altogether. His review was a showpiece with respect to language and content; Wolfgang Schwarz characterized it in these words:

> It is astonishing with what empathy, insight, and linguistic sympathy Felix Hausdorff, who was not particularly involved with number theory, approaches this by no means easy task of reviewing Landau's *Handbuch* [· · ·], a seminal work whose importance for analytic number theory can hardly be exaggerated. (*HGW* IV, 519)

Hausdorff concludes his review with a prognosis: while it is hopeless to speculate about which discoveries in which mathematical fields will promote prime number theory,

> [· · ·] we may perhaps expect, however, that the author of the present manual will play a leading part in the future development of this discipline, and that all further progress will take its starting point from this work as the furthest station reached so far. [H 1911, 97; IV: 518]

Hausdorff turned out to be absolutely right with this prognosis, for Edmund Landau's position as one of the leading exponents of analytic number theory remained undisputed up until his death in February 1938. In their obituary for Landau, Godfrey Harold Hardy and Hans Heilbronn offered this opinion:

> The *Handbuch* was probably the most *important* book he wrote. In it the analytic theory of numbers is presented for the first time, not as a collection of a few beautiful scattered theorems, but as a systematic science. The book transformed the subject, hitherto the hunting ground of a few adventurous heroes, into one of the most fruitful fields of research of the last thirty years. Almost everything in it has been superseded, and that is the greatest tribute to the book. [Hardy/Heilbronn 1938, 307–308]

During the summer of 1912, Hausdorff was hard at work on his opus magnum: the monograph *Grundzüge der Mengenlehre* (*Basic Features of Set Theory*) [H 1914a], his single most influential work. A first indication of how he proceeded comes from the manuscript for his course on set theory from the summer semester of 1912. Therein he presented a new approach to point sets, departing from the ideas in his lectures on set theory from 1901 and 1910. For this, he formulated four properties for open spherical neighborhoods U_x of points $x \in \mathcal{E} = \mathbb{R}^n$:

> (α) Each U_x contains x and is contained in $\mathcal{E}$.
>
> (β) For two neighborhoods of the same point, either $U_x \subseteq U'_x$ or $U'_x \subseteq U_x$.
>
> (γ) If y is in U_x, then there is also a neighborhood U_y contained in U_x ($U_y \subseteq U_x$).
>
> (δ) If $x \neq y$, then there are two neighborhoods U_x, U_y without a common point ($\mathfrak{D}(U_x, U_y) = 0$).
>
> The following remarks are initially based only on these properties. Therefore they apply in general if $\mathcal{E}$ is a point set $\{x\}$ whose points x are assigned point sets U_x with these 4 properties. (Hausdorff estate: Capsule 9: Fascicle 34, p. 21)

Hausdorff's remarks thus explicitly emphasized the axiomatic character of the four fundamental properties he established. In fact, except for (β), these are precisely the axioms Hausdorff introduced for his definition of a topological space in the *Grundzüge*. There he replaced (β) by the following axiom:

> (B) If U_x, V_x are neighborhoods of x, then there exists a neighborhood W_x of x with $W_x \subseteq U_x \cap V_x$.

The beginning of his actual work on the book can be dated quite precisely. On February 27, 1914, Hausdorff wrote to Hilbert, offering to send him a paper on the sphere paradox [H 1914b] for the *Annalen*. There he indicated that his "book on set theory, on which I have been working for $1\frac{1}{2}$ years" would soon be published (*HGW* IX, 338–339). This means that he must have started working on the manuscript around August 1912. This dating is also consistent with the information Hausdorff imparted to Friedrich Engel in a letter written a year earlier than the one to Hilbert. After complaining about the general noise level in Bonn and particularly the racket made by their neighbors, he writes:

> In spite of all this, I have pulled myself together in the last six months and have largely completed a book on set theory, which I hope will appear in the course of this year and, as I imagine, will turn out quite decently. Perhaps you will be so kind, though, not to talk about it (only Study, London and Schoenflies know of it), because I would find it fatal to announce a deed before it is completely done. (*HGW* IX, 223)

Thus, Hausdorff had nearly completed his book while in Bonn, though he put the final touches on it in Greifswald before it came out in April 1914. The contents of Hausdorff's *Basic Features of Set Theory* are discussed in detail later in Section 7.6. Next, however, we turn to the circumstances that led to his appointment in Greifswald.

7.4 Full Professor in Greifswald

In 1912, Friedrich Engel obtained an offer from Kiel to replace the recently deceased Georg Landsberg. That same year, however, he received almost simultaneously a call to Giessen, where Eugen Netto was about to retire. Engel opted to accept the offer from Giessen. This created a vacancy in Greifswald, which the faculty hoped to fill as quickly as possible. Already on February 15, 1913, they submitted their proposal to the Ministry:

> For the first place, we recommend Dr. Felix Hausdorff, Associate Professor at the University of Bonn. His numerous works are all distinguished by extraordinary acumen. He is one of the leading experts on set theory, which was founded by G. Cantor and has had a profound influence on all of modern mathematics. He has critically illuminated this difficult field from all sides and has substantially promoted the further development of the entire theory through his own work. He can look back on many years of successful teaching in Leipzig and Bonn. [Ilgauds 1985, 67]

The others names on the list were Heinrich Jung, for second place, and for third, Paul Koebe and Oscar Perron.

Engel was, of course, responsible for nominating Hausdorff, and he informed Study about this in a telegram immediately after the faculty meeting. After learning about these developments in Greifswald from Study, Hausdorff sent Engel an emotional letter, dated February 15, 1913, thanking him for his support and reconfirmed confidence:

Study has shown me your long telegram, according to which you have once again recommended me for the first place. I would like to tell you how deeply honored and pleased I am by this reconfirmed proof of your confidence, and how sincerely grateful I am to you for it. The good opinion that you and Study have of me has elevated my self-confidence, which is not very well developed and in Leipzig was already about to fall below any positive ε, bringing it back to the lower limit needed for undertaking scientific research. If the Ministry gives its approval to your repeated promotion of me, I will do my utmost to ensure that the gulf between you and your successor is not too deep.

By the way, I very much hope that this time something will come of it. Of course, I would have preferred even more to have been appointed to work together with you two years ago; now, if this case arises, we will feel very lonely in Greifswald at first. You may be surprised that I seem so happy to leave Bonn: the academic relations here are excellent, and we will certainly not find such friendly and cordial relations as those with Studys and Londons again, after the good Engel [good Angel] of Greifswald has left. (*HGW* IX, 223)

Hausdorff originally contemplated taking over the Engels' apartment at least temporarily. Since he did not want to decide that question alone, he and his wife traveled to Greifswald on March 18 for a visit. After their return to Bonn, Hausdorff wrote to Engel, thanking him and his wife for their hospitality during the visit. This letter also reveals his deep sense of relief that finally, at nearly 45 years of age, he had achieved the academic standing of a full professor, a position he long deserved:

Dear and honored Herr Engel!

Today it has been eight days since we said goodbye to you. We stayed in Berlin for several days in order to enjoy a bit of theater, etc., in anticipation of our future provincial existence. Now that we have returned home, the first thing we would like to do is thank you and your wife most sincerely for the kind welcome you have given us in our new home. Not only have you done something wonderful and decisive for me, for which I will always be grateful, you have also added to this numerous gifts and help of a secondary nature, so to speak; we have enjoyed your hospitality but have occupied your time to a considerable degree. All this has done us a great deal of good and has made it much easier for us to adjust to the foreign milieu. If there is one thing I regret, though, it is that I will come to Greifswald as your successor and not as your colleague, so that the friendly affection with which you both met us could only be a swan song and not a prelude. (*HGW* IX, 227)

Hausdorffs then actually lived in Engel's apartment at 11 Arndtstrasse for a few months before they found a suitable house to purchase.

7.5 The Greifswald Seminar from 1913 to 1921

At the time of Hausdorff's arrival, Greifswald was a small provincial Prussian university with between 1300 and 1400 students. Of these, about 120 were enrolled in mathematics and natural sciences, but during the war that number fell to just under 100, though many of these were in the army. For the winter semester 1914/15, out of 95 enrolled students only 32 were at the university; all the others were engaged in war service. Thus, the number of students in Hausdorff's courses during the war years was surely very low, but afterward the numbers rose again. During the summer semester of 1919, enrollments in mathematics and natural sciences rose to a maximum of 145, but soon they dropped again to less than 100.

Only two other mathematicians were working in Greifswald when Hausdorff took up his post there. Theodor Vahlen was the other full professor and the Austrian Wilhelm Blaschke was a private lecturer with a regular teaching contract. However, Blaschke already left Greifswald after the end of the summer semester to take up a position at the German Technical University in Prague. Nevertheless, he and Hausdorff clearly got along well; even when he was in Prague, Blaschke offered his assistance with the completion of *Grundzüge der Mengenlehre.* In fact, he drew several figures that appear in it and he also helped with reading and correcting the final page proofs, for which Hausdorff thanked him in the preface of the book. Their two families must have had some contacts, because in a card Hausdorff sent to Blaschke in Prague on October 29, 1913, he wrote:

> Thank you very much for what you have done so far and for what you will do in the future! Many greetings from both of us to you and your wife.
>
> How does the Moldau reflect the *cima* [zenith] of the moon?
>
> Yours faithfully, F. Hausdorff
> (*HGW* IX, 159)

Starting the winter semester of 1913/14, Blaschke's lectureship was assumed by Clemens Thaer. Vahlen and Thaer remained Hausdorff's only mathematical colleagues in Greifswald until he left for Bonn in 1921.

As noted earlier (7.1), Vahlen came to Greifswald as an associate professor in 1904 and then gained a full professorship in 1911. Until about 1907 he published a considerable number of papers in fields ranging from differential geometry, algebraic geometry, number theory, and algebra, but they – except for his investigations on hypercomplex number systems following Clifford – attained little importance. In 1905, his book *Abstract Geometry. Investigations on the Foundations of Euclidean and Non-Euclidean Geometry* was subjected to a scathing critique by Max Dehn [Dehn 1905]. In a thoroughly matter-of-fact tone, Dehn devoted more than two pages to various errors and unclear passages in the book. Vahlen tried to rebut these criticisms, but could do little to overcome the impression that he lacked the competence to write about the foundations of geometry.

One can easily imagine that Vahlen's later openly anti-Semitic views were not unrelated to the fact that Dehn happened to be of Jewish background. In any case, from 1907 onward, Vahlen turned to applied mathematics, suggesting that this disaster contributed to his reorientation. In 1938, after he had become a bigwig in the Nazi administration, he obtained a seat in the Berlin Academy. In his inaugural address, he referred to this shift in his research interests, which he justified ideologically: "My Abstract Geometry brought about a change, and after its completion I was drawn to the natural concrete way of thinking of our race" [Siegmund-Schultze 1984, 20]. In his new field, however, Vahlen published only a few research papers containing results of rather meager significance.

One can only assume that Hausdorff and Vahlen had a distant relationship limited to official business. Moreover, for seven semesters, from the winter semester 1914/15 to 1917/18, Vahlen was not in Greifswald, since he was engaged in war service as an officer in the German army. Vahlen's subsequent immersion in political developments was unusual for a full professor. If Hausdorff took note of these at all, or even Vahlen's steep career in the National Socialist Party, then he did so only from a distance [Siegmund-Schultze 1984]. Only one statement by Hausdorff about Vahlen seems to have survived. This appears in a letter to Engel, dated May 26, 1913, thus shortly after Hausdorff arrived in Greifswald. There he wrote:

> Hopefully I will come into somewhat closer intercourse with him [Thaer], for there seems to be not much with Vahlen. We were, however, spoiled in this respect by the Studys and Londons; and the kindness with which you have received us aggravates our grief that you are no longer here. (*HGW* IX, 231–232)

Hausdorff's second mathematical colleague, Clemens Thaer, had received his doctorate in 1906 under the well-known mathematician Moritz Pasch. Following a brief stay in Göttingen, he took a position as assistant in Jena in 1908 and habilitated there the following year. In 1911 he married Gertrud Pasch, the daughter of his doctoral supervisor. In the course of transferring his Habilitation to Greifswald, Thaer had to deliver an inaugural lecture, which Hausdorff wrote about in the same letter to Engel:

> Thaer gave his inaugural lecture the day before yesterday, to my mind very fine and thoughtful, though I would disagree with him (existence of irrational numbers). Hopefully I will come into some closer contact with him [···] (*HGW* IX, 231)

Some contact did take place, including in family circles, as is evident from the following passage that appears in Thaer's autobiographical notes:

> The Greifswald mathematicians, Vahlen and Hausdorff, both approached us in a friendly manner, though in their temperamentally different ways. I was more attracted to Vahlen's distinguished, reserved nature, also to his gentle, fine, quiet wife; your mother was more drawn to the intellectually very impressive Hausdorff family, in which the conversation

> about scientific and artistic questions was always witty, though often sarcastic. Friends like in Jena, I did not find here.

As the last sentence suggests, a more intimate friendly relationship did not develop between Thaer and Hausdorff; their contacts appear not to have gone much beyond formalities. Clemens Thaer later gained a reputation in the field of history of mathematics. His most significant achievement was a German translation of Euclid's *Elements* following the standard Greek-Latin edition by Heiberg; Thaer's translation appeared in Ostwald's Classics in 1933. He also translated the *Data* of Euclid, one of his twelve publications on the history of mathematics.

During Hausdorff's first three semesters in Greifswald, his two colleagues covered most of the more elementary courses. This gave Hausdorff the opportunity to teach higher-level courses on such topics as differential geometry, algebraic numbers, elliptic functions, and integral equations; he also taught the basic lecture course on function theory. However, starting from the winter semester 1914/15, Vahlen was absent and later Thaer, too, after he was drafted. For two semesters, Hausdorff had to do all the teaching, since the Mathematics Seminar consisted of him alone! He thus gave several basic lecture courses during this period, but he also taught set theory, mechanics, and a two-semester course on elliptic functions. The situation after the war improved very little, since most of the teaching had to concentrate on elementary subjects. During this time, Hausdorff gave only one higher-level course beyond the one on elliptic functions, namely on Fourier series. Only after he returned to Bonn in 1921 did this unsatisfactory teaching situation change for the better; there he often offered lecture courses that dealt with topics of current research.

7.6 Grundzüge der Mengenlehre

Hausdorff had nearly completed his *Grundzüge der Mengenlehre* (*Basic Features of Set Theory*) in Bonn, so his first priority in Greifswald was to finish the text and see the book through the printing process. In view of this, he asked for Engel's advice about what he might teach since, as he admitted in a letter from March 11, 1913:

> First, I have no idea what I ought to teach in the summer semester; I would like to note right away, though, that I cannot this time in the beginning load myself up with too much in order finally to finish my book. (*HGW* IX, 224)

As it turned out, it took more than a year before Hausdorff's monograph finally appeared. This was mainly because he decided to revise many parts of the existing manuscript in Greifswald, as can be seen from documents in his estate. *Grundzüge der Mengenlehre* [H 1914a, II: 93–576] came out no later than April 1914, since on April 20 Hausdorff sent thanks to Hilbert for "your kind words about my book ..." (*HGW* IX, 340). He had, of course, also sent Engel a copy with a personal dedication. In reply to Engel's letter of thanks, Hausdorff wrote him in May, expressing satisfaction with his work – though not without a dose of irony and understatement:

> Dear and honored Herr College,
>
> thank you very much for your kind words about my book. I hope that it really is a relatively decent achievement and will as such offer you, who have promoted me to into the exalted circle of full professors (*Ordinarien*), subsequent vindication. (*HGW* IX, 234)

Set theory, as it was then understood, included three areas that are today considered distinct fields:

1. general set theory with its core subject matter: the theory of order types along with transfinite ordinal and cardinal numbers;
2. the theory of point sets in $\mathbb{R}^n$;
3. the theory of content and measure, including the theory of real functions.

This is not the place to give an overview of earlier works that dealt with these three aspects of set theory. Interested readers with knowledge of German can refer to Purkert's historical introduction to Hausdorff's *Grundzüge* (*HGW* II, 1–89). Summarizing what one finds there, we note that before the first decade of the twentieth century there was no independent study of set theory that dealt systematically and comprehensively with the whole field as it was understood at that time. Arthur Schoenflies hoped to publish such a work by revising and expanding his two earlier reports [Schoenflies 1900], [Schoenflies 1908]. For this project, he had enlisted the support of Hans Hahn, an outstanding expert on measure theory and real analysis. In 1913, thus a year before the publication of Hausdorff's *Grundzüge*, Schoenflies brought out his *Entwickelung der Mengenlehre* [Schoenflies 1913]. This was a completely revised and greatly expanded version of the first part of his report on set theory from 1900, containing – as noted in the title – general set theory and the theory of point sets. However, the second part, which Hahn was preparing, could not be completed due to the outbreak of the First World War. This was to have contained measure theory and the theory of real function, but also a chapter on the axiomatics of set theory.

As he emphasized in the preface, Schoenflies considered his book to be a "mixture of an historical account and methodological textbook" [Schoenflies 1913, IV], though, in fact, it essentially had the character of a textbook. It also undoubtedly benefited, in terms of precision, from the assistance the author received from Hahn and especially from Brouwer, as Schoenflies emphasized in the preface. Hausdorff evidently thought of Schoenflies' book as a competitor to his *Grundzügen*, as in a letter to Hilbert from February 27, 1914, he wrote, "[u]nfortunately, Schoenflies has beaten me to it; his second edition is after all considerably better than the first"

Bessel · Hagen

GRUNDZÜGE

DER

MENGENLEHRE

VON

FELIX HAUSDORFF

O. PROFESSOR DER MATHEMATIK AN DER UNIVERSITÄT GREIFSWALD

MIT 53 FIGUREN IM TEXT

LEIPZIG

VERLAG VON VEIT & COMP.

1914

Figure 7.2: Title Page of Hausdorff's *Grundzüge der Mengenlehre*

(*HGW* IX, 339). Schoenflies' book was without doubt the most comprehensive book on set theory up to that time, providing a good overall view of the theory. However, it did not introduce new methodological or conceptual points of view, nor did it present any important new results. Mainly, though, Schoenflies was simply unlucky, since his work appeared only months before Hausdorff's *Grundzüge*, a far superior textbook that ultimately overshadowed his book completely.

Hausdorff's *Grundzüge der Mengenlehre* was, indeed, the first comprehensive presentation of all three aspects of set theory that treated each systematically and provided complete and comprehensive proofs of all results. Most importantly, Hausdorff completely renovated the theory of point sets by creating a theory of general topological spaces that included the special case of metric spaces, within which the topology of $\mathbb{R}^n$ represented only a very special case. Beyond this essential innovation, the book contains a number of other original contributions by its author. Indeed, *Basic Features of Set Theory* had tremendous importance for the development of mathematics in the twentieth century, as reflected in Volume II of the Hausdorff edition, which is devoted entirely to this single work, supplemented by numerous notes and a series of essays that analyze its many novel features.

In his preface, Hausdorff noted that the *Grundzüge* was written as a textbook, not as a report. It thus presumes no prior knowledge, as everything is presented "with fully developed proofs." Regarding this last point, Hausdorff underscored its importance due to the peculiarities of set theory, a field in which

> nothing is self-evident, and what is correct often seems paradoxical, whereas what appears plausible is false; there is hardly any means, apart from unbroken chains of deduction, to save oneself and the reader from deceptions. [H 1914a, V; II: 97].

Although he considered that anyone could read his book "who possessed some [ability for] abstraction of thought," he also hoped to offer some new things to his professional colleagues as well, "at least in methodological and formal respects." Hausdorff dedicated his book to "Georg Cantor, the creator of set theory, in grateful reverence." Indeed, the entire work is based on Cantor's so-called "naive" theory of sets.

In surveying the contents of Hausdorff's *Grundzüge der Mengenlehre*, we take up the three components of set theory noted above, as these appear in the ten chapters of the book. Its first six chapters present a number of general concepts, but they are otherwise devoted to set theory in the narrower sense. Point set topology dominates the next two chapters, but in Chapter 7 Hausdorff first develops the theory of general topological spaces. In Chapter 8, "Point Sets in Special Spaces," he mainly focuses on those spaces later identified with descriptive set theory. Chapters 9 and 10 then enter the third branch of contemporary set theory, namely theories of content and measure along with their applications in real analysis.

7.6.1 General Set Theory

In the first chapter, "Sets and Their Connections: Sum, Intersection, Difference," Hausdorff defines the difference $B - A$ as the complement of A in B, assuming $A \subseteq B$. Following Cantor, he calls the union of sets a sum. After proving the propositions linking sums and intersections, he develops the elements of the algebra of sets. This begins with de Morgan's rules, from which it follows that an equation between sets leads to a related equation if one replaces all sets by their complements and interchanges the operations $\cup$ and $\cap$. Hausdorff calls this important property in the algebra of sets the *duality principle.*

Hausdorff also introduced a number of new notions of set algebra which – in modified form or sometimes under other names – still belong to the foundations of various disciplines, including descriptive set theory, general topology, theory of real functions and measure theory. He calls a system of sets a ring (field) if for any two sets A, B the sets $A \cup B$, $A \cap B$ ($A \cup B$, $A - B$ in case $B \subseteq A$) belong to the system. A set field then always contains $A \cap B$; such fields in Hausdorff's sense are today called Boolean rings. A set system is called a σ-system (δ-system) if it is closed with respect to countable unions (countable intersections). A ring or field that is also a σ-system is called a σ-ring or σ-field, respectively; δ-ring and δ-field are defined accordingly. A σ-field is also a δ-field, but not vice versa. Hausdorff later gave applications of these notions to the domains mentioned above.

In the preface, one of the novelties he noted was the concept of limits of set sequences. Let $X_1, X_2, \ldots$ be a sequence of sets, the following constructions turn out to be of special interest:

$$\limsup_i X_i = \bigcap_{k=1}^{\infty} \bigcup_{j=k}^{\infty} X_j \,, \quad \liminf_i X_i = \bigcup_{k=1}^{\infty} \bigcap_{j=k}^{\infty} X_j \,.$$

The set $\limsup X_i$ consists of all $x \in \cup_i X_i$ contained in infinitely many of the X_i, $\liminf X_i$ the set of all $x \in \cup_i X_i$ contained in almost all X_i, i.e., in all but finitely many. A set sequence is then called convergent if $\limsup X_i = \liminf X_i$; this set is then denoted $\lim X_i$.

In Chapter 2, "Sets and their Operations: Function, Product, Power," Hausdorff made a decisive contribution in support of his earlier claim that set theory is "the foundation of all mathematics" by showing that the concept of function (i.e., a single-valued mapping) can be viewed in purely set-theoretic terms. Hausdorff thus created the modern notion of function, which is omnipresent in mathematics today, even though one rarely finds any mention of where it originated (see Ulrich Felgner's essay "Der Begriff der Funktion" in *HGW* II, 621–633.) .

Hausdorff's basic idea defines a mapping or function of A into B as a suitable subset of $A \times B$, thereby clarifying the notion of mapping, which had not been precisely defined earlier. This approach is based on a purely set-theoretic definition of the notion of "ordered pair" (a, b), where the Cartesian product $A \times B$ is the set of all ordered pairs (a, b) with $a \in A$ and $b \in B$. Hausdorff defines the ordered pair (a, b) as the set $\{\{a, 1\}, \{b, 2\}\}$. It should be noted that the definition does not really depend on the notion of natural number; one only needs two different objects, say $\emptyset, \{\emptyset\}$. In 1921 Kazimierz Kuratowski gave the definition generally accepted today, namely, $(a, b) = \{\{a\}, \{a, b\}\}$, which has the advantage of not requiring any other object except a, b. Hausdorff then introduces a *single-valued function* from A into B *as a set* $P \subset A \times B$ with the property that every $a \in A$ occurs in exactly one $(a, b) \in P$ as the first element. The set P then realizes the desired property, namely that for each a there is a uniquely associated b; one then traditionally writes $b = f(a)$. If, in addition, P is such that every $b \in B$ occurs in exactly one pair $(a, b) \in P$, then P represents a one-to-one mapping (bijection) from A to B. This notion then leads directly to the notion of equivalence of sets, which serves as the basis for the later definition of cardinal numbers. By means of the notion of function, one can now also define the Cartesian product of any number of A_α, $\alpha \in J$, J an arbitrary index set, as the set of all single-valued functions f of J in $\cup_\alpha A_\alpha$ with $f(\alpha) \in A_\alpha$. If all $A_\alpha = A$, this yields the power set A^J as the set of all single-valued functions of J with "values" in A.

In Chapter 3, "Cardinal Numbers or Powers," Hausdorff first assigns to each set A a "thing" or "character" $\mathfrak{a}$, its cardinal number or power, in such a way that equivalent sets and only such sets have the same cardinal number. In comparing cardinal numbers, he must first allow the possibility that two cardinal numbers are incomparable, since he only brings in the well-ordering theorem much later. The Cantor-Bernstein equivalence theorem thus follows. Besides Bernstein's proof, Hausdorff also gives Zermelo's, which was based on ideas of Dedekind. This is followed by definitions of sum, product and power of cardinal numbers, theorems on inequalities, in particular König's theorem, and expositions on the cardinal numbers $\aleph_0$, $2^{\aleph_0} = \aleph$ and $2^{\aleph}$, together with examples of sets of numbers, points and functions having power $\aleph_0$ or $\aleph$ or $2^{\aleph}$.

Chapter 4 is entitled "Ordered Sets. Order Types." To define an order on a set A, Hausdorff utilizes ordered pairs, as in the notion of functions. Thus, an order on A is a subset P of $A \times A$ with the following properties:

1) Linearity: for $a, b \in A$ with $a \neq b$, either $(a, b) \in P$ or $(b, a) \in P$.
2) Irreflexivity: for any $a \in A$, $(a, a) \notin P$ holds.
3) Transitivity: From $(a, b) \in P$, $(b, c) \in P$ follows $(a, c) \in P$.

The notion of order (today one usually says linear order) is thus – like the notion of function in Chapter 1 – based on the set concept.

Hausdorff calls two order-isomorphic sets similar, which establishes an equivalence relation. The definition of order type suffers from the same weakness as the definition of the cardinal number, namely the introduction of an extraneous symbol.

> We assign namely to each ordered set A a sign α such that the same sign corresponds only to similar sets [...]. This sign α is called the order type (or type) of the set A. [H 1914a, 73; II: 173]

Hausdorff later defines the ordinal numbers as the order types of well-ordered sets. Thus, in *Grundzüge der Mengenlehre* it was not yet possible to treat the notions of cardinal number, order type and ordinal number strictly in terms of sets alone. This "blemish," so to speak, does not affect the theories developed in any way, as these lie within the scope of what can be derived from the system **ZFC**. In this chapter Hausdorff also gave a summary account of the foundations of the theory of ordered sets. Parts of this already originated with Cantor, but Hausdorff had to develop much himself in his work on ordered sets, including basic results on ordered sums, ordered products (with finitely many factors), segments, cofinality and coinitiality, as well as dense, continuous and discontinuous sets.

Chapter 5 on "Well-Ordered Sets. Ordinal Numbers" presented a systematic, yet concise exposition of the core ideas of Cantor's set theory. This drew on Gerhard Hessenberg's efforts to present a systematic theory of well-ordered sets [Hessenberg 1906], enriching what was already there with original ideas. After the definition of well-ordering and propositions about sums and products of well-ordered sets, Hausdorff demonstrates the comparability of ordinal numbers; he showed further that every set of ordinal numbers, ordered by size, is a well-ordered set. This is followed by the laws for calculating with ordinal numbers, limit formulas, the Euclidean algorithm, and the principle of transfinite induction (both as a principle for proofs as well as definitions).

His general theory of normal functions and their critical numbers is new. To each ordinal number α, let there be an associated ordinal number $f(\alpha)$. Then $f(\alpha)$ is called a *normal function* if it is monotone and continuous, i.e.

$$\text{for } \alpha < \beta \text{ one has } f(\alpha) < f(\beta) \text{ and for } \alpha = \lim_{\mu} \xi_{\mu}, \quad f(\alpha) = \lim_{\mu} f(\xi_{\mu}).$$

Because of monotonicity, $f(\alpha) \geq \alpha$. Hausdorff shows that there must be infinitely many numbers α with $f(\alpha) = \alpha$. These numbers he calls the *critical numbers* of the normal function f. From the critical numbers κ_{α} of a normal function one can obtain by transfinite induction a new normal function $g(\alpha)$, and from its critical numbers another one, and so on.

In anticipation of the well-ordering theorem, proved at the end of the chapter, Hausdorff here takes cardinal numbers to be powers of well-ordered sets. From the theorem that every set of cardinal numbers, ordered by size, is a well-ordered set, it follows that for every set K of cardinal numbers there is a next larger cardinal number, thus, the smallest cardinal number after K. Now Hausdorff introduces the aleph series by means of transfinite induction: Each ordinal number α is associated with a cardinal number $\aleph_{\alpha}$ so that $\aleph_0$ is the smallest transfinite cardinal number and $\aleph_{\alpha}$ for $\alpha > 0$ is the smallest cardinal number following $K = \{\aleph_{\xi},\ \xi < \alpha\}$. Each aleph is thus given an index, the type of all preceding alephs. The function $\aleph_{\alpha}$ has properties analogous to those of a normal function:

$$\aleph_{\alpha} < \aleph_{\beta} \text{ for } \alpha < \beta; \quad \aleph_{\alpha} = \lim \aleph_{\xi} \text{ for } \alpha = \lim \xi\,.$$

The number class $Z(\aleph_\alpha)$ represents the set of all ordinal numbers of power $\aleph_\alpha$. The smallest number ω_α of $Z(\aleph_\alpha)$ is called the initial number of this number class. The initial numbers ω_α play a prominent role in the whole theory. Hausdorff then proves that $Z(\aleph_\alpha)$ has type $\omega_{\alpha+1}$ and power $\aleph_{\alpha+1}$. He also gives a simplified proof for the aleph relation $\aleph_\alpha^2 = \aleph_\alpha$ first proved by Hessenberg.

In a section on initial numbers, Hausdorff summarized his own earlier results. Distinguishing between regular and singular ordinal numbers, he underscored that every transfinite regular ordinal number is an initial number ω_α and that the ω_α with α a non-limit ordinal are regular. He then asks, as he did earlier in [H 1908], whether there are regular initial numbers with a limit number as index. A plausibility consideration shows that such numbers would need to have an unimaginably large cardinality. Later researchers who worked on the theory of inaccessible cardinal numbers mainly relied on the *Grundzüge* rather than Hausdorff's earlier work dealing with this theory. The main theorem then follows that every ordered set is cofinal with exactly one regular ordinal number. If a set has no final element, it is cofinal with exactly one regular initial number. This theorem – as noted in 6.2 – is the basis for the study of ordered sets by means of their element and gap characters.

The last section of this chapter then takes up the well-ordering theorem. Hausdorff first hints at Zermelo's argument in his first proof from 1904, before reproducing the second proof from 1908: He then concludes:

> With the well-ordering theorem, the desired simplicity in the structure of set theory has finally been achieved. All infinite powers are now alephs and all powers are recognized as pairwise comparable. [H 1914a, 138; II: 238].

Up to this point, it was still conceivable that, for instance, $\aleph_1$ might be one among many incomparable powers immediately succeeding $\aleph_0$; but now it is the unique successor of $\aleph_0$. Hausdorff never seriously considered the possibility of developing set theory without the axiom of choice.

The sixth chapter, "Relations between Ordered and Well-Ordered Sets," is devoted almost entirely to summarizing Hausdorff's own results on ordered sets. First, he introduces the notion of partially ordered sets before proving by means of the well-ordering theorem the maximum chain theorem that today bears his name. He then treats element and gap characters, giving an existence proof for sets of prescribed species. Hausdorff next turns to general product and power notions, showing how these fit into the general theory of ordinal numbers. By highlighting special cases, such as the η_α-sets, he indicates the enormous potential for forming new order structures. In a section on "Rational Ordinal Numbers," he introduces the continued fraction algorithm for ordinal numbers, and in another section on "Initial and Final Ordering" he presents various results from [H 1909a]. Finally, Hausdorff establishes a connection between his notion of powers and non-archimedean systems of magnitude, including a proof of Hahn's embedding theorem (*HGW* II, 606–607).

7.6.2 General Topology

Chapters 7 and 8 on point sets were really devoted to general topology and, as such, they breathe the spirit of a new age. In Chapter 7, "Point Sets in General Spaces," Hausdorff develops a systematic theory of topological spaces on an axiomatic basis grounded in set theory. The classical point set theory of $\mathbb{R}^n$ here becomes a special case within the framework of this newly created theory. Its fundamental notion is that of a topological space:

> By a *topological* space we understand a set E in which the elements (points) x are associated with certain subsets U_x, which we call neighborhoods of x in accordance with the following *neighborhood axioms* [H 1914a, 213; II: 313].

Then follow the four axioms (α) - (δ) given already in 7.3 in connection with his lecture course from 1912. Before introducing the general notion of a topological space, Hausdorff also presents the axioms for a metric space:

> By a *metric* space we understand a set E in which every two elements (points) x, y determine a real nonnegative number, their *distance* $\overline{xy} \geq 0$; moreover, we require the validity of the following *distance axioms*:
>
> (α) (symmetry axiom): $\overline{yx} = \overline{xy}$.
>
> (β) (coincidence axiom): $\overline{xy} = 0$ if and only if $x = y$.
>
> (γ) (triangle axiom): $\overline{xy} + \overline{yz} \geq \overline{xz}$. [H 1914a, 211; II: 311]

Hausdorff then immediately shows that every metric space is a topological space by means of the definition $U_x = \{y \in E;\ \overline{xy} < \rho\}$ (ρ arbitrary > 0). Spaces satisfying the three conditions (α)–(γ) had already been introduced in 1906 in the dissertation of Maurice Fréchet under the name "classes (E)" [Fréchet 1906]. Fréchet's work, however, initially met with little response, whereas in chapter 8 of the *Grundzüge* Hausdorff developed the theory of these "metric space" as part of the more general theory of topological spaces. Since the publication of his book, metric spaces have become part of the common stock of knowledge among mathematicians, as Hans Freudenthal once remarked:

> It cannot be said that Fréchet's notions had much appeal until Hausdorff's book. The first to really think them through anew with all their consequences and the one who popularized them, though with other names, was Hausdorff (Freudenthal estate, Rijksarchief Noord-Holland, Haarlem. Inv. no. 557, fol. 6.)

Hausdorff drew from several historical lines of development for his axiomatic construction of the theory of topological spaces. These may be considered as lying at the roots of general topology:

1. the theory of point sets of $\mathbb{R}^n$ and its applications in the theory of real functions, measure and integration theory, function theory, and geometry;

2. the theory of ordered sets, in particular the numerous attempts to apply topological notions to order structures;
3. the extensive generalization of the concept of space in geometry in the course of the nineteenth century, in particular the study of manifolds and the disputes over the space problem; finally,
4. the formation of the first abstract concepts of space (Fréchet, Friedrich Riesz, and the American school around E. H. Moore).

Hausdorff was an important early representative of a new kind of mathematics, which came to be called the modern mathematics of the twentieth century. Characteristic for this new conception were theories of fundamental structures based on axiom systems. As in Hausdorff's theory of topological spaces, one identified a common structural core by extracting this from known examples or fields. Then, on the basis of suitably chosen axioms, one deductively developed a general theory for this core. This general theory not only applied to examples stemming from various special fields, it also opened many new possibilities for further theorizing and for applications. Such theories not only bring about a gain in generality and transparency but also in simplification, unification, and ultimately in economy of thought. The chapters on topology in Hausdorff's *Grundzüge* represent one of the classic texts of modern mathematics and thus represent a pioneering achievement not only in terms of content but also methodologically.

Before further discussing the contents of Chapter 7, some remarks on Hausdorff's fourth topological axiom (δ) will not be out of place. This so-called separation axiom is today usually called Hausdorff's axiom or axiom T_2. In the twenties and early thirties of the last century, both weaker and stronger axioms of separation were soon introduced. Hausdorff himself had not yet taken the step of characterizing spaces by different separation properties. Fréchet later introduced the weaker T_1-axiom in [Fréchet 1928a]. This states that if $x \neq y$, then there exist neighborhoods U_x and U_y with $y \notin U_x$ and $x \notin U_y$. To distinguish these T_1-spaces from those of Hausdorff, Fréchet called the latter "espaces de Hausdorff," terminology that quickly caught on. Clearly, every Hausdorff space is also a T_1-space, but the converse does not hold in general. The stronger separation axioms T_3 (regular space) and T_4 (normal space) were introduced, respectively, by Leopold Vietoris [Vietoris 1921] and Heinrich Tietze [Tietze 1923].

Let now X be a topological space and $A \subseteq X$ an arbitrary subset. Hausdorff then constructs a general topology on X by defining and systematically studying a set of sets arising from A by first considering characteristic properties of points with respect to A. Thus he forms the set A_i of interior points of A and the set A_r, the boundary points of A (points of A which are not interior points). Those sets A for which $A_i = A$ are called open sets in today's terminology (Hausdorff called them *domains*). In a footnote, he emphasizes their fundamental importance for the whole theory: "The notion of a set without boundary points, however, is so fundamental that it decidedly deserves its own name" [H 1914a, 215; II: 315]. He then proved that the union of arbitrarily many and the intersection of finitely many open sets will be an open set. In the mid-1920s, Paul Alexandroff invoked these two properties as axioms along with the axiom $\emptyset \in \mathfrak{O}$, $X \in \mathfrak{O}$ to introduce a topology on a set X by means of a system $\mathfrak{O}$ of open sets [Alexandroff 1925].

Taking any set $A \subseteq X$ Hausdorff distinguishes between three types of limit points. In modern terminology, $x \in A$ is called an accumulation point (ω-accumulation point, condensation point) of A, if in each neighborhood U_x there is at least one point (infinitely many points, uncountably many points) of A. He then introduces three symbols to define the sets A_α of accumulation points, A_β of ω-accumulation points, and A_γ of condensation points of A. The set A_β is the derivation A' of A already studied by Cantor. Following Cantor, Hausdorff calls a set $A \subseteq X$ *closed* if $A_\beta \subseteq A$, *dense in itself* if $A \subseteq A_\beta$ and *perfect* if $A = A_\beta$. He then shows that A is closed exactly if $A = A_\alpha$ or if $X - A$ is open.

Hausdorff then proves the following fundamental theorems: I. The intersection of arbitrarily many and the union of finitely many closed sets is closed. II. The union of arbitrarily many sets that are dense in themselves is dense in itself. These propositions serve as the basis for defining the kernel and hull for collections of sets, and it is noteworthy that Hausdorff introduces these notions in a very general way. Thus, if a system of sets M (such as a system of open sets) has the property that the union of any number of sets of the system belongs to the system again, then for any set A containing at least one set in M, one can define the *largest set in M contained in A*. Likewise, for systems (such as with closed sets) that allow for arbitrary intersections, one can define the *smallest set in M containing A* [H 1914a, 217–218; II: 317–318]. In the first case, this set is the union of all $M \subseteq A$, in the second case the intersection of all $M \supseteq A$. According to this scheme, Hausdorff searches for the smallest closed set containing A, which is just A_α. At the end of these considerations, he shows that by taking unions and intersections all these types of sets can be reduced to α-sets, i.e. to closed hulls.

In the next section, Hausdorff introduces compactness, a notion of fundamental importance for topology. Implicitly, mathematicians had long employed compactness when studying the properties of closed bounded sets of $\mathbb{R}^n$ and exploited this property in analysis. However, this definition of compactness requires making use of the metric on $\mathbb{R}^n$, so it is unsuitable for general topological spaces. Hausdorff adopts the definition of Fréchet: $A \subseteq X$ is called *compact* if every infinite subset of A has at least one accumulation point (finite sets and the empty set are also considered to be compact) [Fréchet 1906]. For compact closed sets in this sense, Cantor's intersection theorem holds: a descending sequence of nonempty compact closed sets has a nonempty intersection. Furthermore, one has a countable version of the Heine-Borel covering theorem: If a compact closed set A is contained in the union of countably many open sets, then A is covered by finitely many of these open sets. Hausdorff wanted to transfer "two main notions of elementary analysis to the theory of point sets, [$\cdots$] the notions of limits and convergence" [H 1914a, 232; II: 332]. A point x is called a limit point of the infinite set A if each of its neighborhoods U_x contains all but finitely many points of A. A set A is then called convergent if it has a limit point, in which case it has only one. Convergence can then be expressed in terms of compactness: A set A is convergent exactly if it is a compact set with a single accumulation point.

In current terminology, Hausdorff's notion of compactness is equivalent to countable compactness, whereas a set A is called compact if *every* open covering of A has a finite subcovering. The latter property was introduced and studied in the early 1920s by Alexandroff and Urysohn, two young Russians, who wrote to Hausdorff about their work in April 1923 (*HGW* IX, 5–9). They actually called such sets bicompact, reserving the term "compact" for those that were compact in Hausdorff's sense. In 1930 Andrei Tychonoff showed that the general property of bicompactness is preserved under arbitrary products [Tychonoff 1930]; since then this terminology was dropped in favor of the general notion of compactness familiar today.

In the section following on "Relative Concepts," Hausdorff introduces the relative topology on a subset $M \subset X$ derived from the topology of X. As he emphasized in the preface, "the systematic implementation of the relative concepts" was one of the book's many novelties [H 1914a, VI; II: 98]. Although Hausdorff does not speak of subspaces of a topological space, this fundamental notion is clearly prefigured in the *Grundzüge*, whereas sums, products and quotients of topological spaces do not appear. Hausdorff takes any fixed subset M of a topological space X and calls a set A *closed in* M (relatively closed) if there exists F closed in X such that $A = M \cap F$. Likewise, A is called *a relative domain of* M (i.e., an open set in M) if for some set G open in X, $A = M \cap G$. It turns out that relative domains in M and closed sets in M are complements of each other (with respect to M). The characteristic properties for unions and intersections of open and closed sets are then easily proved, as are several other fundamental topological results. That Hausdorff here had in mind the notion of topological subspaces is clear from the following:

> The notions of relative domain and relatively closed set are still capable of a somewhat more systematic conception by developing a whole *relative theory* which, instead of referring to the set E [as the ambient space] refers instead to a subset of it as the underlying space. [H 1914a, 241; II: 341]

Since Hausdorff defines a topological space by a system of neighborhoods V_y for points $y \in M$, he obtains this immediately by virtue of $V_y = U_y \cap M$, where U_y are the neighborhoods in the space E. Having established that V_y satisfy the four axioms for a topological space, he notes:

> After the neighborhoods are hereby fixed, all else depends only on the set M, and to every such (non-vanishing) set M corresponds a *relative theory* in which M instead of E has assumed the role of the ambient space. [H 1914a, 242; II: 342]

The theory of connectivity developed by Hausdorff in §7 of Chapter 7 had a lasting influence; indeed, his basic definition of a connected set of points soon gained acceptance in general topology. Other definitions had been given earlier, for example by Cantor, Jordan, Schoenflies, and Study. Hausdorff prefaces his own definition with the following remark:

> Various definitions are common for the context; the one given in the text does not coincide with any of them, but seems to us the most natural and general. [H 1914a, 244; II: 344]

He calls a nonempty subset A of a topological space X connected if it is impossible to find a decomposition $A = A_1 \cup A_2$, where A_1, A_2 are *disjoint nonempty subsets closed in* A. Thus, a topological space X is connected if it is not the union of two disjoint nonempty closed (or open) sets. Hausdorff's definition of connectivity has prevailed, however, not only because it is the "most natural and general" but also because he developed a whole theory that a great number of researchers built upon. Of particular importance for general topology was the recognition that Hausdorff's notion of connectivity expresses a topological invariant: the homeomorphic image of a connected set $A \subseteq X$ will always be connected [H 1914a, 363; II: 463].

7.6.3 Special Topological Spaces

Hausdorff begins Chapter 8, "Sets of Points in Special Spaces," by defining "equivalence" for two systems of neighborhoods $\{U_x\}, \{V_x\}$. These are called *equivalent* if they generate the same topology (i.e., the same system of open sets), which is exactly the case if every U_x contains a V_x and every V_x contains a U_x as a subset. A subset $\{U'_x\}$ equivalent to the entire neighborhood system $\{U_x\}$ is called a basis of neighborhoods. Hausdorff's goal in Chapter 8 was to further specialize spaces by means of certain constraints on the size of such a basis. For this purpose he uses two *countability axioms*, which he calls the first and second countability axioms.

> First countability axiom: *Every point x of the space X has a countable neighborhood basis.*
> Second countability axiom: *The entire space X has a countable neighborhood basis.*

Hausdorff first examines the consequences of the weaker first countability axiom. His main result is that the accumulation points of a set A are identical with the limit points of convergent sequences of points in A. Thus, in Hausdorff spaces satisfying the first countability axiom, the topological structure can be completely described by means of convergent sequences. By means of this result Hausdorff was able to relate his theory of spaces to Fréchet's theory of limit spaces [H 1914a, 265–268; II: 365–368].

After this, he explores the consequences of the second countability axiom, in particular with respect to the existence of condensation points. For a countable set A in an arbitrary Hausdorff space, one can show that the set A_γ of condensation points is empty. If, however, the second countability axiom holds, then the converse is also true: if $A \cap A_\gamma = \emptyset$, then A is at most countable. Even more, the set of condensation points of A belonging to A is dense in itself and hence a subset of the kernel of A, which is dense in itself. Furthermore, in a space X satisfying the second countability axiom, the following bounds hold: the set of all open (closed) sets of X has at most the cardinality $2^{\aleph_0}$ of the continuum, hence the set X itself (the set of points of the space) also has at most the power of the continuum. A set of pairwise disjoint open sets of X, on the other hand, is at most countable (for $\mathbb{R}^n$, these theorems had been found by Cantor).

Hausdorff was able to extend various other results on Euclidean spaces – theorems found by researchers such as Cantor, Bernstein, Young and Lindelöf – to topological spaces with second countability axiom. In particular, for such a space X, a set $A \subseteq X$ is compact if and only if it is countably compact, i.e., compact in the sense of Hausdorff. Already at the beginning of Chapter 7, Hausdorff formulated the distance axioms for metric spaces, the topic elaborated in sections 6–9 of Chapter 8. Taken as topological spaces, these satisfy the first countability axiom, whereas metric spaces satisfying the second countability axiom are precisely those that are separable, i.e. contain a countable dense subset. Fréchet had introduced separable spaces in his dissertation from 1906.

Hausdorff next introduced a concept that would later become important in various fields of mathematics, even in computer science: the so-called Hausdorff metric. Let X be a metric space with metric $d(x, y)$ and A, B be two nonempty compact subsets of X. Hausdorff showed that the function

$$d_H(A, B) = \max\left\{\sup_{a\in A}\inf_{b\in B} d(a,b),\ \sup_{b\in B}\inf_{a\in A} d(a,b)\right\}$$

satisfies the three axioms for a metric space. Thus, with $d_H(A, B)$ as distance, the set of nonempty compact subsets of X becomes a new metric space $H(X)$, which became the starting point for the modern theory of hyperspaces. Already in 1905, the Romanian mathematician Dimitrie Pompeiu had introduced the above distance function for non-empty compact subsets of the Euclidean plane, as Hausdorff himself mentioned in his notes [H 1914a, 463; II: 563]. Hausdorff was able to show that if X is separable, then so is $H(X)$, and later researchers proved that these hyperspaces inherit other important properties, including completeness and compactness. Mikhail L. Gromov used the Hausdorff metric to measure how far two compact metric spaces X, Y are from being isometric. This distance is called the Gromov-Hausdorff metric $d_{GH}(X, Y)$, and is given by the smallest possible $d_H(X, Y)$ when X and Y are embedded isometrically in a metric space Z:

$$d_{GH}(X, Y) = \inf\left\{d_H(f(X), g(Y)),\ f : X \to Z,\ g : Y \to Z\right\},$$

where f and g are isometric embeddings of X and Y, respectively, into Z. The

Gromov-Hausdorff metric and its convergence properties play an important role in differential topology [Gromov 1999].

In turning to Borel sets in metric spaces, Hausdorff entered the arena now called "descriptive set theory." He would later investigate this theory in detail in his book *Mengenlehre* ([H 1927a] in *HGW* III). In a modern account of classical descriptive set theory, this field was characterized as follows:

> Descriptive set theory is the study of "*definable sets*" in *Polish* (i.e., separable completely metrizable) *spaces*. In this theory, sets are classified in hierarchies, according to the complexity of their definitions, and the structure of the sets in each level of these hierarchies is systematically analyzed. [Kechris 1995, XV]

The hierarchies of sets alluded to here are generally indexed by transfinite ordinal numbers. An early example of such generalized "set sequences" was studied by Cantor, namely, the sequence $\{P^{\alpha}\}$ of successive derivatives of a point set in $\mathbb{R}^n$ (α is here a number of the first or second number class). The transfinite ordinals provide an indispensable tool for descriptive set theory, especially for the study of higher classes of points.

Descriptive set theory has now found applications in numerous branches of mathematics, e.g., measure theory, probability theory, topology, functional analysis, harmonic analysis, potential theory, ergodic theory, representation theory of Lie groups, combinatorics, and mathematical logic. The term "descriptive set theory" originated with Alexandroff and was first used by him in a letter to Hausdorff dated July 4, 1926 (*HGW* IX, 44). After Alexandroff and Hopf introduced it in their classic *Topologie I*, the terminology soon became generally accepted [Alexandroff/Hopf 1935, 19–20].

The roots of this theory originated in the "descriptive function theory" of the French school associated with Borel, Baire and Lebesgue. Their most important contributions were the book [Borel 1898] and two extensive essays, [Baire 1899] and [Lebesgue 1905]. This descriptive theory of functions had arisen not least from philosophical considerations concerning the problem of the existence of mathematical objects. For Baire, Borel and Lebesgue, mathematical objects had to be "describable" (from décrire) by certain accepted procedures. All three were firmly convinced that the recent work undertaken by German mathematicians failed to meet this standard. Thus, in their eyes, Cantor's theory of arbitrary sets of arbitrarily large cardinality, but also Dirichlet's arbitrarily defined functions, and especially the arbitrary choices employed in Zermelo's proof of the well-ordering theorem, did not belong to the domain of "describable" objects [Moore 1982, 92ff.].

In the *Grundzüge* Hausdorff succeeded in presenting the most important concepts and results of the French school in a new systematic form, but also enriched in content. In their analysis of his achievement, Vladimir Kanovei and Peter Koepke remark:

> In the *Grundzüge* Hausdorff builds up descriptive theory for the first time as a theory of *Mengen* and not as a theory of *Funktionen*. This view has prevailed historically. (*HGW* II, 773)

In the appendix to the *Grundzüge*, Hausdorff clearly distanced himself from the philosophically motivated restrictions imposed by the French researchers [H 1914a, 450; II: 550]. For him, the results of the French theory of real functions were interesting because of their mathematical substance: functions and sets which arise by certain countable procedures have interesting structural properties, which required investigation.

Here the class of Borel sets is of special interest. Let $\mathfrak{A}$ be a given set system and let $\mathfrak{A}_{(\sigma\delta)}$ denote the smallest set system which encompasses $\mathfrak{A}$ and is closed with respect to countable unions and intersections. Hausdorff first shows, with recourse to the set algebra developed in Chapter 1, how $\mathfrak{A}_{(\sigma\delta)}$ can be successively constructed. The sets A_σ (A_δ) are the countable unions (intersections) of sets from $\mathfrak{A}$; the A_σ (A_δ) form the smallest σ-system (δ-system) $\mathfrak{A}_\sigma$ ($\mathfrak{A}_\delta$) over $\mathfrak{A}$. After forming $A_{\sigma\delta}$ ($A_{\delta\sigma}$) we can continue in the same manner. By means of this process $\mathfrak{A}_{(\sigma\delta)}$ can be built up as follows: Let $\varphi(\mathfrak{A}) = \mathfrak{A}_{\sigma\delta}$, then for each $\eta < \omega_1$ we define:

$$\mathfrak{A}_0 = \mathfrak{A}, \quad \mathfrak{A}_\eta = \bigcup_{\xi<\eta} \varphi(\mathfrak{A}_\xi); \quad \text{finally, we obtain } \mathfrak{A}_{(\sigma\delta)} = \bigcup_{\eta} \mathfrak{A}_\eta \quad (\eta < \omega_1).$$

Hausdorff now considers a metric space X and denotes its closed sets by F (*ensemble fermé*) and open sets by G (*Gebiet*, i.e domain). The F_δ are sets F, the G_σ are sets G, so nothing new arises in each case. The first stage in the above process then takes G_δ and F_σ to construct $F_{\sigma\delta}$ and $G_{\delta\sigma}$, etc. F and G, F_σ and G_δ, $F_{\sigma\delta}$ and $G_{\delta\sigma}$ are complements of each other, respectively. Let $\mathfrak{G}$ be the system of open sets, $\mathfrak{F}$ that of closed sets. One can then apply the above process for $\mathfrak{A} = \mathfrak{G}$ or $\mathfrak{A} = \mathfrak{F}$ and obtain with $\mathfrak{G}_{(\delta\sigma)}$ ($\mathfrak{F}_{(\sigma\delta)}$) the smallest set system which contains all open (closed) sets and is closed under countable unions and intersections. Hausdorff then shows that every closed set F is a G_δ and every open set G is a F_σ. Therefore, it does not matter whether one starts the construction with $\mathfrak{G}$ or $\mathfrak{F}$; finally, the resulting system of Borel sets is the smallest set system which contains all open and closed sets and is closed under countable unions and intersections. The name "Borel sets" in this general sense is due to Hausdorff; Schoenflies had employed the term Borel sets only for countable intersections of open sets, that is G_δ-sets in Hausdorff's terminology. Hausdorff's names for the sets of Borel's hierarchy (G_δ, F_σ, etc.) long prevailed and served as the standard terminology for decades; only in more recent times was it partially replaced by other symbolism in the context of the development of mathematical logic and descriptive set theory. Hausdorff's σ-algebra of Borel sets and its systematic structure thus became standard tools for descriptive set theory and other fields, including measure theory and general topology.

Hausdorff next turns to complete spaces, which he defines as metric spaces in which every Cauchy sequence converges. Fréchet had already dealt with these under the rubric of spaces which admit "une généralisation du théorème de Cauchy" [Fréchet 1906], but the term "complete" has become generally accepted. Hausdorff showed that any metric space can be isometrically embedded in a complete metric space. This so-called "Hausdorff completion" is a generalization of the procedure by which Cantor and Charles Méray had obtained the field of real numbers from the rationals. This leads naturally to separable complete metric spaces (called Polish spaces today), which are those studied in classical descriptive set theory.

Cantor had been able to settle the question of the power *of closed* sets of $\mathbb{R}^n$ by means of the theorem now named after him and Ivar Bendixson . This states that a closed set in $\mathbb{R}^n$ is either at most countable or has the power $\aleph$ of the continuum. This result marks the beginning of an important line of development in descriptive set theory, to which Hausdorff contributed significantly. William Henry Young had shown for G_δ-sets of $\mathbb{R}$ that they are at most countable or of the power $\aleph$, and Hausdorff was able to extend this result to G_δ-sets in Polish spaces. Finally, in the appendix to the *Grundzüge*, he showed that in a Polish space the cardinality of a $G_{\delta\sigma\delta}$-set is either countable or equivalent to the continuum. Thus, in Polish spaces, the continuum problem was solved for the sets G, G_δ, $G_{\delta\sigma}$, $G_{\delta\sigma\delta}$, $G_{\delta\sigma\delta\sigma}$, F, F_σ, $F_{\sigma\delta}$, $F_{\sigma\delta\sigma}$, i.e. all were either at most countable or had the power $\aleph$. To this, Hausdorff added the following remark: "[$\cdots$] it does not seem hopeless to prove the same for all Borel sets, i.e., those arising from domains or closed sets by unions and intersections over sequences" [H 1914a, 466; II: 566]. In fact, Hausdorff (and independently Alexandroff) succeeded in proving this in 1916, as will be discussed at the outset of Section 7.7.

The last two sections of Chapter 8 were devoted to Euclidean spaces and the Euclidean plane especially. Among other goals, Hausdorff aimed to link specific properties of Euclidean spaces, such as the existence of line segments and polygonal paths with the general theory, in particular connectivity properties. A central topic was the Jordan curve theorem in the Euclidean plane and its proof following Brouwer. No really new results were brought forth, but like the entire book, the presentation is a model of clarity and methodological skill.

7.6.4 Mappings and Real Analysis

In Chapter 9, "Mappings or Functions," Hausdorff begins by considering single-valued mappings of one topological space X onto another Y. He distinguishes here for the first time between pointwise and global continuity with respect to such mappings: $f: X \to Y$ is continuous at the point x if for every neighborhood V_y of the point $y = f(x)$ there is a neighborhood U_x of x with $f(U_x) \subseteq V_y$; if $f(x)$ is continuous for all points of X, then it is globally continuous. Hausdorff then shows that the latter holds exactly when the pre-image $f^{-1}(F)$ of every closed set $F \subseteq Y$ (or open set $G \subseteq Y$) is closed (open) in X. Some important properties then follow, e.g.: the continuous image of a connected set is again connected; the continuous image $f(A)$ for any set A cannot have more connected components than A itself. If a space X satisfies the first axiom of countability, then a continuous function f on X is determined by its values on a dense subset P in X. This enables estimations for the cardinality of the set of continuous functions when the power of P is known.

In the section following, Hausdorff discusses curves by considering continuous mappings $f : A \to B$, where A, B are point sets in Euclidean spaces X, Y. Of special interest are *continuous images of line segments*, though Hausdorff finds it problematic "to call such a set a *continuous curve*" [H 1914a, 369; II: 469] in view of the "curves" constructed many years before by Peano and Hilbert, which involve mappings that raise dimension. The "Peano curve" helped to launch investigations of numerous topics in point set topology, such as Peano continua, arcs, simple closed curves, various types of continua (sets that are compact and connected), and ultimately led to a suitable notion of curves (for an historical overview see [Charotnik 1998]). Documents in Hausdorff's estate show that he took a lively interest in this subject, especially during the 1920s and 1930s (*HGW* III, 798–825). In particular, he was very involved with the first monograph on curve theory [Menger 1932].

Another topic Hausdorff treated in Chapter 9 was the topological invariance of dimension. In 1878 Georg Cantor showed that the sets $[0,1]$ and $[0,1]^n$ have the same cardinality [Cantor 1878]. Confronted with this surprising finding, Dedekind remarked in a letter to Cantor that (when $n > 1$) every bijective mapping between these two sets will "necessarily be discontinuous" [Cavaillès/Noether 1937, 38]. Thereafter, a number of attempts were made to prove the invariance of dimension in Euclidean spaces, i.e. the impossibility of altering the dimension of a set under a homeomorphism (a bijective, bicontinuous mapping) [Johnson 1979/1981]. Although this seemed intuitively obvious, it was not until 1911 that Brouwer managed to publish a stringent proof of this fundamental result [Brouwer 1911].

Hausdorff made the following general remark before taking up the problem of dimension:

> The term *analysis situs* refers to the study of those properties and relations of point sets that remain invariant under reversibly single-valued bicontinuous mappings; if one attributes the same *geometric type* to sets which are [$\cdots$] images of each other [under such mappings], then just as one attributes the same order type to similarly ordered sets, so is *analysis situs* concerned with the theory of geometric types, with their enumeration, classification, etc. But this theory is, in our opinion, not yet at all in the desired state of simplicity and completeness; we cannot penetrate into it very deeply here [$\cdots$]. [H 1914a, 376; II: 476]

Hausdorff here hints at a problem that has long preoccupied topologists, namely, the connection between the concepts of general topology he helped to create and the older geometric topology, earlier known as "*analysis situs*." It would take a long time before the language and methods of general topology became accepted as a conceptual foundation in this more established field, in which algebraic methods dominated more and more (*HGW* II, 70–75). Hausdorff only proves the topological invariance of dimension in the Euclidean plane by using Jordan's curve theorem; for higher dimensions he refers to Brouwer.

In the remaining sections of Chapter 9, Hausdorff treats the foundations of the theory of continuous and Baire functions. In terms of results, he does not go beyond those obtained by the French school, but Hausdorff's findings are more general in that they deal with functions on metric spaces that take "values" in metric spaces or real-valued functions on metric spaces. Hausdorff develops the theory from a set-theoretic perspective by linking properties of functions to set classes in Borel's hierarchy. For example, the points of continuity of a function form a G_δ set, whereas the points of discontinuity comprise a F_σ.

Hausdorff's tenth and last chapter, "Content of Point Sets," gave the first detailed presentation of Lebesgue's theory of measure and integration in the German language [Lebesgue 1902], [Lebesgue 1904]. But here he not only masterfully presented an already existing theory, he also made contributions of far-reaching importance (*HGW* II, 788–800). First and foremost among these was the sphere paradox, later named after Hausdorff, probably the most spectacular single result in the entire book. It appears in the appendix to the tenth chapter, but he also published it separately in *Mathematische Annalen* (*HGW* IV, 11–18).

The starting point for this counter-example was the so-called "equal measure problem": to find a non-negative real-valued set function μ such that for any bounded set A in $\mathbb{R}^3$, the content $\mu(A)$ satisfies the following three conditions:

(α) Congruent sets have the same content.

(β) The unit cube has content 1.

(γ) For disjoint sets A, B, $\mu(A \cup B) = \mu(A) + \mu(B)$.

Hausdorff showed that, without imposing further conditions, no such μ exists that could apply to all subsets of Euclidean 3-space. From this also follows the negative solution of the equal measure problem for all dimensions > 3; for dimensions 1 and 2 the problem remained at first open, but was answered positively by Stefan Banach in 1923 [Banach 1923]. Hausdorff had already recognized why his decomposition method could not work in dimensions 1 and 2, as he remarked: "For the straight line and the plane the question must remain open, since the structure of the group of motions in these cases does not admit the above procedure" [H 1914a, 472; II: 572].

Hausdorff's proof was based on a paradoxical decomposition of a sphere K into four pairwise disjoint subsets $K = Q \cup A \cup B \cup C$. The paradox arises from the fact that A is congruent to B, B is congruent to C, but A is also congruent to $B \cup C$. The set Q is countable and thus can be neglected in the consideration of content. Now if there were a content μ that applied to these subsets, then $\mu(A) = \frac{1}{3}\,\mu(K)$, whereas, on the other hand, $\mu(A) = \mu(B \cup C) = \frac{1}{2}\,\mu(K)$, so additivity obviously cannot hold in this case. To produce such a spherical decomposition, Hausdorff needed Zermelo's axiom of choice, which he used without mentioning, just as he had explained already in [H 1908]. Whether a paradoxical decomposition like

Hausdorff's might be obtainable with a weaker variant of the axiom of choice appears to be an open question.

Banach and Tarski published an important generalization of Hausdorff's paradoxical sphere decomposition in [Banach/Tarski 1924]. They showed that in Euclidean spaces of dimension $n \geq 3$, any two bounded sets A and B with interior points can be decomposed into an equal finite number of mutually disjoint pieces

$$A = A_1 \cup A_2 \cup \cdots \cup A_m, \quad B = B_1 \cup B_2 \cup \cdots \cup B_m,$$

such that A_i is congruent to B_i for $i = 1, 2, \ldots m$. This theorem has become famous as the Banach-Tarski paradox; it was proved on the basis of a generalization of the Schröder-Bernstein equivalence theorem originating with Banach.

Hausdorff's paradox led to developments in two interrelated fields, the study of paradoxical decompositions and the theory of amenable groups. The latter theory has its origin in work of John von Neumann. Inspired by the paradoxes of Hausdorff and Banach/Tarski, von Neumann introduced "measurable groups" (the precursors of "amenable groups") in [Neumann 1929]. These are topological groups G that admit a finitely additive measure on subsets of G. The theory of amenable groups has since found applications in functional analysis and in a number of other areas such as graph theory and automata theory [Wagon 1993].

Hausdorff's Chapter 10 also pointed out a noteworthy connection between measure theory and probability theory. Although only formulated as a brief hint, this was a truly visionary look into the future in view of the contemporary state of probability at that time:

> We remark in addition that some theorems on the measure of point sets perhaps reveal a more familiar face if they are expressed in the language of the *probability calculus*. [H 1914a, 416; II: 516]

Hausdorff takes about half a page to explain this remark, the upshot being that one can often translate probabilistic statements to measure theory and vice versa. This new approach to probability theory, laid out here only very briefly, later served as the basis for Hausdorff's lecture course on probability, which he taught in Bonn during the summer semester of 1923; this course will be discussed in Section 8.3.

In Section 4 of Chapter 10, "Examples and Applications," Hausdorff presented one particularly important result for probability theory. This involves a theorem first set forth by Émile Borel, which can be interpreted as a special case of the strong law of large numbers [Borel 1909]. Borel considered the decimal fraction expansions for the numbers $x \in (0, 1)$, and then, taking some fixed digit, he let $p_n(x)$ be the number of times this occurs in the first n digits of the expansion for x. His theorem asserts that the subset of x for which $\lim_n \frac{p_n(x)}{n} = \frac{1}{10}$ has Lebesgue measure 1. His proof, however, was completely inadequate, a problem remedied here by Hausdorff, who considers dyadic fractions instead of decimals. Since the dyadic rationals are countably infinite, they have measure zero, so Hausdorff restricts himself to the irrational numbers $x \in (0,1)$. Letting μ be Lebesgue measure, the result analogous to Borel's theorem now reads: If $p_n(x)$ is the number of zeros in the first n dyadic places of x, then

$$\mu\left(\left\{x;\ \lim_{n\to\infty} \frac{p_n(x)}{n} = \frac{1}{2}\right\}\right) = 1\,.$$

With these remarks on probability, we conclude this cursory survey of the contents and most novel features of Hausdorff's *opus magnum.* A few remarks about its reception will now be added.

7.6.5 Reception of the Grundzüge, 1914–1920

Hausdorff's *Grundzüge* appeared in the spring of 1914, when the great powers of Europe were already in a state of high tension approaching the eve of the First World War. After war broke out in August, it dramatically affected scientific life in Europe, and as the fighting dragged on few paid close attention to new works in mathematics. Under these circumstances, it should come as no surprise that, even five or six years after its publication, Hausdorff's book went almost unnoticed. On the other hand, one should not overlook the fact very few established mathematicians were pursuing research in set theory. Indeed, even the leading mathematical journals in Germany took no notice of Hausdorff's book until well after the war. Scanning, for example, Hilbert's *Mathematische Annalen* up to Volume 82 (1921), only one article cited Hausdorff's *Grundzüge*, namely a paper published by the author himself, his now-famous *Dimension und äusseres Mass* [H 1919a].

Perhaps even more remarkable, Hausdorff's important new book never received a review in the official journal of the German Mathematical Society. In 1917, its readers were only offered the meager information Hausdorff himself had inserted, mainly a few excerpts from his preface (*HGW* II, 829). Moreover, the review of the *Grundzüge* that appeared in the standard abstracting periodical, *Yearbook on the Progress of Mathematics*, was not written by a prominent mathematician but rather by its editor, the aged Emil Lampe. Although he was a personal friend of Cantor, Lampe had little familiarity with set theory. His synopsis essentially only repeated information in Hausdorff's preface and then merely named the chapter headings in the book. In 1920, an almost identical review by Lampe appeared in *Archiv der Mathematik und Physik*, probably because the *Yearbook* volume presenting works from 1914/15 only came out in 1922.

To be sure, a scattered series of mainly shorter reviews of the *Grundzüge* did appear (*HGW* II, 827–855). One came out already in 1914 in the *Literarisches Zentralblatt für Deutschland*; it discussed Hausdorff's book, alongside the 1913 study by Schoenflies and another by Julius König. The part dealing with the *Grundzüge*, however, was scanty in content and essentially confined to the enumeration of topics. In 1915 three reviews appeared: one by Heinrich Wieleitner gave a brief synopsis; another by Wilhelm Gross emphasized the work's originality without going into great detail: the third by Giulio Vivanti was published in Italian in *Bollettino di bibliografia e di storia delle science matematiche e fisiche.* Vivanti was a leading authority on set theory, and his review does much greater justice to Hausdorff's achievement than the other two. Much of it focused on the topological chapters in the *Grundzüge*, from which he even quoted the topological axioms verbatim. Vivanti concluded with the conviction that Hausdorff's book "is an important step forward both in terms of generalization and rigor" (reprinted in German translation in *HGW* II, 837–840). Nor surprisingly, given the state of Franco-German relations during the war, the *Grundzüge* found very little response in France. In 1916, a nine-line review appeared anonymously in *L'enseignement mathématique*; while asserting that Hausdorff's book offers an excellent introduction to the study of set theory, it otherwise only repeats, *mutatis mutandis*, passages from its introduction. The *Bulletin des Sciences Mathématiques*, which regularly reviewed numerous books, took no notice of the *Grundzüge* at all. Even Maurice Fréchet, who must have had a vital interest in Hausdorff's work, only read it some years later. According to his own testimony: "Ce n'est qu'après la guerre que j'ai pu lire l'intéressant Livre de F.H." [Fréchet 1921, 367].

Unlike the situation in France, mathematicians in the United States were made aware of for Hausdorff's *Grundzüge* soon after the war ended. This circumstance was due to a very thorough and competent review that appeared in the December 1920 issue of the *Bulletin of the American Mathematical Society.* Its author, Henry Blumberg, had taken his doctorate in 1912 under Edmund Landau in Göttingen and worked mainly in the field of real functions. One can still enjoy reading this review today, as it reflects a deep understanding of and admiration for Hausdorff's work (see *HGW* II, 844–853). Blumberg's review appeared at a time when a younger generation of mathematicians was preparing to pick up and pursue new lines of research, like those richly contained in Hausdorff's book. Topology stood at the center of their interests, as will be described in Section 8.4 along with the intensive phase in the reception of Hausdorff's *Grundzüge.*

7.7 From Borel Sets to Fractal Dimension

After he published the *Grundzüge*, but especially after 1916, Hausdorff's research interests shifted away from set theory and general topology in favor of topics in analysis. We can see this not only from his publications, but even more so when looking at the many other studies documented in his unpublished posthumous papers. Examining the many manuscripts he produced between 1914 and 1921, we can see this shift as marking two periods, the first dating from publication of his *Grundzüge* in the spring of 1914 to the end of 1916, the second from January 1, 1917 until his departure from Greifswald at the end of September 1921. During the first period, 51 dated studies are preserved in his papers, 23 of them devoted to set theory or topology, 25 to analysis, and 3 to other areas such as geometry and number theory. For the second period, there are even more: 63 dated studies, but only 5 are on topics in set theory or topology, whereas 52 fall in the area of analysis, and 6 deal with topics in geometry, probability, and algebra.

One particular unpublished study, written immediately after completion of the *Grundzüge*, deserves mention here: Hausdorff's "Metrische und topologische Räume" (Metrical and Topological Spaces) (*HGW* III, 750–751). The text begins with the following straightforward sentence: "In metric spaces various propositions hold which are not true in topological ones." In the *Grundzüge*, Hausdorff had proved that every countably compact metric space is separable, a theorem that does not hold for topological spaces in general. In his manuscript, he shows this by constructing a counterexample that would soon become famous: the "long half-line" $H = \lambda + (1 + \lambda)\omega_1$, equipped with the order topology. Here λ has the order type of the real line $\mathbb{R}$ and ω_1 is the initial number of the third number class. H is then a countably compact Hausdorff space, but *it is not separable.* Hausdorff did not publish this interesting construction, but he communicated it to Heinrich Tietze. The latter then published H as an example of a countably compact space that "does not satisfy the second axiom of countability, [$\cdots$] an example which I owe to a kind communication of Herr F. Hausdorff" [Tietze 1924, 218]. Independently of Hausdorff and Tietze, Alexandroff discovered the "long line" $(1+\lambda)\omega_1^* + \lambda + (\lambda + 1)\omega_1$ [Alexandroff 1924], which one can obtain from the long half-line by reflection about the origin. Today this counterexample is often referred to as "Alexandroff's long line."

In his paper "Die Mächtigkeit der Borelschen Mengen" (The Power of the Borel Sets) [H 1916], Hausdorff proved the correctness of the conjecture he had formulated in the *Grundzüge*, namely that his power theorem, which he had proved for $G_{\delta\sigma\delta}$ sets, could be extended to arbitrary Borel sets. Thus, he now had proved the following theorem:

Every Borel set is either finite, countably infinite, or has the power of the continuum [H 1916, 433; III: 434].

His proof followed the same idea used before by Cantor, Young, and Hausdorff himself. Since by Cantor's theorem, every perfect set of $\mathbb{R}^n$ has the power of the continuum, it suffices to show that every uncountable Borel set contains a perfect subset. Hausdorff's proof is very technical, however, and occupies $3\frac{1}{2}$ printed pages (see *HGW* III, 440–441). The continuum problem for Borel sets was thus solved. However, Hausdorff was very skeptical (and rightly so, as we know today) about whether one could come closer to solving the continuum problem in this way. As he put the matter:

> This clarifies the issue of cardinality for a very comprehensive category of sets; but this can hardly be regarded as a step towards the solution of the continuum problem, since the Borel sets are still very special and form only an infinitesimally small subsystem (with the power $\aleph$ of the continuum) in the system of all sets (which has the power $2^{\aleph} > \aleph$). [H 1916, 437; III: 438]

Hausdorff concludes this paper with the remark that his theorem on the power of Borel sets applies not only to Euclidean spaces but also to complete separable metric spaces.

Independently of Hausdorff and also in 1916, young Paul Alexandroff also solved the continuum problem for Borel sets [Alexandroff 1916]. Alexandroff's argument was based on ideas which later enabled him to prove an analogous theorem for the still more general class of Suslin sets. In summing up, one can certainly assert that these and other works by Alexandroff and Hausdorff provided a decisive impulse for the further development of descriptive set theory, particularly in Russia and Poland.

7.7.1 Dimension and Outer Measure

Among all of Hausdorff's journal articles, "Dimension und äusseres Mass" ("Dimension and Outer Measure") [H 1919a, IV: 21–43] has been by far the most influential. Here he introduced notions which today play a vital role in numerous mathematical fields: Hausdorff measure, Hausdorff dimension, and fractal geometry. As an indicator of its long-term impact, one need only look through section 28A, "classical measure theory," in *Mathematical Reviews*. Today this contains a separate subfield 28A78, "Hausdorff and packing measures," which the American Mathematical Society decided to add to its classification system. It is surely no exaggeration to say that Hausdorff's "Dimension and Outer Measure" is currently one of the most cited papers from the decade 1910 to 1920.

Recognition of its importance was long in coming, however, and Hausdorff's own introductory sentences are very modest:

> Carathéodory has developed an impressively simple and general measure theory that contains Lebesgue's as a special case. In particular, one can define with it the p-dimensional measure of a point set in q-dimensional space. We give a small contribution to this [theory] in the following. [H 1919a, 157; IV: 21].

It seems that Hausdorff never imagined at all the importance this "small contribution" would one day attain.

In Lebesgue's measure theory, as set forth in his dissertation from 1902 [Hawkins 1975, 120–124], he constructs for sets $A \subset [0,1]$ an *outer measure* $\mu^*(A)$ as follows: Letting I_n be intervals of length $l(I_n)$,

$$\mu^*(A) = \inf \left\{ \sum_n l(I_n); \;\; A \subset \bigcup_n I_n \right\}.$$

Lebesgue defines the *inner measure* of A as $\mu_*(A) = 1 - \mu^*(\overline{A})$ with $\overline{A} = [0,1] \backslash A$. A set A is then called *measurable* (Lebesgue measurable) if $\mu_*(A) = \mu^*(A)$. Since all Borel sets are Lebesgue-measurable, this measure theory is an extension of Borel's theory. In fact, the extension is considerable in view of the resulting set systems, since the system of Lebesgue-measurable sets has power $2^{\aleph}$, whereas the system of Borel sets only has the power $\aleph$. Carathéodory's theory of measurement [Carathéodory 1914], which Hausdorff adopts, requires only an outer measure to define measurable sets; it thus simplifies Lebesgue's theory considerably.

Carathéodory introduced an outer measure in Euclidean spaces by means of the following axioms:

1) Every point set $A \subseteq \mathbb{R}^q$ determines a unique extended real number $L(A)$ with $0 \leq L(A) \leq \infty$.

2) If $B \subseteq A$, then $L(B) \leq L(A)$.

3) The set function $L(A)$ is σ-subadditive.

4) For sets A, B with distance $\delta(A, B) > 0$, $L(A \cup B) = L(A) + L(B)$.

These axioms can be used to define $L(A)$ on the subsets of any metric space. A set function L for which 1)–4) holds is called a *metric outer measure.*

Carathéodory next defines a set A to be *L-measurable*, if for each $W \subseteq \mathbb{R}^q$ the equation

$$L(W) = L(A \cap W) + L(W \backslash (A \cap W))$$

holds. Because of 4), Borel sets are L-measurable for any metric outer measure.

As a final requirement, Carathéodory places the following condition on a set function L:

5) $L(A) = \inf\{L(B);\ B \supset A,\ \ B \text{ is } L\text{-measurable}\}$.

The measure theory developed in chapter 1 of Carathéodory's paper from axioms 1)–5) turns out to be completely equivalent to Lebesgue's; thus, no measurable quantity derived from Lebesgue's theory will be lost. In handwritten notes to his paper, Carathéodory enumerated the "great advantages" of his approach to Lebesgue measure theory:

> 1. It can be used for the linear measure.
> 2. It holds in the Lebesgue case, even if $L(A) = \infty$.
> 3. The proofs of the main theorems are incomparably simpler and shorter.
> 4. But the main advantage is that the new definition [of measure] is independent of the notion of inner measure.
>
> (*Collected Mathematical Writings* IV, 276.)

The advantage formulated here as 1. refers to the actual aim of Carathéodory's paper [Carathéodory 1914], as expressed in its title: *Über das lineare Maß von Punktmengen.* This was to define a *linear (one-dimensional) measure* in q-dimensional space ($q > 1$) in such a way that the measure of a rectifiable curve was equal to its length. This posed a completely new problem for Lebesgue's theory, which originally only provided a linear measure in $\mathbb{R}^1$, an area measure in $\mathbb{R}^2$, a volume measure in $\mathbb{R}^3$, etc., but no p-dimensional measure in $\mathbb{R}^q$ ($p < q$).

Carathéodory constructed an outer measure in the following way: Let $\mathfrak{U}$ be a system of bounded subsets of $\mathbb{R}^q$ such that for any $\varepsilon > 0$ each $A \subset \mathbb{R}^q$ can be covered by finitely or countably many $U_k \in \mathfrak{U}$ with diameters $d(U_k) < \varepsilon$. Such a system would be, for example, the system of open spheres or that of closed spheres with diameters $< \varepsilon$. Carathéodory then introduces

$$L_\varepsilon(A) = \inf \left\{ \sum_{n \geq 1} d(U_n);\;\; A \subset \bigcup_{n \geq 1} U_n,\;\; U_n \in \mathfrak{U},\;\; d(U_n) < \varepsilon \right\} \tag{7.1}$$

and defines

$$L(A) = \lim_{\varepsilon \downarrow 0} L_\varepsilon(A)\,.$$

He then proves in the following that this set function satisfies conditions 1) to 5) and that for a rectifiable curve A, $L(A)$ equals its length.

Hausdorff follows Carathéodory's approach for introducing outer measures as set functions $L(A)$ on the subsets A of $\mathbb{R}^q$ that satisfy conditions 1) – 5). He then seeks a way to compare such outer measures, first by designating $L(A), M(A)$ to be *of the same order* if for every set A they are simultaneously finite, infinite, or zero. This is the case, in particular, if for positive constants h, k one always has

$$hL(A) \leq M(A) \leq kL(A).$$

If $M(A) = 0$ when $L(A)$ is finite, or $L(A) = \infty$ when $M(A) > 0$, then *L is of lower order with respect to M* or *M is of higher order with respect to L*. In particular, this is the case if for any positive ρ one has

$$M(A) \leq \rho L(A)\,.$$

Hausdorff's fundamentally new idea was to modify Carathéodory's (7.1) as far as possible, and then to study the resulting outer measures. He does this by letting $\mathfrak{U}$ be a set system as above and then by defining for each set $U \in \mathfrak{U}$ an associated nonnegative real number $l(U)$. Hausdorff then forms (analogous to (7.1)):

$$L(A) = \lim_{\varepsilon \downarrow 0} \inf \left\{ \sum_{n \geq 1} l(U_n);\;\; A \subset \bigcup_{n \geq 1} U_n,\;\; U_n \in \mathfrak{U},\;\; d(U_n) < \varepsilon \right\} \tag{7.2}$$

This $L(A)$ satisfies Carathéodory's conditions 1) – 4) without any restrictions on $l(U)$; moreover, if the U are Borel sets, then condition 5) holds as well.

In the following, Hausdorff considers only monotone and continuous set functions $l(U)$. To define continuity, he surrounds each point of U with a sphere of radius β and forms U_β as the set of interior points of all these spheres. $l(U)$ is called *continuous* if for any given $\rho > 0$ there exists a β such that for all U holds: $l(U_\beta) \leq l(U) + \rho$. He calls $l(U)$ *monotone* if $U \subseteq V$ implies $l(U) \leq l(V)$. Hausdorff then gives seven concrete examples, two of which are:

$l(U) = \sup \Delta_p$, where Δ_p is the p-dimensional volume of a simplex with vertices in U and the supremum extends over all such simplices. For $p = 1$, this is identical with Carathéodory's linear measure.

$l(U) = q$-dimensional volume in the sense of Lebesgue's convex hull U_c of U.

Hausdorff's seven examples hardly convey an idea of the immense variety of outer measures that can be constructed in this way, as he himself emphasized. Having generalized Carathéodory's length measure and described the general scope of the new theory, Hausdorff could now tackle his real goal, which he had outlined at the beginning of his paper:

> [$\cdots$] we set forth an explanation for p-dimensional measure which immediately extends to non-integral values of p and makes sets *with fractional dimension* appear possible, in fact, even sets whose dimensions fill out the scale of positive numbers [$\cdots$] . Dimension thus becomes a graduated feature, like the "order" in approaching zero, the "strength" of convergence, and related notions. [H 1919a, 157; IV: 21]

For $\mathfrak{U}$ Hausdorff chooses the system of open spheres of $\mathbb{R}^q$. $A \subset \mathbb{R}^q$ can then always be covered with finitely or countably many balls U_n with diameters $d_n < \varepsilon$. Let $\lambda(x)$ be a real monotonically increasing continuous function defined on $[0, \infty)$ with $\lambda(0) = 0$. Then

$$L^\lambda(A) = \lim_{\varepsilon \downarrow 0} \inf \left\{ \sum_{n \geq 1} \lambda(d_n); \ \ A \subset \bigcup_{n \geq 1} U_n, \ \ d_n < \varepsilon \right\}$$

yields an outer measure or, on the sets which are Carathéodory L^λ-measurable, a measure. Outer measure (or measures) of this kind are today named after Hausdorff; they can be similarly defined on general metric spaces.

Hausdorff then introduces his notion of dimension: If $0 < L^\lambda(A) < \infty$, "we say A is of dimension $[\lambda(x)]$" [H 1919a, 166; IV: 30]. If $\lambda(x), \mu(x)$ are two functions with the above properties and if L^λ, L^μ *are of the same order*, then the dimensions $[\lambda], [\mu]$ *will equal each other*, since every set A of dimension $[\lambda]$ will be of dimension $[\mu]$ and vice versa. Moreover, since L^λ depends only on the behavior of the function $\lambda(x)$ in a small neighborhood of 0, it follows from

$$0 < a \leq \frac{\mu(x)}{\lambda(x)} \leq b < \infty \ \text{ for } \ x \in (0, \rho)$$

(ρ sufficiently small) that $[\lambda] = [\mu]$.

If L^μ is of *higher order* than L^λ, then $[\mu]$ is called the higher, $[\lambda]$ the lower dimension (in characters $[\mu] \succ [\lambda]$). From the earlier definition of order then follows the characteristic property: a set A of a certain definite dimension has for each higher dimension the measure zero, for each lower dimension the measure infinity. If

$$\lim_{x\downarrow 0}\frac{\mu(x)}{\lambda(x)}=0\,,$$

then $L^\mu \leq \rho\, L^\lambda$ for any $\rho > 0$, i.e., for $L^\lambda(A)$ finite then $L^\mu(A)$ vanishes; $[\mu] \succ [\lambda]$. In modern terms, $[\lambda] = [\mu]$ is an equivalence relation in the set of admissible "Hausdorff functions" $\lambda(x)$ and $[\lambda] \prec [\mu]$ is a partial ordering on the set of equivalence classes.

The *most important special case* is $\lambda(x) = x^p$, p an arbitrary positive real number. Hausdorff denotes the associated dimension $[x^p]$ by (p), the associated outer measure by $L^{(p)}$. Then $(q) \succ (p)$ exactly when $q > p$, so the scale of possible dimensions in this case is all positive real numbers. Today, when speaking of Hausdorff measures and Hausdorff dimensions, one usually means this special case. The Hausdorff dimension H-dim(A) of a set A is then that real number $p > 0$ for which:

$$0 < L^{(p)}(A) < \infty,\quad L^{(q)}(A) = 0 \text{ for } q > p,\quad L^{(q)}(A) = \infty \text{ for } q < p\,.$$

The central question for the entire theory is now, whether for a given $\lambda(x)$ there really exist sets $A \subset \mathbb{R}^q$ with dimension $[\lambda]$. Hausdorff does not solve this problem in full generality, but he proves the following: If $\lambda(x):\ [0,\infty) \to [0,\infty)$ is a strictly monotonically increasing concave function with $\lambda(0) = 0$, $\lim_{x\to\infty}\lambda(x) = \infty$, then there exists $A \subset \mathbb{R}$ with dimension $[\lambda]$. His construction is analogous to the one for the classical Cantor set and it yields a set A which, like the Cantor set, is perfect and nowhere dense. Hausdorff first fixes positive numbers $\xi_0, \xi_1, \ldots, \xi_n$ such that $\xi_0 > 2\xi_1$, $\xi_1 > 2\xi_2, \ldots, \xi_{n-1} > 2\xi_n$ and $2^n\lambda(\xi_n) = 1$. From the interval $[0, \xi_0]$ one now removes a middle open interval of length $\xi_0 - 2\xi_1$ such that the two remaining closed intervals have the same length ξ_1. From these, the middle open intervals of length $\xi_1 - 2\xi_2$ are removed, leaving four closed intervals of length ξ_2, and so on. After the nth step, one has 2^n closed intervals of length ξ_n; let the union of these 2^n intervals be A_n. Then

$$A = \bigcap_{n=1}^{\infty} A_n$$

will be a perfect nowhere-dense set, and $L^\lambda(A) \leq 2^n\lambda(\xi_n) = 1$. The most difficult and technical part of the proof involves showing that $L^\lambda(A) \geq 1$ [H 1919a, 169–172; IV: 33–36]. So with $L^\lambda(A) = 1$ the dimension of A is therefore equal to $[\lambda]$.

After clarifying the existence question, Hausdorff presents various examples. First he wants to construct a set of dimension p with $0 < p < 1$. So now $\lambda(x) = x^p$, $0 < p < 1$. One obtains the numbers ξ_n used for the construction of a perfect nowhere dense set $A \subset \mathbb{R}$ with $L^{(p)}(A) = 1$ and with the required properties by setting $\xi_n = \xi^n$ with $\xi = \left(\frac{1}{2}\right)^{\frac{1}{p}}$. We then have $\xi_0 = 1$, $\xi_i > 2\xi_{i+1}$

and $2^n\,\lambda(\xi_n) = 2^n\,\xi^{np} = 1$. For example, a set with Hausdorff dimension $\frac{1}{2}$ is obtained by choosing $\xi = \frac{1}{4}$, i.e., by removing in each case the middle half of the closed intervals remaining in the construction of A. Leaving $\xi < \frac{1}{2}$ initially undetermined, the condition $2^n\,\xi^{np} = 1$ for p yields:

$$p = \frac{\ln 2}{\ln \frac{1}{\xi}}\,.$$

For the Cantor set $\xi = \frac{1}{3}$, so Hausdorff obtains its dimension as $p = \frac{\ln 2}{\ln 3} = 0.630929753\cdots$. With an appropriate choice of ξ, sets $A \subset \mathbb{R}$ of any Hausdorff dimension between 0 and 1 can thus be constructed.

Hausdorff left open the question of determining the weakest conditions that must be imposed on $\lambda(x)$ in order to prove the existence of a set $A \subset \mathbb{R}$ with dimension $[\lambda]$. This natural question was finally answered by Aryeh Dvoretzky in 1948 when he published the following theorem: A necessary and sufficient condition for a set $A \subset \mathbb{R}$ of dimension $[\lambda]$ to exist is

$$\liminf_{x \downarrow 0} \frac{\lambda(x)}{x} > 0\,.$$

Similarly, for the existence of $A \subset \mathbb{R}^q$ of dimension $[\lambda]$, the condition

$$\liminf_{x \downarrow 0} \frac{\lambda(x)}{x^q} > 0$$

is necessary and sufficient [Dvoretzky 1948].

Hausdorff also dealt with the problem of constructing plane sets of prescribed dimension. Here he succeeded in showing that if $A \subset [0, \xi_0]$ is a perfect nowhere-dense set of dimension $[\lambda]$, constructed as above, and B a second such set of dimension $[\mu]$, then $A \times B \subset \mathbb{R}^2$ has dimension $[\lambda\mu]$. So, in particular, for the Hausdorff dimension ($\lambda(x) = x^p$, $\mu(x) = x^q$), for these two sets

$$\text{H-dim}(A \times B) = \text{H-dim}(A) + \text{H-dim}(B)\,.$$

Hausdorff thus succeeded in showing that for every p with $1 < p < 2$ there exists a set $C \subset \mathbb{R}^2$ with $\text{H-dim}(C) = p$. He then goes on to suggest how to prove more specifically that there exist "plane Jordan curves with any dimension between 1 and 2" [H 1919a, 179; IV: 43]. Hausdorff concludes his paper with the remark that for three sets A, B, C of the above kind, one can also form a spatial set $A \times B \times C$ of dimension $[\lambda\mu\nu]$.

Hausdorff began his considerations of plane sets by remarking that "the construction of plane sets of prescribed dimension is not without difficulty" [H 1919a, 177; IV: 41]. Indeed, the general problem of what can be said about the dimension of $A \times B$ when A has dimension $[\lambda]$ and B has dimension $[\mu]$ is very difficult, and many researchers have written about it. For the Hausdorff dimension, in general,

$$\text{H-dim}(A \times B) \geq \text{H-dim}(A) + \text{H-dim}(B).$$

Here one would like to know, for example, under which conditions equality holds or by how much H-dim$(A \times B)$ can exceed the sum on the right-hand side [Steffen 1996, 195–197].

7.7.2 Reception of "Dimension and Outer Measure"

Hausdorff's now famous paper from 1919 went virtually unnoticed for many years. A decade afterward, though, the Russian-born mathematician Abram Samoilovitch Besicovitch published an article entitled *On linear sets of points of fractional dimension* [Besicovitch 1929], the first of many studies that opened up this field. The fact that a full ten years had to pass before mathematicians began to take notice of Hausdorff's "Dimension and Outer Measure" can perhaps be taken as an indication of how far ahead of his time he was with his bold idea of non-integral dimension. In his first paper, Besicovitch noted the significant step beyond Lebesgue that Carathéodory took by introducing p-dimensional measure in $\mathbb{R}^q$ $(q > p)$, before adding:

> But there are sets which from the point of view of the compactness of their points represent something intermediate between sets of finite positive measure of two consecutive numbers of dimensions, *e.g.* between sets of finite linear measure and of finite plane measure. *These are Hausdorff's fractional dimensional sets.* [Besicovitch 1929, 161]

Besicovitch calls a set $A \subset \mathbb{R}$ with H-dim$(A) = s$ and Hausdorff measure $L^{(s)}(A)$ $(0 < s < 1)$ an s-set. He then introduces the notions of upper density $D(x, A)$ and lower density $d(x, A)$ of a point x:

$$D(x, A) = \limsup_{h \to +0} \frac{L^{(s)}(A \cap (x-h, x+h))}{(2h)^s},$$

$$d(x, A) = \liminf_{h \to +0} \frac{L^{(s)}(A \cap (x-h, x+h))}{(2h)^s}.$$

Then x is called a regular point of A if $D(x, A) = d(x, A)$, i.e. regular points have unique density, whereas all other points of A are called irregular. Besicovitch's main result then asserts that almost all points of an s-set are irregular. Furthermore, he gives upper and lower estimates for $D(x, A)$ and $d(x, A)$.

Besicovitch and his students further developed Hausdorff's ideas and applied them to various mathematical fields. In particular, they contributed to a considerable extent to the development of a new field which has been developing briskly since the 1930s alongside abstract measure theory, namely geometric measure theory. This area of research investigates the geometric properties of sets, mappings and functions in Euclidean spaces by combining geometric and differential geometric methods with those of measure theory. Much of the motivation behind this field stems from problems in analysis, e.g. the theory of minimal surfaces (Plateau's problem), but also calculus of variations in general as well as potential theory and integration theory. In addition to classical Lebesgue measure, various specifically adapted measures are used. Hausdorff measures, in particular, are among the basic tools of geometric measure theory. In the classical monograph by Herbert Federer [Federer 1969], seven different measures are introduced using Hausdorff's construction (7.2) and by making different choices for the functions $l(U)$. The Hausdorff measures $\mathcal{H}^p(A)$ thus have special importance in this theory, as Federer emphasized in the introduction to his book.

We can only give a few examples of how Hausdorff's approach to dimension has permeated different areas of mathematics. Beyond geometric measure theory, the earliest application of his ideas was in the field of number theory, in particular Diophantine approximations. In 1929, the same year that Besicovitch followed up on Hausdorff's work, the Czech mathematician Vojtěch Jarník published "Diophantine Approximations and Hausdorff Measure" [Jarník 1929]. There he considers approximations of a real number $x \in [0,1]$ by rational numbers, while pursuing the following question: what can be said about the set

$$M_\alpha = \left\{ x \in [0,1];\ \left| x - \frac{a}{b} \right| \leq \frac{1}{b^\alpha} \text{ for infinitely many integers } a, b \right\}?$$

Alexander J. Khintchine had proved in 1926 that the Lebesgue measure of M_α for $\alpha > 2$ is zero. Jarník commented on this as follows:

> But one will still suspect that, for example, the set M_3 must be "much larger" than the set M_4 in some respect. For such a classification of sets of Lebesgue measure zero, Hausdorff's notion of measure and dimension is well suited; indeed, the aim of this note is to show in a particularly simple case – namely, precisely for the sets M_α – the importance of this notion for such questions. [Jarník 1929, 373]

The main result of his note proves that H-dim$(M_\alpha) = \frac{2}{\alpha}$, though soon thereafter he obtained substantially stronger findings in [Jarník 1931].

Another remarkable number-theoretic application of Hausdorff's ideas is due to Besicovitch. In the dyadic representation of $x \in (0,1)$, let $p_n(x)$ be the sum of the first n digits (i.e., the number of ones). Then, as Hausdorff had proved in the *Grundzüge*, the following strong law of large numbers holds (μ is the Lebesgue measure):

$$\mu\left(\left\{x;\ \lim_{n\to\infty} \frac{p_n(x)}{n} = \frac{1}{2}\right\}\right) = 1 \tag{7.3}$$

In Besicovitch's paper "On the sum of digits of real numbers represented in the dyadic system" [Besicovitch 1934], he devotes himself to the problem of characterizing the sets

$$E_p = \left\{ x \in (0,1);\ \limsup_n \frac{p_n(x)}{n} \leq p < \frac{1}{2} \right\}.$$

Since because of (7.3) $\mu(E_p) = 0$ for any $p < \frac{1}{2}$, Lebesgue measure cannot provide any information about these sets, but as Besicovitch states: "The solution of this problem is given in terms of sets of fractional dimensions" [Besicovitch 1934, 321]. He thus computes the Hausdorff dimension of E_p and thereby determines to some extent the size of E_p as a function of p, i.e., as a function of the deviation from the "normal case." Similarly, for the Cantor sets constructed by Hausdorff: the Hausdorff dimension gives a kind of size measure for these sets (all of which have Lebesgue measure 0) as a function of the parameter ξ, just as in Jarník's studies. S.D. Chatterji in his commentary on [H 1919a] characterizes this utility of the Hausdorff dimension as follows:

> This is a typical service rendered by Hausdorff dimensions in general and they are used in this sense in many refined investigations; for example, many curves that arise in various studies of dynamical systems or stochastic processes have no tangents anywhere and their Hausdorff dimension functions indicate the degree of their fractal nature. (*HGW* IV, 52)

Using $I(p) = -(p \log p + q \log q)$; $p + q = 1$, the Hausdorff dimension of E_p is:

$$\text{H-dim}\,(E_p) = \frac{I(p)}{\log 2}\,.$$

$I(p)$ can be interpreted as Shannon information on the alphabet (discrete probability space) $Z = \{0,1\}$ with $P(\{1\}) = p$, $P(\{0\}) = q$. The connections between Hausdorff dimension and Shannon information lead to further applications of Hausdorff's ideas in ergodic theory and statistical mechanics.

We now turn to some analytical applications. Otto Frostman was a student of Marcel Riesz, who worked at Lund University. In his dissertation [Frostman 1935], Frostman linked the notion of capacity in potential theory with Hausdorff's concept of measure by means of a suitable Hausdorff function $\lambda(x)$. He then proved for any compact set $A \subset \mathbb{R}^q$ that the Hausdorff dimension $\inf\{p > 0;\ L^{(p)}(A) = 0\}$ equals its capacitance dimension (*dimension capacitaire*). This opened the way for a powerful new analytic method for computing the Hausdorff dimension of compact sets.

The solutions of nonlinear elliptic systems of partial differential equations will, in general, possess singularities, as Ennio de Giorgi showed in 1968. Numerous subsequent studies have been undertaken to estimate the Hausdorff dimension of these sets of singular points (see [Giaquinta 1983]). Similarly, Hausdorff measures have come to play an important role in the theory of trigonometric series. Of special importance are the sets A of Cantor set type, constructed by Hausdorff using suitable number sequences $\{\xi_n\}$. Some authors call these "ensembles parfaits symétriques" and use them and suitable generalizations to solve problems in harmonic analysis, e.g., in investigations of uniqueness sets.

Hausdorff measures and dimensions also stand in close relation to packing measures and packing dimensions. Whereas the outer Hausdorff dimension $L^\lambda(A)$ is constructed by covering A as tightly as possible by spheres, the packing dimension $p_\lambda(A)$ is constructed dually by exhausting A from the inside by filling it with spheres. Although packing measure can be defined on arbitrary metric spaces, here we restrict ourselves to Euclidean spaces $\mathbb{R}^n$. By a *ball packing* $\mathcal{P}$ *for* $A \subset \mathbb{R}^n$ we mean a set of closed pairwise disjoint balls K with centers in A and radii $r(K)$. Taking a Hausdorff function $\lambda(x)$, one forms

$$P_\lambda(A) = \inf_{\varepsilon>0} \sup_{\mathcal{P} \text{ for } \mathcal{A}} \left\{ \sum_{K\in\mathcal{P}} \lambda(r(K));\;\; r(K) \le \varepsilon \right\}.$$

Using the quantity function $P_\lambda(A)$, the packing measure is then defined as follows:

$$p_\lambda(A) = \inf \left\{ \sum_n P_\lambda(A_n);\;\; A \subseteq \bigcup_n A_n \right\},$$

which is a metric outer measure. Let the packing measure derived from $\lambda(x) = x^\alpha$ be denoted by $p_\alpha(A)$. For this special case, the packing dimension $\mathrm{Dim}(A)$ is defined as

$$\mathrm{Dim}(A) = \inf\{\alpha > 0;\;\; p_\alpha(A) < +\infty\}.$$

One can then show that H-dim$(A) \le \mathrm{Dim}(A)$, with equality holding for the self-similar sets of Cantor set type constructed by Hausdorff. As remarked above, numerous investigations have been devoted to determining H-dim$(A \times B)$, a problem Hausdorff realized was difficult. The lower bound given by H-dim(A) + H-dim(B) has already been mentioned. Using the packing dimension, one can also find good upward bounds. Following a series of preliminary works by various authors, Claude Tricot proved the following result [Tricot 1982]: For $A, B \subset \mathbb{R}^n$,

$$\text{H-dim}(A \times B) \le \text{H-dim}(A) + \mathrm{Dim}(B) \le \mathrm{Dim}(A \times B) \le \mathrm{Dim}(A) + \mathrm{Dim}(B).$$

Fractal geometry represents a broad field that traces its origins to Hausdorff's "Dimension and Outer Measure" [Falconer 1990]. The basic notion of a *fractal object* was set forth quite intuitively by Benoît Mandelbrot [Mandelbrot 1977], who conceived of these as high irregular subsets of Euclidean spaces. Fractals cannot be grasped by properties such as smoothness conditions, etc. They typically manifest a detailed fine structure on arbitrarily small scales and often possess strict or at least approximate self-similarity. Typical examples date back more than a century, as for example, classical Cantor sets and Hausdorff's sets of Cantor type, von Koch's snowflake curve, Sierpiński's triangle, and the Menger sponge.

The word "fractal" was chosen by Mandelbrot because sets of this kind generally have non-integral ("fractional") Hausdorff dimension. However, since some sets with integral Hausdorff dimension would intuitively be considered as fractal sets, he opted for another definition. In his principal work, *The Fractal Geometry of Nature* [Mandelbrot 1982], he thus defined a fractal as a subset of $\mathbb{R}^n$ whose Hausdorff dimension is larger than its topological dimension [Mandelbrot 1982, 15]. Nevertheless, there seems to be no general consensus among mathematicians about what is meant by a "fractal set," a circumstance that led Kenneth J. Falconer to comment:

> [Mandelbrot's] definition proved to be unsatisfactory in that it excluded a number of sets that clearly ought to be regarded as fractals. Various other definitions have been proposed, but they all seem to have this same drawback. [Falconer 1990, xx]

Owing to this, Falconer has suggested that mathematicians should employ the term fractal much as biologists do when speaking of life forms, i.e., one should consider certain properties – such as those mentioned above – as sufficient for a fractal, but without insisting that all of these properties necessarily be present in each individual case.

Mandelbrot wrote his book for a wide circle of readers. As its title suggests, he wanted to raise awareness of fractals among scientists from many different disciplines, including physicists, astrophysicists, chemists, biologists, physicians, geographers, meteorologists, economists, and statisticians. The text is accompanied by numerous illustrations, some of structures found in nature, but mostly computer-generated images visualizing certain types of fractals. In demonstrating the significance of fractal phenomena in a variety of fields, Mandelbrot focused especially on the property of self-similarity or scale invariance. Hausdorff's name already appears in the preface, where Mandelbrot emphasizes the importance of what he calls Hausdorff-Besicovitch dimension. This attempt to pay tribute to both of these pioneers did not catch on, however.

The Fractal Geometry of Nature triggered a great deal of interest, especially among researchers in mathematics and computer science, but also for experts from other fields. This explosion of activity was documented a decade later in the volume *Chaos and Fractals. New Frontiers of Science* [Peitgen et al. 1992]. Unfortunately, this boom in "fractal theory" led to unrealistic expectations and even sheer nonsense, especially in popular scientific accounts. Still, this did not detract from the positive impact these ideas had for serious interdisciplinary research. Of course, mathematical fractals do not "exist in nature" any more than do exact circles or perfect ellipses. Yet as mathematical models for certain physical phenomena, fractal objects have the same justification as do ellipses for approximating the motion of the planets. As Christoph Bandt and Hermann Haase aptly summarized:

> Hausdorff has widened our eyes for new analogies and models for understanding the world around us. From this point of view the extra-

> mathematical use of the concept of Hausdorff dimension seems altogether justified. [Bandt/Haase 1996, 177]

The possibilities opened by modern computer graphics contributed greatly to the popularity of fractals, which has extended far beyond the realm of professional mathematicians and natural scientists. Images of Julia sets, the Mandelbrot set, and strange attractors are aesthetically pleasing figures that happen to illustrate how fractals arise in the study of dynamical systems. These can be either discrete (as in the simple iteration of the mapping $z \to z^2 + c$ in the complex number plane) or continuous (e.g., in turbulence theory) [Peitgen/Richter 1986]. Indeed, the theory of dynamical systems constitutes a major field in which Hausdorff's concepts play a central role.

A dynamical system is a triple (T, X, Φ) consisting of a period T, a state or phase space X, and a mapping $\Phi(t, x) : T \times X \to X$ with

$$\Phi(0, x) = x; \quad \Phi(s, \Phi(t, x)) = \Phi(s + t, x),$$

where $T = \mathbb{N}_0$ or $\mathbb{Z}$ (discrete systems) or $T = [0, \infty)$ or $\mathbb{R}$ (continuous systems). For X one can choose $\mathbb{R}^n$, but X can be, more generally, a Riemannian manifold, a metric space, or even a Hausdorff space. For a discrete system, let $\Phi(1, x) = \varphi(x)$, then $\Phi(2, x) = \Phi(1, \Phi(1, x)) = \varphi(\varphi(x))$, etc., i.e., the dynamical system is generated by iterating the function $\varphi(x)$. Continuous systems are often generated by nonlinear systems of ordinary differential equations; in such cases, the $\Phi(t, x)$ are then, as functions of t, the solution curves.

For various nonlinear dynamical systems, the state space typically contains sets of non-integral Hausdorff dimension that exhibit chaotic behavior, meaning that small differences in their initial states lead to large differences in the system over the course of time. Such sets thus manifest an extraordinarily high sensitivity with respect to the initial data of the system. With this in mind, let us consider the discrete case in somewhat more detail; the continuous case is conceptually analogous.

Let $\varphi^k(x)$ denote the k-th iterate of φ. A closed set $F \subset X$ invariant under φ (thus, $\varphi(F) = F$) is called an *attractor* for φ if there exists an open set $V \supset F$ (the basin of attraction of F) such that

$$\lim_{k \to \infty} \text{dist.}(F, \varphi^k(x)) = 0$$

for any $x \in V$. A closed invariant set F is called a *repeller* for φ if it is an attractor for the (not necessarily unique) inverse function φ^{-1}. An attractor is called a *fractal* or *strange* attractor if the following holds:

1) There exists a $\delta > 0$ such that for any $x \in F$ there exist points $y \in F$ arbitrarily close to x with $|\varphi^k(x) - \varphi^k(y)| \geq \delta$ for suitable k (highly sensitive dependence on initial conditions).

2) The orbit $\{\varphi^k(x)\}$ is dense in F for $x \in F$.

3) The periodic points of φ in F (i.e., the points $x \in F$ with $\varphi^j(x) = x$ for some

j) are dense in F.

Gaston Julia published the first examples of fractal repellers in [Julia 1918], where he studied iterations of rational functions $\varphi(z)$ in the complex plane. He found that the points $z \in \mathbb{C}$ form two classes, $F(\varphi)$ and $J(\varphi)$, depending on how the initial value z determines – graphically speaking – the dynamics of the process. The set $F(\varphi)$ (later called the Fatou set of φ) consists of those z for which the orbit $\{\varphi^k(z)\}$ depends continuously on z; thus for z' in a small neighborhood of z the orbits will remain close to the orbit of z (stable dynamics). For points in $J(\varphi)$ (the Julia set of φ), the orbit is very sensitive to small deviations from z (chaotic dynamics). Julia studied the properties of the sets $J(\varphi)$ in detail and found, for example, that they are perfect sets and thus uncountable. From today's point of view, they are usually viewed as fractals. Even for the simple functions $\varphi_c(z) = z^2 + c$, the study of Julia sets $J(\varphi_c)$, where the Hausdorff dimension depends on the parameter c, leads to many interesting problems.

The notion of strange attractors was first introduced by David Ruelle and Floris Takens in their studies on turbulence [Ruelle/Takens 1971]; a summary of relevant results can be found in [Ruelle 1989]. The important role played by Hausdorff dimension in turbulence theory was taken up by Mandelbrot, Kahane and Takens at the "Turbulence and Navier Stokes Equations" conference held in Orsay in 1975. The earliest investigation involving strange attractors (though the term had yet to be coined) was undertaken by the mathematician and meteorologist Edward N. Lorenz in [Lorenz 1963]. Lorenz studied a simplified mathematical model of atmospheric convection, which led to the following system of differential equations:

$$\frac{dx}{dt} = \sigma(y - x), \quad \frac{dy}{dt} = rx - y - xz, \quad \frac{dz}{dt} = xy - bz.$$

He found that certain values of the parameters σ, r, and b lead to chaotic behavior, thus minute changes in the initial conditions produce very large changes in the states of the system. This implied that long-term weather forecasting was, in principle, impossible due to the inevitability of measurement errors in trying to determine initial conditions [Sparrow 1982]. Visualizations of the Lorenz attractor and numerous other types that arise in continuous dynamical systems can be found in [Holden 1986, 15–34]. More recent developments in the theory of dynamical systems have shown that Hausdorff dimension and invariants derived from it also play a crucial role in connection with dynamical concepts, such as entropy and Lyapunov exponents [Bothe/Schmeling 1996, 239–241].

Regarding the relevance of Hausdorff's ideas for stochastics, we briefly mention just two examples: random fractals and Brownian motion. One standard type of random fractal can be illustrated by constructions of Cantor sets, where the determination of which middle third to remove might be based on rolling dice. Likewise, one might remove middle intervals from those preceding but do so in such a way that the lengths of the remaining intervals were random (so-called statistical self-similarity).

Brownian motion, named after the Scottish botanist Robert Brown, is an irregular teeming motion of very small particles suspended in liquids. Brown discovered this phenomenon in 1827 when observing pollen grains suspended in water. Over the course of the nineteenth century numerous researchers dealt with it and contributed accurate experimental findings, but it was not until the next century that the mechanism that caused Brownian motion was finally clarified. Up until then, scientists lacked a theoretical model from which they could determine the mean displacement of Brownian particles [Purkert 1983]. This situation changed when Albert Einstein published his famous paper [Einstein 1905] which gave a one-dimensional model $X(t)$ that measured the random location of a particle. Under certain physically motivated assumptions for stochastic independence, Einstein could show that $X(t)$ is normally distributed with variance $2Dt$ (where D is the diffusion coefficient) and mean displacement $\sigma_X = \sqrt{2Dt}$. Einstein was concerned in this work with proving the physical reality of atoms and molecules. In 1905 he left it open whether the theoretically predicted fluctuation phenomena had something to do with Brownian motion, though he would later emphasize this connection. This story is one of the more famous in the history of physics, but it also had great importance for the history of probability theory with remarkable ties to Hausdorff's theory of measure and dimension.

In 1908, the French physicist and later Nobel laureate Jean Baptiste Perrin experimentally confirmed Einstein's calculations, a key breakthrough for atomic theory [Perrin 1908]. Later, in the early 1920s, Norbert Wiener investigated the general stochastic process underlying Einstein's model for Brownian motion, processes that now bear Wiener's name [Wiener 1921], [Wiener 1923]. For the Hausdorff dimension of the trajectories of Wiener processes, these two theorems hold:
1) If $X(t)$ is a one-dimensional Wiener process, then the graph $\{(t, x(t)) \in \mathbb{R}^2;\ t \in [0, c)\}$ of almost all trajectories $x(t)$ has Hausdorff dimension $1\frac{1}{2}$.
2) For $n \geq 2$, almost all trajectories $x(t) = (x_1(t), \cdots, x_n(t))$ of an n-dimensional Wiener process have Hausdorff dimension 2.

Strangely enough, Hausdorff never returned to the ideas in "Dimension and Outer Measure," nor did he react to related works by other mathematicians. There is nothing in his estate that deals with this topic, nor does the name Besicovitch appear anywhere in it, even though one finds references to hundreds of papers by other mathematicians in Hausdorff's manuscripts. The Polish mathematician Edward Szpilrajn found a remarkable connection between Hausdorff dimension and topological dimension in 1936. (From 1940 onward, Szpilrajn called himself Edward Marczewski to escape persecution as a Jew under German occupation.) In 1936, he lectured on "La dimension et la mesure" at the International Congress of Mathematicians in Oslo [Szpilrajn 1937]. For a separable metric space (X, ρ), let $d(X, \rho)$ denote its small inductive dimension (Menger-Urysohn dimension). If ρ' is another metric on X generating the same topology, i.e. (X, ρ) and (X, ρ') are homeomorphic as topological spaces, denoted by $\rho' \sim \rho$, it follows that $d(X, \rho) =$

$d(X, \rho')$ holds, and Szpilrajn's theorem then states:

$$\inf_{\rho' \sim \rho} \{\text{H-dim}(X, \rho')\} = d(X, \rho).$$

In Hausdorff's estate, Szpilrajn's name appears in five different fascicles, but "La dimension et la mesure" is nowhere mentioned.

Thus geometric measure theory lay far outside Hausdorff's main fields of interest, which certainly included various theories of topological dimension. Indeed, numerous studies on this topic can be found in his estate. One is an extensive manuscript on "Dimension Theory" of 200 pages, written between 1930 and 1936 (Hausdorff estate, Capsule 47, Fascicle 986). From its structure, one might well imagine he composed this as a manuscript for a book. In any case, it reveals that Hausdorff was an outstanding connoisseur of this entire field. If he had initially thought of publishing this text, after 1935 he would have had no chance of finding a publisher, as a Jew living in National Socialist Germany.

7.8 Other Analytic Works

This section begins with brief remarks on some of Hausdorff's minor works in analysis before turning to a more detailed discussion in 7.8.4 of his work on divergent series.

7.8.1 Numerical Range of Bilinear Forms

Otto Toeplitz, who later became Hausdorff's colleague and friend in Bonn, was a leading expert on integral equations. Beginning in 1906, when he came to Göttingen, he took up investigations dealing with infinite matrices and bilinear forms with infinitely many variables. This work was inspired by Hilbert's ongoing research in this fast-breaking field. In Toeplitz's paper [Toeplitz 1918], he introduced the notion of the value set ("Wertvorrat") of a bilinear form

$$C(x, y) = \sum_{i,j=1}^{n} c_{ij} x_i y_j$$

with $c_{ij} \in \mathbb{C}$ and $x, y \in \mathbb{C}^n$. Toeplitz calls the set $\mathfrak{W} = W(C) = \{C(x, \overline{x}) \in \mathbb{C};\ \sum x_i \overline{x}_i = 1\}$ the *value set* of the bilinear form (or matrix C). The central theorem in this paper asserts: *The outer edge of the domain* $\mathfrak{W}$ *is a convex curve.* He then wrote: "The question whether $\mathfrak{W}$ fills the complete interior of its outer convex edge, or has holes, remains $[\cdots]$ open" [Toeplitz 1918, 195].

Hausdorff picked up this question in his note "Der Wertvorrat einer Bilinearform" [H 1919b], published in *Mathematische Zeitschrift*. His result is that $\mathfrak{W}$ has no holes in the interior, while the value set for each C is a convex subset of $\mathbb{C}$. The results of Toeplitz and Hausdorff can be applied to much more general situations.

For example, let H be a complex Hilbert space with scalar product (x, y) and $T : H \to H$ a bounded linear operator. Then the set $W(T) = \{(Tx, x) \in \mathbb{C};\ |x| = 1\}$ (the analogue of the value set for Toeplitz and Hausdorff) is now called the *numerical range* of T. The following holds: *the numerical range of values is a convex set in* $\mathbb{C}$. This theorem is now called the Toeplitz-Hausdorff theorem.

7.8.2 Power Series with Analytic Continuation

Together with [H 1919b], Hausdorff also published a second short paper in *Mathematische Zeitschrift*, "On the Distribution of Continuable Power Series" [H 1919c, IV: 67–72]. This concerned the subject of analytic continuation of power series beyond their circle of convergence as formulated by George Pólya in [Pólya 1916]. Pólya there offered the following conjecture: "If a power series with integer coefficients has radius of convergence 1, only two cases are possible: either it represents a rational function, or the series cannot be continued beyond the unit circle" [Pólya 1916, 510].

As can be seen from a later letter he sent to Pólya, Hausdorff immediately tried to solve this problem, but was unsuccessful (*HGW* IX, 548). In the meantime, Pólya published [Pólya 1918], in which he introduced the concept of neighborhoods for sets of power series with radius of convergence 1, thereby bringing topological considerations to bear on these questions. In this way, he was able to give precise expression to the view, widespread among function theorists, that the impossibility of analytic continuation of a power series beyond its circle of convergence was, so to speak, the standard case, whereas continuation was only possible in exceptional cases. Pólya proved, among other results, the following three theorems: I. The set of non-continuable power series is everywhere dense. II. The set of non-continuable power series is open. III. The set of continuable power series has no isolated point.

Hausdorff found an elementary proof of Pólya's Theorem III and communicated this to Pólya on September 15, 1917; he later incorporated this as §1 in his note [H 1919c]. After September 1917, Hausdorff tried to prove Pólya's conjecture that a continuable function must necessarily be rational. By November, he wrote to Pólya again, communicating a partial result he had found. This indicated

> that there are at most countably many functions $f(x) = \sum a_m x^m$ with integer power series which are regular for $|x| < 1$ and continuable beyond this domain (which speaks for the correctness of your theorem), but I have not yet figured out whether these are rational functions. (*HGW* IX, 549)

On December 25,1917, Hausdorff sent Pólya the proof of this theorem (*HGW* IX, 550–552), which forms the second paragraph of his note [H 1919c]. Pólya's conjecture was first proved in 1919 by Fritz Carlson (published in 1921); another proof was given in 1922 by Gabor Szegö.

7.8.3 On Semi-Continuous Functions and their Generalization

Hausdorff's paper [H 1919d, IV: 79–96] was the fourth published that year. It deals with five recently proved theorems on real-valued functions in metric spaces. Hausdorff offered these comments about the first section:

> In §1 I give new and, as it seems to me, particularly simple proofs of some well-known theorems about continuous and semi-continuous functions, while noting a remarkable logical connection [between them]. [H 1919d, 292; IV: 79]

These five theorems were due, respectively, to Baire (I.), Hahn (II.), and Tietze (III.–V.):
I. (Baire's approximation theorem) Every lower continuous function is a limit of an increasing sequence of continuous functions.
II. (Hahn's interpolation theorem) If $g(x)$ is lower continuous, $h(x)$ is upper continuous, and if $g(x) \geqq h(x)$, then there exists a continuous function $f(x)$ with $g(x) \geq f(x) \geq h(x)$.
III (Tietze's extension theorem) A continuous function $\varphi(x)$ defined on the closed set $A \subset X$ can be extended to a continuous function $f(x)$ on the entire space X, i.e., there exists a continuous function f on X with $f(x) = \varphi(x)$ for all $x \in A$.
IV. Let $\varphi(x)$ be bounded below and continuous on the closed set A. Then there exists a continuous function $f(x) \leq \varphi(x)$ with $f(x) = \varphi(x)$ for all $x \in A$.
V. Every lower continuous function $\varphi(x)$ which is continuous on the closed set A is the limit of an increasing sequence of continuous functions $f_n(x)$ with $f_n(x) = \varphi(x)$ on A.
Here $f(x)$ is called upper continuous on x (lower continuous on x) if

$$\limsup_{y \to x} f(y) \leq f(x) \quad (\liminf_{y \to x} f(y) \geq f(x)).$$

The "remarkable logical connection," which Hausdorff mentions at the beginning of his paper, consists in the fact that the proofs of each of these propositions (excepting I.) are based in each case on the preceding result, i.e. they follow the scheme I. → II.→ III. → IV. → V. Moreover, Hausdorff's new proofs represent a major simplification, as Chatterji emphasized in his commentary:

> [$\cdots$] it is hard to imagine easier proofs even when $X = \mathbb{R}^n$. Since these theorems are of wide use in many areas of analysis, Hausdorff's proofs have been taken over by many authors. (*HGW* IV, 98)

Tietze's extension theorem had a special significance for Hausdorff. He dealt with the subject of the extension of continuous mappings in metric and topological spaces again and again over a period of about two decades, as shown by numerous studies in his estate as well as his publications [H 1930b] and [H 1938]. In 9.2.6, we sketch this development in context while describing his final paper [H 1938]. In the second, more extensive part of [H 1919d], Hausdorff generalizes results of

Lebesgue on real-valued Borel and Baire functions in $\mathbb{R}^n$ to the case of metric spaces. These investigations, extended by new insights from descriptive set theory (Suslin sets), were taken up in Chapter 9 of Hausdorff's *Mengenlehre* [H 1927a]; this will be briefly discussed in the treatment of this work in 8.5. Later authors usually cited his *Mengenlehre* rather than [H 1919d].

7.8.4 Works on Divergent Series

Felix Hausdorff's contributions to the summation theory of divergent series exerted a lasting influence on the subject. He set forth this work in two extensive papers, *Summation Methods and Moment Sequences I, II* [H 1921], *HGW* IV, 107–162, both of which first appeared in Volume 9 of *Mathematische Zeitschrift.* Hausdorff's central discovery involved showing the connection between the moment problem and a large class of summation methods, now called Hausdorff summation methods, or *Hausdorff methods* for short. In the development of summation methods, the aim is to assign a generalized limit to the sequence of partial sums $s_0, s_1, \cdots$ of a divergent series $\sum_k a_k$ in a meaningful way. In doing so, one usually seeks a procedure that assigns the sum to a convergent series as a generalized limit.

Already in the eighteenth century, Leonhard Euler worked with divergent series and assigned numerical values to them as "sums" in a generalized sense. Because inconsistencies can often occur in this process, many mathematicians in Euler's time rejected the use of divergent series. However, Euler himself achieved many significant results working with them; his immense experience and unerring intuition saved him from errors. From about 1830 until the last two decades of the nineteenth century, divergent series were largely banned from mathematics. For example, the 108-page bibliography in the monograph [Zeller/Beekmann 1970] begins with the year 1880.

Otto Hölder published the summation method known today by his name in [Hölder 1882]. If $s_0, s_1, \cdots$ is the sequence of partial sums, one forms iterated means as follows:

$$H_n^{(1)} = \frac{s_0 + s_1 + \cdots + s_n}{n+1}, \;\cdots\; H_n^{(k)} = \frac{H_0^{(k-1)} + H_1^{(k-1)} + \cdots + H_n^{(k-1)}}{n+1}$$

If $\lim_{n\to\infty} H_n^{(k)} = s$, then the series H^k is summable; the $H_n^{(k)}$ are then called the kth-order Hölder means. Ernesto Cesàro introduced a similar procedure in [Cesàro 1890], starting with:

$$A_n^{(1)} = s_0 + \cdots + s_n, \;\cdots\; A_n^{(k)} = A_0^{(k-1)} + \cdots + A_n^{(k-1)}, \;\; C_n^{(k)} = \frac{A_n^{(k)}}{\binom{n+k}{k}}\,.$$

If $\lim_{n\to\infty} C_n^{(k)} = s$, then the series C^k is summable; the $C_n^{(k)}$ are then the kth-order Cesàro means.

Hausdorff's approach belongs to the large class of matrix methods. A real-valued number sequence is an element of the vector space $\mathbb{R}^{\mathbb{N}}$. The limiting methods (summation methods) considered are special linear mappings $\mathbb{R}^{\mathbb{N}} \to \mathbb{R}^{\mathbb{N}}$ described by infinite matrices $A = (a_{nm}), n, m = 0, 1, 2, \cdots, a_{nm} \in \mathbb{R}$. A sequence of numbers $\{s_m\}$ is called *A-limitable to value t* if for each n the series $\sum_{m=0}^{\infty} a_{nm} s_m$ converges with limit t_n and if $\lim_{n\to\infty} t_n = t$. The matrix A (or the limiting procedure given by it) is called *convergence-preserving* if every convergent sequence $\{s_m\}$ is A-limitable. One of the outstanding early results on limit theory was obtained by Otto Toeplitz in [Toeplitz 1911], where he gave necessary and sufficient conditions for A to be convergence-preserving, namely that three conditions hold:

$$(1) \sum_{m=0}^{\infty} |a_{nm}| \leq M \text{ for all } n; \quad (2) \lim_{n\to\infty} a_{nm} = l_m\,; \quad (3) \lim_{n\to\infty} \sum_{m=0}^{\infty} a_{nm} = l\,.$$

The procedure is called *permanent* (or *regular*) if A is convergence-preserving and every convergent sequence limits to $\lim_{m\to\infty} s_m = s$, i.e., when $t = s$. This is the case exactly when for all m, $l_m = 0$ in (2) and when $l = 1$ in (3). This theorem is now known as the Toeplitz Permanence Theorem.

The set of all sequences $\{s_m\} \in \mathbb{R}^{\mathbb{N}}$ which are limitable to a finite value by a given process A is called the *effective field* (which Hausdorff calls the *convergence field*) $\mathfrak{A}$ of A. If for two matrices A, B the convergence fields $\mathfrak{A}, \mathfrak{B}$ are equal, then the corresponding methods are called equivalent. If $\mathfrak{A} \subset \mathfrak{B}$ ($\mathfrak{A} \supset \mathfrak{B}$), then A is called weaker (stronger) than B. If none of these three cases occurs, then the processes are called incomparable. Prior to Hausdorff's work, an important result in limiting theory was the equivalence theorem of Konrad Knopp and Walter Schnee. This states that for any $k \in \mathbb{N}$, Hölder's method H^k and Cesàro's method C^k are equivalent. Hausdorff substantially generalized this equivalence theorem of Knopp and Schnee in [H 1921].

In the construction of Hausdorff methods, the nth-order differences $a_{m,n}$ arising from a sequence of numbers $\{a_m\}_{m=0,1,\cdots}$ play an important role. These are defined as:

$$\overset{n}{\nabla} a_m \doteq a_{m,n} = a_m - \binom{n}{1} a_{m+1} + \binom{n}{2} a_{m+2} - \cdots + (-1)^n a_{m+n}.$$

For $m = 0$, one obtains, instead of the double sequence $\{a_{m,n}\}$, an ordinary number sequence $\{b_n\}_{n=0,1,\ldots}$, the *difference sequence* for $\{a_m\}$,

$$b_n = a_{0,n} = a_0 - \binom{n}{1} a_1 + \binom{n}{2} a_2 - \cdots + (-1)^n a_n.$$

With these coefficients, Hausdorff forms the infinite triangular matrix

$$T = (\rho_{n,m}) = \left\{ \begin{array}{ll} \binom{n}{m}(-1)^m, & 0 \le m \le n \\ 0, & m > n \end{array} \right\} = \begin{pmatrix} 1 & 0 & 0 & 0 & 0 & \cdots \\ 1 & -1 & 0 & 0 & 0 & \cdots \\ 1 & -2 & 1 & 0 & 0 & \cdots \\ 1 & -3 & 3 & -1 & 0 & \cdots \\ \vdots & \vdots & \vdots & \vdots & \vdots & \cdots \end{pmatrix}$$

Since $a_{m,0} = b_{0,m}$, the a_m form the difference sequence of the b_n; it follows from this that $T^{-1} = T$ or $T^2 = E$.

A Hausdorff procedure is then given by a matrix of the form $T^{-1}\mu T = T\mu T$, where μ is a *diagonal matrix* and the elements along the main diagonal are $\mu_0, \mu_1, \mu_2, \ldots$. Thus, the sequence $\{\mu_n\}_{n=0,1,\cdots}$ determines the procedure. Matrices of the type $H = T\mu T$ always commute, and they are usually called Hausdorff matrices. Let H be a Hausdorff matrix with distinct μ_n and K another matrix interchangeable with H. Forming $\mu' = TKT$, one easily sees that μ' must be a diagonal matrix. So $K = T\mu'T$ is also a Hausdorff matrix. Since the matrix C^1 associated with the first-order Cesàro means is a Hausdorff matrix with $\mu_n = \frac{1}{n+1}$, Hausdorff could state the following theorem: *The class of our [i.e. Hausdorff] matrices coincides with the class of matrices interchangeable with* C^1.

Hausdorff was also able to show further that the classical methods of Hölder and Cesàro belonged to his general scheme. The Hölder process with kth-order H^k is a Hausdorff process with $\mu_n = (n+1)^{-k}$, whereas the Cesàro method with k-th-order C^k corresponds to Hausdorff's with

$$\mu_n = \frac{1}{\binom{n+k}{n}}.$$

Classically, these two methods were defined only for $k \in \mathbb{N}$, but Hausdorff's approach was, in this respect, completely unconstrained. He could thus introduce $\mu_n = (n+1)^{-\alpha}$ or $\mu_n = \binom{n+\alpha}{n}^{-1}$ and also H^α for any $\alpha \in \mathbb{R}$ and C^α for $\alpha \in \mathbb{R}\backslash\{-1, -2, \cdots\}$. This meant he could consider finer scales for the order.

The question that now naturally arises is this: which sequences $\{\mu_n\}$ lead to permanent summation procedures in the sense of Toeplitz's theorem? To address this question, Hausdorff introduces the following terminology. He calls a convergence-preserving matrix A a *C-matrix*, which will be a *pure C-matrix* if it limits zero sequences to zero sequences, and furthermore a *normalized C-matrix* if it provides a permanent procedure. Accordingly, he calls a sequence $\{\mu_n\}_{n=0,1,\cdots}$ a *C-sequence* (*pure C-sequence, normalized C-sequence*) if the Hausdorff matrix $T\mu T$ is a C-matrix (pure C-matrix, normalized C-matrix).

Now if $T\mu T$ is a Hausdorff matrix, $a = \{a_m\}$, and $(T\mu T)a = \{A_n\}$, then using the properties of the differences, one finds

$$A_n = \sum_{m=0}^{n} \binom{n}{m} \mu_{m,n-m}\, a_m$$

with $\mu_{m,n-m} = \nabla^{n=m} \mu_m$. From the Toeplitz Permanence Theorem for the associated $T\mu T$ matrices, one can then show that a Hausdorff process with the sequence $\{\mu_n\}$ is exactly then permanent, i.e. a normalized C-matrix, when

$$(1)\ M_n = \sum_{m=0}^{n} \binom{n}{m} |\mu_{m,n-m}| \leq M \text{ for all } n,$$

$$(2)\ \lim_{n\to\infty} \binom{n}{m} \mu_{m,n-m} = 0 \text{ for all } n, \quad (3)\ \lim_{n\to\infty} \sum_{m=0}^{n} \binom{n}{m} \mu_{m,n-m} = 1\,.$$

Since $\sum_{m=0}^{n} \binom{n}{m} \mu_{m,n-m} = \mu_0$, condition (3) simplifies to $\mu_0 = 1$.

By using the notion of total monotonicity of a sequence due to Issai Schur, Hausdorff determined a necessary and sufficient criterion for the validity of (1). A sequence $\{\mu_n\}$ is *totally monotone* if all its differences are nonnegative, that is, if $\nabla^n \mu_m \geq 0$ for $m, n \in \mathbb{N}$. Then (1) holds exactly when $\mu_n = \alpha_n - \beta_n$, where $\{\alpha_n\}, \{\beta_n\}$ are totally monotone sequences, and it turns out that condition (1) *alone* is necessary and sufficient for a Hausdorff matrix to be a C-matrix. Thus $\{\mu_n\}$ is a C-sequence if and only if it is the difference of two totally monotone sequences. In particular, every totally monotone sequence is a C-sequence.

For the Hölder method, $\mu_n = (n+1)^{-\alpha}$. This can be expressed as follows:

$$\mu_n = \frac{1}{\Gamma(\alpha)} \int_0^1 u^n \left(\ln \frac{1}{u}\right)^{\alpha-1} du\,; \quad \alpha > 0\,.$$

Hausdorff also found a corresponding integral representation for the Cesàro process:

$$\mu_n = \binom{n+\alpha}{n}^{-1} = \alpha \int_0^1 u^n (1-u)^{\alpha-1}\, du; \ \ \alpha > 0\,.$$

So in both cases, the μ_n are the momenta of a function $\chi(u)$ given on $[0,1]$, as Hausdorff explained:

> Following the hint given by the integral representations [$\cdots$ the two representations given above], we form with a function $\chi(u)$ of bounded variation on the interval $0 \leq u \leq 1$, the Stieltjes integrals
>
> $$\mu_n = \int_0^1 u^n\, d\chi(u) \ \ (n = 0, 1, 2, \cdots)$$
>
> and call this a *moment sequence*. [H 1921, 84; IV: 117]

The central theorem of Hausdorff's limiting theory can now be stated: *A sequence $\{\mu_n\}$ is a C-sequence if and only if it is a moment sequence. Moreover, it yields a permanent process, i.e., it corresponds to a normalized C-sequence,* *exactly when* $\chi(u)$ *satisfies:*

$$\chi(0) = \lim_{u \to +0} \chi(u) = 0 \ \textit{and} \ \chi(1) = 1\,.$$

That moment sequences are C-sequences follows from the fact that they can be written as the difference of two strictly monotone sequences. More specifically, the moment sequence of a non-decreasing function $\chi(u)$ is strictly monotone. The converse, that every C-sequence is a moment sequence, requires solving the moment problem: Given a sequence $\{\mu_n\}_{n=0,1,\cdots}$ representing the difference of two strictly monotone sequences, does this uniquely determine a function $\chi(u)$ of bounded variation on $[0,1]$ such that

$$\mu_n = \int_0^1 u^n \, d\chi(u) \ \ (n = 0, 1, 2, \cdots)?$$

In the last paragraph of [H 1921], Part I, Hausdorff gave a solution independent of Stieltjes' theory, thereby answering the question in the affirmative. If $\{\mu_n\}$ is itself strictly monotone, then $\chi(u)$ turns out to be monotonically increasing. Hausdorff later returned to his original solution of the moment problem in detail in [H 1923b].

Hausdorff also obtained decisive results concerning the comparisons with his summability methods. Let $\{\mu_n\}$, $\{\mu'_n\}$ be sequences which yield permanent Hausdorff methods, i.e., normalized C-sequences, and let $\mathfrak{A}, \mathfrak{A}'$ be their effective fields. He then proved that $\mathfrak{A} \subseteq \mathfrak{A}'$, *exactly when* $\{\frac{\mu'_n}{\mu_n}\}$ *is a normalized C-sequence.* Thus, the two procedures are equivalent if and only if $\{\frac{\mu'_n}{\mu_n}\}$ and $\{\frac{\mu_n}{\mu'_n}\}$ are normalized C-sequences, so that both yield permanent Hausdorff procedures. By means of this criterion Hausdorff found a beautiful new proof of the equivalence theorem of Knopp and Schnee. In fact, he proved much more, namely: H^α and C^α are equivalent for any real $\alpha > -1$; he later gave an extension of this result in [H 1930a].

Chatterji gave this final assessment of Hausdorff's limiting theory:

> The Hausdorff matrices provided a first general class of summability methods which included a number of important methods as special cases and provided a general theory, based on a solution of Hausdorff's moment problem, which seems really satisfactory. Hence, almost all the monographs after 1930 dealing with summability theory devote an important place to Hausdorff matrices [$\cdots$] (*HGW* IV, 170).

In [H 1921], Part II, Hausdorff generalized his moment problem. Let $t_0 = 0 < t_1 < t_2 < \cdots$ be an *unrestricted* sequence of real numbers. $\{\mu_n\}$ is called a $\{t_n\}$-*moment sequence* if there exists a monotonically nondecreasing function $\chi(u)$ on $[0,1]$ with

$$\mu_n = \int_0^1 u^{t_n} \, d\chi(u) \ \ (n = 0, 1, 2, \cdots)\,.$$

In the so-called *divergence case* $\sum_{n=1}^{\infty} \frac{1}{t_n} = \infty$, Hausdorff gave a complete solution of the t_n-momentum problem. For this purpose he introduces the notion of strict monotonicity of a sequence $\{\mu_n\}$ *with respect to the sequence* $\{t_n\}$ and shows that $\{\mu_n\}$ is a $\{t_n\}$-moment sequence if and only if it arises as the difference of two sequences strictly monotone with respect to $\{t_n\}$. Hausdorff undertook an extraordinary amount of work to obtain results for the *convergence case* $\sum_{n=1}^{\infty} \frac{1}{t_n} < \infty$. Numerous studies on this can be found in his estate, seven of which are printed in (*HGW* IV, 339–373). Later functional-analytic investigations have shown that there can be no unique solution of the moment problem in this case.

Beginning in the 1930s, Hausdorff's summability theory became a lively field of research. The bibliography in [Zeller/Beekmann 1970], which covers the literature up to 1968, lists 93 papers that refer to Hausdorff already in the title; certainly his ideas played a role in numerous others. One can gain an impression of the development of the theory in more recent times from the monograph [Boos 2000/2006] as well as the anthology [Dutta/Rhoades 2016], in particular by perusing the extensive bibliographies in these works.

7.9 Family Life in Greifswald

As mentioned earlier, when the Hausdorffs first moved to Greifswald they rented the apartment formerly occupied by Friedrich Engel and his family. For Hausdorff, the decisive issue when it came to housing matters – besides having enough space for his library and Bechstein grand piano – was always to find a peaceful setting where he could quietly work. In this respect, the Engels' former apartment building was far from ideal. However, a favorable opportunity soon arose to acquire a relatively new house (built in 1904) in a pleasant location, "Am Graben 5," in a quiet middle-class neighborhood.

This house went up for sale after the physician Anton Steyrer departed for Innsbruck. The address book for Greifswald in 1914 indicates the professions of the owners of the 12 houses on "Am Graben": six were university professors and two were pensioners; the other four houses were occupied by a doctor, a teacher, a railroad administrator, and a municipal official. Hausdorff's immediate neighbors were the pensioner Albert Ruge and the railroad administrator Robert Jenssen. The house is currently in good condition; a plaque installed by the University of Greifswald reminds us that Hausdorff lived there from 1913 to 1921. Otherwise, our knowledge of the Hausdorffs' family life in Greifswald is very sketchy, based as it is on more or less accidentally preserved sources and a few letters from that time.

At the beginning of this chapter, we noted that Hausdorff had quite distant personal relationships with the other two Greifswald mathematicians. However, he was on friendly terms with three other colleagues from the faculty. Probably his closest relationship, also in terms of the families, was with Theodor Posner. Born on February 18, 1871 in Berlin as the son of the Jewish merchant Isidor

Figure 7.3: The Hausdorffs' Home, Am Graben 5 in Greifswald
The Family is standing on the Balcony with Daughter Lenore in the Middle.
Source: Hausdorff Estate, Capsule 65, Nr. 26.

Figure 7.4: Visit of the Posner Family in Bonn, March 1927
Felix Hausdorff, Theodor Posner, Else Posner, Charlotte Hausdorff, Lenore König. Source: König Estate, Nr. 161.4.

Posner, the younger Posner abandoned Judaism, perhaps already prompted by his parents; in the Greifswald records he is listed as "Protestant." In Berlin, he studied chemistry, receiving his doctorate in 1893. Four years later, he habilitated in Greifswald in 1897 and became a private lecturer and assistant at the Chemical Institute. In 1901, while still a private lecturer, he assumed the directorship of the institute, and in 1903 he published a widely used textbook on organic chemistry. Academic recognition was slow in coming, however. Posner was promoted to associate professor in 1907, but it was not until 1921 that he received a full professorship in chemistry, eight years before his death in 1929.

To a certain extent, Posner and Hausdorff were kindred spirits. They shared an affinity for looking at the ironic sides of life with humor, and both were passionate lovers of music, though they also took a lively interest in all sides of cultural life. Posner was remembered in his university obituary with these words:

> For over 30 years he has worked at the University as a quiet, but self-confident and characterful personality of rich intellectual culture and deep kindness of heart. Beyond his scholarly activities, he has rendered the greatest services to the musical life of the city.

A special indication of the strong friendship that drew Hausdorff and Posner together is reflected in the poem Hausdorff composed for Posner's 50th birthday. He included this handwritten dedication in his birthday gift, a copy of *Sant' Ilario.*[1] In the poem he addresses, among other things, the ominous social situation in Germany after World War I, whereas the final two verses pay touching tribute to his good friend. After Posner's death, the Hausdorffs continued to maintain friendly relations with his widow Else; she later stood by the Hausdorffs during the last hours of their lives (see 9.4.4).

In her conversations with Egbert Brieskorn, Hausdorff's daughter Nora also reported friendly relations with the family of the mineralogist and petrologist Ludwig Milch. Like Hausdorff, Milch came from a Jewish family in Breslau; he took his doctorate in Heidelberg in 1889 before returning to habilitate in Breslau. He arrived in Greifswald in 1907 to assume a post as associate professor of mineralogy, crystallography and petrology, and five years later he was promoted to full professor. Milch left, however, in 1917 and returned to his home town as professor of mineralogy, conducting research on the geology of the Giant Mountains in the Czech Republic. Nora described him as broadly educated and a great enthusiast for classical music. He was also a friend of the playwright Gerhard Hauptmann.

Another friendly acquaintance of Hausdorff during his years in Greifswald was the theoretical physicist Gustav Mie. After taking his doctorate in mathematics in Heidelberg in 1891, Mie habilitated six years later in physics at the Technical College in Karlsruhe. He then came to Greifswald in 1902 as associate professor for theoretical physics, gaining promotion to full professor three years later. Mie left in 1917 to take a position in Halle, before moving on to Freiburg, where he worked until his retirement in 1935; he died there in 1957. A quiet critic of National Socialism, Gustav Mie became a member of the "Freiburg Council," an opposition group consisting mainly of professors and church representatives formed in response to the November 1938 pogrom.

Mie made significant contributions to optics (Mie scattering) and he developed a well-known theory of matter based on electromagnetism. For a short time, Mie's theory attracted the interests of Max Born and Hermann Weyl, but especially David Hilbert, who attempted to combine it with Einstein's general theory of relativity. Hausdorff no doubt learned about these fast-breaking developments through his physicist colleague. He must have taken a deep interest in the debates sparked by general relativity, since the space problem had occupied his attention since his student days. We gain a glimpse of conversations between Hausdorff and Mie on questions of relativity from a letter Hausdorff wrote to Hilbert on March 14, 1916:

> Very Honored Herr Geheimrath!
>
> I have had several detailed discussions with Mie about your work "The Foundations of Physics," which I thank you most obligingly for having sent to me. The mathematical part, which naturally I have

[1] The poem is printed in its entirety in (*HGW* VIII, 255–256). Apparently Posner's wife Else returned the book with the poem to Hausdorff sometime after Posner's death; both are in the Hausdorff estate (Capsule 52, Fascicle 1142).

Figure 7.5: Felix Hausdorff and his Daughter Lenore, ca. 1920
Source: König Estate, Nr. 161.3.

> taken up, causes me some not inconsiderable difficulties [· · ·]. (*HGW* IX, 341)

In fact, Hausdorff's difficulties in understanding Hilbert's text were shared by many others. Emmy Noether, who was then assisting Hilbert, was one of the few who understood it completely, as Einstein came to appreciate when he tried to penetrate Hilbert's text [Rowe 2021, 70–73] (for an overview of Hilbert's efforts to axiomatize physical theories, see [Corry 2004]).

The living conditions of the Hausdorff family during the war years were certainly not easy – especially beginning in 1916 when the situation regarding foodstuffs and other necessities in the German Reich deteriorated dramatically. Obtaining food to eat was certainly not as difficult in a small city like Greifswald, surrounded by vast rural regions, as it was in larger cities and industrial areas. Still, the country faced widespread privation during the winter of 1917. Shortly before Christmas of that year, Hausdorff had to send an apology to a friend, explaining that he could not send her a goose because "our food distributors let us down" (*HGW* IX, 195). Heating fuel posed a far more difficult problem, however. In a later letter, Charlotte Hausdorff reminded her daughter of those times in Greifswald when her father had to fetch their own wood in a handcart. On one occasion, Ludwig Milch had just arrived as a guest, and he helped Hausdorff in lugging the heavy blocks into the house (König Estate, Universitäts- und Landesbibliothek Bonn).

What Hausdorff most missed throughout the war and for several years afterward were the summer trips, especially to southern France and Italy. He wrote about this to Franz Meyer on January 4, 1916: "That now the landscape of Zarathustra remains forbidden to us for a decade takes away an essential charm from life!" (*HGW* IX, 510). A certain compensation – though not remotely equivalent – was offered by the region surrounding Greifswald, especially the island of Rügen. During the summer months, the family also traveled regularly to Bad Reichenhall, where Charlotte Hausdorff's sister Sitta lived. On one of these visits in the summer of 1916, the Hausdorffs got to know the famous actress Louise Dumont and her husband, the theater director Gustav Lindemann. In 1905, Dumont and Lindemann had founded the Schauspielhaus Düsseldorf, which they directed together for many years when it became one of Germany's leading theaters. Hausdorff's one-act play, *Der Arzt seiner Ehre*, was performed there eleven times during the 1905/06 season and another nine times six years later. However, apparently no personal meeting took place on those occasions between the playwright and the house management.

After they met, Hausdorff wrote openly to Louise Dumont about political matters. His letters contain several passages that show his decidedly critical attitude toward the war, in particular, regarding the jingoism that still dominated public attitudes at the end of 1916. In the first preserved letter, dated December 28, 1916, he wrote to wish Gustav Lindemann a speedy recovery from an illness, before going on to say:

> In these dreadful times, one can hardly have more than such individually limited hopes and wishes; precisely because the fate of the individual, as preached to us daily in the newspapers, has become completely nil and insignificant today, one can only shout our personal petitions into the ear of our dear God, who has by now become deaf through gun fire, for ourselves and those bound to us by sympathy. Hopefully, you will repay like with like and, when you address the deaf old man, you will also include us in your prayers! (*HGW* IX, 187)

In the summer of 1917, they met once again in Bad Reichenhall, and after returning to Greifswald Hausdorff wrote to Louise Dumont:

> In a letter of Heine from the year 48 I found the following words: "About the events of the time I say nothing; that is universal anarchy, world muddle, God's madness become visible! The old man must be locked up, if this goes on in such a way." – What must have happened to "the old man" today? And there are still people (like our Reich Chancellor), who consider him to be a reasonable and benevolent old gentleman. (*HGW* IX, 191–192)

Hausdorff also strongly opposed the Russian October Revolution under Lenin, as seen from a letter to Louise Dumont, written on November 14, 1917:

> Incidentally, the world is as dark as it can be, and even in your beloved

> Russia the day never dawns, rather only twitches hell's flaming light. (Well said, almost Pan-German!). (*HGW* IX, 193)

Dumont felt a strong sympathy for Russia and its culture. She had celebrated great successes there on tours as a guest performer. Later she endeavored to bring Russian authors, such as Gogol and Gorky, to German audiences. The bracketed remark is a contemptuous and ironic allusion to the Pan-German League, founded in 1891, which had a particularly strong following during the period of World War I. Its radical program for creating a "Grossdeutsches Reich" – expansionist aims and militarism spiced with large doses of antisemitism – later served as a model for the National Socialists.

The Hausdorffs' daughter Nora grew up to be a lovely young woman in Greifswald. She was 13 years old when the family moved there, and she went on to complete the last grades of her schooling at a local private lyceum. In 1917 she trained to become a nurse, a job she temporarily undertook toward the end of the war. From April 1920 to September 1921, she prepared to work in the field of infant and toddler care. After completing her training and passing the appropriate examination, she received a certificate as a state-certified infant nurse.

Although Felix Hausdorff held strong political opinions, Nora's conversations with Brieskorn suggest that he only rarely expressed his views outside small private circles. She thus indicated that, even though politics was often discussed in the family, her father was not really a political personality. For him, freedom and the rights of the individual were always of paramount importance. He strongly rejected the notion that the individual had to subordinate him- or herself to a "higher destiny," but he also distrusted do-gooders with their ideologies, and he thoroughly despised demagogic propaganda. Nevertheless, he did become a member of a political party in Greifswald, if only for a relatively short period of time. In a personnel questionnaire filled out after November 7, 1934, Hausdorff entered the following in the "political activity" column: "Member of the German Democratic Party, 1919–1921, in Greifswald (non-active)."

In November 1918, after the abdication of Kaiser Wilhelm II and the collapse of the entire monarchical system that followed, Germany's political system was in a state of chaos. As socialists and communists vied for power, liberals hoped to found a new party out of the wreckage of the older ones. Thus, the German Democratic Party (DDP) came about largely by uniting the left-liberal Progressive People's Party with the liberal wing of the National Liberal Party. The party's program professed individual freedom, social responsibility and support for the new parliamentary democracy established through the constitution that created the Weimar Republic. At the outset, the DDP enjoyed considerable success with voters, and in the Reichstag elections of January 1919 it became the third-strongest force (with 75 seats, behind the Social Democrats with 165 and the Center Party with 91 seats). The DDP attracted liberal intellectuals, many of whom were of Jewish background, and it had a distinctly elitist clientele. Among its more prominent members were Ernst Cassirer, Theodor Heuss, Friedrich Naumann,

Walter Rathenau, Ludwig Quidde, Max Weber and Theodor Wolff.

The DDP's main opponent in the Reichstag was the German National People's Party (Deutschnationale Volkspartei, or DNVP, also founded in November 1918), which staunchly opposed parliamentary democracy and sought restoration of the monarchy. Many of its members were hyper-nationalists and outspokenly anti-Semitic. In the Reichstag elections at the beginning of 1919, the DNVP came in well behind the DDP with 44 seats. In Western Pomerania and especially in Greifswald, however, the balance of power was somewhat different from that in the German Reich as a whole. This applies in particular to the University of Greifswald. Among its full professors who, like Hausdorff, arrived before the war, 19 joined political parties afterward: 12 were members of the DNVP, 3 joined the right-wing Stahlhelm, 2 were with the moderately conservative Deutsche Volkspartei, and only 2 belonged to the DDP. The other member of the DDP besides Hausdorff was the art historian Max Semrau, who also came from Breslau. These two representatives of a liberal democratic position were thus quite alone among the group of professors who aligned themselves with one of the postwar political parties.

Unfortunately, the DDP and the liberal democratic movement it represented lost momentum very quickly during the early years of the Weimar Republic, even though the party participated in several coalitions. Over the years, the DDP gradually drifted ever further to the right. By the time of the 1930 Reichstag elections, it formed an alliance with the short-lived Volksnationale Reichsvereinigung, an outgrowth of the nationalist and anti-Semitic Jungdeutscher Orden. Whether Hausdorff followed the course of his former party's decline after he left it in 1921 remains an open question.

Chapter 8

Full Professor in Bonn, 1921–1933

Mathematics in Bonn underwent some significant changes both during the war and in its aftermath. Shortly before the death of Franz London in 1917, Hans Hahn was called as his successor. In early 1921, though, Hahn accepted an attractive offer from the University of Vienna, where he became one of the leading members of the famous Vienna Circle. In this new situation, Eduard Study hoped to bring Hausdorff back to Bonn as Hahn's successor, much as he had largely engineered his earlier appointment more than a decade earlier. In the meantime, however, circumstances had changed, even though Study remained the key figure in Bonn mathematical circles. By now, younger talents had emerged, mathematicians whose accomplishments were in some respects comparable to Hausdorff's own. Another novel development in Bonn, entirely absent during the years when London taught there, concerned internal frictions and outright factionalism. Both of these circumstances played a role in the process of choosing candidates to succeed Hans Hahn, and although Study's favorite ultimately prevailed, Hausdorff's appointment signaled the end of harmonious relations between the mathematicians in Bonn. To appreciate how the atmosphere changed, though, first requires a glance backward to events in Bonn during the period after Hausdorff had departed for Greifswald in 1913.

8.1 Returning to Bonn

When Hausdorff gave up his associate professorship in Bonn, his position was briefly occupied by the algebraist Issai Schur. As a Russian Jew, Schur faced many obstacles in pursuing an academic career, despite his talents and accomplishments. His mentor, Georg Frobenius, hoped to find a place for him in Berlin, but Schur spent a decade there as a private lecturer before gaining Hausdorff's position in Bonn in 1913. By 1916, though, Frobenius finally managed to orchestrate Schur's appointment as an associate professor [Biermann 1988, 178]. In the meantime,

E. Brieskorn, W. Purkert, *Felix Hausdorff*, Vita Mathematica 21,
https://doi.org/10.1007/978-3-031-52135-5_8

Franz London had become too ill to continue teaching just when Schur was getting ready to return to Berlin. This left Study with the unenviable task of having to fill two vacancies at once. Finding suitable candidates in the middle of the Great War, when able-bodied younger men were scarce, proved none too easy.

Study brought Hans Hahn over from Czernowitz to assume London's position, and he decided to nominate Hans Beck as Schur's succesor. Beck had studied in Greifswald during Study's tenure there, after which time he took a position as senior teacher (*Oberlehrer*) at a Berlin Gymnasium. Since he had never habilitated, this appointment clearly reflected the desperate circumstances, and Study would later suffer the consequences. He surely never suspected that his former student, who owed his position in Bonn to him alone, would later oppose him when it came time to choose Hahn's successor.

These deliberations began quietly at first, when the appointed faculty commission convened in January 1921 and Study offered the mathematicians' preliminary recommendations:

> Herr Study reported on the prospective candidates considered by representatives of the discipline: 1. Prof. Hausdorff – Greifswald and Prof. Koebe – Jena; 2nd Prof. Tietze – Erlangen. The next meeting is to be held as soon as [external] information arrives.[1]

Ten external colleagues had been asked to express their opinions, and among these seven eventually replied. When the commission reconvened, Study moved for approval of the following list: 1. Koebe, 2. Hausdorff, 3. Tietze and Radon pari passu. This list was adopted by a 4:1 vote, the lone dissent coming from Beck, who evidently expressed his views in a provocative way. A heated internal dispute then unfolded that must have shocked some who witnessed it, as afterward a special commission was appointed by the faculty with the aim of "restoring collegial relations" between the parties. Minutes from a meeting held in February state that its purpose was "to attempt mediation in the conflict which arose between Herr Study and Herr Hahn, on the one hand, and Herr Beck, on the other, during the deliberations on the professorship terminated by the departure of Herr Hahn." While Beck was reprimanded on three substantive points, "Study and Hahn stated that they regretted that in their agitation they had not chosen the proper place for their argument with Herr Beck." In other words, Hahn and Study certainly saw no reason to amend or qualify their views.

On February 9, 1921, the Mathematical and Natural Sciences Section of the Faculty met and voted on the names that should appear on their list of recommendations. The results were: "Koebe unanimous; Hausdorff 10:1; Tietze unanimous; Radon 8:(3 abstentions)." Again, it can be assumed that the dissenting vote on Hausdorff came from Beck, who had been a member of the DNVP since 1919. As a former Greifswald resident, Beck may well have known that Hausdorff had joined the DDP after the war. Anti-Semitism, one of the ideological foundations

[1]This quotation and those following are from file PF-PA 180, Bonn University Archives.

of the DNVP, could also have been a motive, or possibly Beck's growing animosity toward Study, who clearly favored Hausdorff's candidacy. At this meeting it was also decided that the commission formed at the beginning of the procedure should draw up an expert opinion for the minister. This report begins as follows:

> For the professorship vacated by the departure of Prof. Hahn, the Philosophical Faculty proposes:
> I. Dr. Paul Koebe, full professor at the University of Jena.
> II. Dr. Felix Hausdorff, full professor at the University of Greifswald, and Dr. Heinrich Tietze, full professor at the University of Erlangen.
> III. Dr. Hans Radon, associate professor at the University of Hamburg.

Before briefly describing each candidate, the faculty emphasized that the person to be appointed should have demonstrated "versatility that is highly desirable for representing analysis as a fundamental discipline for higher geometry and theoretical physics." The part of the report referring to Hausdorff clearly confirms that he possessed the desired qualities:

> He has above all rendered services to set theory, in which he himself has largely created the theory of ordered sets. His Grundzüge der Mengenlehre (1914) is the best textbook on this subject. He has also worked successfully in geometrical optics, probability theory, group theory, and, more recently, in the theory of real functions. He is thus one of our most versatile mathematicians. His teaching activities in Bonn are well remembered by the faculty, and make him seem particularly suitable for the position to be filled.

Nevertheless, the faculty ranked him equal with Heinrich Tietze, a topologist, and behind the analyst Paul Koebe, both of whom were more than ten years younger than Hausdorff. Particularly noteworthy is the praise accorded to Koebe:

> Dr. Paul Koebe [...] is the most outstanding among the younger generation of German mathematicians and even one of the most important mathematicians working today. His investigations on function theory and conformal mapping are highly original and of imperishable value. We have only favorable news about Koebe's teaching activity.

Koebe suddenly became famous in 1907 by proving the general uniformization theorem for Riemann surfaces, thereby solving Hilbert's 22nd problem. In the same year, Henri Poincaré also proved the uniformization theorem by a different method. Koebe's other related results on conformal mappings and the Riemann mapping theorem were undoubtedly achievements of "imperishable value," but one may doubt that many informed contemporaries – other than Koebe himself – would have agreed in calling him "one of the most important mathematicians working today." From a present-day perspective, the influence of Hausdorff on the development of mathematics as a whole was considerably greater than that of Koebe.

In any case, the Prussian Ministry of Culture passed over Koebe and decided to call Hausdorff instead. Considering how Hausdorff himself had been passed over when Engel originally recommended him for the vacancy in Greifswald, this was truly an ironic turn of events. Again, though, we can only speculate about the motives behind this decision. One possibility might have been an assessment of Koebe submitted by the Berlin faculty in 1917 during the search for candidates to fill the chair formerly occupied by Hermann Amandus Schwarz. They placed Koebe third on that list behind Erhard Schmidt and Issai Schur. While praising Koebe's work, Friedrich Schottky wrote: "However, his thoughts all go in one direction, though important problems lie in that direction" [Biermann 1988, 329]. This single comment suggests narrowness rather than versatility, thereby directly contradicting the main criterion that the Bonn faculty explicitly sought to fulfill. Hausdorff's tenure as full professor in Bonn began officially on October 1, 1921. Before settling in, he bought a house located at Hindenburgstrasse 61, today aptly renamed Hausdorffstrasse.

8.2 The Bonn Mathematics Seminar

When Hausdorff took up his post in 1921, he joined three other mathematicians working in Bonn: the full professors Eduard Study and Hans Beck as well as the associate professor Johann Oswald Müller. Beck's promotion to a personal full professorship in 1920 had nothing to do with his accomplishments since arriving in Bonn, but rather was part of a general policy implemented at the Prussian universities under the new Social Democratic state government. This new formal status no doubt boosted his self-confidence and sense of legitimacy in challenging Study, who had been full professor in Bonn since 1904. Moreover, unlike his colleagues, Beck and Müller, Study was an important mathematician, whose career began when he habilitated under Felix Klein in Leipzig in 1885. In addition to his studies on the invariant theory of the general projective group, Study had developed invariant theories for other important transformation groups. His aim was to derive from a few "fundamental invariants" and the "main identities" existing between them, geometrical theorems associated with the given group, an approach originally sketched in Klein's famous "Erlangen Program." Perhaps due to its abstract presentation, his first book on invariant theory, *Methoden zur Theorie der ternären Formen* (1889), enjoyed very little success. Many copies had to be pulped when it went out of print, but during the twentieth-century renaissance of classical invariant theory a new edition was issued in 1982 (!).

Study was not only one of the founders of geometry over the field of complex numbers, he also made important contributions to the theory of hypercomplex number systems (algebras of finite rank over the real or complex numbers). In a number of his geometrical works he searched for a hypercomplex system that would enable him to express the formulas related to a particular problem as transparently as possible. Thus, for the treatment of metric line geometry in his principal

Figure 8.1: The Hausdorffs' Home in Bonn
Source: König Estate, Nr. 161.3.

work *Geometrie der Dynamen* (1903), he introduced "dual numbers" $z = x + \varepsilon y$ with $\varepsilon^2 = 0$, $\varepsilon \neq 0$ as the appropriate tool. Another among his favorite research methods involved mapping various classes of geometric objects onto manifolds in projective spaces. Thus, in *Geometry of Dynamen*, Study studied the kinematics of rigid bodies by means of a quadratic hypersurface in 7-dimensional projective space. Study made an important contribution to algebra with his work on systems of linear equations over the domain of quaternions, in which he generalized the concept of determinants to non-commutative cases. Here Study's "nabla function" plays a fundamental role; Hausdorff followed up on this in [H 1927c]. In striving for rigor, precision and, in a certain sense, completeness for geometrical methods, Study did not shy away from public criticism and polemics. Thus, although he appreciated Sophus Lie as one of the most ingenious mathematicians of the previous century, he criticized his theory of contact transformations and Lie's sphere geometry, while pointing out ways to overcome some of its weaknesses.

Study also pursued wide-ranging intellectual interests and wrote several books on epistemological questions and philosophical problems related to mathematics. He also enjoyed esteem as an entomologist, especially for his work on mimicry in insects, and his private collection of butterflies was one of the largest in Europe. Although Study was known as a prickly personality, Hausdorff had already established a friendly relationship with him dating from his first period in Bonn; their warm relations continued throughout the 1920s. As Hausdorff's daughter reported, they enjoyed an unclouded friendship right up until Study's death on January 6, 1930. He died after suffering from a severe form of cancer. Hausdorff eulogized him for a secular funeral service held at a crematorium in Mainz on 9 January 1930. His words stand as a touching testimony to their friendship and the spiritual affinity that drew them together (*HGW* VI, 343–344).

As already noted, Study and Beck had crossed swords over Hausdorff's candidacy in 1921. Predictably, the faculty's efforts to restore "collegial relations" between these two antagonists proved completely futile. In fact, their mutual hostility only grew deeper, even leading to public disputes in professional journals. These first arose when Beck attempted to insert himself into discussions of Lie's sphere geometry, a theory Study had criticized, in part because Lie had neglected to employ oriented spheres. Beck tried to create the impression that his own publications obviated Study's critique, since, as he claimed, his earlier work had already clarified these matters. Around this time, Study wrote a detailed letter to the faculty explaining why he would "no longer in the future participate in the evaluation of dissertations for which Herr H. Beck serves as referee" (PF-PA 536 Eduard Study. University Archives Bonn).

In August 1926, Study undertook a frontal attack on Beck's work in the official journal of the German Mathematical Society [Study 1926]. Not only did he belittle Beck's contributions to sphere geometry, he also went so far as to accuse him of intellectual dishonesty, both with regard to Beck's treatment of his own work as well as that of Study's deceased predecessor, Rudolf Lipschitz. In the next issue, Beck responded to several technical points Study had raised, but avoided addressing the larger points bearing on his character. At the same, Beck apparently also filed a complaint with the faculty. An investigation turned to professors from other universities, asking for their opinion regarding the dispute. One of those contacted was Issai Schur, though obviously not because of his expertise on sphere geometry, but rather as Beck's immediate predecessor on the Bonn faculty. Schur responded on 11 May, 1927 by offering these general remarks:

> I certainly have the impression that Herr Study is by far the superior mind here, and he knows how to judge things from quite a different height than Herr Beck. [⋯] The sharpness of tone in Study's publication is in individual places [⋯] certainly to be regretted. But this can be explained by Herr Beck's entire conduct toward Study, an outstanding researcher to whom he owes so infinitely much in scientific respects. (*ibid.*)

Hausdorff did not get involved in this controversy, though it was obvious whose side he was on. His daughter later reported that her father did not think much of Hans Beck. He surely tried to stay as far from him as possible in order to cope with this unpleasant situation. Toward the end, he must have felt the effects of living in an atmosphere dominated by the tensions between these two colleagues.

Another major conflict arose over the case of Ernst August Weiss, a protégé of Study who applied to habilitate in Bonn [Segal 2003, 190–191]. Weiss had studied mathematics in Hanover, Hamburg and finally in Bonn, taking his doctorate in 1924 under Study with a geometrical thesis. Study clearly had a good opinion of Weiss, as he made him his assistant at the beginning of 1923. He also wrote a very positive report on his postdoctoral thesis on Weddle's surface, which Hausdorff simply endorsed "in all points." Beck, however, criticized it as narrow and out of fashion (thereby implying that Study's research was dated as well). He tried to block or at least delay Weiss's habilitation, arguing to postpone the decision on it for two semesters. Study sharply rejected this argument, supported by Hausdorff, though probably more owing to his friendship with Study than out of conviction. Hausdorff probably cast the decisive vote that broke this deadlock, though Beck fought on until the bitter end.

On May 19, 1926, Weiss completed the official procedure for appointment as a private lecturer. After spending three semesters in France on a Rockefeller scholarship, he returned to Bonn and in 1932 gained an appointment as a non-tenured associate professor. Weiss worked in the field of algebraic geometry and invariant theory in the tradition of Study. After 1933, he played an inglorious role as a dedicated and highly active National Socialist (see 9.1.1 and [Segal 2003, 188–191]). Weiss never achieved prominence as a researcher; in fact, in his obituary for him, Wilhelm Blaschke devoted only two sentences to his research, while paying tribute to him instead as a successful teacher and author of two textbooks.

Throughout these years, Beck clearly had few substantive interactions with the other mathematicians in Bonn. Confirmation of this can be seen from the fact that Beck's name never appeared in announcements for the Mathematical

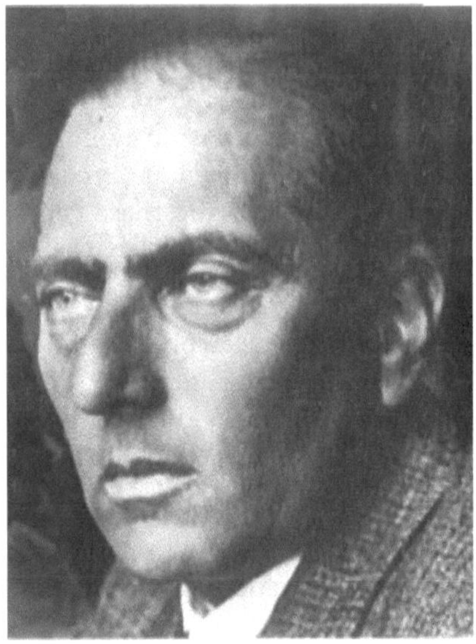

Figure 8.2: Hausdorff's friend and colleague Otto Toeplitz
Source: Stefan Hildebrandt, Peter Lax: Otto Toeplitz, *Bonner Mathematische Schriften*, Nr. 319, Bonn 1999.

Seminar printed in the course catalog. This was attended by advanced students, who presented lectures on special topics. From the winter semester of 1922 up until Study's retirement, these announcements read: "Prof. Study in association with Professors Hausdorff and Müller." Later Hausdorff co-directed seminars with Otto Toeplitz, and he always got along well with Müller.

Eduard Study retired at the end of the summer semester of 1927. No doubt the deliberations in the faculty over his successor set off more fierce disputes, but what transpired will never be known, since no documentary evidence survives. We only know that the process dragged on for a long time, so long, in fact, that the Minister threatened "to appoint a successor to Prof. Study even without [the faculty's report]." This threat was evidently effective, as one week later the list of candidates was complete. The faculty's first choice was a familiar name: Gerhard Kowalewski, Hausdorff's predecessor when he first came to Bonn as associate professor; in the meantime, he was now full professor at the Technical College in Dresden. For second place, the faculty named *pari passu* Hans Mohrmann, full professor at the University of Basel, and Otto Toeplitz, full professor in Kiel. Among these three, Toeplitz was certainly by far the most impressive, and the Ministry again took the liberty of passing over the faculty's first choice in order to appoint him.

Toeplitz was born in 1881 and came from a family of mathematicians: both his grandfather and father were high school mathematics teachers. He grew up in Breslau and studied mathematics there and in Berlin, taking his doctorate in 1905 with a thesis in algebraic geometry. Toeplitz then continued his studies with Klein and Hilbert in Göttingen, where he habilitated and became a Privatdozent in 1907. Six years later he gained an associate professorship in Kiel, where he was promoted to full professor in 1920. His appointment as Study's successor in Bonn began in the summer semester of 1928.

Toeplitz's most influential single work, still often cited today, is a barely two-page note on Fourier expansions published in the *Rendiconti del Circolo Matematico di Palermo* in 1911. There he showed that a real periodic function f is positive exactly when the quadratic forms $\sum_0^N a_{k-i} x_k \overline{x_i}$, $0 \leq i, k \leq N$ arising from its Fourier coefficients a_l are positive definite for all N. The matrices and corresponding operators that arise in this context are known today under Toeplitz's name; they play an important role in the theory of partial differential equations, statistical mechanics, and in X-ray crystallography.

Toeplitz continued to work on Hilbert's theory of linear integral equations, and with his friend Ernst Hellinger he elaborated on the theory of infinite linear, bilinear and quadratic forms and their associated infinite matrices. Their classic survey article on integral equations and equations with infinitely many unknowns was published in the German Encyclopedia of the Mathematical Sciences in 1927. The rich results it contains helped impart strong impetus to the emerging field of functional analysis, particularly after this older theory was translated into the language of abstract Hilbert space theory. In the 1930s, Toeplitz worked closely with Gottfried Köthe on certain infinite-dimensional non-normalizable coordinate spaces, so-called perfect spaces. This work constituted an important preliminary stage for the later development of the theory of locally convex spaces [Köthe 1969].

Toeplitz also had deep interests in the history of mathematics, especially the classical works of ancient Greek geometers. He conducted seminars on the history of mathematics in Kiel (together with Heinrich Scholz and Julius Stenzel) and in Bonn (with Oskar Becker and Erich Bessel-Hagen). Together with Stenzel and Otto Neugebauer, he founded the journal *Quellen und Studien zur Geschichte der Mathematik.* In teaching, he advocated the "genetic method," which traced the historical course of development of a particular subject. His book *The Calculus: A Genetic Approach* [Toeplitz 1963/2007] (*Die Entwicklung der Infinitesimalrechnung* [Toeplitz 1949], published by Köthe from his estate) was based on an elementary course taught according to this method. Together with Hans Rademacher, Toeplitz published the book *Von Zahlen und Figuren* [Rademacher/Toeplitz 1930]. As one of the best popular scientific books on mathematics, it was later translated into English (*The Enjoyment of Math* [Rademacher/Toeplitz 1966]) as well as numerous other languages. One of Toeplitz's main concerns was elevating mathematics education in schools and improving the quality of training for teachers. To promote this effort, he and Heinrich Behnke founded the journal *Mathematische Semester-Berichte*, which continues to flourish today.

Hausdorff was delighted by the Ministry's somewhat surprising decision to appoint Toeplitz, to whom he wrote:

> Having returned the day before yesterday from a journey, where I was cut off from time and news of it, I learn of your appointment and congratulate both you and me: you on your resettlement in a milder climate, and me, surrounded by geometers, on now being able to press an analytical breast against my own. I hope and believe that we will get along quite well. (*HGW* IX, 621)

Hausdorff's hopes were soon to be exceeded, as he and Toeplitz got along spledidly, as did their families, in part thanks to their mutual love for music-making. Toeplitz's wife Erna, *née* Henschel, was very musical; already at the age of 19 she had sung the part of Isolde in a private performance of Wagner's *Tristan und Isolde.* Their son Erich, who later became a professional musician, also performed with Hausdorff in his youth.

Toeplitz also hoped to establish a historical-didactic department within the Bonn Mathematical Seminar, and he succeeded in gaining an appointment for Erich Bessel-Hagen to support this effort. Bessel-Hagen had received his doctorate in 1920 under Constantin Carathéodory with a thesis on the calculus of variations. He then left for Göttingen, where from 1921 to 1923 he served as a private assistant to Felix Klein, who was then preparing his Collected Works in three volumes. After habilitating in Göttingen in 1925, he transferred to Halle, where he worked under Helmut Hasse.

While in Göttingen, Bessel-Hagen contributed not only to the Collected Works of Klein but also to the much larger edition of Gauss's *Werke.* Toeplitz was well aware of Bessel-Hagen's ability to work with diverse source materials when he brought him to Bonn, where they regularly held a seminar on the history of mathematics. Although no one could have anticipated the important role he would later play, Bessel-Hagen was in a unique position to preserve written materials bequeathed to him by various mathematicians. Thus, after Toeplitz's emigration to Palestine, he salvaged parts of his estate, just as he did after Hausdorff's death; among the latter's important documents was his farewell letter, partially reproduced in 9.4.4. In 1931, Bessel-Hagen was appointed associate professor in Bonn. Four years later, following Toeplitz's dismissal, he took over as head of the historical-didactic department of the Mathematical Seminar. He remained a loyal friend of the Hausdorff family to the last.

8.3 Hausdorff's Teaching Activity in Bonn

As described earlier in Section 7.5, Hausdorff carried a heavy teaching burden in Greifswald. During the war years, shortages in staff forced him to cover the introductory courses, leaving him little opportunity to teach more advanced material. In this respect, as in others, his appointment in Bonn brought about a significant improvement for his working conditions. Up until the winter semester 1932/33, Hausdorff offered regular cycles of introductory lectures courses. Two of these – differential and integral calculus and ordinary differential equations – he

Figure 8.3: Group Photo from the Annual Meeting of the German Mathematical Society in Leipzig, 17–24 September 1922
From left to right: Felix Hausdorff, Regina Schur, Issai Schur, Charlotte Hausdorff. Photo: Stella Pólya, *Pólya Picture Album*, Birkhäuser, Basel 1987

Figure 8.4: Felix Hausdorff in his Home Office in Bonn, 1924
Source: Hausdorff Estate, Capsule 65, Nr. 29

taught over four semesters, whereas function theory was covered in three. Alongside these courses in analysis, Hausdorff also occasionally taught a one-semester course on elementary number theory. This was a relatively light load of elementary lectures compared with the situation in Greifswald, giving him the opportunity to teach advanced courses, which often dealt with specialized topics at the frontiers of current research.

Hausdorff's teaching career spanned 78 semesters, during which time he prepared carefully handwritten manuscripts of nearly all his lectures. These form the lion's share of his estate in Bonn, and in most cases they are in a polished state, practically ready for publication. As such, they represent a treasure trove for documenting Hausdorff's teaching activity, covering his lecture courses from first to last. Such a complete and authentic documentary record for such an important mathematician is a true rarity. Of course, only a small selection from his more advanced courses can be described here.

In the winter semester of 1921/22, Hausdorff taught a four-hour course on "Set Theory and Theory of Real Functions." He began by following the exposition of the algebra of sets in the *Grundzüge*, while also incorporating more recent work by Suslin and Lusin, in particular Suslin's construction of a set system $\mathfrak{M}_s$ over a given set system $\mathfrak{M}$. Following this came general set theory (cardinal numbers, order types, ordinal numbers), after which he presented point set theory as an axiomatically grounded theory of metric spaces. This lecture course served as the basis for Hausdorff's textbook *Mengenlehre* (*Set Theory*) [H 1927a]. In fact, the contents of his manuscript were taken over almost completely, and in some places entire paragraphs from it appear practically verbatim in his book. Considering that Hausdorff's *Mengenlehre* was reprinted in 1991 (the 4th edition of the English translation), thus 70 years after he wrote the original manuscript, the significance of this lecture course can hardly be overstated. Indeed, it sheds considerable light on what Hausdorff's appointment in Bonn meant for the quality of teaching there.

Another striking example can be seen from a three-hour lecture course on probability theory that Hausdorff taught in the summer semester of 1923. We consider it here only from the foundational standpoint, in which respect his approach pointed far into the future. Because of its special historical interest, the entire manuscript appears in volume V of the Hausdorff edition (*HGW* V, 595–723). Its fundamental significance stems from the axiomatic treatment of probability on the basis of set theory and measure theory. Hausdorff gradually introduced his audience to the general theory via the special case of classical Laplacian probability. For the general theory, he set forth the following framework: Given a set M representing all possible cases (today called the set of elementary events), Hausdorff introduced $\mathfrak{M}$ as a Borel system of subsets of M (in modern terminology a σ-algebra of subsets of M). He then calls the sets $A \in \mathfrak{M}$ "measurable," and to each $A \in \mathfrak{M}$ he associates a real number $w(A)$ satisfying the following axioms:

1. $0 \leq w(A) \leq 1$.
2. $w(M) = 1$.
3. $w(A)$ is a σ-additive set function.

It then follows immediately that $w(\overline{A}) = 1 - w(A)$. The events A, B are then called *independent* if $w(A \cap B) = w(A) \cdot w(B)$.

When dealing with continuous cases, one cannot calculate probabilities by counting the ratio of favorable to possible cases (as, e.g., in the theory of measurement errors); instead, a distribution or density function must be introduced. Hausdorff dealt with such cases in his axiomatic system by means of an appropriate class of these functions. He thus showed that for every monotonically non-decreasing left-continuous function $F(x) : \mathbb{R} \to [0,1]$ with $F(-\infty) = 0$, $F(\infty) = 1$ there exists a σ-additive non-negative set function $w(A)$ on the σ-algebra generated by the intervals, such that $w([\alpha, \beta)) = F(\beta) - F(\alpha)$ for $-\infty < \alpha < \beta < \infty$. He also handled the case of $\mathbb{R}^2$ in such a way that the extension to the general case of $\mathbb{R}^n$ was obvious. However, he did not develop the foundations of measure theory, but rather took its results for granted, as he did with the theory of measurable functions and Lebesgue-Stieltjes integration theory, which one needs for the treatment of random quantities (the terms "random quantity" or "random variable" do not yet occur).

Probability theory stood in an awkward place toward the end of the nineteenth century. On the one hand, there was an already quite considerable body of theory – with the Law of Large Numbers and the Central Limit Theorem at its core – along with sophisticated analytical methods and numerous applications in error theory, statistics, physics, and insurance mathematics. On the other hand, the central concept of the theory, the notion of probability, was highly ambiguous. It was based on the intuitive idea of "equally possible cases," i.e. it referred to facts outside mathematics, for example to assumed symmetries in ideal dice or in games like roulette. Attempts to understand probability as an inner-mathematical concept inevitably led to a vicious circle: for what could "equally possible" mean other than "equally probable"? Hausdorff pointed out this circularity *expressis verbis* in his lecture (*HGW* V, 597–598). The successes in statistical mechanics, but also the logical inconsistencies occurring in this field and elsewhere, triggered a lively debate over the foundations of probability theory at the end of the century whose participants included philosophers and natural scientists as well as mathematicians. Thus, it seems hardly surprising that David Hilbert highlighted the need to axiomatize probability theory as the sixth among his famous list of 23 problems.

Hausdorff's lecture course provided an axiomatic foundation of probability theory, which coincided, except for its formal presentation, with the theory Andrei Kolmogorov published ten years later. Since Hausdorff's text remained unknown, however, historians date the birth of modern, axiomatic probability theory to the appearance of Kolmogorov's *Grundbegriffe der Wahrscheinlichkeitsrechnung* (*Fundamental Concepts of Probability Theory*) [Kolmogoroff 1933]. In its preface,

Kolmogorov himself noted that the measure-theoretic structure of probability theory "had been familiar in the relevant mathematical circles for some time." Still, he was the first to deal with probabilities in function spaces, while developing the theory of conditional probabilities and conditional expectations.

In the summer semester of 1925, Hausdorff offered a 2-hour lecture course on "Divergent Series," drawing on his groundbreaking work on this topic (see Subsection 7.8.4). After an historical introduction, he first dealt with summation methods based on studies of power series, such as those of Abel and Borel. He then turned to a general discussion of matrix methods, which Hausdorff calls "discrete limit methods." Central aspects included the Toeplitz Permanence Theorem and investigations of the field of action of a summation method. After detailed treatment of the Cesàro and Hölder methods with applications to Fourier series, Hausdorff discusses "methods exchangeable with C^1" and his own theory from "Summation Methods and Moment Sequences I" [H 1921]. He also takes up further results on limit theory from recent work of other authors. The final topics in the course concerned inverse theorems of Tauberian type as well as remarks on two procedures of Borel. Although his manuscript for this course contained no important new results, its publication certainly would have had a desirable effect, since no comparable overview of current methods in this field was then available.

Hausdorff taught Divergent Series again in the winter semester of 1933/34, at which time he reworked his earlier lectures. In this new version, the theory of matrix methods was already clearly placed within the context of functional analysis. Certain key words and concepts in his course manuscript illustrate this more abstract approach, e.g.: vector space of sequences; linear functionals; a general concept of limiting method, permanent methods; continuation theorem of Hahn-Banach with implications for the existence of permanent procedures; Toeplitz's Permanence Theorem. What was said above about the unfortunate circumstance that Hausdorff neglected to publish his earlier text of lectures applies a fortiori to this more mature and updated version.

Hausdorff's interest in topics now classified as belonging to the field of functional analysis date back much earlier. Already in the summer semester of 1914, when he began teaching in Greifswald, he gave a two-hour course on integral equations, which covered large parts of what was then known about the theory of operators in infinite-dimensional spaces. He repeated that course in Bonn in the winter semester of 1923/24 and offered a completely revised version of it a decade later. One can see from the manuscripts in Hausdorff's estate that from about 1926 onward he dealt more and more with topics from functional analysis. During the summer semester of 1931 he taught a three-hour course, entitled "Point Sets," that had a strong analytic flavor. After presenting two introductory sections on metric spaces and injective mappings between them, he began a lengthy treatment of the "Theory of linear metric spaces." In it, he gave an exposition of the fundamental concepts and results of linear functional analysis, an area then just emerging as a new field. These lectures also served as the starting point for Hausdorff's paper "Zur Theorie der linearen metrischen Räume" [H 1931], a work not without influence on developments in functional analysis, as will be discussed in 8.6.5.

Two years later, during the summer semester of 1933, Hausdorff taught a two-hour "Introduction to Combinatorial Topology." This course is particularly remarkable when compared with contemporary monographs and treatises, because it contains a mature and conceptually clear introduction to homology theory. Moreover, the tools needed for building the theory of finitely generated abelian groups are developed within the course itself. Up until the mid-twenties, Hausdorff had been skeptical of combinatorial topology, which he regarded as resting on insecure foundations. After exchanging ideas with Paul Alexandroff, however, he became increasingly interested in the field. Due to its historical importance, Hausdorff's manuscript for these lectures was printed in its entirety in (*HGW* III, 893–953). Erhard Scholz's introduction to the text discusses in detail how Hausdorff turned to this older field, as reflected both in his critical views on earlier works, but also in the approaches he took in his lectures as well as in a number of studies available in the estate. Excerpts from two of these other studies are printed in the edition (*HGW* III, 954–980).

Unfortunately, there are very few testimonials from former students that shed light on Hausdorff as an academic teacher. A relatively detailed account comes from Magda Dierkesmann, who studied in Bonn in the late twenties and early thirties. In her article, "Felix Hausdorff. Ein Lebensbild" [Dierkesmann 1967], she wrote:

> Felix Hausdorff was a man who exuded great modesty and charm, though he was always held in high esteem and respect by his students, an effect that only a great mind, a personality of extraordinary prominence, can exert on those around him, because it is unconscious and unintentional. [···] He spoke very softly, thereby forcing his listeners to pay strict attention. Speaking altogether freely, his head slightly bent forward, a piece of chalk in his hand, he walked up and down in front of the blackboard writing down the most important things in small letters. The reason why his lectures nevertheless exerted such an intense fascination was something I only understood in later semesters. The captivating thing was that he presented rigorous science in an artistic way. He combined the unconditional clarity of mathematical thinking with artistic ease, surprising us over and again when he presented the solution of profound mathematical problems with consummate mastery. [Dierkesmann 1967, 52–53]

Considering the print-ready quality of his lecture manuscripts, it seems astonishing that Hausdorff lectured freely and did not use these texts when he taught his courses. Magda Dierkesmann's recollection of Hausdorff's lecture style was, in fact, directly confirmed by his former student Günter Bergmann in a conversation from 1995. Bergmann recalled how Hausdorff entered the lecture hall, put the folder containing his manuscript on a desk, and then spoke freely for the entire hour. Afterward, he picked up the folder again and left the lecture hall, never even glancing at his manuscript.

Hausdorff had only three doctoral students. This may seem somewhat surprising for such an active scholar, whose courses introduced students to various areas of current research. The reasons for this small number were, for the most part, personal. On the one hand, a dissertation should make some contribution to a field of mathematical research, and to mentor work of a lower standard, as Hans Beck did in many cases, would have been out of the question for Hausdorff. On the other hand, he felt responsible for a doctoral student and feared nothing more than failure should a topic prove too difficult or insufficiently fertile.

External circumstances also certainly played a role. The situation in Greifswald during the war years and post-war period hardly made it possible to supervise dissertation projects. Bonn also had relatively few students who were interested in research or aspired to an academic career when compared with the larger mathematical centers, especially Göttingen and Berlin. With three full professors to choose from, the more ambitious mathematics students in Bonn had ample opportunities to obtain suitable dissertation topics, and word surely spread that with Beck the hurdles were hardly high. The following data are, indeed, suggestive. During Hausdorff's second period in Bonn, from 30 June 1921 until the end of 1936, 37 students took doctorates in the field of mathematics, 2 of them still under Hans Hahn. Of the remainder, 16 took their degrees under Beck (two of them supervised by Weiss), 8 studied with Toeplitz, 7 with Study, 3 were students of Hausdorff, and one was a pupil of Bessel-Hagen.

Hausdorff's first doctoral student was Karl Bögel, who had earlier studied in Tübingen, Stuttgart, and Göttingen before becoming a mathematics teacher at a Gymnasium. He was in his thirties when he published a paper on real functions in *Mathematische Annalen*, in which he proved a generalization of a theorem of Baire. His dissertation dealt with a similar topic; Hausdorff accepted it in 1924 with a grade of "very good." Bögel continued his work in the teaching profession, during which time he published three more papers related to his dissertation topic as well as eleven on topics in mathematics education. In 1935, National Socialist authorities forced him to relinquish his teaching position. Probably he then took a job in industry or some form of applied research, because after the war he was conscripted to do scientific work in the Soviet Union, where he remained until 1953. After returning to East Germany at the age of 66, he was nevertheless appointed full professor of mathematics at the newly founded College of Electrical Engineering in Ilmenau (Thuringia). There, as Dean, he rendered great service in building up its mathematics department, before retiring two years later as professor emeritus.

Gustav Steinbach spent his entire university education in Bonn. He received his doctorate from Hausdorff in 1930 with a thesis on set theory, a work that Hausdorff also graded "very good" and which was described in some detail in the abstracting journal *Jahrbuch über die Fortschritte der Mathematik*. Its focus was mainly on descriptive set theory, following work by Lusin. Steinbach also

Figure 8.5: Felix Hausdorff in front of the main building of Bonn University, March 1932. Photo taken by Erna Bannow, one of Hausdorff's students.

considered various versions of "Baire's condition" and their mutual relations for arbitrary spaces (in separable spaces these are all equivalent). In preparing the second edition of his monograph *Set Theory* [H 1935a], Hausdorff added a section on "The Baire condition" in which he cited Steinbach's results [H 1935a, 304; III: 348].

Hausdorff's third and last doctoral student was Franz Hallenbach. He had studied in Cologne, Göttingen and Bonn, taking his doctorate in July 1933 with a dissertation on limiting methods applied to double sequences. It seems likely, judging from documents in the estate, that Hausdorff assigned this rather difficult topic because of his own interest in the problem. He left behind 151 pages dealing with limiting theory and the moment problem for double sequences. Hausdorff also provided considerable assistance to Hallenbach in writing his dissertation, in which he succeeded to a large extent in transferring Hausdorff's limiting theory from *Summation Methods and Moment Sequences* to double sequences. Hausdorff gave this thesis the grade "good." It received an unusually detailed summary in the *Jahrbuch* and was later cited by various authors.

Franz Hallenbach subsequently took up a career in geophysics, joining the Prussian Geological Survey in 1936. His specialty was geoelectric investigations for projects in hydrogeology. After the war, he rose to the position of head of the Geoscientific Joint Tasks of the Federal Republic. In this position, he also held lectureships at the University of Bonn. Hallenbach published a number of important papers on applications of geophysics, especially for efforts in water exploration.

8.4 Reception of Grundzüge der Mengenlehre

The initially very weak response to Hausdorff's *Grundzüge*, as reflected in the dearth of serious reviews of the book, has already been described in 7.6.5. There we also highlighted the single exception, a review by Henry Blumberg (*HGW* II, 844–855) that only appeared in December 1920, but which stands out as particularly rich in content and highly astute in its appraisal of the work. Three quotations from this review may here serve to introduce our account of the belated, but all the more lasting reception of Hausdorff's *opus magnum*.

Blumberg began as follows:

> If there are still mathematicians who hold the theory of aggregates under general suspicion, and are reluctant to grant it full recognition as a rigorous, mathematical discipline, they will find it hard to retain their doubts under fire of the logic of Hausdorff's treatise. It would be difficult to name a volume in any field of mathematics, even in the unclouded domain of number theory, that surpasses the Grundzüge in clearness and precision. (*HGW* II, 844)

After summing up the contents of the chapters on general set theory, Blumberg described the topological and analytic chapters in greater detail, beginning with this remark:

> The remaining chapters of the book (VII–X) will prove of more general interest because they are concerned with the applications of the abstract theory to the study of space relations. It is in these chapters especially that Hausdorff impresses you with his masterful exposition. The theory of point sets is cast into a new and more general mold, and the resulting treatment is characterized throughout by originality, naturalness, and beauty. (*HGW* II, 849)

At the end of the review, Blumberg mentioned a few misprints and minor oversights and noted that the only thing one might actually criticize had less to do with the book itself than its potential impact on future readers. Blumberg imagined its high quality could evoke more admiration than motivation, thus suggesting that Hausdorff's presentation was so complete that others might not feel inclined to pursue work of their own. About this he then added: "But such remonstrance would be like quarreling with Beethoven for having written symphonies instead of operas" (*HGW* II, 853).

Blumberg's perhaps not entirely serious concern turned out to be completely unfounded. From the beginning of the 1920s, in fact, a new generation of mathematicians picked up on numerous suggestions contained in Hausdorff's *Grundzüge*, particularly in topology, a field of research that soon moved to center stage during the next two decades. Among these mainly younger researchers, a great many came from Poland, where set theory, general topology, and functional analysis flourished during the 1920s and 30s. Among those who contributed to these related fields were Stefan Banach, Karol Borsuk, Witold Hurewicz, Zygmunt Janiszewski, Bronisław Knaster, Kazimierz Kuratowski, Stefan Mazurkiewicz, Wacław Sierpiński, Hugo Steinhaus, Alfred Tarski and Stanislaw Ulam. The upsurge of interest in topology was also strongly represented by Russia's younger generation of mathematicians, including Paul Alexandroff, Lev Pontryagin, Lev Tumarkin, Andrei Tychonoff and Paul Urysohn. Among German-speaking mathematicians, leading topologists and functional analysts included Hans Hahn, Karl Menger and Leopold Vietoris from Austria, together with Heinrich Tietze in Germany. During these years, reception of the *Grundzüge* was thus particularly strong in Poland and Russia, whereas only a few mathematicians from Hausdorff's own country were deeply influenced by his book.

A special role in the reception of Hausdorff's ideas was played by the journal *Fundamenta Mathematicae*, founded in Poland in 1920 [Duda 1996]. Already in 1917, Zygmunt Janiszewski wrote a memorandum with suggestions for how mathematics in a newly emergent Poland could attain international prominence. He proposed concentrating the future nation's intellectual forces on research in a promising new modern area, supported by a special journal devoted exclusively to work in related fields. Two years later, when Janiszewski, Waclaw Sierpiński and Stefan Mazurkiewicz were united as professors at the newly resurrected Warsaw University, they began putting these plans into place; they chose set theory and its applications as the locus for this future research area, which would be showcased in a new journal entitled *Fundamenta Mathematicae.* From its founding by Sierpiński and Mazurkiewicz[2] as one of the first specialized mathematical journals in the world, *Fundamenta Mathematicae* soon attained an international reputation, especially in the field of general topology, but in many related fields as well. These included set theory proper, theory of real functions, measure and integration theory, functional analysis, logic and foundations of mathematics. From the first volume onward, the journal's readers were exposed to Hausdorff's *Grundzüge* with remarkable regularity. Of the 558 papers (omitting three by Hausdorff himself) published in the first 20 volumes, from number 1 (1920) to 20 (1933), no fewer than 88 cited his *Grundzüge.* Moreover, Hausdorff's terminology and conceptualizations had become so commonplace that many other authors used them without explicitly mentioning his name. The most frequent citations refer to concepts and results from the topological part of the *Grundzüge.*

Hausdorff's relations with two talented Russian topologists were particularly close. Through his extant correspondence with Paul Alexandroff, both before and after the early death of his friend Paul Urysohn, we are well informed about how

[2]Janiszewski had succumbed to the then rampant viral influenza, the Spanish Flu, shortly before the first volume was published.

this friendship evolved. Around the time these two young Russian mathematicians first met in Moscow, they undertook a thorough study of the *Grundzüge* (*HGW* IX, 3–134). In a first letter from Alexandroff and Urysohn to Hausdorff, dated April 18, 1923, they made an oblique reference to this: "For quite a long time we have been wanting to communicate to you the results we have obtained in the theory of topological spaces created by you" (*HGW* IX, 5). After this, they delivered a long string of results on compact and locally compact spaces. By the end of 1923, they had together completed a summary account of all these findings, though intervening events prevented their publication until several years later [Alexandroff/Urysohn 1929]. At the end of their letter to Hausdorff, the two Russians asked if they could visit him at a suitable opportunity. Political circumstances made this impossible in 1923, however, because foreigners were not allowed to enter the area around Bonn, which was then under French occupation.

One year later, though, when Alexandroff and Urysohn returned again to Göttingen in the summer, they had a chance to visit Hausdorff in his home for a few days. Afterward they spent a week with Brouwer in Blaricum before going on to France, where they planned to spend August in the village of Batz-sur-Mer on the Atlantic coast. In a letter dated August 3, 1924 from their vacation spot, they thanked Hausdorff "for the so kind and gracious reception we received in your home," adding that they would cherish the memory of their time together. In the meantime, they informed Hausdorff that Urysohn had succeeded in constructing a universal separable metric space, i.e., a separable metric space into which any such space can be embedded. This announcement immediately inspired Hausdorff to search for such a construction, and only eight days later, he wrote them back in Batz-sur-Mer:

> Your communication about the universal metric separable space, which Mr. Urysohn has constructed, interested me very much and, since you wrote nothing more about it, served to challenge me to find such a space myself. (*HGW* IX, 22)

Then he briefly shared his construction of such a universal space, noting that he "found it only yesterday" (first published from his estate in *HGW* III, 762–765).

This idyllic vacation in Batz-sur-Mer ended tragically for the two Russians. While bathing in the Atlantic Ocean on August 17, 1924, Urysohn was caught by a large wave that threw his head against a rock. Alexandroff swam out to save his friend, but he could only recover his dead body. One day later, he informed Hausdorff about the accident. Deeply saddened, Hausdorff replied on August 23 in a sensitive and touching letter:

> With a shudder before the senselessness of fate, I hold the letter in my hands that Urysohn wrote to me on August 16, probably the last letter of his life, written one day before his death. I wanted to answer him that his proof of the theorem on the metrizability of normal spaces with the second axiom of countability is so beautiful and simple and that he had

> thus far surpassed [my] long and complicated work on the metrization of compact spaces that I no longer consider my proofs worth publishing. (*HGW* IX, 25–26)

In fact, Hausdorff never published his certainly interesting results on the problem of metrization. They have been preserved in his estate and are now printed in (*HGW* III, 755–758).

Alexandroff subsequently spent many months of efforts preparing Urysohn's posthumous results for publication. On September 2, 1924, shortly before returning to Moscow, he wrote to Hausdorff:

> You will be truly amazed at how deeply he was able to penetrate into the most hidden secrets of topological space structure when you cast your eyes on his principal work "Mémoire sur les multiplicités Cantoriennes." (*HGW* IX, 28)

This work [Urysohn 1925/1926], which later appeared in two parts in *Fundamenta Mathematicae*, could have easily filled a book. In it, Urysohn introduced a theory of continua in compact metric spaces along with the notion of small inductive dimension, later named after him and Karl Menger. He relied heavily on the *Grundzüge* throughout, as seen from the fact that he cites Hausdorff's book no fewer than 60 times in his text.

In the years afterward, a strong friendship developed between Hausdorff and Alexandroff, despite their difference in age (the Russian was 28 years younger). Alexandroff spent considerable time in the West during the twenties. Except for the year 1929, he returned to Göttingen every summer, often teaching as a guest lecturer. He also enjoyed lengthy visits with L.E.J. Brouwer in Blaricum while lecturing in Amsterdam, and on two occasions stayed for a few months in Princeton. He usually went to France for summer vacations, and whenever possible he visited Hausdorff in Bonn for a few days. After the first trip to Bonn with Urysohn in August 1924, he made at least three subsequent visits: in November 1925, August 1927, and late October 1930; in the fall of 1932 they also met at Lake Maggiore, where both were vacationing.

After Urysohn's death, his professorship was transferred to Alexandroff, who soon began to introduce a circle of students in Moscow to general topology in the spirit of Hausdorff's *Grundzüge*. As an enthusiastic and talented teacher, Alexandroff held high hopes that his efforts would bear fruit. Writing to Hausdorff about this promising group, he thought it likely that from them "a certain number of new workers in our field might arise" (*HGW* IX, 30). These hopes were eventually fulfilled, as the Moscow school of topology gradually took form. Although he never visited Moscow, Hausdorff might well be regarded as the "spiritual father" of this school [Purkert 2015a]. At any rate, Alexandroff always emphasized how important *Grundzüge der Mengenlehre* was for him.

In 1926, Hausdorff was putting the final touches on his *Mengenlehre* (*Set Theory*), which he and others called the second edition of the *Grundzüge*, even though this was actually an entirely new book. While he was in Göttingen, Alexandroff received the proofs along with a request from the author for his critical remarks. On May 13, 1926, he wrote back to Hausdorff:

> I thank you again very much for giving me the opportunity to become one of your first readers. By the way, I notice when presenting my lectures in Göttingen that I am already quoting from your first edition by heart (which is how good conductors conduct, e.g., Beethoven's symphonies even without a score!). (*HGW* IX, 42)

When it came to the readership of his *Grundzüge*, Hausdorff was certainly very well aware of those countries whose mathematicians were responding with the greatest enthusiasm. He also hoped that his *Mengenlehre* would find a similarly responsive readership. Shortly after it appeared, he wrote to Alexandroff on May 29, 1927 (not without a strong dose of self-irony): "Hopefully Russia and Poland will swallow immense quantities of my book, so that I can impress my publisher!" (*HGW* IX, 51). Meanwhile in Moscow, Alexandroff had founded a Topological Society, whose members unanimously elected Hausdorff as an honorary member; this pleased him very much when he heard about it. Russian mathematicians actually helped to spread awareness of Hausdorff's *Grundzüge* in Weimar Germany, as can be seen from several articles that appeared in *Mathematische Annalen*. Of the 34 papers published in volumes 83 (1921) through 106 (1932) that cite the *Grundzüge*, 16 were written by Russians.

Alexandroff's letters to Hausdorff often mention results obtained by members of the Moscow School, but he also wrote about his own larger vision for topology. His goal was to create a synthesis joining general topology with the older branches of the subject. As he put this in a letter to Hausdorff from July 4, 1926, he hoped "to fill the deep chasm existing up to now between general (set-theoretic) and classical topology" (*HGW* IX, 44). In pursuing this objective, he struck up a collaboration with Heinz Hopf, whom he had met in Göttingen. Alexandroff had virtually internalized Hausdorff's *Grundzüge*, but he also worked closely with Brouwer and, together with Hopf, he spent the winter of 1927/28 with Alexander, Lefschetz, and Veblen in Princeton. During the summer afterward, Alexandroff began work on a monograph that aimed to overcome the "deep chasm." He was joined by Hopf, with whom Alexandroff had held a "quite extensive topological seminar" that summer before departing for Batz (*HGW* IX, 69). Alexandroff described this work again and again in his letters to Hausdorff. Yet it took many more years before Alexandroff and Hopf were able to complete the first volume, and *Topologie I* [Alexandroff/Hopf 1935] could finally appear.

Originally, the authors had planned on publishing three volumes, but the others never materialized due to increasing restrictions on travel and deteriorating relations between Stalin's Soviet Union and Western countries. Hausdorff, already experiencing difficult times himself, read a large portion of the manuscript in page proofs. In a letter from March 9, 1935, the last in their preserved correspondence, Alexandroff wrote

> The two authors are really extraordinarily grateful to you for your participation in reading the proofs. I am very pleased that you find at least the first chapters good. The other chapters have undergone a thorough reorganization since the first stage of corrections; I very much hope that you will like these chapters better in the second stage than in the first! In particular, the former chapter VI, though more in its second half, has been completely reworked. (*HGW* IX, 131)

The textbook by Alexandroff and Hopf exerted a great influence on subsequent topological research, even if the "deep chasm" Alexandroff had spoken of could not be overcome. This remained a challenge for decades to come.

Not surprisingly, Hausdorff's *Grundzüge* eventually received due acknowledgment in the textbook literature on set theory and general topology. A.A. Fraenkel's *Einleitung in die Mengenlehre* (*Introduction to Set Theory*) [Fraenkel 1919] was the first important account from this time. In the preface, he contrasted his elementary book with more demanding studies:

> Whoever wants to become thoroughly acquainted with the set-theoretical building from the mathematician's point of view must not, therefore, be content with the present little book, but must reach for one of the works listed at the end (preferably Herr Hausdorff's "Grundzüge"). [Fraenkel 1919, IV].

In 1923, Fraenkel published a second revised edition of his book, which appeared in Springer's yellow series [Fraenkel 1923]. Compared with the first, it treated the basic questions in more detail and, in particular, presented Zermelo's axioms for set theory together with Fraenkel's own additions. The above quotation in the preface was also repeated. Fraenkel brought out his third edition in 1928, again expanded and updated. Although this edition can be considered a comprehensive textbook on general set theory, Fraenkel noted in it: "For penetrating studies, the excellent textbook by Hausdorff alone comes into consideration."

Also in 1928, Erich Kamke's *Set Theory* appeared in the Göschen collection, which aimed to provide short introductory material for students. Accordingly, this little volume presented general set theory along with some elements of point set theory in $\mathbb{R}^n$. For deeper studies in the whole field of set theory, Kamke recommended Hausdorff's *Grundzüge* and his *Mengenlehre* together with Fraenkel's book for the axiomatic foundations. This booklet served its purpose very well; after the Second World War, it went through six more editions.

Among the textbooks on general topology, we shall only mention the single most important one: Kazimierz Kuratowski's *Topologie I* [Kuratowski 1933]. In the preface, Kuratowski mentions works from which he benefited in writing his own, first Hausdorff's *Grundzüge* and *Mengenlehre*, then [Sierpiński 1928/1934], and finally [Fréchet 1928a]. Sierpiński was Kuratowski's teacher, but after him Hausdorff received by far the most citations in [Kuratowski 1933]. *Topology I* was without doubt the most influential book on general topology of the "second generation"; after 1933, it was increasingly cited in articles published in *Fundamenta*

Mathematicae. In a paper on Kuratowski's mathematical work, Ryszard Engelking called Kuratowski's *Topologie I* "a masterpiece of mathematical literature" [Engelking 1998, 449]. The impact of Kuratowski's classic was due in no small part to the fact that he showed the manifold relations that tied general topology to measure theory, real functions, descriptive set theory, and functional analysis, thereby arousing interest in general topology among wide circles of mathematicians.

Although Hausdorff did not live to witness the postwar reception of the *Grundzüge*, it soon attracted influential readers in both France and the United States. General topology occupied a central place in Bourbaki's structural conception of modern mathematical theories, a viewpoint already made evident in Hausdorff's book. When Chelsea Publications began reprinting German mathematical textbooks in New York, North American mathematicians gained the chance to own a copy of *Grundzüge der Mengenlehre*, which came out in 1949 and again in 1965.

In closing these remarks on the longer-term reception of the *Grundzüge*, we cite an assessment from 1972, thus 60 years after Hausdorff first began work on his masterpiece. That same year also marked the 30th anniversary of his tragic death. Both events were the occasion for publication of an anthology, *Theory of Sets and Topology. In Honour of Felix Hausdorff (1868–1942)*, whose 40 authors included not only young researchers but also a number of prominent senior mathematicians from several countries [Asser/Flachsmeyer/Rinow 1972]. In the preface, the editors quoted from a lengthy passage in Hausdorff's *Grundzüge* before commenting:

> Whole generations of young mathematicians have gained essential knowledge and inspiration from this book, and it is to be hoped that even today many students may reach for this work, which is of a timeless modernity. It is fair to say that the "Grundzüge der Mengenlehre" and the individual works of Hausdorff had a decisive influence on the development of set theory, general topology, measure theory, and thus on mathematics as a whole. [Asser/Flachsmeyer/Rinow 1972, 6]

8.5 Hausdorff's Mengenlehre

By mid-1923 Hausdorff's *Grundzüge* was sold out, and in the meantime its publisher, Veit & Co. in Leipzig, belonged to the firm of Walter de Gruyter in Berlin. In 1921, De Gruyter launched a new series, "Göschens Lehrbücherei," a textbook library with volumes ranging from 260 to 320 pages. When the publisher approached Hausdorff in mid-1923 about preparing a new edition of the *Grundzüge* for this series, it was clear that considerable cuts would need to be made (the first edition ran to 476 pages). Hausdorff agreed to this and signed a contract that called for him to submit the finished manuscript by the end of 1924. As it turned out, however, this "second edition" did not appear until 1927 [H 1927a, III: 41–351]. What probably accounts for this delay was Hausdorff's decision not to simply shorten the text and here and there smooth some transitions, but rather to write an entirely new book.

In the preface to this so-called second edition of the *Grundzüge*, Hausdorff explained why he felt it was necessary to completely revise the earlier version. He omitted large parts of the theory of ordered sets, which he himself had created, as well as the material on Euclidean spaces. Furthermore, he decided to sacrifice completely the final chapter on the theory of measure and integration, since "there is no lack of other presentations." Probably the most serious change, however, concerned the abstract theory of topological spaces presented in the *Grundzüge*, about which Hausdorff wrote:

> More than these deletions will perhaps be regretted that, in order to save further space in point set theory, I have abandoned the *topological* standpoint, through which the first edition apparently has won many friends, in order to confine myself to the simpler theory of *metric* spaces, for which the cursory survey (§40) of topological spaces is not a sufficient substitute. [H 1927a, 5–6; III: 45–46]

Although only six pages long, this §40 is nevertheless of special interest. There Hausdorff considers different axiomatic possibilities for imposing a topology on a set. Following this approach, any one of the following undefined basic terms can be chosen: open set, closed set, closure or interior of a set, neighborhood of a point. He then sketches how a system of separation axioms with ever more stringent conditions leads to a hierarchy of topological spaces with different separability properties. These included the axioms introduced by Fréchet and Hausdorff (later called T_1, resp. T_2), as described already in 7.6.2, as well as the axiom for so-called normal spaces, which requires that disjoint closed sets can be separated by disjoint open sets containing them. As Hausdorff remarks, metric spaces are always normal. Although Kuratowski had already adopted a similar axiomatic approach, Hausdorff's §40 was historically important: it gave the first textbook presentation of axiomatic methods for classifying topologies based on their separability properties. Previously, this approach could only be found in various forms scattered throughout the research literature, whereas after the appearance of Hausdorff's *Mengenlehre* this became the canonical approach taken in nearly all textbooks on general topology.

Hausdorff knew from Alexandroff's letters that the latter was unhappy about the suppression of material on general topological spaces; nor was his the only voice of regret. Arthur Rosenthal, in a review for the *Deutsche Literaturzeitschrift* wrote:

> Among what has now been omitted are some things one wishes not to miss. Precisely those theories, for which we are indebted to H. himself, have been eliminated in the 2nd edition; namely, the higher parts of the theory of ordered sets and – what is especially regrettable – the theory of topological spaces created by H. in the 1st edition, which so many

Mengenlehre

Von

Dr. F. Hausdorff
o. Professor der Mathematik
an der Universität Bonn

Zweite, neubearbeitete Auflage

Mit 12 Figuren

Walter de Gruyter & Co.
vormals G. J. Göschen'sche Verlagshandlung
J. Guttentag, Verlagsbuchhandlung — Georg
Reimer — Karl J. Trübner — Veit & Comp.
Berlin W 10 und Leipzig
1927

Figure 8.6: Title Page of Hausdorff's Monograph from 1927 *Mengenlehre*

other mathematicians have pursued in the meantime. [Rosenthal 1928, 294].

Yet Hausdorff did not merely want to abridge his earlier book; nor was it "space saving" alone that induced him to concentrate on metric spaces. As one textbook among many in Göschen's Lehrbücherei, his *Mengenlehre* aimed to introduce readers to the most useful topics in general topology. The theory of metric spaces had been intensively studied during the 1920s, having already found applications in many varied branches of mathematics. For his new text, he naturally wanted to include important advances in the field of set theory made since the appearance of the *Grundzüge*. Such progress occurred primarily in two areas:
1) axiomatic foundations of set theory, as first set forth earlier by Ernst Zermelo, had been refined and completed by A.A. Fraenkel and Thoralf Skolem;
2) developments in the field later called descriptive set theory (Alexandroff, Hausdorff, Lusin, Suslin, Sierpiński, Kuratowski and others).

Problems connected with the axiomatic foundations of set theory were never particularly close to Hausdorff's heart, so he was happy to leave this to others. He even expressed a sense of relief to Fraenkel after the appearance of the latter's books [Fraenkel 1919], [Fraenkel 1923]. Matters stood altogether otherwise, on the other hand, when it came to descriptive set theory. In his *Mengenlehre*, Hausdorff endeavored to incorporate all the important progress made in this field, especially that of the Russian school around Lusin. The overwhelmingly positive response to Hausdorff's book was thus largely due to the fact that it gave *the first monographic presentation of the then current state of descriptive set theory.*

In the following we give a short overview of the contents of Hausdorff's book. Its first four chapters present, in somewhat more concise form than in the *Grundzüge*, general set theory in the tradition of Cantor, Dedekind, Bernstein, Zermelo, and Hessenberg (Hausdorff also mentions these authors in his references to Chapters 1–4). From the standpoint of modern axiomatic set theory, Hausdorff's naive theory corresponds to an extension of the system **ZFC** (Zermelo-Fraenkel plus the axiom of choice) by the addition of primal elements, a theory equivalent to **ZFC**.

Chapter 5 presents "Set Systems," where a *set system* $\mathfrak{M}$ is simply a set of sets. One then studies the smallest or largest set systems which remain closed with respect to certain algebraic operations on sets, e.g., rings or fields of sets and the σ- and δ-systems, as already presented in *Grundzüge*. The Borel system $\mathfrak{B}$ over a given set system $\mathfrak{M}$ is the system $\mathfrak{M}_{(\sigma\delta)}$. Thus, here Borel's hierarchy is considered purely from the standpoint of the algebra of sets, whereas in the *Grundzüge* $\mathfrak{M}$ was the system $\mathfrak{G}$ of open (or $\mathfrak{F}$ of closed) sets of, say, a metric space.

Hausdorff's most significant innovation in this chapter was the introduction of the δs-operation (also later called the Hausdorff operation). A δs-operation assigns to each sequence $\{M_i\}_{i=1,2,\cdots}$ of sets $M_i \in \mathfrak{M}$ a set X as follows: Let N be a set of increasing sequences $\nu = (n_1, n_2, \cdots)$ of natural numbers. For each such sequence ν, one constructs the intersection

$$M_\nu = M_{n_1} \cap M_{n_2} \cap M_{n_3} \cap \cdots$$

of the sets associated with ν from the set sequence $\{M_i\}$. Building from the (in general, uncountable) set N, one then forms the set

$$X = \Phi_N(M_1, M_2, \cdots) = \bigcup_{\nu \in N} M_\nu \, .$$

Hausdorff calls this operation a δs-function (later usually called δs-operation), because it first takes a countable intersection and then an uncountable union. He then proves the following theorem:
For every ordinal number $1 \leq \xi \leq \omega_1$ there exists a set N_ξ of sequences of increasing natural numbers such that the δs-function

$$X = \Phi_{N_\xi}(M_1, M_2, M_3, \cdots), \quad M_i \in \mathfrak{M}$$

runs through exactly the Borel sets of class $\mathfrak{B}^\xi$ generated by $\mathfrak{M}$.

In the final section of Chapter 5, Hausdorff deals with Suslin sets, a major topic studied in the Russian school around Nikolai Lusin. The Russians called these analytic sets (*ensembles analytiques*) or A-sets defined by the so-called A-operation. In his book, Hausdorff chose to name them after the discoverer of these sets, Michael Suslin, who died at age 25 in 1919 [Lorentz 2001]. An A-operation starts from the set $\mathbb{N}^\mathbb{N}$ of all sequences $\nu = (n_1, n_2, \cdots)$ of natural numbers (so the n_i need not be pairwise distinct as before). Given a sequence ν, one assigns to the finite sections $(n_1), (n_1, n_2), (n_1, n_2, n_3), \ldots$ the sets $M_{n_1}, M_{n_1,n_2}, M_{n_1,n_2,n_3}, \cdots \in \mathfrak{M}$. One then forms

$$M_\nu = M_{n_1} \cap M_{n_1,n_2} \cap M_{n_1,n_2,n_3} \cap \cdots \, .$$

If for each sequence ν the $M_{n_1}, M_{n_1,n_2}, \cdots$ are taken from the system $\mathfrak{M}$ in all possible ways, then the sets

$$X = \bigcup_{\nu \in \mathbb{N}^\mathbb{N}} M_\nu$$

form the system $\mathfrak{M}_S$ of all Suslin sets generated by the system $\mathfrak{M}$. Hausdorff notes that the A-operation is also a δs-operation. He then shows $\mathfrak{M}_{(\sigma\delta)} \subseteq \mathfrak{M}_S$ and $(\mathfrak{M}_S)_S = \mathfrak{M}_S$. His main theorem states the following:
There exists a fixed set N of sequences $\nu = (n_1, n_2, \cdots)$ of increasing natural numbers such that the δs-function

$$X = \bigcup_{\nu \in N} M_\nu = \Phi_N(M_1, M_2, \cdots), \quad M_i \in \mathfrak{M}$$

exactly represents the Suslin sets generated by $\mathfrak{M}$.

Hausdorff's Chapter 6, "Point Sets," corresponds to Chapters 7 and 8 in the *Grundzüge*, but here restricted to the case of metric spaces rather than topological T_2-spaces (Hausdorff spaces). This restriction makes some things much simpler, of course. For example, in the *Grundzüge* Hausdorff investigated T_2-spaces that satisfy either the first or second axioms of countability, the first of which holds in every metric space. Those that satisfy the second countability axiom are precisely the separable metric spaces, to which Hausdorff devotes a special section in Chapter 6.

In the *Grundzüge*, Hausdorff introduced a topology on the set of nonempty, closed and bounded subsets F of a metric space X by means of the so-called Hausdorff metric, denoted by $\mathbf{F}(X)$. These represent special types of hyperspaces in today's terminology; Hausdorff calls them *set spaces* and he devotes a separate section to these in Chapter 6. The terminology "set spaces" made clear that interesting new types of spaces arise by introducing a metric (or topology) on suitable classes of subsets of a space X. For example, considering the subspace $\mathbf{K}(X)$ of $\mathbf{F}(X)$, whose elements are nonempty *compact* subsets of X, one obtains a "set space" that inherits important properties of X, such as completeness, compactness, and separability. The space $\mathbf{K}(X)$ is also of special interest for descriptive set theory.

After publication of his *Grundzüge*, Hausdorff became deeply interested in descriptive set theory; moreover, he also made an essential contribution to it with [H 1916]. The lively developments in this field, especially in the Russian school of Lusin and Sierpiński's circle in Poland, continued to draw Hausdorff's attention well into the 1930s, as can be seen from documents in his estate. Chapters 7 to 9 – entitled "Point Sets and Ordinal Numbers," "Mappings of Two Spaces," and "Real Functions" – provided a systematic, well-rounded, and generally complete account of the current state of research on descriptive set theory. The propositions of this theory, as originally developed by Suslin, Lusin, Sierpiński, and others, were valid for sets in Euclidean spaces or in the Baire space of irrational numbers. Hausdorff tried to extend these results to more general cases, though he generally worked with Polish spaces, i.e. complete separable metric spaces.

Chapter 7 begins with a section on "Closures and Interiors." Although Hausdorff had introduced these concepts in a more specific way in the *Grundzüge*, here he gives a general method for carrying out these important topological constructions. Let each subset A of the space X be associated with a set A_φ, and let the set function φ be monotone, i.e., if $A \subset B$ then $A_\varphi \subseteq B_\varphi$. Then the union V (intersection D) of arbitrarily many sets $A \subseteq A_\varphi$ ($A \supseteq A_\varphi$) will again be such a set, i.e., $V \subseteq V_\varphi$ ($D \supseteq D_\varphi$). For any $M \subseteq X$, one can now define:

$$\underline{M} = \bigcup_{A \subseteq M} A, \quad A \subseteq A_\varphi\,; \quad \overline{M} = \bigcap_{A \supseteq M} A, \quad A \supseteq A_\varphi\,.$$

$\underline{M}$ is called the φ interior, $\overline{M}$ is called the φ closure of M. For example, if $A_\varphi = A_i =$ set of interior points of A, then $A \supseteq A_\varphi$ always holds, i.e., $\overline{M} = M$

and $\underline{M}$ provides the open interior of M. For $A_\varphi = A_\beta =$ set of cluster points of A, then the sets A with $A \subseteq A_\beta$ are dense in themselves and the sets A with $A \supseteq A_\beta$ are closed, i.e. $\underline{M}$ is the dense in itself interior and $\overline{M}$ is the closed exterior of M.

A short section follows with some applications, where Hausdorff proves the existence of certain maximal or minimal sets using induction on the ordinal numbers. As examples, he shows the existence of a Hamel basis for $\mathbb{R}$ as well as the existence within a compact continuum of at least one irreducible (i.e. minimal) subcontinuum connecting any two points x, y. The remainder of Chapter 7 is devoted to Borel and Suslin sets, the core structures of descriptive set theory. Let X be a metric space, $\mathfrak{F}$ its system of closed sets, and $\mathfrak{G}$ its system of open sets. Then the set operations discussed in Chapter 5 yield the system $\mathfrak{B}$ of Borel sets and the system $\mathfrak{S}$ of Suslin sets of X: $\mathfrak{B} = \mathfrak{F}_{(\sigma\delta)} = \mathfrak{G}_{(\delta\sigma)}$, $\mathfrak{S} = \mathfrak{F}_S = \mathfrak{G}_S$. As in the *Grundzüge*, Hausdorff now constructs the σ-algebra $\mathfrak{B}$ by means of transfinite recursion over the ordinal numbers of the second number class (Borel's hierarchy).

An important theorem about the power of Suslin sets in a Polish space now follows: such a set is either at most countable or has the power of the continuum. (This theorem had been proved first by Suslin for the space $X = \mathbb{R}$.) From this and the fact that $\mathfrak{B} \subseteq \mathfrak{S}$ follows immediately the result proved earlier in [H 1916]. Nothing is gained for the continuum problem, however, because in such a space $\mathfrak{S}$ has the power $\aleph$ of the continuum as does X itself (if its perfect kernel is nonempty), so the set of all subsets of X has power $2^\aleph > \aleph$. As Hausdorff observed: "For the vast majority of point sets, then, the power question and hence the continuum problem remain unsolved" [H 1927a, 181; III: 225].

In the following section, "Existence Theorems," Hausdorff proves two fundamental results:

> *Let X be a complete space with nonempty perfect interior, then:*
> *1.) For every $\xi < \omega_1$ there are elements in the Borel class $\mathfrak{B}^\xi$ which do not belong to any $\mathfrak{B}^\eta$ with $\eta < \xi$; thus the length of the Borel hierarchy in such a space has the maximum value ω_1.*
> *2.) In such a space there exist Suslin sets which are not Borel sets, i.e. $\mathfrak{B} \subset \mathfrak{S}$.*

These results lead naturally to the question: under what conditions is a Suslin set a Borel set? Unlike Borel sets, the complement of a Suslin set (a co-Suslin set in today's terminology) is not generally a Suslin set. Suslin had himself already conjectured that a Suslin set will be a Borel set exactly when its complement is also a Suslin set.

A second criterion, conjectured by Lusin, asserts that a Suslin set S is a Borel set exactly when it can be represented by means of disjoint summands, i.e. S has at least one representation

$$S = \bigcup_{\nu \in \mathbb{N}^\mathbb{N}} F_\nu$$

with *disjoint* F_ν, where $F_\nu = F_{n_1} \cap F_{n_1 n_2} \cap \cdots$ $(F_{n_1}, F_{n_1 n_2}, \cdots \in \mathfrak{F})$ is the summand belonging to the number sequence ν. Hausdorff proved the necessity of the second criterion for any metric space, i.e., if S is a Borel set, it has a representation with disjoint summands. Sufficiency is much more difficult and requires Lusin's theory of indices, which Hausdorff expounded in concise form. He was able to prove the sufficiency of both criteria for the case of Polish spaces, whereas the original theorems of Lusin and Sierpiński were only proved for Euclidean spaces. The general proofs he gave for these results constitute one of Hausdorff's most original achievements in his *Mengenlehre.*

Hausdorff divided Chapter 9 of the *Grundzüge* on "Mappings or Functions" into Chapters 8 and 9 in the *Mengenlehre.* In Chapter 8, he treats continuous mappings $X \to Y$ of metric spaces, continuous injective mappings (he calls these *schlicht* mappings), and homeomorphisms. Much of this appears in the *Grundzüge* for topological spaces, some only for metric spaces. New here is especially the study of the behavior of Borel and Suslin sets under such mappings. Since Y is a continuous image of X if and only if the pre-image of every closed set in Y is closed in X, the following holds for arbitrary metric spaces:

> *If Y is a continuous image of X, then the pre-images of Borel or Suslin sets in Y are Borel or Suslin sets in X, respectively.*

Statements about the images of Borel and Suslin sets require additional assumptions on the spaces.

> *If X and Y are Polish spaces, then:*
> *The continuous image of a Suslin set in X is a Suslin set in Y.*
> *The injective continuous image of a Borel set in X is a Borel set in Y.*

Hausdorff also gave a theorem he had proved in [H 1924, III: 445–447]:

> *Every set G_δ in a complete space is homeomorphic to a complete space.*

Chapter 9 deals with real-valued functions on metric spaces. Hausdorff here essentially presents the completion of his earlier results published in [H 1919d]. In the opening section, "Functions and Pre-Image Sets," he assumes nothing at all about the domain space. So letting X be an arbitrary set and $f : X \to \mathbb{R}$ a real function on X, he calls the pre-images

$$[f > y] = \{x \in X,\ f(x) > y\} \text{ and } [f \geq y] = \{x \in X,\ f(x) \geq y\}$$

the *Lebesgue sets* of f. Let $\mathfrak{M}, \mathfrak{N}$ be two given set systems. If $[f > y]$ for each y is an $M \in \mathfrak{M}$ and $[f \geq y]$ for each y is an $N \in \mathfrak{N}$, then f is of class (M, N). For example, continuous functions on a metric space X are of class (G, F), where G belongs to the set $\mathfrak{G}$ of open sets and F to the set $\mathfrak{F}$ of closed sets of X.

A system $\{f\}$ of real functions is called an *ordinary function system* if every constant function is an f, the absolute value of an f is an f, and the sum, difference, product, and quotient (with nowhere vanishing divisor) of two fs is again an f. An ordinary function system $\{f\}$ is called *complete* if the limit of a uniformly convergent sequence of functions $f_n \in \{f\}$ is again an $f \in \{f\}$. From properties of the systems $\mathfrak{M}, \mathfrak{N}$, one can infer those of the system of all functions of class (M, N); namely, the following class theorem holds:

> *(K) If $\mathfrak{M}$ contains the sets X and $\emptyset$ and is a σ-ring, and if $\mathfrak{N}$ consists of the complements $N = X\backslash M$ of sets from $\mathfrak{M}$ (thus is a δ-ring), then the functions f of class (M, N) form a complete system of functions.*

Hausdorff then shows that this class theorem is invertible in the following sense. Let $\{f\}$ be an ordinary function system and let $M = [f > y]$, $N = [f \geq y]$ be its Lebesgue sets, then the class theorem (K) can be inverted as follows:

> *Let $\{f\}$ be a complete system of functions. Then the M form a σ-ring, the N form a δ-ring, and the functions of the class (M, N) are identical to the functions of the function system $\{f\}$.*

The remainder of Chapter 9 is devoted essentially to the theory of Baire functions, where the underlying space X is assumed to be a metric space. Hausdorff introduces the hierarchy of Baire functions, deviating somewhat from the usual procedure at that time. A system $\{f\}$ of functions $f : X \to \mathbb{R}$ is called a *Baire system* if the limit of every convergent sequence of functions in $\{f\}$ belongs to $\{f\}$. If Φ is a given system of functions, then the intersection of all Baire systems $\{f\}$ with $\{f\} \supseteq \Phi$ is called the system of Baire functions generated by Φ. Just as Baire had done, this system can be constructed inductively as a hierarchy of function classes via the ordinal numbers $\xi < \omega_1$.

If X is a metric space and we choose as the initial system Φ^0 of the Baire hierarchy the set of continuous functions $f^0 : X \to \mathbb{R}$, then the functions of the whole Baire system over Φ^0 are called the *Baire functions of space X*. Then the following theorem, already dating back to Lebesgue, holds:

> *The Borel functions of the space X are identical to the functions of the class (B, B) where B ranges over the Borel sets of the space X.*

Hausdorff next considers connections between Suslin sets and Borel functions. For this purpose he considers the Cartesian product $X \times \mathbb{R}$, which is again a metric space, for which he derives the following two theorems:

> *A real function $f : A \to \mathbb{R}$ on a Suslin set A in a Polish space X is a Baire function exactly when its graph*
>
> $$\Gamma_f = \{(x, f(x)); x \in A\}$$
>
> *is a Suslin set in $A \times \mathbb{R}$.*
> *If A is a Borel set in X, then f is a Borel function on A if and only if Γ_f is a Borel set in $A \times \mathbb{R}$.*

Hausdorff calls the projection of Γ_f onto $\mathbb{R}$, when $f(x)$ is a Baire function, a *Baire image of A*. The following theorems then hold for Baire images:

> *The Baire image of a Suslin set in a Polish space is a Suslin set in* $\mathbb{R}$.
> *The injective Baire image of a Borel set in a Polish space is a Borel set in* $\mathbb{R}$.

Finally, Hausdorff generalized the Baire classification to maps $f : X \to Y$ of a metric space X into a metric space Y. He succeeded in transferring a number of results from the theory of real functions (though not all) to Baire mappings of X into a Polish space Y.

Having sketched the contents of Hausdorff's *Mengenlehre*, we conclude with some remarks about its reception. As we have seen, primarily due to the First World War the initial reception of *Grundzüge der Mengenlehre* was very weak and only picked up several years after its publication. By contrast, Hausdorff's sequel monograph was immediately received with considerable interest and for two quite obvious reasons. First, external conditions for scientific work had improved greatly and were far more favorable in 1927 than during the Great War and the immediate post-war years. Second, by this time Hausdorff had become a world-renowned mathematical authority, not least because of his *Grundzüge*. No fewer than eleven reviews of his *Mengenlehre* appeared in 1927 and 1928, five in Germany, two in the United States, and one each in Holland, Norway, Spain, and Czechoslovakia. Arthur Rosenthal wrote at the end of his review:

> The 2nd edition will certainly be studied as much and as eagerly as the 1st edition; and it is to be expected that it will be stimulating to the same high degree as the first. [Rosenthal 1928, 395]; (*HGW* III, 419)

This prediction was borne out fully. Taking, for instance, the 20 volumes of *Fundamenta Mathematicae* that appeared after its publication – i.e., volumes 9 (1927) to 28 (1937) – we find *Mengenlehre* was cited in 70 papers out of a total of 590, thus in 11.9 %. For the *Grundzüge*, the rate had been 15.8 % for the first 20 volumes of *Fundamenta* (volumes 1(1920) to 20(1933)). This impressive success reflects the fact that Hausdorff's *Mengenlehre* – in accordance with the aims of Göschen's Lehrbücherei – was an excellent textbook. It provided a clear and cogent introduction to general set theory, the topology of metric spaces, and the foundations of the theory of real functions. Although sales data are no longer available, variant printings reveal that De Gruyter issued at least one unaltered reprint, and in 1935 another version, expanded by two paragraphs, came out as the third edition.

Yet *Mengenlehre* was not merely a standard textbook for novices, it also attracted considerable interest among active researchers. Not only did it provide the first summary account of descriptive set theory, it also presented a number of new ideas as well as generalizations of earlier results. This was clearly appreciated by several reviewers. The first monograph devoted exclusively to descriptive set theory appeared three years later in Paris: Nikolai Lusin's *Leçons sur les ensembles analytiques* [Lusin 1930]. In a review of this book, no less an authority than John von Neumann wrote:

> The theory of analytic sets created by the author and *M. Suslin* is one of the most important stages in the progress of point set theory and a theoretical structure that already today stands in its main features finished and complete. Since the original literature is vast, the need for a summarizing exposition of the subject is urgent. (In German, the excellent summarization in Hausdorff's Mengenlehre has been available for some years; [· · ·]). (*Jahrbuch über die Fortschritte der Mathematik* **56** (1930), 85)

Vladimir Kanovei, by way of comparing the works of Lusin and Hausdorff, wrote the following:

> 1) Hausdorff presented descriptive set theory in a form applicable to all separable metric spaces. This general form is one of Hausdorff's original achievements in his *Mengenlehre.* In his *Leçons*, Lusin considered only the case of real numbers and mainly only the Baire space of irrational numbers.
> 2) Hausdorff substantially modernized Lusin's somewhat prolix style. Lusin favored a more geometric style of exposition, which did not always harmonize with the content and led to unnecessarily long and cumbersome argumentation, both in the *Leçons* and in his earlier works. Hausdorff's presentation of the then current state of descriptive set theory in [H 1927a] was much more modern and compact. (*HGW* III, 28–29)

Even after 1930, Hausdorff's *Mengenlehre* was cited with respect to descriptive set theory in *Fundamenta Mathematicae* nearly as often as Lusin's *Leçons.* Indeed, a number of researchers took up questions or ideas that Hausdorff first addressed in his book (see *HGW* III, 22–25).

We cannot conclude this discussion of Hausdorff's *Mengenlehre* more appropriately than by citing the final words from Hans Hahn's review, which perhaps already hinted at the rising dangers of antisemitism in Germany:

> If a summary judgment is to be made about this book, it can only be: An exemplary account in every respect of a difficult and thorny field; a work of the type that has carried the fame of German science around the world and of which the author and all German mathematicians may be proud. (*Monatshefte für Mathematik und Physik* **35** (1928), 58).

On the fate of *Mengenlehre* in Germany during the Nazi era as well as its translations into Russian and English, see Section 9.3.

8.6 Other Mathematical Works

During his second period in Bonn, Felix Hausdorff published a number of papers on a wide variety of topics. In this section, we describe his more important works from the years 1921 to 1933; those he wrote thereafter are taken up in Section 9.2 of the final chapter.

8.6.1 The Hausdorff-Young Inequality

On May 20, 1922, Hausdorff submitted a short paper to *Mathematische Zeitschrift* entitled "Eine Ausdehnung des Parsevalschen Satzes über Fourierreihen" ("An Extension of Parseval's Theorem on Fourier Series") [H 1923a, IV: 175–181]. Although only seven pages long, this publication had far-reaching influence.

Let $f(t)$ be a Lebesgue integrable function $f : [0, 2\pi] \to \mathbb{C}$ with Fourier coefficients

$$a_n(f) = \frac{1}{2\pi}\int_0^{2\pi} f(t)e^{-int}\,dt; \quad n = 0, \pm 1, \pm 2, \cdots .$$

Furthermore, let p, q be positive real numbers, both exceeding 1 and with

$$\frac{1}{p} + \frac{1}{q} = 1.$$

The substance of Hausdorff's paper lies in proving the following two theorems:

I. *If $p \leq q$ and if for a sequence $\{a_n\}$, $n = 0, \pm 1, \cdots$ the series $\sum_{n=-\infty}^{\infty} |a_n|^p$ converges, then the a_n are the Fourier coefficients of a function $f(t) \in L^q([0, 2\pi])$ for which*

$$|f|_{L^q} = \left(\frac{1}{2\pi}\int_0^{2\pi} |f(t)|^q\,dt\right)^{\frac{1}{q}} \leq \left(\sum_{n=-\infty}^{\infty} |a_n|^p\right)^{\frac{1}{p}}.$$

II. *If $p \leq q$ and $f(t) \in L^p([0, 2\pi])$ has Fourier coefficients $a_n(f)$, then the series $\sum_{n=-\infty}^{\infty} |a_n(f)|^q$ converges and furthermore*

$$\left(\sum_{n=-\infty}^{\infty} |a_n(f)|^q\right)^{\frac{1}{q}} \leq |f|_{L^p}.$$

Before proving these two theorems, Hausdorff noted an anomaly, namely, they fail to hold for $p > 2$, i.e. for $p > q$. Indeed, in 1914 Hardy and Littlewood gave a sequence $\{a_n\}$ for which $\sum |a_n|^p$ converges for every $p > 2$, but nevertheless there is no function f with $a_n = a_n(f)$. In 1918, Carleman exhibited a continuous function $f(t)$ (i.e., with $f \in L^p$ for all $p > 2$) for which $\sum |a_n(f)|^q$ diverges for every $q < 2$.

For the case $p = q = 2$, I. and II. correspond to the classical Riesz-Fischer Theorem, proved independently by Friedrich Riesz [Riesz 1907] and Ernst Fischer [Fischer 1907]: $f \in L^2([0, 2\pi])$ if and only if the sequence of its Fourier coefficients

$a_n \in \ell^2$, i.e., the series $\sum |a_n|^2$ converges. The equality of the norms in this case

$$\left(\frac{1}{2\pi}\int_0^{2\pi} |f(t)|^2\, dt\right)^{\frac{1}{2}} = \left(\sum_{n=-\infty}^{\infty} |a_n|^2\right)^{\frac{1}{2}}$$

is called Parseval's equation or Parseval's theorem in Fourier series theory, which accounts for the title of Hausdorff's paper.

The Riesz-Fischer Theorem and Parseval's equation served as the starting point for numerous papers by William Henry Young (some jointly written with his wife Grace Chisholm Young) with significant results for the theory of trigonometric series. Young had already proved a special case of Hausdorff's Theorems I and II in 1912. Let $a_n(f)$ denote the Fourier coefficients of the function f and choose $q = 2k$, $p = \frac{2k}{2k-1}$, $k = 1, 2, \cdots$, then Young's theorem states: If $f \in L^p([0, 2\pi])$, then $\{a_n(f)\} \in \ell^q$, and when $\{a_n\} \in \ell^p$, there exists $f \in L^q([0, 2\pi])$ such that $a_n = a_n(f)$. However, the inequalities formulated by Hausdorff in I and II do not occur in Young's work. Hausdorff writes that Young proved both theorems "in a slightly different form" for the pairs of values of p and q given above. This theorem of Young underlies Hausdorff's proof of his theorems I. and II, making it quite justified to call Hausdorff's inequality in II the Hausdorff-Young inequality.

A second result of Young was needed by Hausdorff for his proof, namely Young's convolution inequality. For two sequences $a = \{a_n\}, b = \{b_n\}$ the convolution $a * b$ is defined as the sequence $c = \{c_n\}$ with $c_n = \sum_{k \in D} a_k b_{n-k}$, D a finite subset of $\mathbb{Z}$. Then for $1 < \alpha, \beta < \infty$ and $\frac{1}{\alpha} + \frac{1}{\beta} - 1 > 0$ follows

$$|a * b|_{\ell^\gamma} \leq |a|_{\ell^\alpha} \cdot |b|_{\ell^\beta} \text{ with } \gamma = \frac{\alpha\beta}{\alpha + \beta - \alpha\beta},$$

an inequality due to Young.

Hausdorff also required a result proved by Friedrich Riesz in 1910, which gives a sufficient condition for a real measurable function on $[a, b]$ to be in $L^q([a, b])$ [Riesz 1910]. In 1922, the year Hausdorff wrote his paper, he also inspired Riesz to generalize his results to arbitrary orthonormal systems of uniformly bounded functions. This came about through their encounter at the annual meeting of the German Mathematical Society, held that year in Leipzig from September 17 to 24. Friedrich Riesz was among those in the audience when Hausdorff gave a lecture, "Über Fourierkonstanten," that featured Theorems I. and II. For the proofs, he referred to his forthcoming paper, but Riesz subsequently sought to prove these theorems on his own, and he succeeded in a surprising way. For he found thereby – as he remarked in a later publication – "that those results hold not only for the trigonometric orthogonal system, but also for any bounded orthogonal system" [Riesz 1923, 118].

Before submitting this paper for publication, however, Riesz sent his results to Hausdorff, from whom he received an interesting reaction. In a letter from

November 18, 1922, Hausdorff expressed his appreciation for Riesz's new proof in these words:

> So your method is easily extendable to bounded orthogonal functions, whereas mine, as you rightly remember, is only suitable for trigonometric series; my supporting point, Young's results, turned out at the same time to be an obstacle, which made me overlook the simpler and more far-reaching line of reasoning. [⋯]
>
> In any case, you must publish your finding; it is very beautiful, so beautiful that I am angry I did not find it myself because of my clinging to Young. (*HGW* IX, 569)

Other significant developments related to the Hausdorff-Young inequality resulted from work by Marcel Riesz, Friedrich's brother, and his student Olof Thorin. Their findings helped to launch an extensive new branch of functional analysis, the theory of interpolation spaces [Bergh/Löfström 1976]. Another line of research, due to André Weil, concerned the extension of harmonic analysis to locally compact abelian groups. In 1940, Weil found the most general version of the Hausdorff-Young inequality known up to that time. Since then, harmonic analysts have developed further generalizations for non-abelian groups. A look at the entries in Wikipedia under the keyword "Hausdorff-Young inequality" indicates that the developments which began with Young and especially with Hausdorff and the Riesz brothers continues to play a role in present-day research.

8.6.2 Moment Problems

Hausdorff's immediately preceding work on moment problems in connection with divergent series was discussed earlier in 7.8.4. These ideas were set forth in his two-part study, "Summation Methods and Moment Sequences I, II" [H 1921], published in *Mathematische Zeitschrift.* In "Momentprobleme für ein endliches Intervall" ("Moment Problems for a Finite Interval") [H 1923b, IV: 193–221], he elaborated further on this theme. Today, the "Hausdorff moment problem" is usually formulated as follows: Given a sequence $\{\mu_m\}_{m=0,1,2,\cdots}$ of real numbers, determine necessary and sufficient conditions for the existence of a Borel measure χ on $[0,1]$ so that μ_m are the moments belonging to χ, i.e.:

$$\mu_m = \int_0^1 x^m \, d\chi(x), \quad m = 0, 1, 2, \cdots .$$

Furthermore, if $\{\mu_m\}$ is a given moment sequence, then the corresponding measure χ shall be determined.

This general version of the moment problem differs somewhat from Hausdorff's formulation. In his case, $\chi(x)$ was a suitable function defined on $[0,1]$ and the integral defining the moments was a Riemann-Stieltjes integral. For the interval $[0,\infty)$ with $\chi(x)$ monotone, Thomas Jean Stieltjes had already solved the moment problem in 1894. It was not until 1920 that Hans Hamburger solved the corresponding Stieltjes moment problem for the interval $(-\infty,\infty)$. Stieltjes, for his part, required extensive techniques from the theory of functions and continued fractions. Hausdorff, on the other hand, noted already in [H 1921] that the moment problem "can also be treated in an almost elementary way without that extensive algebraic and function-theoretic apparatus" [H 1923b, 221; IV: 194].

Hausdorff considered the following three types of "density function" (*Belegungsfunktion*) $\chi(x)$:

1) $\chi(x)$ is a real monotonically nondecreasing function, normalized as follows: if u is a discontinuity point of χ, let $\chi(u) = \frac{1}{2}[\chi(u-0)+\chi(u+0)]$, and further $\chi(0) = 0$.
2) $\chi(x)$ is a real function of bounded variation, i.e. $\chi(x) = \chi_1(x)-\chi_2(x)$, where χ_1, χ_2 are functions of type 1).
3) $\chi(x)$ has a density $\varphi(x)$, i.e. $\chi(x) = \int_0^x \varphi(t)\,dt$, so $d\,\chi(x) = \varphi(x)\,dx$, and in addition $\varphi \in L^\alpha([0,1])$, $\alpha > 1$.

As in [H 1921], the higher differences of the sequence $\{\mu_m\}$ play a decisive role:

$$\overset{n}{\nabla}\mu_m \doteq \mu_{m,n} = \mu_m - \binom{n}{1}\mu_{m+1} + \binom{n}{2}\mu_{m+2} - \cdots + (-1)^n\mu_{m+n},$$

as do also the quantities

$$\lambda_{p,m} = \binom{p}{m}\mu_{m,p-m}\,.$$

A sequence $\{\mu_m\}$ is called *totally monotone* if $\mu_{m,n} \geq 0$; this is the case exactly when $\lambda_{p,m} \geq 0$. Hausdorff next proves the following theorems:
(I) *A sequence $\{\mu_m\}$ is the sequence of moments of a covering function $\chi(x)$ of type* 1) *exactly when it is totally monotone.*
(II) *A sequence $\{\mu_m\}$ is the sequence of moments of a covering function $\chi(x)$ of type* 2) *exactly when for all p:*

$$\sum_{m=0}^{p} |\lambda_{p,m}| \leq L < \infty\,.$$

(III) *A sequence $\{\mu_m\}$ is the sequence of moments of a covering function $\chi(x)$ of type* 3) *exactly when for all p:*

$$(p+1)^{\alpha-1}\sum_{m=0}^{p} |\lambda_{p,m}|^\alpha \leq M < \infty\,.$$

Now let $\mathfrak{P}$ be the vector space of all real polynomials on $[0,1]$. Hausdorff's basic idea for proving (I) and effectively constructing $\chi(x)$ – an idea that also motivated Marcel Riesz's work on the moment problem [Riesz 1921/1922/1923] –

was to assign to each sequence $\{\mu_m\}$ a linear functional $Mf : \mathfrak{P} \to \mathbb{R}$, the *moment* of f. Thus, for $f(x) = a_0 + a_1 x + \cdots + a_n x^n$, one has $Mf = a_0\mu_0 + a_1\mu_1 + \cdots + a_n\mu_n$. Hausdorff then proved that a necessary condition for the existence of a monotone function $\chi(x)$ with moments μ_m is that Mf be positive, i.e. for $f \in \mathfrak{P}, f \geq 0 \Longrightarrow Mf = \int_0^1 f(x)\, d\chi(x) \geq 0$. Now, if the $f(x)$ are the Bernstein polynomials $f(x) = x^m(1-x)^n$, then $M[x^m(1-x)^n] = \mu_{m,n}$, from which the necessity of total monotonicity of the sequence $\{\mu_m\}$ immediately follows. Hausdorff gave several proofs showing that this condition is also sufficient for the existence of a monotone χ with moments μ_m.

A moment problem is called *definite* if for a given moment sequence $\{\mu_m\}$ the measure generating this sequence is uniquely determined; otherwise it is *indefinite*. The Stieltjes and Hamburger moment problems are, in general, indefinite. Thus, for these types of moment problems, one searches for further conditions that will guarantee definiteness, whereas the Hausdorff moment problem proves to be definite in every case.

8.6.3 G_δ Sets in Complete Spaces

Hausdorff's first paper in *Fundamenta Mathematicae* was a short note, "Die Mengen G_δ in vollständigen Räumen" ("G_δ Sets in Complete Spaces") [H 1924, III: 445–447]. This was written in reaction to a note Alexandroff had published in the *Comptes Rendus* of the Paris Academy [Alexandroff 1924]. Hausdorff began as follows:

> P. Alexandroff has found the interesting theorem that G_δ sets in complete separable spaces are themselves topologically nothing other than complete separable spaces, i.e., homeomorphic with them. Since the proof (only sketched there) seems rather difficult according to the author's own statement, I would like to give here a short and simple proof, moreover without the restriction to separable spaces. The theorem reads thus: *Every G_δ set in a complete space is homeomorphic to a complete space.* [H 1924, 146; III: 445].

Prior to publication, Hausdorff had sent his proof to Alexandroff and Urysohn. The accompanying letter is no longer extant, but they replied thus:

> We have read your letter with the greatest interest. Your so extraordinarily simple and elegant proof of the theorem on G_δ surprised us all the more, because it does not use the second axiom of countability [i.e., separability]: although we were always convinced that the theorem also holds in general, it was clear to us that the old method [i.e., that of Alexandroff] could not be freed from the second axiom of countability. (*HGW* IX, 16)

This letter also contains Urysohn's proof of the converse theorem, namely: if a complete metric space E is homeomorphic to a subset A of a complete metric space R, then A is a G_δ in R (*HGW* IX, 17–18).

Hausdorff proved his theorem by introducing a cleverly chosen change in the original metric. He let E be a complete metric space with metric $d(x,y)$ and let $A \subset E$ be a G_δ. The complement $B = E \setminus A$ is then an F_σ, i.e. $B = F_1 \cup F_2 \cup \cdots$ with closed sets F_n. Hausdorff then introduces the new metric on A: $d'(x,y) = \max[d(x,y), \rho(x,y)]$ with

$$\rho(x,y) = \sum_{n=1}^{\infty} c_n \sup_{t \in F_n} \frac{d(x,y)}{d(x,y) + d(x,t) + d(y,t)}.$$

Here $\sum\{c_n\}$ is a convergent series of positive numbers. He then shows by means of relatively simple calculations: 1) $d'(x,y)$ is a metric on A; 2) $d'(x,y)$ and $d(x,y)$ generate the same topology on A; and 3) A with metric $d'(x,y)$ is a complete metric space, which proves the theorem.

Hausdorff had introduced the concept of completeness for metric spaces in his *Grundzüge*. He also showed there that one can always complete a given metric space E, i.e. construct a mapping that embeds E as a dense subset of a complete metric space $\overline{E}$. This technique is essentially a generalization of the method employed by Cantor and Meray for constructing the real numbers from the rationals. Hausdorff also proved a number of fundamental theorems on complete metric spaces in the *Grundzüge* [H 1914a, 315–328; II: 415–428].

Alexandroff's work and [H 1924] mark the beginning of general studies of completeness in the topological sense, as they gave a topological characterization of completely metrizable spaces. Hausdorff continued to study such questions, for example, by introducing the notion of a closed basis. A subsystem $\mathfrak{B}$ of the system $\mathfrak{G}$ of open sets of a topological space is called a *closed basis* if from $G_1 \supseteq G_2 \supseteq \cdots$ with $G_n \in \mathfrak{B}$ follows $\cap_{n=1}^{\infty} G_{n\alpha} \neq \emptyset$, where $G_{n\alpha}$ is the closure of G_n. This notion enabled Hausdorff to prove the following theorem:
A metric space E is an absolute G_δ (i.e., a G_δ in any space containing E) if and only if the system of its open sets has a closed basis. He did not publish this result, however, because he learned from Alexandroff that one of his students, Nikolai Vedenissoff, had found the same result independently.

8.6.4 Four Notes on Topics in Analysis

On December 10,1924 Hausdorff submitted a short note to *Mathematische Annalen*, "Zum Hölderschen Satz über $\Gamma(x)$" [H 1925, IV: 239–242]. Its first two lines already convey its purpose:

> The following proof that $\Gamma(x)$ does not satisfy any algebraic differential equation would seem to be much simpler than the previous ones.

Hausdorff here referred to three earlier papers: [Hölder 1887], [Moore 1897] and [Ostrowski 1919].

A function $f(z) : \mathbb{C} \to \mathbb{C}$ satisfies an algebraic differential equation if there exists a polynomial $G(x, x_0, x_1, \cdots, x_n)$ over $\mathbb{C}$ with $G(z, f(z), f'(z), \cdots, f^{(n)}(z)) \equiv 0$. Otto Hölder presented the first proof that the gamma function $\Gamma(z)$ *does not* satisfy an algebraic differential equation. Hölder mentions at the beginning of his paper that Karl Weierstrass had already conjectured this theorem and had set the task of proving it.

Hausdorff's proof was based on the following general theorem:
Let $\varphi(x)$ be a rational function, such that $\varphi(x) \to 0$ when $x \to \infty$ and where $\varphi(x)$ has no pair of distinct poles that differ by an integer. If a function $y(x)$ satisfies an algebraic differential equation and at the same time satisfies the difference equation $y(x+1) - y(x) = \varphi(x)$, then $\varphi(x) \equiv 0$.
Hausdorff then observes that the logarithmic derivative of the gamma function $y(x) = \frac{\Gamma'(x)}{\Gamma(x)}$ satisfies the difference equation $y(x+1) - y(x) = \frac{1}{x}$, proving that it cannot satisfy any algebraic differential equation. From this it is then easy to conclude that $\Gamma(x)$ itself has the same property.

In his paper "Proof of a Theorem of Arzelà" [H 1927b, IV: 249–251], Hausdorff gives a new and – as he thinks – "short and natural proof, suitable also for elementary lectures" for what is now usually called Arzelà's limit theorem [Arzelà 1885]: If *uniformly bounded* Riemann integrable functions $f_n(x)$ defined on $[a, b]$ converge to a *Riemann integrable* function $f(x)$, then the limit and integral are interchangeable, i.e.

$$\lim_{n \to \infty} \int_a^b f_n(x)\, dx = \int_a^b f(x)\, dx\,.$$

A major drawback of the Riemann integral arises from having to assume the integrability of the limit function, as reflected in this theorem. Lebesgue was able to circumvent this limitation in his theory. Thus, for the Lebesgue integral he was able to prove: if $\lim f_n(x) = f(x)$ almost everywhere on $[a, b]$ and if $|f_n(x)| \le g(x)$ holds for a Lebesgue-integrable function $g(x)$, then $f(x)$ is Lebesgue-integrable and the integral and limit are interchangeable. This theorem was of fundamental importance for modern analysis, for example, in the theory of L^p-spaces. However, in this paper Hausdorff does not even mention the Lebesgue integral at all, though he was, of course, well aware of these issues. Here, though, he was merely interested in a "proof suitable for elementary lectures," and at that time the Lebesgue integral was only taught in "advanced lecture courses." In his *Grundzüge*, which for the first time gave mathematicians in German-speaking countries an overview of Lebesgue's measure and integration theory, he had even emphasized this "fundamental advantage of Lebesgue's integral over Riemann's" [H 1914a, 440; II: 540].

It may seem surprising to mention Hausdorff's name in connection with so-called hypercomplex number systems, which today are known as finite-dimensional associative algebras over $\mathbb{R}$ or $\mathbb{C}$. These had been studied since the first half of the nineteenth century by a great number of eminent mathematicians, including (in alphabetical order): E. Cartan, Cayley, Clifford, Dedekind, Frobenius, Gauss, Hamilton, Kronecker, Molien, and Weierstrass. A major focal point of interest was the classification problem, on which Eduard Study and Georg Scheffers, both working in Leipzig at that time, achieved far-reaching results around 1890. Given these circumstances as well as Hausdorff's wide-ranging curiosity, one can well imagine why he became interested in this topic already as a young man. In fact, he published his own small contribution in [H 1900b], which, however, had no influence on further developments.

In the winter semester of 1926/27, Hausdorff took up this subject again in a lecture course on "Hypercomplex Numbers." The last part of this dealt with "Lipschitz number systems," known today as Clifford algebras. The works of Study from 1923–1926 and this course led to Hausdorff's 1927 paper: "Lipschitz Number Systems and Study's Nabla Functions" [H 1927c, IV: 469–483]. Like [H 1900b], this paper also had no demonstrable influence on further developments, so here it suffices to cite Winfried Scharlau's concluding assessment:

> Overall, Hausdorff's work should be regarded as a contribution to the structural theory of Clifford algebras. Many essential notions and results were in essence already present in Lipschitz's work [$\cdots$], but no systematic exposition of the theory existed using the basic notions of algebra and linear algebra which had become common knowledge by about 1920. (*HGW* IV, 485)

Hausdorff's paper "The Equivalence of Hölder's and Cesàro's Limit Values of Negative Order" [H 1930a, IV: 257–267] follows directly on a topic addressed in the two-part paper "Summation Methods and Moment Sequences I, II" [H 1921] (see Subsection 7.8.4). There he introduced the Hölder and Cesàro matrices H^k and C^k – which until then had only been defined for natural numbers k – now written C^α and valid for real α, excluding the values $\alpha = -1, -2, \cdots$. He also had shown the equivalence of both procedures, i.e., $H^\alpha \sim C^\alpha$, for $\alpha > -1$, thereby substantially generalizing the equivalence theorem of Knopp and Schnee .

The basic idea in [H 1930a] was to replace the matrix C^α by a modified matrix Γ^α, which was then defined for all $\alpha \in \mathbb{R}$. C^α is the Hausdorff matrix associated with the sequence of moments

$$\mu_n = \binom{n+\alpha}{n}^{-1} = \frac{\Gamma(n+1)\Gamma(\alpha+1)}{\Gamma(n+\alpha+1)},$$

which is undefined for $\alpha = -1, -2, \cdots$ because these are singular places for $\Gamma(x)$. Hausdorff therefore introduces the modified moment sequence $\mu_n = \frac{\Gamma(n+1)}{\Gamma(n+\alpha+1)}$ and denotes the corresponding Hausdorff matrix by Γ^α. Then the following theorem holds:

> *If $\{x_n\}$ is H^α-limitable to the value x, then $\{x_n\}$ is also Γ^α-limitable to the value $\frac{x}{\Gamma(\alpha+1)}$.*

However, the converse does not hold in general, since for $\alpha \leq -1$ an additional condition must be satisfied. In contrast to his earlier paper [H 1921], which had a considerable influence on the development of limiting theory, the refinement Hausdorff made in [H 1930a] attracted rather little notice.

A topic that occupied Hausdorff on and off for over twenty years was the extension of continuous mappings. He wrote about this already in [H 1919d], which gives a new and quite transparent proof of Tietze's extension theorem. After many efforts to obtain more general results, he later published "Erweiterung einer Homöomorphie" ("Extension of a Homeomorphism") [H 1930b]. Finally, in 1938, in his last publication, "Erweiterung einer stetigen Abbildung" ("Extension of a Continuous Mapping") [H 1938], he once again returned to this subject. We discuss this whole matter in context in 9.2.6, where we deal with [H 1938].

8.6.5 Linear Functional Analysis

Hausdorff's paper "Zur Theorie der linearen metrischen Räume" ("On the Theory of Linear Metric Spaces") [H 1931, IV: 271–288] appeared in a special volume of Crelle's Journal dedicated to Kurt Hensel on his 70th birthday. Hensel had served as co-editor of Crelle since 1902, and his star student, Helmut Hasse, planned and organized this special issue. Hasse approached a select group of Hensel's students, friends and colleagues, asking them to contribute a paper for this Festschrift. Ultimately, 38 authors agreed to do so, including many distinguished mathematicians. The lion's share among them were experts in algebra and number theory, whereas Hausdorff's reputation in Germany was largely based on his work in modern analysis. Hausdorff had also engaged intensively and early on in areas at the forefront of functional analysis; a number of his lectures in Bonn bear witness to this. In particular, his paper for the Hensel-Festschrift volume [H 1931] emerged directly from his course on "Point Sets," which he taught in the summer semester of 1931. Although the title hardly suggests this, what Hausdorff offered in these lectures amounted to an excellent introduction to the then current state of linear functional analysis.

Hausdorff's main concern in [H 1931] was the theory of linear normed spaces, a new field whose central theorems were scattered throughout the literature. In this connection, he noted important works by S. Banach, H. Hahn, E. Helly, F. Riesz, J. Schauder, E. Schmidt, W. Sierpiński and O. Toeplitz. As he had done in his monographs on set theory and topology, Hausdorff set the goal of giving a self-contained account that could be understood without further study of the literature, while freeing the proofs from unnecessary assumptions. His entire presentation is so transparent as to be suitable for an introductory course even today. For the post-war generation in Germany, [H 1931] occupied an important place in the growing literature on functional analysis. Klaus Krickeberg acknowledged, in a message from 3 November, 2016, how he was "very much influenced by [Hausdorff's] great and really fundamental work on duality in linear spaces [H 1931]. Dieudonné later treated the same material, but all the essential ideas were already in Hausdorff."

In the following we will mainly employ modern terminology, even if in many cases this only began to consolidate during the 1930s and thus often differs from Hausdorff's vocabulary. His §1 introduces the spaces studied in his paper, familiar today as normed vector spaces over $\mathbb{R}$. Such a space X with $|x-y|$ satisfies the properties of a metric space. Hausdorff then defines the concepts of subspace and basis for such spaces, calling the subspace generated by a set $A \subset X$ its linear hull A_λ. Important examples are the spaces ℓ^p $(p \geq 1)$ and ℓ^∞. In §2 he introduces the notion of a linear mapping $s : X \to Y$ (X, Y normed vector spaces) and the properties of continuity and boundedness for such a mapping. A first important theorem states that such a mapping is continuous if and only if it is bounded, i.e., if $|s(x)| \leq M|x|$ for all x. Furthermore, such a continuous linear mapping is completely determined by its values on a basis for X. Thus, from the beginning, Hausdorff links topological properties of the spaces under consideration with the properties of linear mappings.

If the linear mapping s is injective with range $R(s) \subseteq Y$, then $s^{-1} : R(s) \to X$ is a linear mapping which is continuous if and only if s is bounded from below, i.e. $|sx| \geq m|x|$. If s is continuous, injective and s^{-1} is also continuous, then Hausdorff calls the spaces X and $R(s)$ *linearly homeomorphic.* A space that is linearly homeomorphic with a complete space is itself complete. For two given normed vector spaces X, Y, the set S of all linear continuous mappings $s : X \to Y$ forms a normed vector space, by virtue of

$$(\alpha s)x = \alpha \cdot sx, \quad (s_1 + s_2)x = s_1 x + s_2 x, \quad |s| = \sup_{|x|=1} |sx|.$$

Moreover, the following theorem holds:

If Y is complete, then S is also complete, and if Y is a Banach space, so is S.

A particularly important special case for continuous linear mappings arises when $Y = \mathbb{R}$. Hausdorff calls these mappings $u : X \to \mathbb{R}$ linear forms or linear functionals, and the space S is called the space *dual* to X, denoted by X'. Since $\mathbb{R}$ is complete, the dual space X' of a normed vector space X is always complete. Moreover, for each linear continuous mapping $s : X \to Y$ one has a *dual mapping* $s' : Y' \to X'$. This is a linear continuous mapping between dual spaces, defined by

$$v(sx) = (s'v)x \ \text{ with } v \in Y',\ u = s'v \in X',\ x \in X,\ sx \in Y\,.$$

Furthermore, the norms of these two mappings satisfy $|s'| = |s|$.

Following these preparations, Hausdorff could now prove some fundamental theorems in the theory of normed vector spaces in paragraphs 3 and 4. At the beginning of §3, he presents the novel concept of a quotient space. Let X be a normed linear space and F a closed linear subspace, then one has an equivalence relation $x' \equiv x \pmod F \Leftrightarrow x' - x \in F$. The equivalence classes $[x]$ form a linear space, the *quotient space* $X|F$, defined by $\alpha[x] = [\alpha x]$ and $[x+y] = [x] + [y]$. With $|[y]| = \inf_{x\in[y]} |x|$, $X|F$ is a normed linear space, and when X is complete so is $X|F$. Another new concept introduced by Hausdorff is that of a topological homomorphism. Let $s : X \to Y$ be a continuous linear mapping, with $R(s)$ its range of values and kernel $N(s) = \{x \in X; sx = 0\}$, then s induces a continuous linear mapping $\sigma : X|N(s) \to R(s)$. The mapping $s : X \to Y$ is called a *topological homomorphism* (or continuously invertible) if σ is a homeomorphism, i.e. σ^{-1} is continuous. An important first theorem, Hausdorff states that $s : X \to Y$ is a topological homomorphism if and only if the image $s(G)$ of an open set $G \subseteq X$ is open in $R(s)$.

Hausdorff then presents a version of the open mapping theorem central for the entire theory, the "first main theorem," which reads: *Let $s : X \to Y$ be a continuous linear mapping and let X be complete. Then the range space $R(s)$ will be complete or of first category in itself depending on whether s is or is not a topological homomorphism.* The next major result in §3 is familiar today as the uniform boundedness theorem (or Banach-Steinhaus Theorem). Hausdorff calls it "the boundedness theorem": Let $\{s_t\}_{t \in T}$ be a set of continuous linear mappings $s_t : X \to Y$ and let X be complete. If for each x the image points $s_t x$ form a bounded set, i.e. $|s_t x| \leqq a_x$, $a_x > 0$, then the norms $|s_t|$ form a bounded set. Applying the theorem that every vector space has a Hamel basis, Hausdorff proves two more remarkable theorems in §3:

> 1.) *There are normed vector spaces which are incomplete and yet of second category in themselves.*
> 2.) *Every infinite-dimensional normed vector space is the injective continuous linear image of a normed vector space of first category.*

The final paragraph of [H 1931] is essentially devoted to the study of duality in normed vector spaces. Hausdorff begins with important examples: the dual space of ℓ^p, $p > 1$ is ℓ^q, where $\frac{1}{p} + \frac{1}{q} = 1$; the dual space of ℓ^1 is ℓ^∞, and the dual of ℓ^∞ is ℓ^1. There follows another "main theorem" of functional analysis: the Hahn-Banach Theorem, which Hausdorff calls the "Extension Theorem": *a linear functional u defined on the linear subspace $L \subset X$ with bound M* (i.e., $|ux| \leq M|x|$) *can be extended to X with the same bound.* As a corollary, we have the following interesting theorem: if L is closed in X, then there exists a vanishing linear functional u on L which does not vanish at a given point $x_0 \in X \setminus L$.

In the following, again let $s : X \to Y$ be a continuous linear mapping and $s' : Y' \to X'$ the corresponding dual mapping. Hausdorff considers the problem of deciding when $y = sx$ or $u = s'v$ are solvable, i.e., when does there exist for given y (or u) an x (or v) such that these equations hold? This comes down to examining certain properties of $R(s)$ (resp. $R(s')$). Besides the kernels $N(s)$ and $N(s')$ (which are closed sets in their respective spaces), Hausdorff introduces the following closed sets:

$$F_y = \{y \in Y;\, vy = 0 \text{ for } v \in N(s')\}, \quad F_u = \{u \in X';\, ux = 0 \text{ for } x \in N(s)\}.$$

The following are then evidently necessary conditions: the solvability of $y = sx$ for x (resp. $u = s'v$ for v) requires that $R(s) \subseteq F_y$ (resp. $R(s') \subseteq F_u$).

Hausdorff calls $y = sx$ (resp. $u = s'v$) *normally solvable* if $R(s) = F_y$ (resp. $R(s') = F_u$). He then shows that the following three fundamental theorems hold:

> *1.) The equation $y = sx$ is normally solvable exactly when $R(s)$ is closed in Y.*
> *2.) For the normal solvability of $u = s'v$ it is necessary and, if X is reflexive (i.e. $X'' = X$), also sufficient that $R(s')$ be closed in X'.*
> *3.) The equation $u = s'v$ is normally solvable exactly when s is a topological homomorphism.*

If X is complete, this will be the case exactly when $R(s')$ is closed in X'. In summarizing, Hausdorff states that "under the usual restrictions in the literature" – meaning the spaces X and Y are complete, i.e., Banach spaces – both $y = sx$ and $u = s'v$ will be normally solvable exactly when s and s' are topological homomorphisms. Then, according to 3), all four of these properties are equivalent [H 1931, 310; IV: 287].

At the end of his paper, Hausdorff briefly discusses the compact linear mappings (which he calls fully continuous) introduced by F. Riesz in 1918. Here $t : X \to Y$ is called *fully continuous* if the sequence of images $\{y_n\} = \{tx_n\}$ of every bounded sequence $\{x_n\}$ contains a fundamental sequence $\{y_p\}$. Hausdorff then proves the following theorem: *If $t : X \to X$ is a fully continuous linear mapping of the complete space X into itself, then $y = sx = x - tx$ is a topological homomorphism.*

This completes our account of the main results in "Theory of Linear Metric Spaces" [H 1931]. S.D. Chatterji thoroughly investigated its reception, reaching the conclusion that Banach's great work [Banach 1932] completely overshadowed all previous work on linear functional analysis; all later authors refer to Banach in nearly every respect. There is one exception, however, namely Jean Dieudonné's important paper [Dieudonné 1942], which introduced and systematically utilized the non-metrizable weak topologies $\sigma(X, X')$ and $\sigma(X', X)$. This led to the advances in the theory of locally convex topological vector spaces made by him and others associated with Boubaki. About this work and the role of functional analysis within the Bourbaki group, Chatterji writes:

> On reading this paper [Dieudonné 1942] it becomes clear that Bourbaki had started on a systematization of the general theory of locally convex topological vector spaces already around 1938 (the basic definitions of the latter theory having germinated around 1934–35 in the works of Kolmogorov, von Neumann and others). Dieudonné (a founding member of the Bourbaki group) was strongly motivated in writing his 1942 paper by the problem of solving the equations $y = sx, u = s'v$ (in our notation) in the general framework of two vector spaces in duality; indeed [Dieudonné 1942] can be said to be the birthplace of this duality theory. Two sources seemed to have influenced Dieudonné: the book [Banach 1932] and the present paper by Hausdorff. He refers to them

several times, and although his methodology is quite different, his aim was to give the clearest possible formulation of the theorems concerning s, s' presented by Banach and Hausdorff. (*HGW* IV, 296)

8.7 Family, Friends and Social Life

In earlier parts of this book that chronologically predate the last period in Bonn – 2.3 "Family Life," 5.3 "Relations with Women, Ideal and Real," and 7.9 "Family Life in Greifswald" – we have already described Hausdorff's personal life, his relationships with his wife and daughter, and the atmosphere in their hospitable home. An important aspect in Felix Hausdorff's social life was his deep and abiding love of music, which he shared with many others who appreciated his knowledge of musical works and his talents as a pianist. To round out the picture, this section describes his happier years as a full professor in Bonn up until the seizure of power by the National Socialists. That disastrous event had immediate repercussions for academic life, after which everything in Bonn changed for the worse, as recounted in Chapter 9, Section 9.1.

As already mentioned, Hausdorff bought a spacious single-family house soon after his appointment in Bonn. At that time it was located in a particularly quiet area, which had been a decisive factor for the purchase. The house also came with a beautiful garden filled with apple trees. Moreover, it was only about a mile away from the main building of the university at Hofgarten, where mathematics classes were held, so Hausdorff could easily reach his place of work on foot. The building survived the war undamaged and thus still stands today; a historical photograph is reproduced in Figure 8.1. Felix and Charlotte Hausdorff enjoyed entertaining there when they were not visiting family or traveling to warmer climes. They often spent time in Bad Reichenhall with Charlotte's sister Sitta. During the 1920s and early 1930s, her younger sister Edith Pappenheim along with her daughter Else visited the Hausdorffs in Bonn nearly every year. Edith lived with the Hausdorffs from September 1938 onward, whereas Else managed to emigrate to the United States.

After the end of hyperinflation in late 1923, Hausdorff and his wife often undertook longer trips to Switzerland, Italy or southern France during the vacation breaks between semesters. Occasionally they met colleagues on these trips, which typically sparked a lively exchange of ideas. Thus, in April 1925, Hausdorff wrote to Georg Pólya from a vacation spot:

> Since your departure we have been to Vira-Magadino, Pallanza (Isola Madre), Camedo, Contra, Lugano. The axiom: if Zurich, then Pólya – stands firm for us! – Here there is now a mathematicians' congress: Brouwer, Schoenflies, Ludwig (geometer from Dresden), Faber (unfortunately already left). (*HGW* IX, 556)

Almost exactly one year later, on April 25, 1926, Hausdorff wrote to Pólya: "We have been in Lugano for a week, in the rain and cold; if we stayed any longer we would need to have you come here again" (*HGW* IX, 557).

One also frequently finds mention of trips or travel plans in Hausdorff's letters to Paul Alexandroff. Thus, he writes on June 14, 1928:

> Over the Easter vacation I was in Italy with my wife, four weeks in Levanto, where a mathematicians' congress developed at the end (Hensel, Courant, Neugebauer, H. Lewi), and then three weeks in Rome. (*HGW* IX, 60)

Alexandroff loved to travel as well, so they shared this mutual interest. Their correspondence reveals that the Hausdorffs were in Switzerland in the summer of 1925 and also in early October 1928. In the spring of 1930 they stayed in Menton on the Côte d'Azur. The last time Alexandroff and Hausdorff saw each other was in 1932 during a stay in Locarno on Lake Maggiore. Hausdorff's daughter reported that her father was particularly fond of the Swiss mountains and made such trips often. He could even speak in Swiss-German dialect quite well. Two of his favorite places were Zermatt and Montreux, and he also liked to stay in Chamonix in the French Alps. When he spent his summer vacation near the sea or a lake, he liked to go swimming. Still, he never swam in the Rhine during his many years in Bonn, even though this was still common practice at the time.

The Hausdorffs' daughter Lenore (Nora) lived with them in the same house until she married. Like the vast majority of middle-class young women, Nora grew up imagining she would become a wife and mother. According to her own account, she met her future husband, Arthur König, at a *professorium*, the ball and banquet hosted each year by the rector of the university. Her father knew König, who attended his course on probability theory during the summer semester of 1923; he was already then an older auditor, so their acquaintance may well have led to this initial contact between the two young people.

Arthur König was born in Berlin on October 13, 1896 to the physicist Arthur Peter König. He was only five, however, when his father died in 1901. His mother, a librarian, then moved with little Arthur to Bonn, where she passed away ten years later. Afterward, Arthur was taken in by his mother's siblings until he joined the army in 1915, serving until the Great War ended. In 1919, he began studying astronomy in Bonn, where he received his Ph.D. in 1923 with a thesis on the star system of the Pleiades. Afterward, he took a position as a research assistant at the Bonn Observatory. One year later, Nora Hausdorff and Arthur König decided to marry (see Figure 8.7). Their wedding took place on February 28, 1925 in a Protestant church, after which the young couple moved into their own apartment in Bonn.

In May of 1927, their first child was born, a son whom they named Felix. Ten days later, Hausdorff reported this happy event to Alexandroff: "It will interest you to know that on May 19 I became a grandfather; my daughter has given birth to a baby boy. Mother and child are in excellent condition" (*HGW* IX, 51). In

Figure 8.7: Lenore Hausdorff and Arthur König in the living room of her parents at the time of their engagement, 1924
Source: Hausdorff Estate, Capsule 65, Nr. 31.

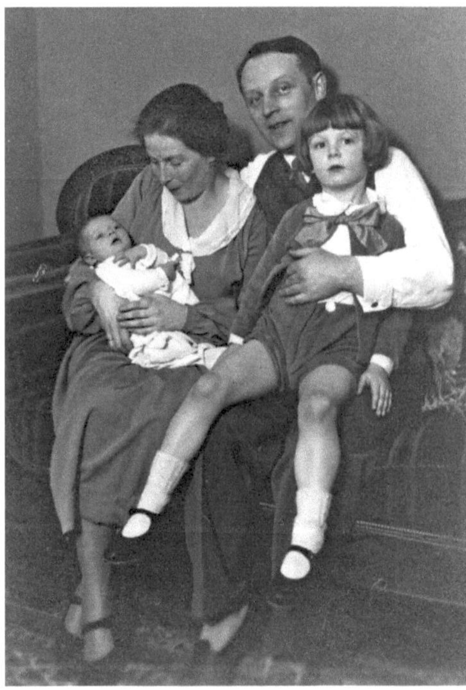

Figure 8.8: Nora and Arthur König with their children Felix and Hermann (1932). Source: Hausdorff Estate, Capsule 65, Nr. 10.

Figure 8.9: Felix and Charlotte Hausdorff with their grandson Hermann (1934)

1929 König received an offer from the firm of Carl Zeiss in Jena, where he would take over management of its astronomy department, so in July of that year, the Königs moved to that city. Three years later, Hermann, their second son was born. After a few years had passed, however, they realized that the child was severely mentally handicapped and would not be able to attend school; his mental state remained at the level of an infant. Only after the war did medical science discover the cause of his disability. A genetic defect had led to phenylketonuria, a metabolic disease, which when untreated causes irreparable brain damage.

Nora König spoke about her father's warm friendships with his sisters and their families in Prague. His older sister Martha was married to Anton Brandeis. She died already in her early 60s in the spring of 1932, half a year after her husband's passing. Although very few relevant documents survive that attest to these relations, the Hausdorffs received a very warm letter from Felix's nephew Ludwig Brandeis and his wife Hana, written in 1931. They had just moved into a new domicile, which prompted the Hausdorffs to send them a house-warming present for their library: a luxurious edition of Nietzsche's works. In thanking them profusely for this gift, Ludwig and Hana Brandeis expressed their hopes "to be able to welcome you soon in our new home" (*HGW* IX, 163).

Felix's younger sister Vally was the wife of Anton Glaser, who was nine years older than his wife and three years older than her brother. Glaser ran an iron construction company in Prague. From him only a single letter is preserved, written on 8 November 1926 to congratulate his brother-in-law on his birthday. In it, he reports on attending an interesting, rather informal lecture by the controversial political journalist Theodor Lessing, which drew a large crowd from Prague's Jewish community. The message ends:

> Now accept again my best wishes with the same generous warmth and affection as always. Many, many greetings! Your Anton
> (*HGW* IX, 299)

While few personal visits with these relatives from Prague are documented, it seems very likely that the Hausdorffs saw them on special occasions. One such family gathering was mentioned in a letter Hausdorff wrote to Alexandroff on January 4, 1929, where he alluded to a "Christmas visit of my siblings." Else Pappenheim remembered another such family meeting in Prague in 1935. One year later, in June 1936, the Hausdorffs were in Prague again. At that time, Hausdorff sent Erich Bessel-Hagen a postcard showing a picture of the Wenceslas Square (*HGW* IX, 147); this was probably their last visit in Prague to see his relatives. At a time when antisemitism had become state doctrine in Germany, one can easily imagine that these stays with dear people in this (still) free city must have done the Hausdorffs a world of good.

Before 1933, they also enjoyed the company of a stable circle of friends in Bonn over the years, many of them notable for their high intellectual standing and integrity. In describing some of them here, we begin with those friends who have already been mentioned in earlier sections of the book. The oldest still active

Figure 8.10: Felix Hausdorff with his wife Charlotte and sister Vally Glaser (left) on a Trip to Interlaken in the Bernese Highlands (1934). Source: Hausdorff Estate, Capsule 65, Nr. 10.

friendship was with Gustav Kirstein, his wife Cläre, and their daughter Marianne. In 1930, Hausdorff took part in the celebration of Gustav Kirstein's 60th birthday by offering him an elaborate acrostic on that occasion (reprinted in *HGW* IB, 438, and in *HGW* VIII, 261). The Hausdorffs also traveled to Leipzig to assist Cläre Kirstein after her husband's unexpected death on 14 February 1934, (see Section 5.1).

In 1929, Marianne Kirstein married the mathematician Reinhold Baer, who came to Halle that year as a Privatdozent. Baer knew, of course, that Georg Cantor had long taught in Halle, though curiously Cantor never taught a course on set theory; in fact, the first to do so was Reinhold Baer, when he arrived there from Freiburg in 1929. As recounted in 6.3.3, Hausdorff's important study on ordered sets [H 1908] came about after speaking with Cantor in Halle. Immediately after the Nazis came to power, however, the Baers emigrated to the United States via England.

Theodor Posner and his wife Else were older friends of the Hausdorffs from their days together in Greifswald (see Section 7.9 and Figure 7.4). After his death in 1929, Posner's widow remained a loyal friend to the Hausdorff family; it was she who assisted the Hausdorffs in the last hours of their lives. They also maintained close relations with the widow of Franz London up until her emigration. Moreover, in sharp contrast with the situations in Leipzig and Greifswald, Hausdorff enjoyed very friendly relations in Bonn with the majority of his mathematical colleagues. This was true from the beginning for Eduard Study, Franz London and Johann Oswald Müller, but also continued after 1927, when Otto Toeplitz and Erich Bessel-Hagen joined the faculty.

Beyond his own discipline, Hausdorff also had a number of close colleagues in other fields at the university. First and foremost among them were the neurologist and psychiatrist Otto Löwenstein and his wife Marta, *née* Grunewald. Löwenstein habilitated in Bonn in 1920 and in 1926 he became head physician of the newly founded Provincial Children's Institution for the Mentally Abnormal. In 1931, following heated disputes among the faculty during which Otto Löwenstein's Jewish background was clearly a factor, he was appointed professor of pathopsychology at the University of Bonn.

Marta Löwenstein was the daughter of Dr. Julius Grunewald, a Jewish physician from Barmen. Her parents were enlightened, liberal, and socially-minded people, highly educated and cultured; for them, music played "an extremely important, even existential role" in their family [Peters 2009, 159]. From early childhood, Marta Grunewald shared in this love for music as a talented singer and pianist; she enjoyed playing four-handed works with her mother. Like her father, she was also drawn to the life sciences; after studying zoology in Munich and Naples she took her doctorate from Munich in 1915. During World War I, she worked in a war hospital as a nurse, and in 1920 she married Otto Löwenstein. The couple had two daughters, Annebeth and Marieli.

The Löwensteins ran a hospitable home in Bonn, in which Marta offered singing performances. Their friendship with the Hausdorffs commenced almost

Figure 8.11: Marta Löwenstein with her baby daughter Marieli.
She was a talented singer, and Hausdorff enjoyed accompanying her on his piano.
Source: König Estate, Nr. 161.4.

from the time when the latter first left Greifswald and returned to Bonn. Annebeth Perls, *née* Löwenstein, knew this because of an important incident in her own life, about which she reported 90 years later:

> I nearly came into the world at the Hausdorff's home. Following an invitation from the Hausdorffs, my parents went there by taxi and had almost arrived when the cab had to turn around as quickly as possible to get to the hospital, where I was then born. (Annebeth Perls to Walter Purkert, 7 March 2012)

Annebeth Perls also described her childhood memories of Felix Hausdorff and Marta Löwenstein making music together during the early 1930s:

> I could not know that Uncle Häuschen was a famous mathematician, philosopher and writer. The musician is unforgettable for me. He was quite a wonderful pianist–and the accompanist of my mother, the singer. They played music together in such a way that every song, every aria was an experience. What luck for us children to grow up in that atmosphere.

The children of Hausdorff's friends got to call him affectionately "Uncle Häuschen;" Marieli Löwenstein later even became Hausdorff's godchild. Marta Löwenstein also referred to their earlier music making in a beautiful letter sent to Felix Hausdorff on the occasion of his 70th birthday (*HGW* IX, 443–444). After emigrating to the USA, her daughter Annebeth Perls was later discovered as a singer with considerable talent. The West Bay Opera in Palo Alto cast her in several leading roles of the opera literature, and she also appeared in numerous concerts as an interpreter of songs.

In the summer of 1932, the Löwensteins bought a house in Endenicher Allee in Bonn. In her autobiography, written in 1954, Marta's mother, Julie Grunewald, recalled their moving-in party:

> One day I was let in on the secret by friends that in the evening there would be a surprise "house-warming" party. Everyone came, most of whom are now no more or scattered to all parts of the world. Music, serious and lighthearted talks, and of course good hospitality, partly provided by us, partly by the friends, made an evening I will never forget. The dear, fine Hausdorffs, the Franks, now our neighbors, Dr. Samuel, who played a trio with the Toeplitz sons. Oh, it was eerily beautiful, and probably no one saw the Mene Tekel, the "sign on the wall." Although our friend Hausdorff wrote in a humorous poem, in which everyone got his verse:
>
> "Man zieht gern um, auch Endenich
> ist noch vielleicht das Ende nicht."
> (One likes to move, even Endenich
> is still maybe not the end.)

> How near that was, we should have been able to guess! Well, we enjoyed this beautiful summer, and it lives in me as a happy memory. [Peters 2009, 140–141].

Hausdorff's whimsical verse later took on tragic significance for his own family; see the excerpt from his farewell letter to the lawyer Hans Wollstein, quoted in Section 9.4.

Hausdorff maintained friendly relations with a number of other colleagues at Bonn University, such as the geographer Alfred Philippson, the orientalist Paul Ernst Kahle, the archaeologist Richard Delbrueck, the theologian Gustav Hölscher, the literary scholar Oskar Walzel and the Egyptologist Hans Bonnet.

Regarding Alfred Philippson, there were already ancestral ties connecting his father, Rabbi Ludwig Philippson, with the families of Hausdorff and his wife. Ludwig Philippson was the most important representative of the liberal Jewish reform movement in the German states. In 1837 he founded the *Allgemeine Zeitung des Judenthums*, which he edited for 52 years until his death in 1889. Philippson was a distinguished scholar, whose translation of the Hebrew Bible was especially influential for Jewish religious practice in the German-speaking world. His friends and comrades-in-arms included Rabbi Abraham Meyer Goldschmidt, the paternal grandfather of Charlotte Hausdorff (see 2.3). In 1869, Ludwig Philippson was involved in the founding of the German-Israelite Community Federation, in which Hausdorff's father Louis Hausdorff served as a member of the governing body from 1878 until his death in 1896 (see 1.3 and 1.5). Together with Abraham Geiger and Salomon Neumann, Philippson established the Hochschule für die Wissenschaft des Judenthums (College for the Science of Judaism) in Berlin, which opened in 1872.

Alfred Philippson studied geography in Leipzig, taking his doctorate in 1886; five years later he habilitated at the University of Bonn. As a private lecturer, he undertook a series of research trips to the eastern Mediterranean and Russia. His reports on these regions attracted the interest not only of geographers and geologists but also archaeologists and historians. After gaining an initial appointment at the University of Bern, he left for a full professorship in Halle in 1906, before returning to Bonn in 1911 to assume the chair in geography. While there, Philippson devoted himself to studying Rhenish geography, about which he wrote a very influential textbook. Through his efforts to expand the library and collections of Bonn's Institute of Geography, he succeeded in making it one of the most modern in Europe before retiring in 1929. The children of the Hausdorff and Philippson families played an important part in strengthening their friendly relations, as Nora Hausdorff was very fond of Dora Philippson, who was four years her senior.

The Orientalist Paul Ernst Kahle was, like Philippson, Hausdorff's colleague in the Faculty of Philosophy. Kahle studied Oriental languages and theology before habilitating in 1909 as a private lecturer for Semitic philology in Halle. Five years later, he received his first professorship in Giessen, before coming to Bonn in 1923 as full professor and director of the Oriental Seminar. Under his leadership, the

Bonn seminar became a world-renowned institution. Even during the first five years of the Nazi dictatorship, it was able to operate relatively unchallenged with an international group of staff members and visitors. Kahle also played a central role in the German Oriental Society. His main field of work was the history of the Hebrew language as well as translations of the Hebrew Bible in antiquity. His research gained him international recognition, and his philological theories regarding the origins of the biblical texts were later confirmed by findings from the Dead Sea Scrolls.

In 1917, Kahle married Marie Grisevius, an elementary school teacher 18 years his junior. She was a lively, socially engaged and fair-minded woman. Like her husband, she detested National Socialist ideology and its antisemitism. The five sons of the Kahle couple were brought up this way so that, despite pressure, they refused join the Hitler Youth. According to Nora Hausdorff's accounts, the Kahle and Hausdorff families were on friendly terms and met on occasion socially. Marie Kahle liked to sing, while accompanied by Hausdorff on the piano. In contrast to Charlotte Hausdorff, she was an emancipated woman who drove a car herself and held rather left-wing views.

Another friend was the Lutheran theologian and Old Testament scholar, Gustav Hölscher who, like Hausdorff, had attended the Nikolai Gymnasium in Leipzig. Hölscher followed in his father's footsteps, as it were, as he had been a pastor at the Nikolai Church. The elder Hölscher's son turned out to have an impressive ability for ancient languages. During the course of his studies in theology and oriental languages at Erlangen, Leipzig, and Berlin, Hölscher became proficient in Hebrew, Syriac, Arabic, Assyrian and Persian. After his habilitation in Halle and appointments as full professor in Giessen and Marburg, the Prussian Ministry of Culture brought him to Bonn in 1929. Prior to his appointment, the theological faculty had four vacant professorships, which placed Hölscher before the immediate task of filling the other three positions in order to move the faculty forward again. He accomplished this brilliantly by nominating a trio of eminent theologians: the New Testament scholar Karl Ludwig Schmidt, the dogmatist Karl Barth and the church historian Ernst Wolf. In just two years, the number of students in theology increased from 99 to 460.

How Hölscher and Hausdorff got to know one another remains unclear, though one possibility is that their relationship came about through Hölscher's second wife, Gertrud, *née* von Meibom, the widow of Axel Schur, who died at age 39 in 1930. Hausdorff had helped arrange Schur's appointment as a private lecturer in mathematics in 1927. Axel Schur and his wife were both highly educated and liberal-minded people who enjoyed socializing with the Hausdorffs.

We know of Hausdorff's friendly interactions with the Germanist and literary scholar Oskar Walzel only from a short passage in the latter's memoirs. Therein Walzel recalled a great number of former colleagues with whom he had been on friendly terms, and noted that "we also consorted with mathematicians, especially with the one who once wrote poetry under the assumed name of Paul Mongré" [Enders 1956, 225]. The Austrian Oskar Walzel had studied in Vienna and Berlin and habilitated in 1894. After professorships in Bern and Dresden, he became a full professor in Bonn and director of the German Language Seminar in 1921. His interdisciplinary approach, explicated in *Wechselseitige Erhellung der Künste* (*Mutual Illumination of the Arts*) [Walzel 1917], is still relevant today. Walzel was editor of the *Handbuch der Literaturwissenschaft*, to which he himself contributed the volumes *Gehalt und Gestalt im Kunstwerk des Dichters* (1923) and *Deutsche Dichtung von Gottsched bis zur Gegenwart* (1927–1930). He became emeritus professor in 1933, but continued lecturing until 1935. Afterward, he came under sharp attack from National Socialist students due to "jüdische Versippung" (his wife was Jewish) and for "political and ideological unreliability," which led to the revocation of his teaching license in July 1936. In the fall of 1944, his wife was deported to Theresienstadt, where she died. Feeling isolated and in despair, Walzel lost his life on December 29, 1944 during a bombing raid on Bonn, under circumstances that are not entirely clear. In our closing Section 9.5 we will describe the further fates of Hausdorff's other friends from Bonn University mentioned above, as well as the important role played by the Egyptologist Hans Bonnet in saving the documents in Hausdorff's estate.

Hausdorff's home was open to younger colleagues from Germany and abroad as well as students who sought his personal or professional advice. His friendship with Paul Alexandroff, including the latter's several visits to Bonn, are well documented in their correspondence. Their letters reveal that, despite their differences in age, they were personally quite close. Both of them described, among other things, their respective health problems, which in Hausdorff's case often involved intestinal problems. In a letter to Alexandroff from October 7, 1931, he reported that a series of thorough examinations "in the course of the last year and a half [had] revealed nothing malignant," but he then went on to say:

> Unfortunately, this does not change the fact that my intestinal complaints – whether they be nervous spasms or the remnants of a catarrhal inflammation – have now lasted, with few interruptions, already a year and thoroughly impair my pleasure for life. (*HGW* IX, 111–112)

On top of such physical ailments came – as his daughter and Else Pappenheim both reported – Hausdorff's disposition to hypochondria, which certainly did not make matters easier. His state of mind, in this respect, is reflected in a letter from November 17, 1930 to Alfred Pringsheim, which ends: "In the hope that you feel very healthy and consequently less old than I do, your very devoted F. Hausdorff sends you his warmest regards" (*HGW* IX, 561). Pringsheim was then 80, so 18 years older than he. Hausdorff also worried about memory loss, though the following passage from a letter to Friedrich Engel, dated May 14, 1927, was perhaps written with more self-irony than real apprehension:

> In general, I find that mathematics is growing over my head and that I know less with each new day, first relatively, since much more is being produced than one can still grasp, but second, absolutely, in that I

> forget more every day than I learn. It is a comfort to me to assume that things go the same way for others, but perhaps this assumption is only self-deception. (*HGW* IX, 236–237)

If Hausdorff really did fear memory loss, then his worries were needless, because his mind remained fresh to the last day.

Following this slight digression, let us return to some of those who were guests in the Hausdorffs' home. Paul Alexandroff reported in a letter to Hausdorff from 1930 that his friend Andrei Kolmogoroff would be staying in Paris during most of the winter and hoped to arrange a visit in Bonn. Whether this plan was ever realized we do not know, but in any case, the Polish mathematicians Kazimierz Kuratowski and Bronislaw Knaster did make a trip to Bonn and were able to visit with Hausdorff. Knaster told this story several times to his student Roman Duda, and in doing so always raved about Hausdorff's hospitality and the excellent wine. This visit probably took place following the International Congress of Mathematicians held in Zurich in 1932.

Younger mathematicians from Bonn, like the aforementioned Axel Schur and his wife, enjoyed socializing with the Hausdorffs. Magda Dierkesmann, who studied mathematics in Bonn, reported that the Schurs regularly consorted with them. Her own first social contact with the Hausdorffs came through Frau Schur, who got along particularly well with Charlotte Hausdorff. After she was introduced to the family, Magda Dierkesmann became a frequent guest as well. Nora Hausdorff also reported on a very friendly relationship with Franz von Krbek and Margarete Flandorffer, whom Krbek married some years later. Von Krbek had studied mathematics, physics, and astronomy in Budapest, receiving his doctorate in 1921 under Leopold Fejér. From 1931 to 1935 he lived in Bonn, where he was an assistant at the Institute of Physics. Margarete Flandorffer came from Greifswald, where she became a good friend of Nora; it was probably in the Hausdorffs' home that she first met her future husband. Von Krbek was witty and charming and, even though they shared no close mutual interests in mathematics, he certainly told Hausdorff some interesting things about physics. His book, *The Fundamentals of Quantum Mechanics and its Mathematics*, was published in Berlin in 1936.

Various testimonials from students indicate that Hausdorff enjoyed good personal relationships with many others. Some, like Magda Dierkesmann and Erna Bannow, frequented his home. Bannow studied in Bonn for two semesters, starting in 1930, after which she left for Göttingen, where she studied under Emmy Noether. She later married the Hamburg mathematician Ernst Witt. Her photograph of Hausdorff from March 1932, showing him standing in front of the main building of the university, is reproduced in 8.5. Frau Witt informed us that Hausdorff's students always presented him with a large bouquet of flowers on his birthday. She herself sent him a bouquet from Göttingen on November 8, 1932, for which Hausdorff thanked her warmly in a letter (*HGW* IX, 142).

Two of Hausdorff's favorite gathering places outside his home were the "Lese" and the Bonner Bürgerverein. The "Lese" refers to the Bonn Reading and Recreation Society, first founded in 1787 by men "of education and standing" committed to the ideals of the Enlightenment; Hausdorff became a member in 1924. Its reading rooms were well stocked with literature and leading newspapers and magazines, so visitors had plenty to read, or they would come to exchange views on current political, scientific and cultural issues. The main building of the Lese was a prestigious city palace at the end of the Hofgarten, less than 5 minutes walk from the main university building. It extended to the banks of the Rhine and housed a large library, a ballroom, social rooms, and – something Hausdorff surely welcomed – its own wine shop. In October 1944, it was completely destroyed during an air raid. Nora Hausdorff and Günter Bergmann both recalled how Hausdorff enjoyed socializing with colleagues and friends over a glass of wine at the Lese.

The Bonner Bürgerverein was founded in 1862 as a private club for "community instruction, sociable entertainment, and cultural education." From the outset, this club had a predominantly Catholic membership, in sharp contrast to the Reading and Recreation Society. It was located at beginning of Poppelsdorfer Allee in an impressive four-story building with a restaurant and terrace garden café; it also housed three bowling alleys. Hausdorff would hardly have considered joining such a Christian-oriented association, but he took advantage of its gastronomic specialties and also enjoyed bowling there with his colleague Johann Oswald Müller.

Nora Hausdorff consistently expressed the view that her father was an apolitical man. It may be true that Felix Hausdorff was not particularly interested in, perhaps even disdained, the day-to-day business of politics and party wrangling. Nevertheless, he was very well aware of overall social developments and their potential dangers. His dedicatory poem for Theodor Posner as well as some of his essays from earlier days bear witness to this. At the beginning of 1932, he still spoke of the National Socialist danger with biting irony. In a letter written to Hilbert on his 70th birthday, January 23, 1932, he considered the problem of finding an appropriate honorary title that he might bestow upon him:

> Since the title *princeps mathematicorum* [C.F. Gauss] is already taken, I would suggest naming you *dux mathematicorum*, if the name dux, duce, Führer would not be so politically discredited today by people who offer themselves to lead the German people on the basis of a self-issued driving license [*Führerschein*]. I would like to call you *lux mathematices* [$\cdots$]. (*HGW* IX, 343)

At that time, Hausdorff probably did not imagine that this "Führer's" self-issued driver's license would one year later become officially valid.

Chapter 9

Hausdorff's Life during the Nazi Dictatorship

As recounted in the preceding chapter, the years in Bonn up until 1933 were happy ones for Felix Hausdorff and his family. Largely sheltered from the hardships of the Great Depression, they went about their lives never really imagining how swiftly everything in their seemingly stable little world would come crashing down. Hitler's National Socialist German Workers' Party (NSDAP or Nazi Party) had a clear vision for the future of Germany. One did not need to have read *Mein Kampf* to recognize that the twin pillars of the Nazi movement were German national expansionism and antisemitism.

9.1 Up to the Pogrom of November 1938

The program adopted by the NSDAP on February 25, 1920 at the Hofbräuhaus in Munich contained 25 points, the fourth of which stated their position on citizenship.

> Only those who are *Volksgenossen* (comrades of the people) can be citizens. Only those of German blood can be *Volksgenossen*, without regard to confession. Therefore, no Jew can be a *Volksgenosse*.

Thus, a necessary condition for citizenship was to demonstrate that "German blood" flowed in one's veins. Since such a blood test had yet to be devised, the Nazis would implement a substitute criterion: a person had to show evidence that there was no "Jewish blood" in their family going back two generations. Such people were then dubbed Aryans, who were in most cases presumed to be *Volksgenossen*, but not always. Social democrats and communists failed that test because they opposed the Nazi Party's cause, which claimed to be a peoples' movement for national renewal after the shameful defeat Germany had suffered

E. Brieskorn, W. Purkert, *Felix Hausdorff*, Vita Mathematica 21,
https://doi.org/10.1007/978-3-031-52135-5_9

in World War I. Many could not bear the idea that German troops were defeated on the battlefield, especially when General Hindenburg himself spoke otherwise. Domestic enemies had undermined the country's great cause, stabbing the military in its back, but who were those enemies? Hitler's Party had a simple answer that millions of Germans would come to accept, even if they did not think of themselves as *Volksgenossen*.

Point 6 of the NSDAP program stated that only citizens could hold public offices. While some points of the program with anti-capitalist overtones gradually faded into obscurity, racist antisemitism became state doctrine after Adolf Hitler was appointed Reich's Chancellor on January 30, 1933. On February 28,1933, one day after the Reichstag fire, Hindenburg issued the Reich President's Decree for the Protection of the People and the State, the so-called "Reichstag Fire Decree." Although enacted legally, in practice it gave the Nazis *carte blanche* to arrest anyone they deemed a threat to their cause. Hindenburg's decree thus served as the ostensible legal basis for the first wave of terror instituted by Hitler's henchmen. They arbitrarily rounded up political opponents, including not a few Jews, placing them under "protective custody" in prisons and concentration camps, where they were mistreated and in some cases even murdered. Their primary target were members of the Communist Party (KPD) in the Reichstag, many of whom fled. This opened the door for Hitler to assume full dictatorial power, and on March 23, 1933 his coalition government steamrolled the opposing Social Democrats in the Reichstag in passing the "Law for the Redress of the Distress of the People and the Reich," better known as the "Enabling Act" (*Ermächtigungsgesetz*). This effectively abolished the separation of powers and, together with the "Reichstag Fire Decree," provided the foundation for the establishment of the National Socialist dictatorship within the formal structure established by the constitution of the Weimar Republic [Broszat 1984], [Evans 2004].

One particular victim of the "Reichstag Fire Decree" was a close friend of the Hausdorffs, namely the psychiatrist Otto Löwenstein (see Section 8.7). Löwenstein soon realized the immediate danger, after his apartment was searched and his clinic occupied by members of the SA, who sought to take him into "protective custody." Warned by a colleague from the clinic, he was able to escape via the Saarland to Switzerland, where his family soon followed him. In Switzerland, he worked from 1933–1939 at a private sanatorium in Nyon, where he also founded and directed a special children's clinic. In 1939, the family moved to New York, where he first worked at New York University and from 1947 onward held a professorship at Columbia University. He died in New York on March 25, 1965. After emigrating to Switzerland, Marta Löwenstein assisted her husband in his work with handicapped children. She died in the same year as her husband, on October 12, 1965, in Houston, Texas.

This first wave of violence reached its peak in early April, when the SA staged "spontaneous" boycotts, while sporadically trashing Jewish businesses throughout the German Reich. A week later, the government issued the Law for the Restoration of the Professional Civil Service (*Gesetz zur Wiederherstellung des Berufs-*

beamtentums), the first of the many racial laws that would soon follow. Since professors who taught at institutions of higher education were in most instances civil servants, this measure had immediate consequences for the German universities. Paragraph 1 of the law read:

> Civil servants who are not of Aryan descent shall be placed in retirement; those who are honorary civil servants shall be dismissed from office.

Paragraph 2 concerned exceptional cases:

> Paragraph 1 shall not apply to those who were already civil servants before August 1, 1914, or to those who fought at the front for the German Reich or its allies in the World War, or whose fathers or sons were killed in the World War. (Reich Law Gazette, Part 1 (No. 34 of April 7, 1933), p. 175).

All civil servants were required to fill out a "Questionnaire for the Implementation of the Law for the Restoration of the Professional Civil Service." They were to give not only personal data but also information about their parents and grandparents, about whom only one line mattered, namely "confession (but also former confession)." In addition, respondents had to give account of earlier membership or activities in political organizations and also whether any of the exceptions in Paragraph 2 applied to them. At the universities and colleges, even those who were not civil servants, such as private lecturers, had to submit this questionnaire.

Section 8.7 briefly touched on the role played by Gustav Hölscher in recruiting three leading theologians to Bonn: Karl Ludwig Schmidt, Karl Barth, and Ernst Wolf. This brilliant period proved to be short-lived, however, once the National Socialists seized power. Schmidt – unusually for a theologian – had been a member of the SPD since 1924; he was already dismissed in September 1933 and emigrated to Switzerland. Karl Barth, one of the co-founders of the Confessing Church, spoke out sharply against the Nazis' persecution of the Jews and refused to take the oath of allegiance to Hitler; he was dismissed in 1934 and returned to Basel. Wolf was another leading member of the Confessing Church in Bonn; he was dismissed in 1934 and transferred to Halle in 1935. Hölscher himself was also dismissed in 1934 for participation in a manifesto against the National Socialist "Reichsbischof" Ludwig Müller in Bonn; he was transferred to Heidelberg one year later.

Regarding Gustav Hölscher's relationship with Hausdorff, his son, the classical scholar Uvo Hölscher, reported the following on 17 December 1991:

> I remember very well from conversations how moved [my father] was by Hausdorff's fate; they must have been close during the Bonn years, from 1930 on. [⋯] I also think I remember a visit to the Hausdorffs with us two sons. (Brieskorn estate, folder no. 20).

In a postscript, Uvo Hölscher refers to a passage at the end of his father's memoirs, written in 1945. There, he says, his father also had the Hausdorffs in mind when he wrote:

> A condemnation of the persecution of the Jews had to be self-evident for every Christian. What help and friendship we were able to offer Jewish friends was unfortunately little enough; some of them chose to escape from the cruelties of a life they found intolerable by taking poison.

9.1.1 Impact on Mathematics in Bonn

At the Mathematics Seminar in Bonn, the Nazi's initial law had no immediate effect, since both Felix Hausdorff and Otto Toeplitz had already been civil servants before 1914, so for the time being they remained in their positions. Neither were any of the associate professors and private lecturers affected by the law. Nevertheless, the atmosphere became tense, owing to the fact that Ernst August Weiss turned out to be a very dedicated and active Nazi (on earlier conflicts in Bonn involving Weiss, see Section 8.2).

Weiss only became a member of the NSDAP in 1937, but he joined the SA in July 1933 and soon worked his way up to the position of *Obertruppführer* and brigade adjutant. Occasionally he even appeared in SA uniform at the Mathematics Seminar. He was especially well known among Nazi loyalists for his role in organizing "mathematical work camps" for students in Kronenburg in the Eifel region. These typically lasted for about 14 days, during which camp life focused on physical training, military-structured guard duties, contacts with local SA figures, and lectures and activities promoting National Socialist education. The students presented mathematical lectures on a daily basis, and occasionally Weiss also gave lectures, e.g., on his experiences connected with mathematical teaching in France. After the first work camp in Kronenburg during the Easter vacation of 1934, others took place on a semi-regular basis up until the fall vacation of 1938 [Segal 2003, 188–197].

Weiss was also a proponent of Bieberbach's obscure ideas about the "racial roots" of mathematical thinking. Ludwig Bieberbach was a well-known function-theorist at Berlin University, who drifted to the right during the 1920s and became an active National Socialist in 1933 [Mehrtens 1987]. Thereafter, in several publications and various other activities he sought to promote a specifically "German mathematics" that was essentially different from other "non-Aryan" approaches, in particular an alleged "Jewish mathematical style" [Segal 2003, 334–418]. Bieberbach tried to legitimize this "völkisch" movement through the journal *Deutsche Mathematik*, which was published by Theodor Vahlen with Bieberbach as its editor. Weiss served as co-editor from the first volume (1936) onward, and he also published in it, including a "Genealogy of Eduard Study" (vol. 1 (1936), 711–715), from which he claimed to have distinguished "clearly two main lines of transmission of valuable hereditary material." The labor camps at Kronenburg and those at other universities inspired by it were also reported in *Deutsche Mathematik*. With regard to Hausdorff, Weiss appears to have behaved in a distant, but proper manner. In Hausdorff's extensive lists of those to whom he should sent offprints, Weiss was named as a potential recipient around 1936.

Among the students, some who were particularly zealous supporters of National Socialism were made "heads of departments" (*Fachabteilungsleiter*). For mathematics in the Bonn student body this student was a certain Hans Peters, who received his doctorate in 1936 under Weiss. On October 31, 1934, Peters sent a report on the individual mathematics instructors to Fritz Kubach, then a student in Heidelberg who was the corresponding national leader in this organization. Peters briefly assessed these mathematicians as follows:

> Beck, Hans: geometry and function theory; sole seminar director; accommodating, but not always clear and transparent;
> Hausdorff, Felix: analyst; Jew; scientifically exceptionally able; reserved, in no way damaging to the student body;
> Toeplitz, Otto: analyst; Jew; to be treated with great caution; man with connections who usually supports his students well;
> Weiss, E. A.: geometer, algebraist; very active in the interests of the student body even before January 30, 1933; S.A. *Rottenführer* [section leader], brigade adjutant; the lecturer in whom I place my full confidence, which he always deserves.

After this, Peters goes on to rant about the need for new scientific journals; he ends by proclaiming:

> The conviction that led us to carry out our two math. work camps, I continue to share: as National Socialists, we must conquer the university; we can conquer it only on the basis of our own scientific achievement.

One method Nazi students used in 1933 and 1934 "to conquer the university" was by disrupting courses or staging boycotts of lectures given by Jewish professors, such as the boycott directed against Edmund Landau in Göttingen [Schappacher 1987], [Schappacher/Scholz 1992]. Hausdorff's lectures were also once disrupted during the winter semester of 1934/35, the last time he was able to teach in Bonn. In the 40-page manuscript for his course "Infinitesimal Calculus III" in Hausdorff's estate, he added a note: "aborted 20.11." During the four decades of his teaching career, Hausdorff only once had canceled a lecture, ironically also due to political turmoil, namely the short-lived Kapp Putsch that rocked Germany in March 1920. From this, one assumes something serious must have happened on November 20, 1934 that caused him to break off his lecture. A search revealed that two days afterward an article appeared in a Nazi newspaper about a "Working Conference of the NSD Student League in Bonn" aimed at the "political education" of the student body, which took place around November 20. There it was emphasized that the Führer himself "had placed the entire political education of German students and technical students under the control of the NSD Student League, as the old political fighting force of the movement at the German universities." The article then added:

> The most important thing for the young student, [Hitler] said, is *political* education. [⋯] This semester the field of political education has the subject of "race and nationality," on which thorough work is to be done. All should help do their part in the great task of shaping the new type of German student who is a National Socialist.

Clearly, some members of the NSD-Studentenbund immediately began this task by disrupting the lecture of a Jewish professor, who happened to be teaching in the same building.

9.1.2 Käthe Schmid, Hausdorff's Cousin

While this episode proved relatively harmless, the Hausdorffs had already experienced among their own relatives the shocking brutality and lawlessness of the National Socialist regime. The victim in this case was the husband of Käthe (Käte) Schmid. She was born in Schwerin in 1899 as Käthe Tietz, daughter of Adolf and Hedwig Tietz; Adolf Tietz was a brother of Hausdorff's mother, Hedwig. Käthe Schmid's biological father was a Baron von Alvensleben, with whom her mother had an affair. Adolf Tietz recognized Käthe as his child and he was listed as her father on the birth certificate; thus, Käthe was considered Hausdorff's cousin.

In 1921 she married Wilhelm Eduard (Willi) Schmid, who had studied Romance languages and literature, art history, musicology, and pedagogy, taking his doctorate in 1923. From 1924 he worked as a music critic and feature writer at the *Bayerischer Kurier* and later at the *Münchener Neuesten Nachrichten*. In addition, he wrote for professional journals and tried his hand as a freelance writer. Schmid played viola da gamba and cello superbly and occasionally toured Germany and especially Italy with a string quintet he founded; they were credited with reviving interest in Renaissance and early Baroque music. The Hausdorffs were very fond of Käthe and Willi Schmid; on one visit to Bonn, they took a beautiful hike together in the Siebengebirge, one among numerous other meetings in Schwerin, Arendsee, Greifswald and Bad Reichenhall (*HGW* IX, 600).

On the evening of June 30, 1934, Willi Schmid was arrested by an SS commando in his apartment and murdered that same night in the Dachau concentration camp. This murder took place in the context of the liquidation of the SA leadership, the so-called Röhm Putsch, ordered by Hitler and carried out by the SS. Willi Schmid had nothing whatsoever to do with the SA, so in all likelihood his arrest involved a case of mistaken identity with the SA leader Wilhelm Schmid (often spelled Schmidt). Although even Rudolf Hess contacted Käthe Schmid to apologize for this "mishap," she had to fight for a long time to receive a pension for their three children from the state (she had refused money from the NSDAP or the SS).

The Hausdorffs were stunned when they received this terrible news from Munich. In a postcard addressed to Käthe Schmid and her mother Hedwig (Hedel) Tietz, who was visiting Munich at the time of her son-in-law's murder, they wrote:

> You poor unfortunates, this blow is so appalling, so terrible, that one freezes & falls silent. Day and night we have only one thought – why, why? Is there anything we can do to help you?
> Your Lotte
>
> Käthe, Hedel, you poor things, we are completely paralyzed by horror, sorrow and compassion. I don't know a word of comfort. – If you have any financial difficulties, I will help as much as I can.
> In deepest sorrow
> Your Felix
> (*HGW* IX, 593)

Hedwig Tietz died in Schwerin on July 25, 1934, less than a month after the murder of her son-in-law. Charlotte and Felix Hausdorff sent a message of condolence, borne of deep sympathy, to Käthe Schmid on the death of her mother, which ended:

> Sometimes we doubt whether you can bear all this; but in your youth, in your soul, in your spirit are such tremendous forces – we hope with all our hearts that they will not let you sink.
> In deepest mutual mourning
> Your Felix
> (*HGW* IX, 596–597)

As Hausdorff had hoped, Käthe Schmid found new tasks, which in addition to caring for the welfare of her three children fulfilled her and gave her the strength to carry on with her life. In the fall of 1934, with the support of Oswald Spengler, she began to compile a book based on her husband's publications and posthumous papers. After submitting the text to the Gestapo for approval, the book was published in 1935 under the title: *Unvollendete Symphonie. Gedanken und Dichtung von Willi Schmid* (*Unfinished Symphony. Thoughts and Poetry of Willi Schmid*). A somewhat modified version was published two years later in Salzburg. The Hausdorffs, of course, also received a copy of the work, but Charlotte wrote to Käthe Schmid: "Felix is not able to read the book, he puts it down again and again, too deeply shaken. Yes, these sounds make all wounds bleed again" (*HGW* IX, 598).

Another important task for Käthe Schmid was the work she did as a volunteer for the German Alpine Club. She and her husband had loved the mountains and made friendships with famous mountaineers, like Willy Merkl. Shortly before Willi Schmid was murdered, Merkl had embarked on the German Himalayan Expedition with the goal of climbing Nanga Parbat, during which time the Schmids set up the expedition's press office in their apartment. In mid-July 1934, just two weeks after her husband's tragic death, Käthe received news of a terrible disaster that claimed the lives of ten climbers, including Merkl. Although she insisted on continuing the press work despite everything, the Alpine Club sent Hermann Hoerlin to help aid her work.

Hoerlin was a physicist who specialized in studies of cosmic radiation by means of photography, often under difficult conditions. That accorded well with his lifestyle, as he was also a devotee of extreme mountaineering. In addition to a number of Alpine mountains, he had climbed Jongsong Peak and had reached the top of the Kanchenjunga massif as part of an international Himalayan expedition in 1930. Two years later, he spent several weeks in Peru at altitudes above 5,000 meters doing research on cosmic radiation. Hoerlin and Schmid gradually grew closer while working together, leading to a long-distance romantic relationship they conducted for several years. Because of antisemitism, which was particularly virulent in Munich, Käthe Schmid and her children moved to Salzburg in March 1937, whereas Hoerlin worked near Dessau as head of the physics laboratory at the Agfa company.

The Nuremberg Race Laws, enacted in 1935, forbade marriages of "Aryans" to Jews; however, marriages between "mongrels of the 1st degree" and "Aryans" were possible, but required special permission. Hoerlin and Schmid fought a long battle with the Nazi bureaucracy before they could obtain this right, since Schmid had to prove that the "Aryan" von Alvensleben was her biological father. By the time of their marriage in Berlin, on July 12, 1938, they had already made plans for leaving Germany, as Hoerlin had in the meantime obtained a similar position with an Agfa/Ansco subsidiary in Binghamton, New York, where their daughter Bettina was born the next year. Many years later, Bettina Hoerlin told this whole story in *Steps of Courage. My Parents' Journey from Nazi Germany to America* [Hoerlin 2011].

Before leaving the country, the Hausdorffs met with Käthe once more in Lucerne, Switzerland, probably in the spring of 1938 on what was likely the Hausdorffs' last trip abroad. On July 16, 1938, he wrote a farewell letter to Käthe before her departure:

> My dear Katja,
>
> All my heartfelt wishes accompany you, your children and your spouse to the new life beyond the Atlantic Ocean. Despite this spatial separation and despite my 70 years, I hope to see you again on earth. (*HGW* IX, 600)

This hope would not be fulfilled, of course, as the Jews in Nazi Germany were about to experience a living hell.

Hermann Hoerlin's later research on the harmful atmospheric effects of nuclear weapons testing, especially on the ozone layer, helped in finally bringing about a moratorium on such testing between the great powers. Hoerlin died on November 6, 1983; Nobel laureate Hans Bethe delivered the eulogy at his funeral. Käthe Hoerlin died two years later, survived by her daughter Duscha Schmid, who married the distinguished physicist Victor Weisskopf. Weisskopf was a co-founder of quantum electrodynamics and served as director of the European nuclear research center CERN in Geneva from 1961 to 1965.

In her old age, Duscha Weisskopf wrote a book about her father, *Willi Schmid: A Life in Germany* [Weisskopf 2004/2012]. It was she who gave Egbert Brieskorn the letters from the Hausdorffs to her mother. In an accompanying letter she wrote:

> I am very glad that these cards and letters are useful to you. They are so loving that I'm sure you get a better picture of the authors' humanity. By the way, my mother burned many letters before she emigrated, because she could not take everything with her. But she carefully kept the Hausdorff letters.

9.1.3 In the Wake of the Nuremberg Racial Laws

On January 21, 1935, the Law on Dismissal and Transfer of University Teachers was enacted as part of the Nazi's reconstruction of the German higher education system. The law stipulated that "civil servant university professors of the German Reich shall be released from their official duties [$\cdots$ when] they reach the age of 65," unless "overriding university interests" militate against this. In the case of a Jewish professor, the Ministry certainly had no motivation to consider such an exception, and so Hausdorff received a letter, dated March 5, 1935, informing him that he had been placed in retirement. This formal communique contained not a single word of appreciation or thanks for his 40 years of faithful service to German higher education. Certainly he had made no plans to stop teaching, as he had already announced his courses for the coming summer semester of 1935: a 4-hour course on elliptic functions and a seminar, offered together with Toeplitz and Bessel-Hagen. Since his retirement had not been voluntary, at least the financial terms seemed satisfactory: he would be offered a regular pension as an emeritus professor.

Later that year, from September 10–16, 1935, the annual Reich Party Congress of the NSDAP took place in Nuremberg. In conjunction with it, the Reichstag, which had by now been reduced to rubber-stamping the government's proposals, was also specially convened in Nuremberg. It then unanimously passed two new laws: the Reich Citizenship Law and the Law for the Protection of German Blood and German Honor; both came to be known as the "Nuremberg Racial Laws." Two months later, the First Ordinance to the Reich Citizenship Law of November 14, 1935 was enacted for the implementation of these laws. It was at this point that the Nazi government set forth its policy regarding who was to be considered a Jew, namely those persons with three or four grandparents who were Jewish by race. However, since determination of a person's "racial characteristics" clearly would have led to a morass of difficulties, the Nazis opted for the more practical method of relying on religious affiliation. Thus, a person's grandparent was considered Jewish "by race" if they belonged to or had belonged to a Jewish religious community.

This ordinance then stipulated the following:

> (1) A Jew cannot be a citizen of the Reich. He shall not be entitled to vote in political matters, nor may he hold public office.
> (2) Jewish civil servants shall enter into retirement on 31 December 1935.

These conditions thus eliminated the exemptions provided by the 1933 Law for the Restoration of the Professional Civil Service. An immediate consequence for Bonn was that Otto Toeplitz had to retire as of January 1, 1936; his retirement pay, set at 35% of pensionable service income, was significantly lower than the pay of an emeritus professor.

Felix Hausdorff received a letter on December 17, 1935 informing him that, as of January 1, he would no longer enjoy the status of an emeritus professor, which paid him 10,676 Reichsmark annually. Instead, his retirement income was to be reduced to only 5,496 Reichsmarks per year, barely more than half of what he had counted on receiving. As it turned out, however, paragraph (2) was not meant to apply to those who already had emeritus status, and the university determined that in such cases the pension payments were to continue as before.

Thus, Hausdorff's material situation was at first still quite comfortable, compared, for instance, with the many Jews who had to live from private means. This circumstance, but also his personal conviction – expressed in the presence of his daughter Nora – "that they probably won't do anything to us old people," helps to explain why Hausdorff made no attempt to emigrate during the years when that might still have been possible. He did not imagine, as few did at that time, that he and his family would witness a series of new decrees and orders, especially from 1939 onward, some of which directly affected Jewish emeriti and retired civil servants. The Nazi state passed these as part of its effort to oust Jews from economic life and gradually expropriate all Jewish property and wealth through "Aryanisation," a process that accelerated after the November 1938 pogrom. Many Jewish business owners felt compelled to close or sell their companies because of the widespread financial pressure and state-sponsored discrimination, even though fair prices could rarely be obtained under these circumstances.

Hausdorff and his co-owners also decided to sell the publishing house "Spinner und Weber. Hausdorff & Co" to the brothers Arthur Gustav and Ludwig Vogel, who resided in Pössneck in Thuringia. Since sales contracts are no longer available, nothing is known about the financial side of this transaction or whether Hausdorff himself was involved in the sales negotiations. His son-in-law Arthur König was a co-owner, so he may well have represented Hausdorff during this sale, especially since he lived in Jena, not far from Pössneck. The first number of *Spinner und Weber* printed there appeared on October 4, 1935.

Although Hausdorff's livelihood was not strongly affected by the Nazi's anti-Semitic legislation, he was deeply aware of the ongoing discrimination, harassment and persecution of Jews, who were confronted with all kinds of harsh laws and decrees. In response to this, he increasingly withdrew from outside life. Having been forced into retirement, he began to devote himself almost exclusively to his mathematical research. Käthe Schmid, who visited the Hausdorffs in October 1935, wrote from Bonn to her partner Hermann Hoerlin:

> I have it very good with the Hausdorffs. I find in them superior, sensitive people, they are very close to me as I have always known them to be. Adversity has not contorted them, nor does it constrict them spiritually. Felix is so entrenched in mathematics that he cannot in any way be pulled out of it – what happiness that is!

Three years later, in July 1938, Erich Bessel-Hagen wrote to Dora Reimann, who had taken her doctorate under Toeplitz:

> Hausdorff lives very secluded with his wife, does not care about the things of the world and works on mathematical problems. Healthwise his condition is unstable (as it always was), but on average better than before. [Neuenschwander 1996, 256]

Hausdorff's friend and colleague, Otto Toeplitz, responded to the persecution of the Jews very differently: he became deeply involved in a wide variety of activities on behalf of the Jewish community. Already in August 1933, he was elected to the board of the Bonn synagogue and in October he became chairman of the newly founded Jewish Cultural and School Association, a post he occupied until 1936. Toeplitz was strongly committed to founding a Jewish elementary school, for which he conducted negotiations with the school board. On May 1, 1934 this school opened with 63 pupils; by 1935 enrollments rose to 84 Jewish children, practically the entire elementary school-age population in the area; they no longer had to be exposed to the disparagement and hostility they would have faced in the public schools. Toeplitz was also head of the higher education department of the "Reich Representation of German Jews" (*Reichsvertretung der deutschen Juden*), and in this capacity he organized scholarships and other aid for the emigration of gifted Jewish students. He also closely followed the situation at German universities, for which he kept a set of index cards recording dismissals and suicides of university members in all disciplines. Although he essentially withdrew into his mathematical work, Hausdorff deeply appreciated the struggles of Otto and Erna Toeplitz against all the adversities of state-sponsored antisemitism. For their silver wedding anniversary, he composed an elaborate poem under the title "Very Difficult Acrostic for the Toeplitz Couple," in which he refers to the Toeplitzes' social activism in an affectionately appreciative manner (*HGW* IB, 956–957).

After Hausdorff's compulsory retirement, Toeplitz hoped to uphold Bonn's established high standards by recruiting a first-rate mathematician. He set his eyes on Helmut Hasse, who was then director of the Mathematics Institute in Göttingen. This effort failed, however, and subsequent negotiations to fill the two vacant positions dragged on for a long time, as the candidates nominated by the faculty turned down the offer. The situation became more difficult, as the Berlin Ministry recommended considering ideologues like Erhard Tornier and

Maximilian Krafft, two representives of Ludwig Bieberbach's Nazi-oriented "German mathematics" (*HGW* IB, 957–958). After both were rejected, ministerial officials took matters into their own hands by transferring Hausdorff's former position to Göttingen. This gave Hasse the opportunity to attract the distinguished number theorist Carl Ludwig Siegel, who had formerly been the only "Aryan" mathematician in Frankfurt, a once leading center since destroyed by Nazi policies. An associate professorship was then created in Bonn, a position that went to the function theorist Ernst Peschl, an NSDAP member since 1933. The Ministry then appointed the eminent algebraist Wolfgang Krull to fill Toeplitz's chair. Krull was also a member of the NSDAP; shortly after his appointment, he offered a course on "mathematical genealogy." From 1941 to 1943 he served as dean of the faculty, but otherwise he appears not to have engaged in party politics. Peschl was not politically active either; on the contrary, he experienced some difficulties with fanatical party comrades [Segal 2003, 461–462]. By 1943, both Krull and Peschl were no longer in Bonn; Krull worked at the Naval Meteorological Service and Peschl at the German Research Institute for Aviation in Brunswick.

After the retirement of Hausdorff and Toeplitz, those who remained on the staff of the Bonn Mathematics Seminar encountered serious problems themselves. Johann Oswald Müller, who was close to Hausdorff, happened to be married to a Jewish woman whom he refused to divorce. As a result, the Ministry revoked his teaching credentials in 1937; three years later, he died of carcinoma of the esophagus. Hans Beck became an emeritus professor in 1941 and died the next year; he had no successor. His younger rival, E.A. Weiss, never had a chance for a professorship in Bonn. Beck had always disliked his deceased former colleague Eduard Study, which likely shaped his attitude toward the latter's former pupil. In any event, Weiss remained in his assistantship until 1939 – albeit with the title of professor – at which point he obtained a salaried lectureship. Although officially appointed full professor in 1941 at the newly created Reichsuniversität Posen, he never took up this post because he had enlisted in the army after the invasion of the Soviet Union. Afterward, Weiss was deployed as commander of an engineering battalion, where he was seriously wounded; he died in a field hospital near Lake Ilmen in February 1942. After Toeplitz's departure, Erich Bessel-Hagen was Hausdorff's only remaining friend at the Mathematical Seminar in Bonn. Bessel-Hagen was appointed associate professor in 1939; his role during the last years of Hausdorff's life will be taken up later in 9.4.4.

The situation of people in Germany who were Jews, as defined by the Nazis, continued to deteriorate in a myriad of smaller and larger ways following implementation of the Nuremberg Racial Laws. Up until the outbreak of the Second World War, the Nazi regime tried to induce as many Jews as possible to emigrate, while confiscating a large part of their assets by means of the Reich Flight Tax and other measures. Between 1933 and 1939, the proportion of Jews in the urban area of Bonn fell by half from 0.8 to 0.4%. [Rey 1994, 233].

Up until 1937, Hausdorff and his wife continued to take trips abroad, as they still possessed valid passports and the necessary financial means to do so.

They visited Locarno in September 1934, Montreux in October 1935, and stayed in Prague in June 1936. Their last vacation trip was in the fall of 1937 when they visited the Ligurian coast, Santa Margherita Ligure and Genoa, the region Hausdorff identified with his Sant' Ilario. He looked back on this trip wistfully in a letter to Franz Meyer, dated December 6, 1938, a time when such a journey was no longer possible:

> Last fall I was once again in S. Margherita with my wife; on October 26, in melancholy farewell weather, we walked up and down the seafront of Nervi. [$\cdots$] Memories can be beautiful, but no substitute for hopes; I am attuned to Dante's "nessun maggior dolore." (*HGW* IX, 511)

In the 5th canto of Hell in Dante's *Divine Comedy* one reads: "Nessun maggior dolore che ricordarsi del tempo felice ne la miseria" ("In misery to remember past happiness is the greatest pain").

Despite all the difficulties, Hausdorff was still able to receive occasional guests from abroad. Harald Bohr visited him in early 1935 (*HGW* IB, 951) and Otto Neugebauer in February 1936 (*HGW* IX, 313). On the other hand, from 1933 onward, Paul Alexandroff could no longer undertake trips to Germany due to Soviet restrictions. In this respect, Alexandroff's letter from July 2, 1933 is particularly revealing. In it, he first cautiously inquires "whether my letters are capable of pleasing you now, or whether it is rather appropriate, in your interest, for me to hold back my friendly feelings toward you a bit after all." Then he hints that it was now virtually impossible for a Soviet scientist to travel to Germany and expresses how shaken he feels by the persecution and expulsion of Jewish colleagues:

> It was a difficult experience for me to come to terms with the fact that a country, which has almost become my second home, is now virtually closed to me. Perhaps even more difficult is the thought that a large circle of my friends and closest colleagues now stand outside its sphere of activity. You can imagine how difficult all this is for me. (*HGW* IX, 130)

The correspondence with Alexandroff ends with a letter of March 9, 1935, which mainly discusses Hausdorff's cooperation in reading the proofs of *Topologie I* [Alexandroff/Hopf 1935] and his suggestions for improving the text. In the Soviet Union, Stalin's purges were soon underway, making postal traffic abroad and cooperation with foreign scientists dangerous for Soviet scientists. As a consequence of these circumstances, the planned volumes II and III were never completed.

Even before the pogrom of November 9/10, the situation for German Jews in 1938 deteriorated considerably. For five years, the focal point of Nazi regulations had been the complete elimination of Jewish interests in German economic life. Even the sham legality of discrimination and oppression of Jews by laws and decrees, which had been practiced after the lawlessness of the early years, was by now quietly abandoned. Thus, in a strictly confidential government directive to local offices from June 1, 1938, one reads:

Figure 9.1: Celebration of Felix Hausdorff's 70th Birthday at his Home in Bonn From left to right: Arthur König, Felix Hausdorff, (*), Inge Köttgen, Lenore König, Charlotte Hausdorff, (*), (*). The three unidentified older women designated (*) are probably Frida, Louise und Anna Köttgen, aunts of Arthur König. Source: Hausdorff Estate, Capsule 65, Nr. 18.

> All Jews who were punished with a prison sentence of more than one month, or with a fine for which a prison sentence of more than one month has been imposed in the case of non-enforcement, shall be arrested and sent to a concentration camp without interrogation. [Walk 1981/1996, Part II, No. 478]

The following drastic announcements and decrees on the stigmatization and isolation of Jews also obviously affected the Hausdorffs:

- Announcement of the Reich Ministry of the Interior of July 23, 1938:

> Jews who are German citizens must apply by 31 December 1938 for the issuance of an identification card indicating their status as Jews. In all oral applications to authorities, they shall present the identification card without being requested to do so; in written applications, they shall refer to their status as Jews and indicate the issuing locality and number of the identification card. [Walk 1981/1996, Part II, No. 506]

– Decree of the Reich Ministry of the Interior and the Reich Ministry of Justice of August 17, 1938:

> Jews who do not have a given name listed as a Jewish given name in the circular issued by the Ministry of the Interior on August 18, 1938 shall, from January 1, 1939 on, adopt as an additional given name the names "Israel" (for males) or "Sara" (for females). [Walk 1981/1996, Part II, No. 524]

– Decree of the Reich Ministry of the Interior of October 5, 1938:

> All German passports whose holders are Jews shall become invalid. Passports issued earlier are to be surrendered. [Walk 1981/1996, Part II, No. 556]

On November 8, 1938, Hausdorff celebrated his 70th birthday (Figure 9.1). The celebration took place within an intimate family circle, whereas many who remembered him on this special day sent their congratulations in writing (*HGW* IX, 712). This birthday celebration was once again a modest reminiscence of the better times now past. Just one day later, on the night of November 9-10, 1938, all hell broke loose for the Jews in Germany and Austria. Before turning to those dramatic events and their aftermath, however, we first briefly describe the serene inner world of Felix Hausdorff's final mathematical works.

9.2 Hausdorff's Mathematical Works

Hausdorff's paper "Zur Theorie der linearen metrischen Räume" [H 1931, IV: 271–288] (8.6.5) was the last he published in a German journal. Beginning in 1933, he published seven more papers, all in Polish journals: six in *Fundamenta Mathematicae* and one in *Studia Mathematica*. *Fundamenta* also carried a special section called "Problems," a forum in which mathematicians could pose open problems to colleagues around the world; Hausdorff twice published solutions: for Problem 58 (1933) and for Problem 62 (1935) (*HGW* III, 481–482, 527–528).

9.2.1 On δs-Functions

In 1933, *Fundamenta* also carried his short note "On the projectivity of δs-functions" [H 1933a, III: 473–477]. Hausdorff had already introduced the notion of a δs-function or δs-operation in his book *Mengenlehre*. This operation plays an important role in his abstract representation of Suslin sets over an arbitrary set system $\mathfrak{M}$. Hausdorff's note follows directly from an extensive study by Leonid Kantorovitch and E. M. Livenson [Kantorovitch/Livenson 1932]. Let N be a fixed set of sequences of natural numbers and $A = \Phi_N(A_1, A_2, \cdots) = \Phi(A_1, A_2, \cdots)$ be a δs-function. If A_n independently run through a set system $\mathfrak{A}$, then A runs through a set system which shall be denoted by $\Phi\mathfrak{A}$. Let $Z = X \times Y$ and $C \subseteq Z$, and let $\pi(C)$ denote the projection of C onto X. If C traverses a set system $\mathfrak{C}$, then let $\pi(\mathfrak{C})$ denote the set system traversed by the $\pi(C)$, $C \in \mathfrak{C}$. Now if $\mathfrak{C}$ is a system of sets of Z and $\mathfrak{A}$ a system of sets of X, then $\mathfrak{C}$ is *δs-projective to* $\mathfrak{A}$, if for every δs-function Ψ there exists a δs-function Φ such that $\pi(\Psi\mathfrak{C}) = \Phi\mathfrak{A}$ holds.

The main result (fundamental theorem) of Kantorovitch and Livenson states that for the product space $Z = X \times Y$ of two metric spaces, where Y is compact and uncountable, the system of sets closed in Z is δs-projective to the system of closed sets in X. Hausdorff's main theorem (Theorem IV) in [H 1933a] generalizes this theorem substantially by sharpening the projectivity, while requiring much weaker conditions on the spaces. It states:

> *Let X be an arbitrary topological space (separation axioms are not needed) and Y a space satisfying the second countability axiom, then the system $\mathfrak{C}$ of sets open in $Z = X \times Y$ is projective to the system $\mathfrak{A}$ of sets open in X.*

From this also follows the analogous statement for closed sets. Moreover, Hausdorff's proof is considerably simpler than that in [Kantorovitch/Livenson 1932].

9.2.2 On Inner Mappings

Hausdorff's three-part paper "Über innere Abbildungen" ("On Inner Mappings") [H 1934, III: 485–497] deals with the relations between inner mappings and completeness properties of topological spaces. Hausdorff calls a continuous mapping $\varphi: X \to Y$ between topological spaces X and Y an *inner mapping* (today an "open mapping") if for every set U open in X its image $V = \varphi(U)$ is open in Y. He calls a space X topologically complete if it is homeomorphic with a complete metric space. He then proves the following theorem:

> I: *Let $\varphi: X \to Y$ be an inner mapping from X to $Y = \varphi(X)$, then if X is topologically complete, so also is Y.*

Hausdorff had already communicated this theorem in a somewhat more specialized form and without proof in [H 1931]. He also had written up a proof, which can be found in an unpublished paper, dated 29 January, 1931. The argument there depends on a result Hausdorff had already found on 28 October, 1926, when he proved that a space X is topologically complete if and only if it has a closed basis.

In the second part of the present paper, Hausdorff referred to "a very astute, but somewhat involved construction" by Stefan Mazurkiewicz, which the latter used to prove this extension theorem for inner mappings [Mazurkiewicz 1932]: Let $y = \varphi(x)$ be a continuous mapping of the topologically complete separable space X onto the topological space $Y = \varphi(X)$, where $\varphi(x)$ is an inner mapping on $A \subset X$. Then there is a G_δ set P with $A \subset P \subset X$ and such that φ is an inner mapping on P. Hausdorff attempted several times to prove this theorem without assuming separability, but without success. Here he offered a much simpler proof of Mazurkiewicz's theorem than the one originally presented. Not until 1981 was it shown that the separability condition is needed in Mazurkiewicz's theorem, so that Hausdorff's efforts in this respect were bound to fail.

In the third and final part of [H 1934], Hausdorff studies open images of subsets in what he calls Baire spaces. By a Baire space, he means a product $X = X_1 \times X_2 \times \cdots$ of a sequence $\{X_n\}$ of sets with the following metric: For $x = (x_1, x_2, \cdots) \in X$, $x' = (x'_1, x'_2, \cdots) \in X$, the distance $d(x, x')$ is defined by $d(x, x') = \frac{1}{n}$ if the n-th position is the first in which x and x' differ, thus where $x_1 = x'_1, x_2 = x'_2, \cdots x_{n-1} = x'_{n-1}, x_n \neq x'_n$. This metric is non-Archimedean, i.e., $d(x, z) \leq \max\{d(x, y), d(y, z)\}$ holds. For $X_n = \mathbb{N}$ for all n, one obtains Baire's null space [H 1927a, 101–102; III: 145–146]. Hausdorff then proves the following theorem:

> *Every metric space Y is an image of a Baire space X by virtue of an inner mapping $y = \varphi(x)$. This mapping can be extended to an inner mapping of a topologically complete Baire space $V \supset X$.*

Hausdorff also suggests on page 290 of his paper that any metric space with nonarchimedean metric is homeomorphic to a subspace of a Baire space. Already in 1927 he had shown that every subspace of a Baire space has Uryson-Menger dimension ind $X = 0$ (*HGW* III, 770–777).

9.2.3 Closure Spaces

In his paper "Gestufte Räume" ("Closure Spaces") [H 1935b, III: 505–521], Hausdorff defines these as spaces E for which the Kuratowski closure axioms hold. Thus, for each $A \subseteq E$ one has an associated $A_\lambda \subseteq E$ such that, *except for idempotency* $(A_\lambda)_\lambda = A_\lambda$, these axioms hold: $\emptyset_\lambda = \emptyset$, $A \subseteq A_\lambda$, $(A \cup B)_\lambda = A_\lambda \cup B_\lambda$. Moreover, Hausdorff also requires the T_1 separation axiom: for every $x \in E \Rightarrow \{x\}_\lambda = \{x\}$. Hausdorff's goal in this paper is to show that every Fréchet limit space (L-space) generates a closure space and every closure space generates a topological space, so that

> the closure spaces, as a link between L-spaces and topological spaces, are not unworthy of a brief investigation, which may perhaps shed brighter light on some known facts about spatial axiomatics *) [the footnote *) refers to the book [Fréchet 1928a]]. [H 1935b, 486; III: 505]

Hausdorff's formulation might be taken as a slight and probably intended dig at Fréchet's book, which was in many respects faulty. Angus Taylor made a detailed study of Fréchet's book and carefully weighed its strengths and weaknesses; he offered this summary assessment:

> Fréchet's book was too late on the scene to have any hope of displacing the influence of Hausdorff's book of 1914. Moreover, it was not

> constructed in a manner to capture the minds of young French mathematicians who might readily have preferred a French book to a German book on Topology. [Taylor 1985, 361]

In §1 "Topological spaces" of [H 1935b], Hausdorff utilizes the axioms for closed sets which he had formulated in §40 of his *Mengenlehre*. If $\mathfrak{F}, \mathfrak{F}'$ are then two subset systems of E satisfying the axioms for closed sets, then following today's terminology in the case $\mathfrak{F}' \subset \mathfrak{F}$, the topology generated by $\mathfrak{F}$ is *finer* than that generated by $\mathfrak{F}'$ and the latter is *coarser*. Hausdorff instead speaks in this situation of $X' = (E, \mathfrak{F}')$ as an *upper space* of $X = (E, \mathfrak{F})$ and X as a *lower space* of X'. The lowest of all possible spaces for the ground set E is then the *discrete space* $(E, \mathfrak{P}(E))$, where $\mathfrak{P}(E)$ is the power set of E. If the closed hull of a set A is denoted by A_α, then for any $A \subseteq E$: $A_\alpha(X') \supset A_\alpha(X)$, because the upper space has fewer closed sets and consequently larger closed hulls.

In a closure space, the system of sets A with $A_\lambda = A$ satisfies the axioms for the system of closed sets of a topological space. Thus, if we define these sets to be closed, we obtain a topological space arising from the closure space with ground set E. For any set $A \subset E$, we obtain the closed hull A_α in this topology by transfinite repetition of the λ-operation:

$$A_0 = A, \quad A_{\xi+1} = (A_\xi)_\lambda, \quad A_\eta = \bigcup_{\xi<\eta} A_\xi \ \text{(where} \eta \text{ is a limit number)}. \tag{9.1}$$

This process leads after at most card(E) steps to a highest level A_α with $(A_\alpha)_\lambda = A_\alpha$. If one has two closure spaces X, X' over the same ground set E with operations λ, λ', then by analogy with the hull relation for topological spaces, one calls X' a closure upper space relative to X (X a closure lower space relative to X') if $A_{\lambda'} \supset A_\lambda$ for all $A \subseteq E$. From $A_\alpha(Y') \supset A_\alpha(Y)$, it follows that for the induced topological spaces Y, Y', Y' is an upper space relative to Y and Y is a lower space relative to Y'.

An L-space over a ground set E arises when among the set of all sequences of points $(x_1, x_2, \cdots)$, $x_n \in E$, certain sequences are distinguished as convergent with limit $x \in E$. In the system $\mathfrak{K}$ of these convergent sequences $x_n \to x$ the following axioms hold:

(1) *The limit of a convergent sequence is uniquely determined.*
(2) *The constant sequence $(x, x, x, \cdots)$ is convergent with limit x.*
(3) *Every subsequence of a sequence convergent to x converges to x.*

For $A \subset E$, let A_λ be the set of all $x \in E$ for which there exists a convergent sequence $x_n \in A$ with $x_n \to x$. This λ-operation satisfies the axioms for a closure space; thus every L-space induces a closure space and hence an associated topological space with the closed hulls formed by (9.1). Also for L-spaces one can compare convergence systems and define upper spaces and lower spaces in Hausdorff's sense. Let $\mathfrak{K}, \mathfrak{K}'$ be two convergence systems on E with $\mathfrak{K}' \subset \mathfrak{K}$, then A has more limit points in $X = (E, \mathfrak{K})$ than in $X' = (E, \mathfrak{K}')$, so $A_\lambda(X) \supset A_\lambda(X')$, and X is an upper space relative to X', X' a lower space relative to X.

To pursue these investigations further, Hausdorff introduces certain minimal and maximal spaces, which can be briefly explained here for the case of topological spaces. Let $\mathfrak{F}_t$, $t \in T$ be a set of systems of closed sets over E such that $X_t = (E, \mathfrak{F}_t)$ are topological spaces. These have common upper spaces and lower spaces. For an upper space X, $\mathfrak{F} \subset \mathfrak{F}_t$ for all t, and among these upper spaces X there is a lowest (minimal) space $\overline{X}$ with $\overline{\mathfrak{F}} = \bigcap_t \mathfrak{F}_t$. For its closed hulls $A_\alpha(\overline{X}) \supset \bigcup_t A_\alpha(X_t)$ holds. Similarly, among lower spaces X with $\mathfrak{F} \supset \mathfrak{F}_t$ there is an uppermost $\underline{X}$ such that $\underline{\mathfrak{F}}$ is the smallest system with $\underline{\mathfrak{F}} \supset \bigcup_t \mathfrak{F}_t$; for this $A_\alpha(\underline{X}) \subset \bigcap_t A_\alpha(X_t)$. Analogous constructions can be carried out with certain modifications for closure spaces as well as for L-spaces.

Here we have merely sketched how Hausdorff introduced closure spaces "as a link between L-spaces and topological spaces." This discussion omits his more detailed study of these relations, in which especially the notion of continuous decomposition of spaces and special minimal and maximal spaces play a role. However, it should be mentioned that closure spaces have recently gained importance again due to the fact that quotient mappings between such spaces are always hereditary, which is not generally the case in topological spaces (see the commentary on [H 1935b] in *HGW* III, 523).

9.2.4 Independent Subsets and Mappings

In his short note, "On two theorems by G. Fichtenholz and L. Kantorovitch [H 1936a, III: 531–532], Hausdorff elaborated on an earlier functional analytic study [Fichtenholz/Kantorovitch 1935]. In that investigation, the two authors required a combinatorial lemma, which Hausdorff formulates in his note in this form as Theorem 1:

> *If A is of infinite power $\mathfrak{m}$, then there are $2^{\mathfrak{m}}$ essentially distinct mappings from A into A.*

A family $\mathfrak{F}$ of mappings $f : A \to A$ is essentially distinct if for any finite subfamily $f_1, f_2, \cdots f_n$ from $\mathfrak{F}$ there exists $a \in A$ with $f_i(a) \neq f_k(a)$ for $1 \leq i < k \leq n$, thus the values of these mappings are pairwise distinct.

This lemma was originally proved by transfinite induction in a lengthy argument that required four pages. Hausdorff's proof took only ten lines! He considers a set M of power $\mathfrak{m}$; the set $\mathfrak{P}_e(M) = \{X \subset M;\ X \text{ finite}\}$ of all its finite subsets then also has the power $\mathfrak{m}$. Hausdorff's trick came by recognizing that it sufficed to consider $\mathfrak{P}_e(M)$ rather than an arbitrary set A. He let T vary over the $2^{\mathfrak{m}}$ subsets of M, and for each fixed T he defined the mapping from $\mathfrak{P}_e(M)$ into itself by $f_T(X) = X \cap T$. These $2^{\mathfrak{m}}$ mappings are essentially different because the $T_1, T_2, \cdots T_n$ differ pairwise, so $T_i \setminus T_k \cup T_k \setminus T_i \neq \emptyset$. If one chooses an element from each of these nonempty sets to form the finite set X, then $X \cap T_i \neq X \cap T_j$, i.e. $f_{T_i}(X) \neq f_{T_j}(X)$ and so the $f_{T_1}, \cdots f_{T_n}$ are pairwise distinct.

Hausdorff also improved a second lemma in [Fichtenholz/Kantorovitch 1935] concerning the subsets of an infinite set C. A subsystem $\mathfrak{M}$ of the power set $\mathfrak{P}(C)$ is called an *independent system of subsets* of C if for finitely many $M_1, M_2, \cdots M_p$ and $M_1', M_2', \cdots M_q'$ from $\mathfrak{M}$ there always exists

$$M_1 \cap M_2 \cap \cdots \cap M_p \cap (C \setminus M_1') \cap (C \setminus M_2' \cap \cdots \cap (C \setminus M_q') \neq \emptyset,$$

i.e., no Boolean combination of finitely many sets from $\mathfrak{M}$ can yield the empty set. Fichtenholz and Kantorovitch considered the set C of all measurable subsets of a closed interval $[a, b]$, which has the power $2^{\aleph_0}$, and proved as "Lemma IV" that there exists an independent system $\mathfrak{M}$ of subsets of C with the power $2^{2^{\aleph_0}}$.

Hausdorff's Theorem 2 generalized this lemma by proving a purely combinatorial theorem, namely:

> *If C is an infinite set with the power $\mathfrak{m}$, then there exists in $\mathfrak{P}(C)$ an independent system $\mathfrak{M}$ of subsets of power $2^{\mathfrak{m}}$.*

Hausdorff's proof is based on Theorem 1 and a trick similar to the one used to prove that result; it is also much shorter than the proof of Lemma IV in [Fichtenholz/Kantorovitch 1935]. In his commentary on [H 1936a], Ulrich Felgner gave an account of various subsequent developments, including a further generalization of Theorems 1 and 2 made by Hausdorff himself. He pointed out, furthermore, that independent subset systems (and mappings) play an important role today in functional analysis, topology, measure theory and algebra, which led him to conclude:

> These remarks may prove that Hausdorff's theorem on the existence of large independent set systems (Theorem 2) is still one of the central theorems of "Infinite Combinatorics." (*HGW* III, 538)

Hausdorff's most influential contribution to general set theory after 1933, however, was his paper "Sums of $\aleph_1$ Sets" [H 1936b]. Since this work is closely related to Hausdorff's earlier study "Graduation by Final Behavior" [H 1909a], we discussed these together already in 6.3.4.

9.2.5 Injective Continuous Images of Zero-Dimensional Space

Hausdorff's article "Die schlichten stetigen Bilder des Nullraums" ("Injective Continuous Images of Zero-Dimensional Space") [H 1937, III: 541–548] was inspired by a paper by Kazimierz Kuratowski published in *Fundamenta Mathematicae* [Kuratowski 1934]. There Kuratowski had introduced the notion of an *(α, β)-homeomorphism* as a refinement of the notion of a *schlicht* (i.e. injective) continuous image. If X, Y are separable complete metric spaces and if $f : X \to Y$ is a bijective mapping from X to Y, then f is called an (α, β)-homeomorphism if

$f[F^0(X)] \subset F^\beta(Y)$ and $f^{-1}[F^0(Y)] \subset F^\alpha(X)$. Here F^α, G^α denote Borel sets of level α in the respective space following Lebesgue's classification: thus, $F^0 = F$ are closed sets, $G^0 = G$ open sets, $F^1 = G_\delta, G^1 = F_\sigma, F^2 = F_{\sigma\delta}, G^2 = G_{\delta\sigma}$, etc. Kuratowski had shown that in a separable complete metric space the $(0,\alpha)$-homeomorphic image of an F is an $F^{\alpha+1}$, and he also had obtained the following results concerning the converse:

> 1. *Every $F^{\alpha+1}$ $(\alpha > 0)$ is a $(0,\alpha)$-homeomorphic image of a zero-dimensional F.*
> 2. *every $F^{\alpha+1}$ $(\alpha > 0)$ is a $(0,\alpha)$-homeomorphic image of a closed set in the Baire null space N. If $F^{\alpha+1}$ is uncountable, then after omitting a countable set, it is a $(0,\alpha)$-homeomorphic image of N.*

Hausdorff calls a set $A \subseteq X$ *clustered (verdichtet)* if every point of A is a cluster point (accumulation point) of A. He then sharpened Kuratowski's results as:

> Theorem I: *Every clustered $F^{\alpha+1}$ $(\alpha > 0)$ in a separable complete metric space X is a $(0,\alpha)$-homeomorphic image of N.*

The case $\alpha = 1$ had caused him special difficulties, so he formulated this case separately as

> Theorem II: *Every clustered $F^2 = F_{\sigma\delta}$ in X is a $(0,1)$-homeomorphic image of N.*

In concluding, Hausdorff characterized the topological structure of the Baire null space:
A space X is homeomorphic to N if and only if it is a separable, topologically complete, 0-dimensional space containing no nonempty compact open set.

9.2.6 Extensions of Continuous Mappings

In Hausdorff's last publication "Erweiterung einer stetigen Abbildung" ("Extension of a Continuous Mapping") [H 1938, III: 557–564] he returned to a topic that had occupied him repeatedly for more than two decades. His first thoughts on the extension of continuous mappings between metric spaces can already be found in the *Grundzügen* [H 1914a, 368–369; II: 468–469]. In 1915, Heinrich Tietze proved the following classical extension theorem: *For a real-valued function $\varphi(x)$ continuous on the closed set A of a metric space X there exists a continuous extension to the whole space, i.e., a continuous function $f : X \to \mathbb{R}$, which coincides with φ on A.* [Tietze 1915] In [H 1919d], Hausdorff, relying on an interpolation theorem of Hans Hahn, gave a particularly simple and transparent proof of this theorem.

In an unpublished study from December 1925, Hausdorff proved the following extension theorem: *If F is closed in a space A and homeomorphic with a bounded set $\overline{F}$, then the homeomorphism can be extended to A and a (suitable) space $\overline{A}$.*

About two years later, Hausdorff succeeded by means of this theorem in proving a more general extension theorem that did not require metrization properties. A similar theorem had been found independently by two of Alexandroff's students, Victor Niemytzki and Andrei Tychonoff [Niemytzki/Tychonoff 1928]. After Hausdorff communicated his result to Alexandroff, the latter encouraged him to send it to a journal, but Hausdorff was reluctant to do so for the following reason:

> Unfortunately, up to now I must assume that $\overline{F}$ is bounded (I *can* presuppose A is bounded); I cannot bring myself to publish as long as I have not eliminated this presupposition, which after all is probably unneeded. (*HGW* IX, 59–60)

In June 1930, Hausdorff finally succeeded in proving his theorem without imposing the boundedness condition, at which point he sent the final manuscript for "Extension of a Homeomorphism" [H 1930b, III: 457–464] to Waclaw Sierpiński, the editor of *Fundamenta Mathematicae.* He formulated the final result of his long efforts as follows:

> We want to prove the following theorem concerning metric spaces F, E, $\overline{F}$, $\overline{E}$: *If F is closed in E, then a homeomorphism between F and $\overline{F}$ can be extended to a homeomorphism between E and a suitable space $\overline{E}$.* [H 1930b, 353; III: 457].

To prove this theorem, Hausdorff introduced a clever "change of metric," which he explained as follows:

> Instead of the old distances uv defining the metric space E, new distances $\overline{uv}$ can be introduced defining a space $\overline{E}$ homeomorphic with E so that pairs of points in F have values prescribed by the metric on $\overline{F}$. [H 1930b, 353; III: 457].

According to Hausdorff, the proof is modeled on his 1919 proof of Tietze's extension theorem, though it "requires considerably more artifice." He uses three types of letters to distinguish the points in different sets: those in E are denoted by u, v, points in F by a, b, c, p, q, r, and those in $E \setminus F$ by x, y. For points $u, v \in E$, he first modifies uv by defining

$$\delta(u) = \inf_{p \in F} pu \quad \text{and} \quad \delta(u,v) = \min[uv, \delta(u) + \delta(v)].$$

The given metric in $\overline{F}$ is denoted by overlined values, which are next used to define:

$$\varphi(x,a) = \sup_{p\in F}\left[\overline{ap} - \overline{cp} - \frac{px}{\delta(x)} + 1\right] + \inf_{p \in F}\left[\overline{cp} + 2\frac{px}{\delta(x)} - 2\right],$$

where c is a fixed point in F. Taking

$$\psi(u,v) = \sup_{a \in F} |\varphi(u,a) - \varphi(v,a)|,$$

we finally have the new metric $\overline{uv} = \max[\psi(u,v), \delta(u,v)]$. Hausdorff's proof that this new metric actually accomplishes what is required is truly a work of art. After this, he shows in a few lines that the theorem of Niemytzki and Tychonoff follows directly from his theorem.

After 1930, Hausdorff continued to work on the extension problem. In preparation for his final publication [H 1938] (*HGW* III, 557–564), he wrote up a manuscript, dated May 21, 1937, in which he proved the following two new theorems:

> I. *A continuous mapping of the closed set F in the metric space E onto the metric space $\overline{F}$ can be extended to a continuous mapping f from E onto a suitable metric space $\overline{E} \supset \overline{F}$.*
> II. *This extension via f can be constructed so that $\overline{F}$ is closed in $\overline{E}$ and $E \setminus F$ is mapped homeomorphically onto $\overline{E} \setminus \overline{F}$.*

As Theorem III, Hausdorff restated his extension theorem from [H 1930b], which he cited in a footnote. Our discussion of [H 1938] has attempted to show how this last publication was part of Hausdorff's long-term interest in the problem of extending continuous mappings. A very nice survey of this topic from the vantage point of modern topology can be found in [Hušek 1992].

9.3 The Fate of Hausdorff's Mengenlehre

On October 2, 1934, an internal meeting took place at the publishing house of Walter de Gruyter in Berlin. In the minutes from that meeting, one finds this note: "Hausdorff, Mengenlehre is coming to an end. Make new edition or anastatic reprint? Decided: reprint 500 copies anastatically" (Archiv de Gruyter, Staatsbibliothek Berlin, 461). Hausdorff was contacted soon thereafter, and he apparently managed to convince the firm that he should add a tenth chapter, while updating the list of sources and bibliography. In this way, he was able "to do at least partial justice to the progress that had been made in the meantime" [H 1935a, 6] (*HGW* III, 41–351). As discussed in 8.5, the main importance of the earlier edition had been due to its presentation of descriptive set theory. Hausdorff's goal, therefore, in adding this new chapter and four short addenda was to describe some of the essential progress made in descriptive set theory since the mid-twenties. This focused on work undertaken by the Polish school (Banach, Hurewicz, Kuratowski, Nikodym, Sierpiński) as well as by members of Lusin's Russian school.

Most reviews of the new edition were brief or referred back to earlier ones for the 1927 edition; some also noted the previous success of Hausdorff's book. For example, in the annual journal of the German Mathematical Society, Erich Kamke wrote:

> Only eight years after the publication of the second edition, a new edition of this standard work of set theory has already become necessary. In view of the by no means always simple subject matter of the work,

> this is such a great success that any further word of recommendation is superfluous. (*Jahresbericht der DMV* 46(1936), p. 25)

There were two more detailed reviews written by Thoralf Skolem and Giulio Vivanti, both eminent experts in the field of set theory (*HGW* III, 425–428).

By mid-1939, only a small number of copies of the new edition was still available. This was noted in a memo in the publisher's archives, dated September 12, 1939, which states:

> Hausdorff, *Mengenlehre* (Göschens Lehrbücherei Vol. 7). The stock of the 3rd edition of this work is coming to an end. However, we have reservations about organizing a new edition, since Hausdorff is Jewish and we fear that a reprint could cause us difficulties.
>
> In the conference held on Aug. 1, 1939, the decision was made to look for a new author; however, since it will certainly take a long time until a new manuscript can be made available, we wanted to contact Professor Bieberbach and request his opinion as to whether we could risk organizing a small reprinting of Hausdorff's book, possibly using the old year. Herr Bieberbach strongly advised us against reissuing the book in this way. (Archiv de Gruyter, Staatsbibliothek Berlin, 227)

Robert Haussner, an emeritus professor from Jena who served as the publisher's advisor for Göschen's Lehrbücherei, was also consulted about potential new authors. Some of his suggestions were downright absurd, including figures like Hans Beck and Erhard Tornier, neither of whom had any expertise in set theory. A suggestion by Bieberbach for a new author was also mentioned, namely Oskar Perron, but the publisher ruled him out, because he would "not impose any constraint on himself and quote Jews to his heart's content." At the end of the memo, Haussner's opinion was quoted: "If you are concerned at all about a [book on] set theory, this could easily be postponed somewhat until it can be clearly seen whether under the present circumstances set theory will find as much attention and appreciation as before" (Archiv de Gruyter, Staatsbibliothek Berlin, 461). Not until the Nazi regime had passed would a new book on set theory be published in Germany.

This situation stands in sharp contrast with that in the Soviet Union, where a Russian edition of Hausdorff's *Mengenlehre* appeared in 1937 under the title *Teoria mnozhestvkh* (Mengenlehre). Alexandroff and Kolmogoroff served as the responsible editors of this book, which represents a rarity in the history of mathematical literature, since some parts of its contents were not written by Hausdorff. These circumstances can best be clarified by quoting from the editors' preface:

> Hausdorff's *Mengenlehre* belongs to the stock of those unique classical works of mathematical literature that not only draw the balance of developments over a whole period for the discipline in question, but also sketch the course of future developments.

> When one speaks of Hausdorff's *Mengenlehre*, one actually has two books in mind: the first edition, published in 1914 under the title *Grundzüge der Mengenlehre* and the second edition, published in 1927 and entitled simply *Mengenlehre*. These two books differ in their contents to such an extent that they must be regarded as two different works rather than as two editions of one and the same book.

Alexandroff and Kolmogoroff then noted that due to length considerations Hausdorff's *Mengenlehre* devoted only a single paragraph to general topology:

> Undoubtedly the abandonment of the *topological* structure of point set theory, which was one of the most glorious achievements of the first edition of Hausdorff's book, is the greatest loss: the author himself speaks in his preface with evident regret of the necessity of proceeding in this way.

Furthermore, the book "carried out the theory of point sets only for metric spaces" and it omitted altogether measure theory and integration as well as the "topology of the Euclidean plane and of n-dimensional space." On the other hand, they emphasized what was new in the *Mengenlehre*, as opposed to the *Grundzüge*, namely Hausdorff's masterly elaboration of Suslin's theory of A-sets.

The preface then briefly describes the changes in the Russian "translation," along with this justification:

> The editors of the Russian translation have made the decision to make good this loss, which the book had suffered for reasons beyond [the author's] control: supported in this respect by the publishing house ONTI, they have decided to return to the topological point of view, on which the excellent reputation of the first edition of the book is justly based. We have thus made an attempt to combine the great merits of the second edition – merits consisting, above all, in the logical perfection and in the polish of the entire material – with the merits of the first. However, the goal set could not be achieved by mechanical transcription of the text of the first edition: this very first edition stimulated a tremendous development in the theory of topological spaces with the result that this theory no longer looks at all like it did in 1914 when the first edition of Hausdorff's book appeared. Thus, when revising those chapters of the first edition which are devoted to general (topological) set theory, we had to adapt them to the present state of the theory of topological spaces. ⋯ Thus we hope, notwithstanding the fact that Hausdorff's book has undergone considerable reworking under our supervision, the true spirit of the original has been fully preserved. Our task has been to place in the hands of the Soviet reader a book which, on the whole, replaces both the first and second editions of Hausdorff's book, but retains that part of the original which could still be kept since it accords with the current state of set theory.

This preface, dated August 7, 1936, was signed by Alexandroff and Kolmogoroff. No evidence exists that would indicate Hausdorff knew anything at all about the Russian edition of his book. Although he still had access to the journals in the Bonn Institute through Erich Bessel-Hagen, none of these mentioned the Russian translation. Whether Hausdorff would have agreed with the very extensive reworking it received is impossible to say. In any case, it seems fairly certain that he was never asked, otherwise Alexandroff and Kolmogoroff would surely have mentioned this in their preface. Nor did they need to consult with the publisher, since at this time the Soviet Union paid no attention to conflicts over copyright issues.

In 1957, thus twenty years after the Russian edition, Hausdorff's book came out in English under the title *Set Theory*, published by Chelsea in New York. This was a literal translation of [H 1935a] prepared by a group of mathematicians, led by John R. Aumann, who was responsible for the final editing. Five years later, it appeared in a second edition with two new appendices by R. L. Goodstein, dealing with Contradictions in Naive Set Theory and the Axiom of Choice. This English edition was highly successful, leading to new printings in 1978 and 1991. After the American Mathematical Society acquired Chelsea, it issued a new printing of Hausdorff's *Set Theory* in 2005, some 80 years after the original book was written.

9.4 The Final Years

9.4.1 The Outbreak of Terror

The pogrom of November 9/10, 1938, the so-called "Night of Broken Glass" (*Reichskristallnacht*) marks the transition from the continuously pursued and expanded discrimination against those deemed Jewish by the National Socialist regime to the brutal persecution and eventual systematic extinction of six million European Jews. A welcome occasion for the Nazi leadership to unleash long-planned pogroms against the Jewish population came on November 7, 1938, when a 17-year-old Polish Jew named Herschel Grynszpan attempted to assassinate Ernst vom Rath, the legation secretary at the German embassy in Paris. Two days later, on November 9, vom Rath succumbed from his wounds, at which point Nazi leaders issued a call for revenge.

The ensuing anti-Jewish riots were portrayed by Joseph Goebbels and other Nazi propagandists as "spontaneous actions" of the German people and "expressions of popular anger." In truth, they were well-organized attacks carried out by Nazi paramilitary forces. The SA group leaders – "Gruppenführer" with high ranks in the SA and SS comparable to generals in the military – issued an order to the staff leaders who oversaw the rioters on November 9, 1938 that left nothing to the imagination:

> All Jewish stores are to be destroyed immediately by SA men in uniform; an SA guard shall ensure that no valuables of any kind can be stolen.

> The press is to be called in. Synagogues are to be set on fire immediately, Jewish symbols are to be seized. Only dwelling houses of Aryans are to be protected by the fire department, but also Jewish adjoining dwelling houses; however, Jews must get out, since Aryans will move in there soon.
> The police are not allowed to intervene. Jews are to be disarmed, in case of resistance then mowed down immediately. Signs are to be put up at the destroyed Jewish stores, synagogues, etc.: "Revenge for murder of vom Rath," "Death to international Jewry," "No understanding with Jewish people." [Walk 1981/1996, 249].

The Jewish death toll on the night of the pogrom alone was about 400. Approximately 30,000 Jews were arrested and deported, including 4,600 from Vienna alone; most were sent to the concentration camps in Dachau, Buchenwald, and Sachsenhausen. Hundreds more died in these horrific places and thousands suffered injuries or became sick due to the inhumane conditions. Most of the survivors were released by August 1939, provided they pledged to emigrate and leave much of their property to the German state. During the pogrom, 1406 synagogues and prayer rooms, as well as more than 7,000 Jewish stores, homes, and community centers were devastated or completely destroyed. On November 12, 1938, Hermann Göring decreed that Jewish shopkeepers, tradesmen and apartment owners were to repair the damaged buildings themselves, except for those used for religious services. In fact, this decree prohibited the reconstruction of destroyed synagogues, nor could Jews convene in the few that remained standing. Göring also issued a decree requiring Jews of German nationality to atone for these damages by paying one billion Reichsmarks to the government [Walk 1981/1996, 255]. Those who were able to emigrate were de facto expropriated almost completely by means of the Reich Flight Tax.

By this time, anyone who dared to criticize the Nazi's antisemitic policies was bound to be ostracized socially if not directly persecuted. The orientalist and theologican Paul Kahle and his wife Marie, introduced in Section 8.7 as good friends of the Hausdroffs, were outspoken critics of Hitler's regime. When in 1933 Paul Kahle was repeatedly asked by the Bonn university administration to provide proof of his status as an "Aryan," he answered:

> Please don't keep sending me this nonsense. You have no right to ask me, a scientist and philologist, to sign such nonsense. I am not an Aryan. It is possible that Indians and Persians are Aryans. I am neither Indian nor Persian. I am a German, and the devil knows what the Germans are. [Kahle/Bleek 2003/2006, 16]

The family had a number of Jewish friends, some of whom Frau Kahle and her sons tried to comfort after the pogrom, for example the Philippsons, who lived only about 50 meters away from them. On the evening after the pogrom of November 9/10, 1938, a policeman in Bonn saw Marie Kahle and her eldest son Wilhelm helping a Jewish shopkeeper clean up her demolished store. A few days later, the "West German Observer" published a four-column inflammatory article attacking Frau Kahle and her son; it appeared under the headline "This is Treason Against the People." Thereafter, the Kahle family was increasingly harassed and threatened. Paul Kahle was suspended from his teaching post and Wilhelm was expelled from the university. Despite various difficulties, the family finally managed to emigrate to England, though their house and Kahle's library with some 8,000 books were confiscated. After a long serious illness, Marie Kahle died in England already in 1948 [Kahle/Bleek 2003/2006, 14–79]. Paul Kahle was able to continue his scientific work, mainly financed by the millionaire Chester Beatty. Most of his work involved describing and cataloging Arabic manuscripts in Beatty's library. In 1963, he moved to Düsseldorf to live with his son Theodor, but he died the next year following an accident in Bonn. He was the recipient of honorary doctorates from three universities, including Oxford, and in 1955 he received the Grand Cross of Merit of the Federal Republic of Germany.

The Hausdorffs could count themselves lucky that their apartment was not destroyed. They lost a quarter of their saved assets, however, on top of which they had to get along on a reduced pension as a result of a decree affecting all retired Jewish civil servants, which went into effect on 1 January 1939. Such financial plundering continued during the ensuing period, as various measures already planned before the pogrom were now implemented. This included the issuance of a special "Kennkarte J" and the compulsory acceptance of the additional first names Israel and Sara. On February 2, 1939, Felix Hausdorff received his identity card J, provided with a passport photograph and – like criminal records – with fingerprints of the index fingers (Figure 9.2).

Up until this time, Hausdorff apparently never considered leaving Germany, though as an internationally recognized scholar he might have been able to do so earlier. His reluctance to pursue such a course remains a matter of conjecture, though it is easy to understand what kept him in Bonn. Here he had his house, his library, his opportunity to work, and some loyal friends. Even with his skeptical attitude, he probably never imagined that he and his wife were living in actual physical danger. The November pogrom, however, and the events that soon followed it made the undisguised brutality of the Nazi regime clear to him. Thus, by January 1939, Hausdorff was ready to make a desperate attempt to emigrate.

This is clear from a letter Richard Courant wrote to him from New York on February 10, 1939:

> Dear Colleague:
>
> I just received your letter of January 31, 1939, and I am dictating my answer in a hurry ... Of course, every mathematician in the world is under a great obligation to you and I certainly always have felt this way. If I could be of any help to you I should be only too glad. However, the circle of my personal influence is extremely narrow and offhand I do not see within it any concrete possibility, but I have immediately

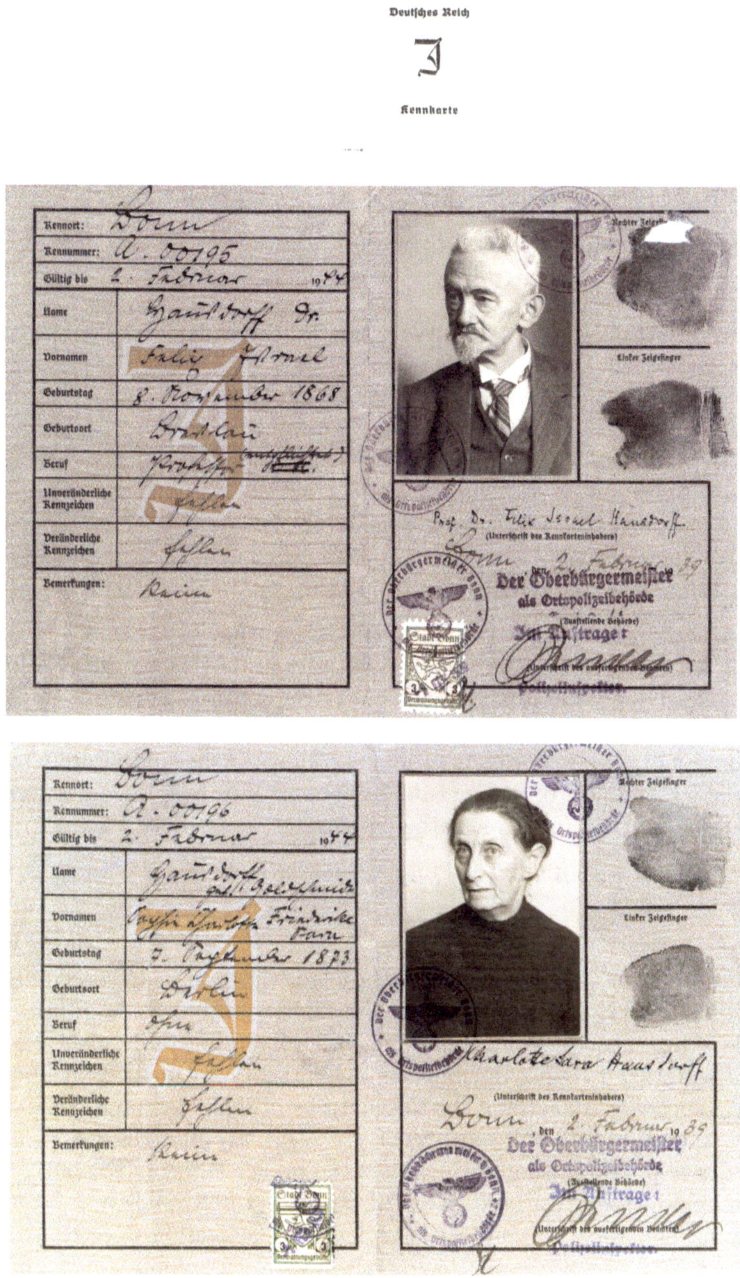

Figure 9.2: "Kennkarten J" for Felix and Charlotte Hausdorff with the compulsory signatures "Felix Israel Hausdorff" and "Charlotte Sara Hausdorff." Source: Hausdorff Estate, Capsule 63, Nr. 10 and 11.

communicated with Weyl, hoping that through a certain connection he has something can be done. Unfortunately, everything here usually develops rather slowly, and therefore please do not think the matter has been forgotten if you should not hear from Weyl for a time.

I remember quite well our meeting in Italy years ago, and it would be a great satisfaction to me if I could see you and Mrs. Hausdorff in this country some time.

With kindest regards to you both, I am,

Sincerely yours, R. Courant[1]

Courant wrote to Hermann Weyl on the same day, asking him to act on behalf of Hausdorff. Weyl was on good terms with Harlow Shapley, director of the Harvard College Observatory. In 1938, Shapley, Weyl and Oswald Veblen established the Asylum Fellowship Plan to support elderly refugees. Soon after receiving Courant's letter, Weyl and others took up Hausdorff's case. Evidence of this can be found in the Veblen Papers, which contains a document, dated February 17, 1939, along with assessments of Hausdorff by Weyl and John von Neumann. These bear witness to the high esteem in which he was held by two of the most important mathematicians from that era:

Hausdorff is known the world over as the author of the classical work on theory of sets in general, and point sets in particular. On this foundation set-theoretic topology has built ever since. Much of his research work is along the same lines. His other important papers are on such diverse subjects as Waring's problem, bi-linear forms of infinitely many variables, problem of momentum, astronomy etc. In spite of his seventy years he is still a creative mathematician.

A man with a universal intellectual outlook, and a person of great culture and charm.

H. Weyl

Hausdorff is a many-sided mathematician who has made contributions in widely varying fields, so that his activities even outside of his main field – set theory – would put him in a very respectable place among mathematicians. His contributions to set theory are of the very first order; especially concerning the foundations of topology, point-set topology, theory of analytic sets, theory of measure, etc. His book on set theory is probably the best ever written on the subject. In spite of his age he still keeps up production of absolutely first quality. I feel that the mathematical community is under great obligation to him.

John von Neumann

[1]The original letter has not been preserved. Prof. Reinhard Siegmund-Schultze (Kristiansand) discovered a carbon copy in the Courant Papers and made it available to us. He also sent us the assessments of Weyl and von Neumann quoted below; he found these in the Veblen Papers. For a thorough discussion of these documents and affairs, see *HGW* IX, 173–177.

These late efforts, however, proved futile. Courant accomplished a great deal in helping mathematicians expelled from Germany reach the US [Reid 1979], as did Weyl and Veblen (on the subject of emigration of persecuted mathematicians, see the fundamental work [Siegmund-Schultze 2009]). Courant continued to promote Hausdorff's cause as late as 1941, but even if he had succeeded, there was no longer any chance for him to leave Germany (*HGW* IX, 176–177).

During the fall of 1938, the Hausdorffs' household took in a new boarder. In early October, Charlotte's sister, Edith Pappenheim, left Vienna and moved into their home in Bonn. Section 2.3 already describes Edith's earlier life and that of her daughter Else, who was born in 1911 and lived with her mother after her parents divorced in 1919. After graduating from a private high school, Else studied medicine in Vienna, specializing in neurology and psychiatry. She received her doctorate in 1935, after which she worked as a secondary physician in a clinic for neurology and psychiatry. Once Hitler's armies had annexed Austria, she was summarily dismissed on April 25, 1938, before she could complete her residency. Afterward, she opened a private practice, but then had to close it when her license was revoked. She then decided to emigrate to the United States via Palestine and France. There, in New York City, she became a well-known psychiatrist; in addition to practicing medicine, she also taught at various universities. Else Pappenheim died of old age on January 11, 2009 (on her life and work, see [Handlbauer 2004]).

When Edith Pappenheim first came to Bonn, she thought this would only be a temporary stay, as she planned to join her daughter in New York. In Vienna, she no longer had a way to earn a living, so she had to give up her apartment. The Hausdorffs' invitation was, as Else later wrote, "a lifesaver until I could get her out." Despite Else's best efforts, though, she was unable to obtain an entry visa for her mother. The latter's letters to her daughter reveal that for a long time she did not give up hope of persuading the American consulate in Stuttgart, but the State Department kept setting ever-higher hurdles for persecuted Germans, especially the many Jews, to enter the United States. By August 1941, at the latest, it must have been clear to Edith that she would probably never see her daughter again.

In the meantime, Charlotte Hausdorff struggled to maintain a home environment that enabled her husband to continue his work, despite continually worsening living conditions. That task had become all the more difficult because she often felt ill, depressed, or just grumpy, while her physical condition deteriorated more and more. Under these circumstances, Hausdorff benefited a good deal from the presence of his sister-in-law, whose practical vein, empathy, and difficult past life had taught her not to despair even under trying circumstances. In a letter to Else Pappenheim, dated May 9, 1939, Felix Hausdorff began by paying acknowledgment to her personal and professional development before writing the following about her mother:

> And there is something else I wanted to write to you, namely how happy I am to have your mother as a housemate and how extraordinarily pleasant her presence is for me. She has a decidedly psychiatric gift, probably drawn from you and your father, and she understands how to deal with me wonderfully; she and the little frog [Hausdorff's wife] do not always have it easy with me! (*HGW* IX, 542)

In the period from the November pogrom up until the outbreak of World War II on September 1, 1939, the anti-Jewish measures of the Nazis concentrated on further plundering the Jewish population, while ousting them from all forms of economic and public life. The property levy and the reduction of pensions already mentioned were only two such measures. Just three days after the pogrom, the president of the Reich Chamber of Culture decreed: "Jews are forbidden to attend theaters, cinemas, concerts, exhibitions, etc." [Walk 1981/1996, 255]. From that day onward, Jews were allowed to withdraw a maximum of 100 Reichsmarks per week from their private bank accounts. On November 18, 1938, the Reich Ministry of Economics implemented the "Exclusion of Jews from Economic Life" by issuing detailed instructions on closing any remaining Jewish businesses and retail establishments. On December 3, 1938, the government ordered the confiscation of all driver's licenses and motor vehicle registrations issued to Jews. On 28 January 1939, Jews were forbidden to sell anything at markets, and by the end of January the licenses of Jewish dentists, veterinarians and pharmacists had expired. Three weeks later, on February 21,1939, the Commissioner for the Four-Year Plan, Hermann Göring decreed:

> All Jews are to deliver within two weeks to the public purchasing offices established by the Reich the objects of gold, platinum, or silver in their possession as well as precious stones and pearls. [Walk 1981/1996, 283]

This amounted to de facto expropriation, as those who relinquished these valuables were not allowed to refuse the financial compensation offered for them! On April 16, 1939, the Reich Ministry of Economics issued a decree placing all Jewish cash assets and any remaining monthly payments under custodianship.

Hitler's larger plans were revealed in his speech to the Reichstag on January 30, 1939. If war came, he threatened, it would lead to "the annihilation of the Jewish race in Europe." Eight months later, on September 1, his warning went into effect when he ordered his armies to invade Poland; in the war propaganda, this invasion was blamed on international Jewry. Already on the first day of war, an evening curfew for Jews was set at 8:00 p.m. (in summer 9:00 p.m.). A decree from September 12 established special stores assigned to Jews for the purchase of foodstuffs; their owners had to pass muster as "flawless" in the eyes of the Gestapo and the Nazi Party. A little later, Jews were no longer allowed to own radio sets, which had to be surrendered without compensation. Again and again, new regulations were issued regarding the use of Jewish property; on July 19, 1940, Jews were denied the possibility of having telephone service. The Hausdorffs had a telephone in their home, so they were directly affected by this decree. Ration cards issued to Jews were marked with a "J" – this entitled them to purchase only reduced rations. With the onset of war, mass murders of Polish Jews had already begun.

By 1941, German society had shifted into a full war economy, and on March 4, the Reich Ministry of Labor, acting on a secret decree issued by Göring, began implementing a program of compulsory labor for all German Jews capable of doing physical work. That month, the Reich Association of Jews received an order to transmit "a complete list of all Jewish apartments in Aryan houses, [⋯ with] addresses, number of rooms, and additional details" [Walk 1981/1996, 338]. Thus began the forced relocation of the Jewish population into so-called "Jewish houses" (*Judenhäuser*) or immediately into local camps, which preceded their deportation to concentration camps. The building in which the Hausdorffs lived was one of these 32 Jewish houses in Bonn. In fact, they were forced to quarter displaced Jews on several occasions, and in the end the three of them were living in a single room.

Elisabeth Hagemann, *née* Reimann, had been Otto Toeplitz's doctoral student and also his assistant; she was well acquainted with the Hausdorff family. In a letter, dated April 27, 1941, Erich Bessel-Hagen wrote to her about their situation:

> Hsdff's [Hausdorffs] are doing reasonably well, even if they cannot escape the anger and excitement over constant new anti-Semitic chicanes. The tax burden and the deductions taken are so high that he can no longer live on his salary alone and must use up his assets; how fortunate that he still has this reserve. Moreover, they have been forced to give up part of their house, which makes things very cramped in terms of space. But I am glad to see that there are more people who care for the Hsdff's, as I occasionally find when I meet one or the other during a visit. The other day, for example, I met a musician who had just been making music with Hsdff. It is really nice that in this way some joy is brought into their house! [Neuenschwander 1996, 257]

In her letters, Edith Pappenheim also reported about a violinist who occasionally played together with Hausdorff. Although unidentified, he was evidently one of those in a small group of upright people who remained on friendly terms with the Hausdorff family and supported them – as best they could – until the end.

9.4.2 Taking Refuge in Mathematics

Another such person was Erich Bessel-Hagen, who was already briefly introduced in Section 8.2. Since he suffered from a walking disability, he took no part in active war service, unlike his colleagues Wolfgang Krull and Ernst Peschl. Their absence meant that, in addition to his own research and teaching duties for history of mathematics, Bessel-Hagen also had to take over some of the other mathematics courses. In 1939 he was appointed associate professor, and he also served as executive director of the Mathematical Seminar during the war years. He died, weakened by illness and poor nutrition, in Bonn on March 29, 1946.

Bessel-Hagen's friendship and regular visits were of existential importance for Hausdorff, who relied on him during these difficult times when his mathematical work truly sustained him. Mathematics carried him away from an increasingly unbearable reality into a familiar world that he loved and which meant a great deal to him. As usual, Hausdorff worked at night, sleeping until around 11:00 a.m. and then recovering, as best he could, in the afternoon and early evening. For his work, he needed access to current literature, especially periodical literature that was not part of his personal library. After his compulsory retirement in 1935, it remains unclear whether Hausdorff was allowed to use the Mathematics Seminar library, but on December 8, 1938, the Reich Minister for Science and Education issued a decree that prohibited Jews from entering the libraries at universities. In any case, probably already after 1935, Bessel-Hagen brought Hausdorff whatever literature he needed for his work at home. This scurrying back and forth was not without danger, especially since it involved circumventing the intended purpose of a ministerial decree, but Bessel-Hagen took that chance. For Hausdorff, this journal literature enabled him to work during the years from 1939 to January 1942 on quite current areas of research, as we will briefly describe here.

Hausdorff's estate from this period contains 64 dated studies (fascicles) with a total of 817 manuscript pages. This does not include a number of undated studies that were presumably written during this time. The principal focus of his work concerned topics in topology: retracts, homotopy, topological curves, Peano continua, homology groups, topological groups, mappings of spheres, dimension theory, intersection and entanglement numbers, and Brouwer's fixed point theorem. One also finds studies in analysis (linear spaces and operators, almost periodic functions), algebra (Boolean algebras and rings, lattice theory, group characters), descriptive set theory, and geometry (projective geometry over Galois fields). Hausdorff wrote out texts for all these studies, some of them quite extensive, although publication of the results was, of course, out of the question during wartime.

Here, just a few words about four of these fascicles. The manuscript in fascicle 742 consist of 101 pages on homology groups, written during the spring and early summer of 1940. This project had an interesting prehistory, as Hausdorff already gave a course on combinatorial topology in the summer semester of 1933 (see 8.3 and *HGW* III, 893–953). In that course, he presented a proof of the topological invariance of homology groups. From the point of view of the development of algebraic topology in the 1930s, however, this earlier proof appeared to him to be in need of improvement. Although he knew in 1940 with certainty that he would never teach again, Hausdorff made a great effort to produce an improved argument for this fundamental theorem. In a letter to his friend and former colleague Johann Oswald Müller, dated June 6, 1940, he indicated the motivation behind this. After briefly mentioning his course from the summer semester 1933, he noted that: "The last paragraph of it, the proof of the topological invariance of Bettian groups, I have now nevertheless simplified in such a way that I have a certain aesthetic pleasure in it" (*HGW* IX, 529).

Hausdorff was also interested in very recent developments in algebra, as can be seen from fascicle 777, "Lattices, Boolean algebras and rings" (63 pages). This was written between October 16, 1941 and January 10, 1942, the period when Jews living in Bonn stood in constant fear of being sent to an internment camp. Hausdorff was busy studying lattice theory, a topic he probably first learned about in the spring of 1938 by reading an important paper by Gottfried Köthe [Köthe 1937]. Not until mid-October 1941, however, did he earnestly take up this new theory, which arose from various sources [Mehrtens 1979]. That the 73-year-old scholar was on top of this material can be seen from pages 18–25, written on December 1, 1941; there he proves that every distributive union is isomorphic to a set union, adding this note: "(according to Köthe p. 131 this has been shown by G[arrett] Birkhoff; I do not know the work in question. The following proof is mine)." After this was written, Bessel-Hagen must have brought him Birkhoff's work, since Hausdorff referred to it on pages 36–39, dated December 6, 1941, noting that Birkhoff's proof was simpler than his own.

The chronologically last two fascicles in Hausdorff's estate (Nrs. 780 and 781) were written on January 16, 1942, ten days before his death. In fascicle 780 he dealt with a problem in a projective geometry over a finite Galois field $GF(p^f)$. For $P_k = P_k(GF(p^f))$, a projective $(k-1)$-dimensional geometry over $GF(p^f)$, Hausdorff calculated the number of points, lines, and planes for the cases P_2, P_3, P_4. It seems he hoped to find formulas for these numbers in the general case P_k by inductive means. Fascicle 781 bears the notation "Question from Arthur," presumably referring to his son-in-law Arthur König, who may have encountered this problem in his work at Zeiss. Hausdorff calculated the improper integral

$$I = \int_{-\infty}^{\infty} \frac{\sin(x-\beta)\sin(x-\gamma)}{(x-\beta)(x-\gamma)}\, dx\,,$$

obtaining $I = \pi\frac{\sin(\beta-\gamma)}{\beta-\gamma}$ (*HGW* IV, 400–402).

In a memorial address, delivered on January 26, 1949, Hans Bonnet described Hausdorff's work ethic during the last months of his life:

> There was the holy, unrelenting seriousness of his research. He did not drive his science, rather it held him in its grip and did not let him go even in the most difficult times. No matter how high the waters rose, he always found hours, day after day, in which he could ponder mathematical problems. His young friend Bessel-Hagen stood faithfully by his side. We also want to remember him gratefully, he who was abruptly taken away by the hardship of the post-war period. (Brieskorn estate, Folder 13)

9.4.3 Loyal Friends

Hans Bonnet studied classical philology, archaeology and Egyptology, first in Breslau and then from 1907 in Leipzig. In 1910 he became an assistant in Leipzig to Georg Steindorff, one of the most important German Egyptologists of his time. During World War I, Bonnet served at the front, returning home severely wounded. In 1916 he received his doctorate under Steindorff and habilitated in Halle/Saale, where he taught as a private lecturer. In 1928 he received a call to the University of Bonn as full professor of Egyptology, a position he held until 1955. Bonnet's main fields of research were Egyptian archaeology and religious history. His *Reallexikon der ägyptischen Religionsgeschichte*, which was printed many times, remains still today an unsurpassed standard work.

Bonnet belonged to the Breathrens' Council in the Bonn Congregation of the Confessing Church, which was critical of the National Socialist system. He saw it as his Christian duty to help oppressed Jews. After the war, his former teacher Georg Steindorff was asked to give his assessment of German Egyptologists and their behavior during the Third Reich as part of the effort to reconstruct the German universities along democratic principles. Steindorff complied with this request in a letter to the American Egyptologist John Wilson in June 1945 (the so-called Steindorff List). About Bonnet he wrote:

> Dr. Hans Bonnet, professor of Egyptology at the University of Bonn, is one of the finest personalities I have ever known. He was my pupil, and later my assistant at Leipzig, and I proved him [sic] as a gentleman without fear and without reproach. During my darkest days at Leipzig, some weeks after the pogrom of November, 1938, he came to our house in Leipzig and invited me and my wife to go with him and find asylum in his house at Bonn, though to give us sanctuary might well have resulted in his confinement in a concentration camp. [Schneider 2011, 231–233]

Although Hausdorff and Bonnet surely met on occasion in the faculty, they apparently knew one another only slightly during Hausdorff's active years. In the memorial address cited above, Bonnet stated: "I only approached Hausdorff when the burden of persecution was already upon him. At that time it had become lonely, very lonely around him." Bonnet visited the Hausdorffs regularly; his last visit took place only a few days before their tragic deaths. In all likelihood, he also arranged for the visits to the Hausdorffs' home made by a pastor of the Lutheran Church, Friedrich Frick, who also belonged to the Confessing Church. Frick's visits were especially comforting to Charlotte Hausdorff and Edith Pappenheim. It was Bonnet who finally saved the manuscripts in Felix Hausdorff's estate (see [Bonnet 1967] and Section 9.5 below).

Among those who remained close to the Hausdorffs during these trying times were three courageous women: Hedwig Cohen, *née* Bouvier, Hulda Heerlein, *née* Fett, and Antonie (Toni) Wilisch, *née* Freiin von Gall. Hedwig Cohen was the widow of the Jewish publisher and bookseller Friedrich Cohen. The Cohen bookstore had existed since 1829 and was an institution in Bonn, as it was closely associated with the university. In better times, Hausdorff had enjoyed attending concerts at the Cohens' house.

After the death of her husband, Hedwig Cohen took over management of the company. Although she was not Jewish (her ancestors were French émigrés), the company encountered economic difficulties in 1933 as a result of the Nazi boycott measures, which in this case were instigated by Nazi students at the university. In order to save the business for her children, Hedwig Cohen resumed using her maiden name, so that in 1937 the company could be known as Buchhandlung H. Bouvier u. Co. Since her three children (two sons and a daughter) were ineligible to take over as company managers, she hired an "Aryan" business manager in 1938; he became co-owner in 1941 and sole owner after the war. Her children emigrated to the USA, where the eldest son, Friedrich Cohen, became a well-known composer. Frau Cohen-Bouvier was, of course, critical of the Nazi regime and sought to alleviate the suffering of persecuted Jews.

Hulda Heerlein was the widow of the Sanitätsrat Dr. Wilhelm Heerlein. She lived in Bonn at Königstrasse 60 and belonged to the close circle of friends of Hedwig Cohen-Bouvier, as did Antonie Wilisch, an artistic woman with many interests. In her younger years, Wilisch was an enthusiastic follower of Friedrich Nietzsche, and she later befriended Carl Spitteler, the Swiss poet and writer who in 1919 won the Nobel Prize for literature. With her husband, Dr. Hugo Wilisch, she lived in a magnificent villa at Fasanenstrasse 25 in Bad Godesberg. The following excerpt from a letter Edith Pappenheim sent to her daughter on June 5, 1939, suggests how eagerly the Wilisch couple wanted to please the Hausdorffs:

> On Saturday I enjoyed the first completely beautiful day here. The whole thing was like a fairy tale. Some time ago Lotte was visited by an unknown lady who brought her flowers. [···] On Friday she came again and invited us [for a visit]. We were picked up by car [the following day] and taken to her magnificent villa (I have never seen anything so beautifully furnished with so much culture). Then there was a wonderful lunch, and in the afternoon we alone, Lotte, Felix and I, were taken on a car tour into the Eifel, to the Aremberg. These people had put a basket with cake, champagne and sandwiches in the car, which we really devoured up there. It was completely secluded and quiet. An old ruin, in the middle of the forest. [···] No houses, no people, only silence and stillness. The whole thing a dream. We drove back over the famous Nürburgring, on a wonderful road. I enjoyed everything to the fullest. That there are still people who think like that, and who strive to bring joy to others, just from the feeling: we are doing well, so we must help others who are not doing so well.

Edith Pappenheim could not give any names because of the censorship. That their hosts were the Wilischs is evident from a later letter, which refers to the hostess's friendship with Carl Spitteler:

> Yesterday we were invited again to the fairy-tale villa. I have hardly ever met such a nice person. Lotte told us that Felix was not quite well and could not go to Vulpera for the cure after all. [Frau Wilisch knew, of course, that Jews could no longer travel to Switzerland.] So she immediately wanted to invite him and Lotte to her place. She knows about a proper diet and how a change would be good for him. She is also very interesting in other ways. She has a lot of famous people as friends. For example, one of her best friends was Spitteler. (Collection of letters from Edith Pappenheim to Else Pappenheim, Manuscript Dept. of the Bonn University Library, Letter No. 79)

Antonie Wilisch stood by the Hausdorffs until the bitter end.

9.4.4 Awaiting the End

With the attack on the Soviet Union, begun on June 22, 1941, the "annihilation of the Jewish race in Europe" announced by Hitler was enacted as one of the principal war aims of his regime, to be pursued with dogged brutality until the very end. As early as July 31, 1941, SS Gruppenführer Reinhard Heydrich was instructed by Göring to make all necessary preparations "substantial and material for a total solution of the Jewish question in the German sphere of influence in Europe" [Walk 1981/1996, 345]. On the conquered territory of the Soviet Union, mass murders of Ukrainian, Belorussian, Russian and Baltic Jews soon began. In October 1941, mass deportations of German Jews from the Reich territory also began with cargo trains departing for Łódź [Gottwald/Schulle 2007]. As early as the spring of 1941, in order to implement Göring's secret decree to harness Jewish slave laborers, the first regional internment camps were built. These would later be increasingly used as holding sites for the deportations to extermination camps in the East.

In Bonn, on April 30, 1941, the head of the Gestapo, Walter Proll, appeared with some men at the convent of the Benedictine nuns Maria Hilf (also called the convent "To Perpetual Adoration"). There, in Bonn-Endenich, he informed the prioress Felicitas Kiese:

> I have orders to inform you that the convent is hereby confiscated with all living and dead inventory. The sisters are to leave within one hour. They may take all personal property with them.

The forced incarceration of Jews from Bonn in the Endenich internment camp began in mid-June 1941 under abominable conditions. At times, as many as 380 people lived in the already small rooms that had previously housed 140 Benedictine nuns. All able-bodied men and women had to perform hard labor, whereas the others helped in the garden, kitchen, or in the laundry. Visitation was strictly regulated and became ever more restrictive over time. The nutritional and hygienic conditions in the camp were entirely inadequate, as noted by Dr. Josef Kill, a physician who reported on the "desolate sight of the elderly and desperate people wandering among the furniture and furnishings they had brought with them." There were no doctors or nurses present, not even a sickroom for special cases. He thus observed, for example, how "in a room occupied by about 10 prisoners, an elderly sick woman, known to me for a long time, had to empty her dysentery-like intestinal catarrh over a cleaning bucket in front of the other people." Moreover, the plumbing system was in such a deplorable state "that for days the water line was released for only a few insufficient hours a day." When Dr. Kill asked Mayor Rickert to undertake immediate and energetic remedial action, the response came from the Gestapo, who interrogated him for two days on suspicion of "friendliness to Jews" (*Judenfreundlichkeit*).

Initially, the Hausdorffs were spared from being sent to the Endenich internment camp, but they clearly recognized the precariousness of their situation. On August 1, 1941, Erich Bessel-Hagen reported to Elisabeth Hagemann about the expulsion of the nuns from their convent and the present threat to their Jewish friends:

> All the Jews still living in Bonn are now being forcibly interned in this stolen building; they must either have their belongings auctioned off or kept in "faithful" hands. In the case of Hdff. and Philippson, it was achieved with difficulty that an exception was allowed. H. has not received an official notice that he may remain in his house, but it is absolutely certain according to the unofficial notice. [Neuenschwander 1996, 257]

Even an official notice would not have been trustworthy, of course, and so the constant fear of being interned in Endenich remained. On top of this came new drastic forms of harassment. On August 28,1941, the Reich Ministry of Economics banned "the free sale of Jewish- owned stocks and shares." These investments were to be exchanged for Reich Treasury notes. Although it is not known to what extent this new ban affected the Hausdorffs financially, they were forced to pay a total of 22,000 Reichsmarks as a Jewish wealth tax ("Judenvermögensabgabe"). As already mentioned, they could only withdraw 100 Reichsmarks per week from their bank account, and now their stock assets were completely out of reach.

On September 1, 1941, the Reich Ministry of the Interior issued a police regulation on the identification of Jews with the following wording:

> As of September 15, 1941, Jews who have reached the age of six are forbidden from appearing in public without wearing a Jewish star. Jews are forbidden to leave their community of residence and to wear medals, badges of honor, or other insignias without written police permission. [Walk 1981/1996, 347]

As a result of such measures, the Hausdorffs rarely left their home. Erich Bessel-Hagen described their situation to Frau Hagemann on October 4, 1941:

> Hausdorffs have had some inconveniences again in the meantime; in particular, Jews must now wear badges when they go out into the street. [· · ·] Furthermore, they have been forced to sell their shares and exchange them for Reich Treasury notes, naturally with losses in exchange. And other things that I cannot write about. [Neuenschwander 1996, 258]

The last sentence likely refers to an event that must have been particularly shocking for the Hausdorffs. Charlotte had made an appointment with the university dental clinic for treatment, and on that day she was accompanied by her husband. She was already sitting in the examination chair with a napkin tied around her, when the senior physician's representative rushed in, tore off her napkin and yelled: "Get out, Jews should hang themselves!" As it happened, this incident would not be forgotten. The perpetrator who uttered these words was one Heinrich Müller, who held the position of physician's representative, even though he had not earned a doctorate, despite several futile attempts.

Müller was nevertheless an important local figure: he held the title of Hauptsturmführer in the SS and he headed its security service in Bonn. In this position, which kept him busy in the afternoons and evenings after he completed his morning duties at the dental clinic, he built up an extensive network of informers and participated – often in a brutal manner – in Gestapo activities directed against Jews. His role as a terrorist was remembered after the war, and in 1948 he was put on trial for a variety of crimes, including participation in burning down the Bonn synagogue. His treatment of the Hausdorffs was also one of the charges filed against him; as the state prosecutor duly elaborated:

> Prof. of Mathematics Hausdorff accompanied his sick wife to the University Dental Clinic for examination. An attending assistant readily accepted the wife for treatment. When she was sitting in the chair, the accused appeared and using rude words demanded she leave the clinic. He shouted at the husband Hausdorff, who was standing nearby, that he should leave and go hang himself. A little later the two spouses poisoned themselves.

As witnesses to this charge, the court obtained statements from Minna Nickol, the Hausdorffs' domestic servant since 1937, as well as from Antonie Wilisch and Arthur König. Müller was sentenced to four years' imprisonment on December 29, 1948, though he was pardoned on November 7, 1951.

On November 8, 1941, Hausdorff turned 73 years old. Edith Pappenheim, in a letter to her daughter, described the atmosphere on his birthday: "Saturday was Lixen's birthday. It was quite nice; a few friends were there and everyone tried to be as cheerful and lighthearted as possible."

On or near January 18, 1942 the Hausdorffs and Edith Pappenheim received orders to report to the Endenich internment camp on January 29. In preparing for this eventuality, they had jointly decided to take their own lives before putting themselves in the hands of Nazi officials. This can be surmised from the coded messages in two letters Edith sent to her good friend Paula Blümel in Vienna. On January 12, 1942, shortly before the internment order arrived, she wrote:

> We spent the holidays quietly and experienced New Year's Eve at least a little more peacefully, because for the time being we do not need to travel. But what the future will bring all of us we do not know. Who knows, indeed, whether I will not meet with your sister-in-law.

Here "need to travel" (*verreisen müssen*) obviously stands for being interned in Endenich or immediately deported. The sister-in-law of Paula Blümel, Resi Blümel, had chosen joint suicide with her husband, Dr. Joseph Blümel, when faced with the persecution of the Jews in Austria, so "meet with your sister-in-law" meant to follow this path as well. In an undated letter, which must have been written soon after receiving the internment order, Edith then wrote:

> Dear Paula!
> We are about to make a big trip and will take the same route as Joseph and Resi. Since we would have to move out in the next few days anyway, we prefer to do it this way. [···] Goodbye and don't forget me, give my regards to Elisabeth as well. Thank you all for your great love and friendship. You should tell Elsele [Else P.] that there was just no other way.

Somehow Else Posner, the Hausdorffs' good friend from Greifswald days and a visitor from time to time since then, was also informed. She traveled from Berlin and stood by until their last hours. On the evening of January 25, 1942, Felix Hausdorff, his wife Charlotte, and Edith Pappenheim took an overdose of Veronal. Felix and Charlotte died during the night, but Edith still showed signs of life the next day, though she did not awaken from a coma. She died on January 29, 1942.

Their lawyer Hans Wollstein was also informed. The son of a Jewish bank director, Wollstein was born in Elberfeld in 1895; his parents had him baptized as a Protestant. After graduating from high school, he volunteered to go to the front in World War I. After the war ended, he studied law and later ran an office in Bonn. He was active in the Protestant church community, especially in programs for the poor. In November 1938 he was disbarred, and in 1939 his home was declared a "Judenhaus." Alfred Philippson, along with his wife and daughter, stayed there for a time. It was probably through Philippson that Hausdorff came to know Hans Wollstein. On July 27, 1942, Wollstein was deported to Theresienstadt, and in the fall of 1944 he was sent to Auschwitz, where he was murdered (the precise date of his death is unknown).

Only hours before taking his own life, Hausdorff wrote a farewell letter to this lawyer friend. Wollstein must have passed this letter on to Erich Bessel-Hagen before his deportation to Theresienstadt, as it has been preserved in the latter's estate. It is published in facsimile and in print in [Neuenschwander 1996] and also reprinted in full in (*HGW* IX, 655–656). Here we present the beginning and end of Hausdorff's letter:

> Dear friend Wollstein!
> By the time you receive these lines, the three of us will have solved the problem in a different way – the way you have been trying to dissuade us from. The feeling of security that you predicted for us, once we had overcome the difficulties of the move, does not want to take hold at all, on the contrary:
>
> even Endenich
> Is still maybe not the end!
> (auch Endenich
> Ist noch vielleicht das Ende nich!)
>
> What has happened against the Jews in the last few months arouses a justified fear that we will not be allowed to live to see a state of affairs that is tolerable for us.

After thanking friends and expressing final wishes regarding burial and wills, Hausdorff continues with great composure:

> Forgive us for causing you trouble beyond death; I am convinced that you are doing what you *can* do (and which may not be very much). Forgive us also for our desertion! We wish you and all our friends to experience better times once again.
>
> Your faithful
> Felix Hausdorff

A few days after the tragic event, Erich Bessel-Hagen wrote to inform Elisabeth Hagemann [Neuenschwander 1996, 258]:

> And now I have some horribly sad news to bring you. In the night from last Sunday to Monday, our dear Hausdorffs made an end to their lives, in order to escape the horrors that threatened them.

He went on to describe the horrific conditions at the Endenich encampment, the constant fear of internment there, the wavering back and forth in Hausdorff's case, finally the internment order in mid-January, and the decision to take their own lives. He continued thus:

> I think you have to approve of the decision of the three. As things stand, it is perhaps best and mildest to clear the field before it is too late. You can hardly imagine how shaken and emotionally depressed I am over this tragedy; and all consideration that it was perhaps better this way cannot, at least at first, help with the pain of having lost such dear friends. It is a horrible feeling to have to watch the demise of such dear people in the wild flood without being able to lift a finger. One feels so

> terribly cowardly and constantly ashamed. And yet I don't know what I could have done. [Neuenschwander 1996, 259]

As all three had wished, their bodies were cremated. Yet there still must have been some difficulties in procuring a place in the municipal cemetery, as Bessel-Hagen wrote about this later to Frau Hagemann:

> At least it has been achieved after some struggle that the family H. got a place for the urns in the Poppelsdorf cemetery; not even that [did they] want to grant them. [Neuenschwander 1996, 259]

The admirable calm with which Hausdorff faced death was reported by Bessel-Hagen in a letter to the logician and mathematician Hans Hermes, who had briefly worked in Bonn before his enlistment in the army. After first describing the events surrounding his suicide, he wrote on April 3, 1942:

> According to everything reported to me by people who have seen or spoken to H. in the last few days, H. ordered everything with stoic calm and deliberateness and until the end never lost his inclination to make humorous remarks. He really "died as a philosopher." It is altogether appalling that a man of his fame and reputation, who was entitled to so much thanks, should have been driven to [his death] by his persecutors, and that no one was strong enough to lift a finger against this from happening. The worst of all is, as must be said in view of what might otherwise have come, that this sad ending was a good thing. [Neuenschwander 1996, 260]

On June 11, 1942, Antonie Wilisch wrote to an acquaintance in Zurich about the death of Edith Pappenheim:

> Else's mother died very gently in her sleep. She was found with a volume of Storm's novellas and the Bible beside her. [· · ·] Around her all the last time I knew her, but especially very shortly before the end, there was an indescribable dignity and – I would like to say majesty. Only when she spoke of her daughter, tears always immediately ran down her face. I already wrote to you that she asked me to tell her, as a final greeting, that she had always experienced only joy, and nothing but joy in her, and she would gladly again experience all the pain she had to bear just for the sake of such a child. – You will tell her, won't you? After all, it's the only loving thing one can still do for the dead.
>
> Now her urn is buried next to those of Lotte and Felix [see Figure 9.3] in a beautiful urn cemetery where the forest grows over the graves. The Protestant minister with whom the family had been friends blessed them up there on the hillside. (Collection of letters from Edith Pappenheim to Else Pappenheim, Manuscript Dept. of the Bonn University Library, Letter No. 228)

How correctly Hausdorff saw things with his play on words – "Auch Endenich/ Ist noch vielleicht das Ende nich!" – soon became apparent. On June 15, 1942, the deportations from Bonn began. From Endenich camp, the innocent victims were first taken to Cologne-Deutz before being deported from there. One transport train went to the concentration camp in Theresienstadt, another to the Lublin district, where the Majdanek concentration camp was located. The latter was initially constructed to intern forced laborers, but later transitioned into an extermination facility. On June 14, 1942, Alfred Philippson, his wife Margarete and his daughter Dora were deported to Theresienstadt. They only survived thanks to Philippson's friendship with the Swedish geographer Sven Hedin, an admirer of Hitler who was respected by leading Nazis. Philippson's family could thus obtain so-called "celebrity status" and a modicum of relief.

In Theresienstadt, Alfred Philippson wrote his memoirs, entitled *How I Became a Geographer* [Böhm/Mehmel 1996/2000]. After the liberation of the Theresienstadt camp by the Red Army, the family returned to Bonn in July 1945. Philippson then resumed teaching at the university, despite his advanced age. He died, highly honored, on March 28, 1953; that same year his wife also died. Dora Philippson worked for reconciliation and Christian-Jewish cooperation until the end of her life (see [Mehmel 1996] and [Mehmel 2014]). Such a survival story was rare. By the end of July 1942, the last transports took place from Bonn to Theresienstadt and elsewhere, after which the internment camp in Bonn-Endenich was closed. Around 470 people had been herded into those quarters before being deported to camps in the east mainly run by SS officers. Among that group, records show that only eleven survived the Holocaust [Rey 1994].

Among the Hausdorffs' Prague relatives who were still alive at the onset of Germany's occupation of Czechoslovakia, all succumbed as victims of the Nazi's persecution of Jews. Hausdorff's nephew Ludwig Brandeis, the son of his sister Martha, was deported to Theresienstadt on 13 July, 1943; on 28 October 1944, he was transported to the Auschwitz death camp and murdered there. His wife Marta and their children Anita and Eva suffered the same fate. Anton Glaser and his wife Vally, Felix Hausdorff's other sister, arrived in Theresienstadt one year earlier, on 9 July, 1942. Vally Glaser died there on 5 July, 1944 and her husband already two weeks later. Their daughter Edith Spitzer, son-in-law Rudolf Spitzer, and the three grandchildren Franz, Hans and Thomas Spitzer were all murdered in Auschwitz.[2]

We end this section with a passage from a letter written by the then 85-year-old Else Pappenheim to Egbert Brieskorn on May 28, 1996. In thanking him for sending [Brieskorn 1996], she went on to write:

> I would like to add – perhaps I have already written this to you – that for me the suicide was much more bearable than the thought that they

[2]The data on the persecution of the Brandeis and Glaser families were researched by lecturer Dr. Leo Boček and Prof. Jiri Vesely, Charles University, Prague, at the Federation of Jewish Communities in the Czech Republic.

would be sent to Auschwitz. At least they died, according to Nora, peacefully asleep in their own beds. I am still amazed that it was possible to bury them. It is still all unbelievable and unfortunately one must fear that similar things could happen again and again. One only has to think of Africa [an allusion to the genocide in Rwanda in 1994]. (Brieskorn estate, Folder 113)

9.5 Fate of Family, Friends, and the Estate

Nora and Arthur König had been living in Jena since 1929. Even under normal circumstances their lives would have been difficult, owing to the fact that their second son, Hermann, was severely handicapped mentally. The Nazi racial laws, however, added a tremendous new burden on the family. According to these, the Königs had entered into a "privileged mixed marriage," since Nora's husband was an "Aryan." She was still required to obtain an identification card "J" and enter her middle name as "Sara" on it, but she could leave their home without wearing a yellow star indicating that she was a Jew. By January 1942, however, the Wannsee Conference had reached a decision about those Jews who had married gentiles: they were to be deported and ultimately exterminated. Of course this was a state secret at the time, but many signs already pointed in this direction. The Holocaust had, in effect, begun in June 1941 with the surprise attack on the Soviet Union, whereas the Nazi officials who met in Berlin-Wannsee devised a deportation plan for removing all Jews from territories under German occupation in western and central Europe. That historic meeting was held only a week before the Hausdorffs and Edith Pappenheim committed suicide.

Arthur König took charge of settling various family affairs, and he presumably arranged for the sale of furniture in the house and its rental afterward. He also consulted with Erich Bessel-Hagen about what to do with Hausdorff's library and the manuscripts in his scientific estate. The Hamburg astronomer Otto Heckmann, who was a personal friend of Arthur König, had the opportunity to purchase some of Hausdorff's books, whereas later the bulk of his library was sold to a bookshop in Leipzig, Koehler's Antiquarium. Heckmann had also suggested that Hausdorff's scientific papers be transferred to Hamburg, where they could either be housed at the astronomical observatory or placed in the care of Wilhelm Blaschke at the Mathematics Institute. Bessel-Hagen objected to this plan, however, fearing that someone might tip off Nazi authorities, thereby risking confiscation of the estate. He then learned from Hans Bonnet that he had adequate room in the basement of his house, and Bonnet thus offered to keep the manuscripts there on a provisional basis. König approved of this plan, and by May 1942 he could report to Bessel-Hagen that the transfer had been completed.

Hausdorff's papers seemed to be safe and secure, but then in December 1944 Bonnet's house was hit by a bomb and destroyed. He had to leave Bonn and did not return until the summer of 1946. Not surprisingly, he found virtually no trace of the furniture he had left behind, but apparently none of the "treasure hunters" had taken an interest in the mathematical papers scattered about in the basement. As Hans Bonnet would later recall:

> Hausdorff's papers were essentially conserved, [$\cdots$ though] losses must certainly have been sustained and the loose pages got completely mixed up more and more. Out of the well-ordered cosmos emerged a chaos. When I gave this collection over to the Mathematics Institute, I had little hope that one could ever recover usable coherent material from it. [Bonnet 1967]

After the war had long ended, Bonnet made several attempts to have Hausdorff's papers removed from his home and placed in the custody of the Mathematics Institute. Finally, in 1963, an initiative to accomplish this was undertaken, largely thanks to Hausdorff's former student, Günter Bergmann (see 8.3). Supported by the German Research Foundation, Bergmann began to order and catalog Hausdorff's manuscripts, which were in a wild state of disarray. He was then working at the University of Münster, and in 1969 he received permission from Nora König, whose husband had died in that year, to transfer Felix Hausdorff's Nachlass from Bonn to Münster.

Returning to the König family's earlier situation in Jena during the war years, in June 1943 they received more bad news. This concerned their elder son Felix, who was then 16; he was attending a Gymnasium in Jena, but as a half-Jew he could no longer continue there. Desperate for help, his father turned to a local optician, who agreed to offer Felix an apprenticeship position. This went well at first, but he later had to leave before completing the program. In the meantime, the racial laws became even more restrictive, and Arthur König's status as "jüdisch versippt" (family ties to Jews) now placed him in a dangerous category. As of 1943, such persons were deemed "unworthy of military service" and many had to perform forced labor in "Sonderkommandos" (work units composed of death camp prisoners), part of the notorious Organization Todt. In October 1944, Arthur and Felix König were brought to such a work camp, but after four months Arthur was released, probably thanks to the intervention of high-level people at Zeiss. His son, however, had to remain in the camp until April 1945.

By the beginning of 1945, Nazi officials were rounding up Jewish spouses from mixed marriages, including those who were formerly considered "privileged," in preparation for deporting them to Theresienstadt. Nora König was now living all alone without any contact with her husband. On 27 January, 1945, she received an official message from the Gestapo in Jena informing her that she was to appear at the local office two days later to receive an oral communication; the written message was to be submitted at that time. Nora had probably already made plans for this eventuality, as she had no intention of putting herself in the hands of the Gestapo. She had a good friend from her days in Greifswald, Anne-Sophie Dumrath, the daughter of a pastor there. In the meantime, this friend had married a clergyman named Wachinger, the pastor of a church in a town near Osterode in the Harz Mountains. Traveling under a false name, Nora was able to reach that destination, where the Wachinger couple courageously agreed to keep her in hiding for the duration of the war.

Some three months later, in late April, the British army had control of the Osterode district and they appointed Konrat Ziegler as regional councilor there. Ziegler had been professor of philology in Greifswald until 1933 and a strong opponent of the Nazi regime. He also had known the Hausdorff family for many years, an added incentive for him to help out their daughter. Through an English officer, he sent a message to the Zeiss firm in Jena on 27 April, informing the company as to the whereabouts of Nora König and requesting information about her husband and son. It took nearly three weeks before Ziegler received a reply, but the news was gratifying. Father and son were well and wished for nothing more than for their wife and mother to travel to Jena as soon as possible. The arrangements for her permission to travel, by foot as well as by car, were soon made, and Nora then arrived to greet her family (except for their son Hermann) on May 26.

Their harrowing adventures were not yet over, however, as the US American troops stationed in Jena and the surrounding region of Thuringia were already in the process of withdrawing. In line with the Allied agreement, the Red Army would soon take over control of this territory. Since the Soviets were keen to take advantage of German scientific expertise, in many cases viewing such experts as war booty, there was due cause for alarm. Many at Zeiss no doubt feared the possibility that they might soon be working in Russia, whereas the Americans were eager to bring them into their own future occupation zone. The Königs were thus one of 77 families that resettled in the area around Heidenheim in eastern Württemberg, where the Zeiss company began building a new factory. Arthur's technical abilities did not fit these plans, however, and so he was forced to search for other work. He taught for a brief time as a substitute teacher at a local school, but his prospects looked so grim that the family even considered settling in the USA. Then, in July 1947, the dark clouds finally began to clear; König would soon be looking through telescopes at the Heidelberg astronomical observatory, where he and his family lived on the Königstuhl with a beautiful view overlooking the city.

In the meantime, Hermann, the Königs' younger son, had returned from Mecklenburg. In Jena he had been staying since 1938 at a special anthroposophical institute for children with disabilities. Three years later, this institution moved to Seewalde in Mecklenburg. It might seem like a miracle that someone like Hermann, a severely handicapped half-Jew, could have survived through the entire Nazi era. His mother, however, provided the answer: several high-level Nazis sent their own children to that same place. After 1953, when Hermann turned 21, he lived in a psychiatric hospital in Wiesloch, a short distance south of Heidelberg. He died in 2010.

His older brother Felix was able to graduate from a Gymnasium in 1948, after which he studied law in Bonn. In 1949, against the will of his parents, he

Figure 9.3: Family Grave in the Poppelsdorf Cemetery in Bonn

converted to Judaism. His studies at the university did not go well, however, and in 1955 he left without a degree. As Bonn was the capital of the newly founded Federal Republic of Germany, Felix managed to get a government job with the Defense Ministry. In 1962 he was promoted to a position with civil service status, which meant he could anticipate a generous pension on retirement. Unfortunately, though, he did not live to benefit from this: he and his wife Ellen were killed in a car crash in 1973. The couple had no children.

A decade earlier, Arthur König had retired from his position as head astronomical observer in Heidelberg. He and his wife then had to move out of their quarters on the Königstuhl. They lived nearby in the city after that, while König continued to teach astronomy in Mainz, where he was appointed honorary professor in 1968. He died the next year and was eulogized by Otto Heckmann in *Astronomische Nachrichten*. Nora König moved the next year to Bad Honnef on the Rhine, before entering the Perthes-Heim in Bonn in 1974, the year after her elder son's death. Around this time she decided to bequeath her father's papers to Günter Bergmann, who had in the meantime published two volumes of facsimiles from the 26,000 pages in Felix Hausdorff's estate. In 1980, Bergmann sold the Hausdorff Nachlass to the Bonn University Library for 18,000 Marks, which he then donated to Nora König, who undoubtedly greatly appreciated this gesture. Bergmann's efforts to restore her father's work were of great importance for the subsequent volumes in the Hausdorff edition, which have been cited throughout this book.

During her last years, Nora König was physically frail, but her mind was completely fresh and clear. The present biography has profited greatly from that circumstance as well, as through her conversations with Egbert Brieskorn she not only imparted many valuable insights into the family history but also provided hints about other people and related matters that served as motivations for further research. Lenore König died on 8 September 1991. She and her husband are interred together with her parents at the family grave located in the Poppelsdorf Cemetery in Bonn.

Bibliography

[H 1891] Zur Theorie der astronomischen Strahlenbrechung (Dissertation). *Ber. über die Verhandlungen der Königl. Sächs. Ges. der Wiss. zu Leipzig. Math.-phys. Classe* 43 (1891), 481–566. *HGW* V, 5–90.

[H 1893] Zur Theorie der astronomischen Strahlenbrechung II, III. *Ber. über die Verhandlungen der Königl. Sächs. Ges. der Wiss. zu Leipzig. Math.-phys. Classe* 45 (1893), 120–162, 758–804. *HGW* V, 93–182.

[H 1895] Über die Absorption des Lichtes in der Atmosphäre (Habilitationsschrift). *Ber. über die Verhandlungen der Königl. Sächs. Ges. der Wiss. zu Leipzig. Math.-phys. Classe* 47 (1895), 401–482. *HGW* V, 215–296.

[H 1896] Infinitesimale Abbildungen der Optik. *Ber. über die Verhandlungen der Königl. Sächs. Ges. der Wiss. zu Leipzig. Math.-phys. Classe* 48 (1896), 79–130. *HGW* V, 315–366.

[H 1897a] Das Risico bei Zufallsspielen. *Ber. über die Verhandlungen der Königl. Sächs. Ges. der Wiss. zu Leipzig. Math.-phys. Classe* 49 (1897), 497–548. *HGW* V, 445–496.

[H 1897b] (Paul Mongré) *Sant' Ilario–Gedanken aus der Landschaft Zarathustras.* Verlag C.G.Naumann, Leipzig. VIII + 379 S. *HGW* VII, 87–473.

[H 1897c] (Paul Mongré) Sant' Ilario–Gedanken aus der Landschaft Zarathustras. Book announcement. *Die Zukunft, 20.11.1897*, 361. *HGW* VII, 477.

[H 1898a] (Paul Mongré) *Das Chaos in kosmischer Auslese–Ein erkenntniskritischer Versuch.* Verlag C. G. Naumann, Leipzig. VI und 213 S. *HGW* VII, 589–807.

[H 1898b] (Paul Mongré) Massenglück und Einzelglück. *Neue Deutsche Rundschau (Freie Bühne)* 9 (1), (1898), 64–75. *HGW* VIII, 275–288.

[H 1898c] (Paul Mongré) Das unreinliche Jahrhundert. *Neue Deutsche Rundschau (Freie Bühne)* 9 (5), (1898), 443–452. *HGW* VIII, 341–352.

E. Brieskorn, W. Purkert, *Felix Hausdorff*, Vita Mathematica 21,
https://doi.org/10.1007/978-3-031-52135-5

[H 1898d] (Paul Mongré) Stirner. *Die Zeit* 213, 29.10.1898, 69–72. *HGW* VIII, 381–390.

[H 1899a] Analytische Beiträge zur nichteuklidischen Geometrie. *Ber. über die Verhandlungen der Königl. Sächs. Ges. der Wiss. zu Leipzig. Math.-phys. Classe* 51 (1899), 161–214. *HGW* VI, 213–266.

[H 1899b] (Paul Mongré) Tod und Wiederkunft. *Neue Deutsche Rundschau (Freie Bühne)* 10 (12), (1899), 1277–1289. *HGW* VIII, 415–429.

[H 1899c] (Paul Mongré) Das Chaos in kosmischer Auslese. Book announcement. *Die Zukunft* 8 (5), (1899), 222–223. *HGW* VII, 811–813.

[H 1899d] L.v.Bortkewitsch, Das Gesetz der kleinen Zahlen (Review). *Zeitschrift für Mathematik und Physik* 44 (1899), 24–25. *HGW* VI, 269–271.

[H 1900a] (Paul Mongré) *Ekstasen.* Gedichtband. Verlag H. Seemann Nachf., Leipzig. 216 S. *HGW* VIII, 39–190.

[H 1900b] Zur Theorie der Systeme complexer Zahlen. *Ber. über die Verhandlungen der Königl. Sächs. Ges. der Wiss. zu Leipzig. Math.-phys. Classe* 52 (1900), 43–61. *HGW* IV, 407–425.

[H 1900c] (Paul Mongré) Nietzsches Wiederkunft des Gleichen. *Die Zeit* 292, 5.5. 1900, 72–73. *HGW* VII, 889–893.

[H 1900d] (Paul Mongré) Nietzsches Lehre von der Wiederkunft des Gleichen. *Die Zeit* 297, 9.6.1900, 150–152. *HGW* VII, 897–902.

[H 1901a] Beiträge zur Wahrscheinlichkeitsrechnung. *Ber. über die Verhandlungen der Königl. Sächs. Ges. der Wiss. zu Leipzig. Math.-phys. Classe* 53 (1901), 152–178. *HGW* V, 529–555.

[H 1901b] Über eine gewisse Art geordneter Mengen. *Ber. über die Verhandlungen der Königl. Sächs. Ges. der Wiss. zu Leipzig. Math.-phys. Classe* 53 (1901), 460–475. *HGW* IA, 5–20. English translation in Plotkin, J. M. (Ed.): *Hausdorff on Ordered Sets.* American Mathematical Society, Providence (Rhode Island) 2005, 11–22.

[H 1902a] (Paul Mongré) Der Schleier der Maja. *Neue Deutsche Rundschau (Freie Bühne)* 13 (9), (1902), 985–996. *HGW* VIII, 453–466.

[H 1902b] (Paul Mongré) Der Wille zur Macht. *Neue Deutsche Rundschau (Freie Bühne)* 13 (12) (1902), 1334–1338. *HGW* VII, 905–909.

[H 1902c] (Paul Mongré) Max Klingers Beethoven. *Zeitschrift für bildende Kunst, Neue Folge* 13 (1902), 183–189. *HGW* VIII, 489–500.

[H 1902d] (Paul Mongré) Public letter against G. Landauer's article 'Die Welt als Zeit'. *Die Zukunft* 10 (37), 14.6.1902, 441–445. *HGW* VIII, 529–533.

[H 1902e] W. Ostwald: Vorlesungen über Naturphilosophie (Review). *Zeitschrift für mathematischen und naturwissenschaftlichen Unterricht* 33 (1902), 190–193. *HGW* VI, 275–278.

[H 1903a] Das Raumproblem (Inaugural lecture at the University of Leipzig, held on 4 July 1903). *Ostwalds Annalen der Naturphilosophie* 3 (1903), 1–23. *HGW* VI, 281–303

[H 1903b] (Paul Mongré) Sprachkritik. *Neue Deutsche Rundschau (Freie Bühne)* 14 (12), (1903), 1233–1258. *HGW* VIII, 551–580.

[H 1903c] *Christian Huygens' nachgelassene Abhandlungen: Über die Bewegung der Körper durch den Stoss. Über die Centrifugalkraft.* Edited by Felix Hausdorff. 79 pp., notes by Hausdorff on pp. 63–79. Verlag W.Engelmann, Leipzig 1903. Reprint: Akademische Verlagsgesellschaft Leipzig (no date given). *HGW* V, 837–915.

[H 1903d] J. B. Stallo: Die Begriffe und Theorien der modernen Physik (Review). *Zeitschrift für mathematischen und naturwissenschaftlichen Unterricht* 34 (1903), 138–142. *HGW* VI, 307–310.

[H 1903e] W. Grossmann: Versicherungsmathematik (Review). *Zeitschr. für mathematischen und naturwissenschaftlichen Unterricht* 34 (1903), 361. *HGW* V, 592.

[H 1903f] M. Kitt: Grundlinien der politischen Arithmetik (Review). *Zeitschrift für mathematischen und naturwissenschaftlichen Unterricht* 34 (1903), 361. *HGW* V, 592.

[H 1904a] Der Potenzbegriff in der Mengenlehre. *Jahresbericht der DMV* 13 (1904), 569–571. *HGW* IA, 31–33. English translation in Plotkin, J. M. (Ed.): *Hausdorff on Ordered Sets.* American Mathematical Society, Providence (Rhode Island) 2005, 31–33.

[H 1904b] Eine neue Strahlengeometrie (Review of E.Study: *Geometrie der Dynamen*). *Zeitschrift für mathematischen und naturwissenschaftlichen Unterricht* 35 (1904), 470–483. *HGW* VI, 323–336.

[H 1904c] (Paul Mongré) Gottes Schatten. *Die neue Rundschau (Freie Bühne)* 15 (1), (1904), 122–124. *HGW* VIII, 663–666.

[H 1904d] (Paul Mongré) Der Arzt seiner Ehre, Groteske. *Die neue Rundschau (Freie Bühne)* 15(8), (1904), 989–1013. *HGW* VIII, 767–799. New edition: Der Arzt seiner Ehre. Komödie in einem Akt mit einem Epilog. Mit 7 Bildnissen, Holzschnitte von Hans Alexander Müller nach Zeichnungen von Walter Tiemann, 10 Bl., 71 S. Fünfte ordentliche Veröffentlichung des Leipziger Bibliophilen-Abends, Leipzig 1910. Reprint: S. Fischer, Berlin 1912, 88 S.

[H 1904e] (Paul Mongré) Max Klinger, Beethoven. Accompanying text for the reproduction of Klinger's sculpture in: *Meister der Farbe. Beispiele der gegenwär-*

tigen Kunst in Europa. Mit begleitenden Texten. E.A.Seemann, Leipzig 1904, Abb.Nr.4. *HGW* VIII, 685–686.

[H 1904f] W.K. Clifford, Von der Natur der Dinge an sich. Übersetzt von H.Kleinpeter. (Review). *Vierteljahrsschrift für wissenschaftliche Philosophie und Soziologie* 28 (1904), 241. *HGW* VI, 312.

[H 1904g] R.Manno, Heinrich Hertz–für die Willensfreiheit? (Review). *Vierteljahrsschrift für wissenschaftliche Philosophie und Soziologie* 28 (1904), 241–242. *HGW* VI, 314.

[H 1904h] R.Schweitzer, Die Energie und Entropie der Naturkräfte, mit Hinweis auf den in dem Entropiegesetze liegenden Schöpferbeweis. (Review). *Vierteljahrsschrift für wissenschaftliche Philosophie und Soziologie* 28 (1904), 242. *HGW* VI, 316.

[H 1904i] C.Stumpf, Leib und Seele. Der Entwicklungsgedanke in der gegenwärtigen Philosophie. (Review). *Vierteljahrsschrift für wissenschaftliche Philosophie und Soziologie* 28 (1904), 242. *HGW* VI, 318.

[H 1904j] Melchior Palágyi, Die Logik auf dem Scheidewege. (Review). *Vierteljahrsschrift für wissenschaftliche Philosophie und Soziologie* 28 (1904), 242–243. *HGW* VI, 320.

[H 1905] B.Russell, The principles of mathematics (Review). *Vierteljahresschrift für wissenschaftliche Philosophie und Soziologie* 29 (1905), 119–124. *HGW* IA, 481–487.

[H 1906a] Die symbolische Exponentialformel in der Gruppentheorie. *Ber. über die Verhandlungen der Königl. Sächs. Ges. der Wiss. zu Leipzig. Math.-phys. Klasse* 58 (1906), 19–48. *HGW* IV, 431–460.

[H 1906b] Untersuchungen über Ordnungstypen I, II, III. *Ber. über die Verhandlungen der Königl. Sächs. Ges. der Wiss. zu Leipzig. Math.-phys. Klasse* 58 (1906), 106–169. *HGW* IA, 41–104. English translation in Plotkin, J. M. (Ed.): *Hausdorff on Ordered Sets.* American Mathematical Society, Providence (Rhode Island) 2005 , 35–95.

[H 1907a] Untersuchungen über Ordnungstypen IV, V. *Ber. über die Verhandlungen der Königl. Sächs. Ges. der Wiss. zu Leipzig. Math.-phys. Klasse* 59 (1907), 84–159. *HGW* IA, 107–182. English translation in Plotkin, J. M. (Ed.): *Hausdorff on Ordered Sets.* American Mathematical Society, Providence (Rhode Island) 2005, 97–171.

[H 1907b] Über dichte Ordnungstypen. *Jahresbericht der DMV* 16 (1907), 541–546. *HGW* IA, 203–208. English translation in Plotkin, J. M. (Ed.): *Hausdorff on Ordered Sets.* American Mathematical Society, Providence (Rhode Island) 2005, 175–180.

[H 1908] Grundzüge einer Theorie der geordneten Mengen. *Math. Annalen* 65 (1908), 435–505. *HGW* IA, 213–283. English translation in Plotkin, J. M. (Ed.): *Hausdorff on Ordered Sets.* American Mathematical Society, Providence (Rhode Island) 2005, 197–258.

[H 1909a] Die Graduierung nach dem Endverlauf. *Abhandlungen der Königl. Sächs. Ges. der Wiss. zu Leipzig. Math.-phys. Klasse* 31 (1909), 295–334. *HGW* IA, 297–335. English translation in Plotkin, J. M. (Ed.): *Hausdorff on Ordered Sets.* American Mathematical Society, Providence (Rhode Island) 2005, 271–301.

[H 1909b] Zur Hilbertschen Lösung des Waringschen Problems. *Math. Annalen* 67 (1909), 301–305. *HGW* IV, 503–507.

[H 1909c] (Paul Mongré) Strindbergs Blaubuch. *Die neue Rundschau (Freie Bühne)* 20 (6), (1909), 891–896. *HGW* VIII, 691–695.

[H 1909d] Semon, Richard, Die Mneme als erhaltendes Prinzip im Wechsel des organischen Geschehens. (Review). *Vierteljahrsschrift für wissenschaftliche Philosophie und Soziologie* 33 (1909), 101–102. *HGW* VI, 339–340.

[H 1910a] (Paul Mongré) Der Komet. *Die neue Rundschau (Freie Bühne)* 21 (5), (1910), 708–712. *HGW* VIII, 723–727.

[H 1910b] (Paul Mongré) Andacht zum Leben. *Die neue Rundschau (Freie Bühne)* 21 (12), (1910), 1737–1741. *HGW* VIII, 743–746.

[H 1911] E.Landau, Handbuch der Lehre von der Verteilung der Primzahlen (Review). *Jahresbericht der DMV* 20 (1911), 2.Abteilung, IV Literarisches, 1. b. Review, 92–97. *HGW* IV, 513–518.

[H 1913] (Paul Mongré) Biologisches. Licht und Schatten. *Wochenschrift für Schwarzweißkunst und Dichtung.* 3.Jg.(1912/13), H.35, Sp.[20a]–[20b]. *HGW* VIII, 757–758.

[H 1914a] *Grundzüge der Mengenlehre.* Verlag Veit & Co, Leipzig. 476 S. mit 53 Figuren. Reprints: Chelsea Pub. Co. 1949, 1955, 1965, 1978. *HGW* II, 93–576.

[H 1914b] Bemerkung über den Inhalt von Punktmengen. *Math. Annalen* 75 (1914), 428–433. *HGW* IV, 5–10.

[H 1916] Die Mächtigkeit der Borelschen Mengen. *Math. Annalen* 77 (1916), 430–437. *HGW* III, 431–438.

[H 1917] Book announcement for Grundzüge der Mengenlehre. *Jahresber. der DMV* 25 (1917), Abt. Literarisches, 55–56. *HGW* II, 829.

[H 1919a] Dimension und äußeres Maß. *Math. Annalen* 79 (1919), 157–179. *HGW* IV, 19–43.

[H 1919b] Der Wertvorrat einer Bilinearform. *Math. Zeitschrift* 3 (1919), 314–316. *HGW* IV, 57–59.

[H 1919c] Zur Verteilung der fortsetzbaren Potenzreihen. *Math. Zeitschrift* 4 (1919), 98–103. *HGW* IV, 67–72.

[H 1919d] Über halbstetige Funktionen und deren Verallgemeinerung. *Math. Zeitschrift* 5 (1919), 292–309. *HGW* IV, 79–96.

[H 1921] Summationsmethoden und Momentfolgen I, II. *Math. Zeitschrift* 9 (1921), I: 74–109, II: 280–299. *HGW* IV, 107–162.

[H 1923a] Eine Ausdehnung des Parsevalschen Satzes über Fourierreihen. *Math. Zeitschrift* 16 (1923), 163–169. *HGW* IV, 175–181.

[H 1923b] Momentprobleme für ein endliches Intervall. *Math. Zeitschrift* 16 (1923), 220–248. *HGW* IV, 193–221.

[H 1924] Die Mengen G_δ in vollständigen Räumen. *Fundamenta Mathematicae* 6 (1924), 146–148. *HGW* III, 445–447.

[H 1925] Zum Hölderschen Satz über $\Gamma(x)$. *Math. Annalen* 94 (1925), 244–247. *HGW* IV, 239–242.

[H 1927a] *Mengenlehre.* Zweite, neubearbeitete Auflage. Verlag Walter de Gruyter & Co., Berlin. 285 S. mit 12 Figuren. *HGW* III, 42–351. Russian translation of chapters 1–4 and 9, with supplements to the other chapters added by N.B. Vedenisov under the supervision of Alexandrov and Kolmogorov: *Teoria mnoshestvch* (Mengentheorie), Moskva 1937.

[H 1927b] Beweis eines Satzes von Arzelà. *Math. Zeitschrift* 26 (1927), 135–137. *HGW* IV, 249–251.

[H 1927c] Lipschitzsche Zahlensysteme und Studysche Nablafunktionen. *Journal für reine und angewandte Mathematik* 158 (1927), 113–127. *HGW* IV, 469–483.

[H 1930a] Die Äquivalenz der Hölderschen und Cesàroschen Grenzwerte negativer Ordnung. *Math. Zeitschrift* 31 (1930), 186–196. *HGW* IV, 257–267.

[H 1930b] Erweiterung einer Homöomorphie. *Fundamenta Mathematicae* 16 (1930), 353–360. *HGW* III, 457–464.

[H 1930c] Akrostichon zum 24. Februar 1930. In: Walter Tiemann (Ed.) *Der Verleger von morgen, wie wir ihn wünschen.* Verlag der Freunde Kirsteins, Leipzig, 1930, S.9. *HGW* VIII, 261.

[H 1931] Zur Theorie der linearen metrischen Räume. *Journal für reine und angewandte Mathematik* 167 (1931/32), 294–311. *HGW* IV, 271–288.

[H 1932] Eduard Study. Worte am Sarge Eduard Studys, 9. Januar 1930. *Chronik der Rheinischen Friedrich Wilhelms-Universität zu Bonn für das akademische Jahr 1929/30.* Bonner Universitäts-Buchdruckerei Gebr. Scheur, Bonn (1932). *HGW* VI, 343–344.

[H 1933a] Zur Projektivität der δs-Funktionen. *Fundamenta Mathematicae* 20 (1933), 100–104. *HGW* III, 473–477.

[H 1933b] Problem 58. *Fundamenta Mathematicae* 20 (1933), 286. *HGW* III, 481.

[H 1934] Über innere Abbildungen. *Fundamenta Mathematicae* 23 (1934), 279–291. *HGW* III, 485–497.

[H 1935a] *Mengenlehre.* Dritte Auflage. Mit einem zusätzlichen Kapitel und einigen Nachträgen. Verlag Walter de Gruyter & Co., Berlin. 307 S. mit 12 Figuren. *HGW* III, 42–351. Reprint: Dover Pub. New York, 1944. English edition: *Set theory.* Translated from the German by J.R. Aumann et al. Chelsea Pub. Co., New York 1957, 1962, 1978, 1991.

[H 1935b] Gestufte Räume. *Fundamenta Mathematicae* 25 (1935), 486–502. *HGW* III, 505–521.

[H 1935c] Problem 62. *Fundamenta Mathematicae* 25 (1935), 578. *HGW* III, 527.

[H 1936a] Über zwei Sätze von G.Fichtenholz und L.Kantorovitch. *Studia Mathematica* 6 (1936), 18–19. *HGW* III, 531–532.

[H 1936b] Summen von $\aleph_1$ Mengen. *Fundamenta Mathematicae* 26 (1936), 241–255. *HGW* IA, 349–363. English translation in Plotkin, J. M. (Ed.): *Hausdorff on Ordered Sets.* American Mathematical Society, Providence (Rhode Island) 2005, 305–316.

[H 1937] Die schlichten stetigen Bilder des Nullraums. *Fundamenta Mathematicae* 29 (1937), 151–158. *HGW* III, 541–548.

[H 1938] Erweiterung einer stetigen Abbildung. *Fundamenta Mathematicae* 30 (1938), 40–47. *HGW* III, 557–564.

[H 1969] *Nachgelassene Schriften.* G. Bergmann, (ed.) 2 vols. Teubner, Stuttgart 1969. Vol. I comprises fascicles 510–543, 545–559, 561–577; Vol. II fascicles 578–584, 598–658 from Hausdorff's estate, rendered in facsimile reproduction.

[Alexandroff 1916] ALEXANDROFF, PAUL: Sur la puissance des ensembles mesurables B. *Comptes Rendus Acad. Sci. Paris* 162: 323–325.

[Alexandroff 1924] ALEXANDROFF, PAUL: Über die Metrisation der im Kleinen kompakten topologischen Räume. *Mathematische Annalen* 92: 295–301.

[Alexandroff 1925] ALEXANDROFF, PAUL: Zur Begründung der n-dimensionalen mengentheoretischen Topologie. *Mathematische Annalen* 94: 296–308.

[Alexandroff/Urysohn 1929] ALEXANDROFF, PAUL; URYSOHN, PAUL: Mémoire sur les espaces topologiques compacts dédié à Monsieur D. Egoroff. *Verhandlingen Koninklijke Akademie van Wetenschappen te Amsterdam, Afdeeling Natuurkunde, Eerste Sectie* 14: 1–96.

[Alexandroff/Hopf 1935] ALEXANDROFF, PAUL; HOPF, HEINZ: *Topologie I.* Springer, Berlin.

[Andreas-Salomé 1894] ANDREAS-SALOMÉ, LOU: *Friedrich Nietzsche in seinen Werken.* Carl Konegen, Wien.

[Andreas-Salomé 1898] ANDREAS-SALOMÉ, LOU: Buchbesprechung von Sant' Ilario, Gedanken aus der Landschaft Zarathustras. *Die Zeit*, Wien, 3. September 1898, 157.

[Arzelà 1885] ARZELÀ, CESARE: Sulla integrazione per serie. *Atti della Reale Accademia dei Lincei. Rendiconti (4)*1: 532–537.

[Asser/Flachsmeyer/Rinow 1972] ASSER, GÜNTER; FLACHSMEYER, JÜRGEN; RINOW, WILLI (Ed.): *Theory of Sets and Topology. In Honour of Felix Hausdorff (1868–1942).* Deutscher Verlag der Wissenschaften, Berlin.

[Aull/Lowen 1998] AULL, E.C.; LOWEN, R. (Eds.): *Handbook of the History of General Topology.* Vol. 2, Kluwer, Dordrecht.

[Bacon 1857] BACON, DELIA: *The Philosophy of the Plays of Shakespeare Unfolded. With a Preface by N. Hawthorne.* London, Boston.

[Bacon 1858] BACON, DELIA: William Shakespeare and his Plays. In: *Putnam's Monthly*, Jan. 1858, 1–19.

[Baeumler 1931] BAEUMLER, ALFRED: *Nietzsche, der Philosoph und Politiker.* Reclam, Leipzig.

[Baire 1899] BAIRE, RENÉ: Sur les fonctions de variables réelles. *Annali di Matematica pura ed applicata*, Ser. III,3: 1–123.

[Baker 1905] BAKER, HENRY FREDERICK: Alternants and continuous groups. *Proc. London Math. Society, Ser. 2*3: 24–47.

[Banach 1923] BANACH, STEFAN: Sur le problème de la mesure. *Fundamenta Mathematicae* 4: 7–33.

[Banach/Tarski 1924] BANACH, STEFAN; TARSKI, ALFRED: Sur la décomposition des ensembles de points en parties respectivement congruentes. *Fundamenta Mathematicae* 6: 244–277.

[Banach 1932] BANACH, STEFAN: *Théorie des opérations linéaires.* Monografje Matematyczne, Warszawa.

[Bandt/Haase 1996] BANDT, CHRISTOPH; HAASE, HERMANN: *Die Wirkungen von Hausdorffs Arbeit über Dimension und äußeres Maß.* In [Brieskorn 1996], 149–183.

[Barbut et al. 2004] BARBUT, MARC; LOCKER, BERNARD; MAZLIAK, LAURENT (Ed.): *Paul Lévy – Maurice Fréchet. 50 ans de correspondence en 107 Lettres.* Hermann, Paris.

[Bauschinger 1898] BAUSCHINGER, JULIUS: Untersuchungen über die astronomische Refraction. *Neue Annalen der Königl. Sternwarte in München* III: 41–229.

[Bayertz et al. 2007] BAYERTZ KURT; GERHARD, MYRIAM; JAESCHKE, WALTER (Ed.): *Der Ignorabimus-Streit.* Volume 3 of *Weltanschauung, Philosophie und Naturwissenschaft im 19. Jahrhundert.* Meiner Verlag, Hamburg.

[Beckert/Purkert 1987] BECKERT, HERBERT; PURKERT, WALTER: *Leipziger mathematische Antrittsvorlesungen. Auswahl aus den Jahren 1869–1922.* Teubner, Leipzig.

[Beckert/Schumann 1981] BECKERT, HERBERT; SCHUMANN, HORST (Ed.): *100 Jahre Mathematisches Seminar der Karl-Marx-Universität Leipzig.* Deutscher Verlag der Wissenschaften, Berlin.

[Bense 1965] BENSE, MAX: *Aesthetica. Einführung in die neue Ästhetik.* Agis-Verlag, Baden-Baden.

[Berding 1988] BERDING, HELMUT: *Moderner Antisemitismus in Deutschland.* Suhrkamp, Frankfurt/Main.

[Berger 1939] BERGER, ALFRED: *Mathematik der Lebensversicherung.* Springer, Wien.

[Bergh/Löfström 1976] BERGH, JÖRAN; LÖFSTRÖM, JÖRGEN: *Interpolation Spaces. An Introduction.* Springer, Berlin Heidelberg New York.

[Bernoulli 1908] BERNOULLI, CARL ALBRECHT: *Franz Overbeck und Friedrich Nietzsche. Eine Freundschaft.* 2 vols. Diederichs, Jena.

[Besicovitch 1929] BESICOVITCH, ABRAM SAMOILOVITCH: On linear sets of points of fractional dimension. *Mathematische Annalen* 101: 161–193.

[Besicovitch 1934] BESICOVITCH, ABRAM SAMOILOVITCH: On the sum of digits of real numbers represented in the dyadic system (On sets of fractional dimensions II.) *Mathematische Annalen* 110: 321–329.

[Biermann 1988] BIERMANN, KURT R.: *Die Mathematik und ihre Dozenten an der Berliner Universität 1810–1933.* Akademie-Verlag, Berlin.

[Boehm 1933–1939] BOEHM, FRIEDRICH: Ueber die Entwicklung und den gegenwärtigen Stand der Risikotheorie in der Lebensversicherung. *Das Versicherungsarchiv. Teil I* 4: 561–584; *Teil II* 5: 241–260, 321–333; *Teil III* 5: 753–777, 913–939; *Teil IV* 9: 161–183, 257–289; *Teil V* 9: 593–629.

[Böhm/Mehmel 1996/2000] BÖHM, HANS; MEHMEL, ASTRID (Ed.): Alfred Philippson: *Wie ich zum Geographen wurde. Aufgezeichnet im Konzentrationslager Theresienstadt zwischen 1942 und 1945.* Bouvier, Bonn, expanded and commented edition.

[Bonnet 1967] BONNET, HANS: *Geleitwort* zu GÜNTER BERGMANN: Vorläufiger Bericht über den wissenschaftlichen Nachlaß von Felix Hausdorff. *Jahresbericht der DMV* 69: 76(*152*).

[Boos 2000/2006] BOOS, JOHANN: *Classical and Modern Methods in Summability.* Oxford Mathematical Monographs, Oxford University Press, New York, Reprint.

[Borel 1898] BOREL, ÉMILE: *Leçons sur la théorie des fonctions.* Gauthier-Villars, Paris.

[Borel 1909] BOREL, ÉMILE: Les probabilités dénombrables et leurs applications arithmétiques. *Rendiconti del Circolo Matematico di Palermo* 27: 247–271.

[Bothe/Schmeling 1996] BOTHE, HANS-GÜNTHER; SCHMELING, JÖRG: *Die Hausdorff-Dimension in der Dynamik.* In [Brieskorn 1996], 229–252.

[Brieskorn 1996] BRIESKORN, EGBERT (Ed.): *Felix Hausdorff zum Gedächtnis. Aspekte seines Werkes.* Vieweg, Braunschweig/Wiesbaden.

[Brieskorn/Purkert 2021] BRIESKORN, EGBERT; PURKERT, WALTER: *Felix Hausdorff: Mathematiker, Philosoph und Literat.* Springer Spektrum, Heidelberg.

[Brinkmann 1995] BRINKMANN, REINHOLD: *Arnold Schönberg–Sämtliche Werke. Abteilung VI: Kammermusik, Reihe B, Vol. 24,1. Melodramen und Lieder mit Instrumenten. Teil 1: Pierrot lunaire op. 21. Kritischer Bericht · Studien zur Genesis · Skizzen · Dokumente.* Schott Musik International, Mainz und Universal Edition, Wien.

[Broggi 1911] BROGGI, HUGO : *Versicherungsmathematik.* Teubner, Leipzig 1911.

[Broszat 1984] BROSZAT, MARTIN: *Die Machtergreifung. Der Aufstieg der NSDAP und die Zerstörung der Weimarer Republik.* dtv, München.

[Brouwer 1911] BROUWER, LUITZEN EGBERTUS JAN: Beweis der Invarianz der Dimensionszahl. *Mathematische Annalen* 70: 161–165.

[Bruns 1891] BRUNS, HEINRICH: Zur Theorie der astronomischen Strahlenbrechung. *Berichte der Königl. Sächs. Ges. der Wiss. zu Leipzig, Math.-phys. Classe* 43: 164–227.

[Bruns 1895] BRUNS, HEINRICH: Das Eikonal. *Abhandlungen der mathematisch-physischen Classe der Königl. Sächs. Ges. der Wiss. zu Leipzig* 21: 323–436.

[Bruns 1897] BRUNS, HEINRICH: Ueber die Darstellung von Fehlergesetzen. *Astronomische Nachrichten* 143: 329–340.

[Bruns 1906] BRUNS, HEINRICH: *Wahrscheinlichkeitsrechnung und Kollektivmaßlehre.* Teubner, Leipzig und Berlin.

[Cantor 1878] CANTOR, GEORG: Ein Beitrag zur Mannigfaltigkeitslehre. *Journal für die reine und angewandte Mathematik* 84: 242–258.

[Cantor 1895/1897] CANTOR, GEORG: Beiträge zur Begründung der transfiniten Mengenlehre, *Mathematische Annalen* 46: 481–512; 49: 207–246; reprinted in [Cantor 1932, 282–351]; English trans. [Cantor 1915/1955].

[Cantor 1915/1955] CANTOR, GEORG: *Contributions to the Founding of the Theory of Transfinite Numbers*, Philip Jourdain, trans., New York: Dover.

[Cantor 1932] CANTOR, GEORG: *Gesammelte Abhandlungen mathematischen und philosophischen Inhalts.* Herausgegeben von ERNST ZERMELO. Springer, Berlin.

[Carathéodory 1914] CARATHÉODORY, CONSTANTIN: Über das lineare Maß von Punktmengen – eine Verallgemeinerung des Längenbegriffs. *Nachrichten von der Königlichen Gesellschaft der Wissenschaften zu Göttingen, Jahrgang 1914*, 404–426.

[Cavaillès/Noether 1937] CAVAILLÈS, JEAN; NOETHER, EMMY (Ed.): *Briefwechsel Cantor–Dedekind.* Hermann & Co., Paris.

[Cesàro 1890] CESÀRO, ERNESTO: Sur la multiplication des séries. *Bulletin des sciences mathématiques (2)* XIV: 114–120.

[Chang/Keisler 1973] CHANG, CHEN CHUNG; KEISLER, HOWARD JEROME: *Model Theory.* North-Holland, Amsterdam.

[Charotnik 1998] CHAROTNIK, J.J.: History of Continuum Theory. In [Aull/Lowen 1998], 703–786.

[Clausius 1865] CLAUSIUS, RUDOLF: Ueber verschiedene für die Anwendung bequeme Formen der Hauptgleichungen der mechanischen Wärmelehre. *Poggendorffs Annalen der Physik und Chemie* 125: 353–400.

[Conrad 1898] CONRAD, MICHAEL GEORG: Rezension von Paul Mongré, Sant' Ilario. Gedanken aus der Landschaft Zarathustras. *Die Gesellschaft. Realistische Monatsschrift für Litteratur, Kunst und öffentliches Leben, Heft3*, 208.

[Corry 2004] CORRY, LEO: *Hilbert and the Axiomatization of Physics (1898–1918): From „Grundlagen der Geometrie“ to „Grundlagen der Physik“.* Kluwer Academic Publishers, Dordrecht.

[Cramér 1936] CRAMÉR, HARALD: Über eine Eigenschaft der normalen Verteilungsfunktion. *Mathematische Zeitschrift* 41: 405–414.

[Cramér 1976] CRAMÉR, HARALD: Half a century with probability theory: some personal recollections. *The Annals of Probability* 4: 509–546.

[Czichowski 1992] CZICHOWSKI, GÜNTER: Hausdorff und die Exponentialformel in der Lie-Theorie. *Seminar Sophus Lie* 2: 85–93.

[Czuber 1900] CZUBER, EMANUEL: Wahrscheinlichkeitsrechnung. *Enzyklopädie der Mathematischen Wissenschaften mit Einschluß ihrer Anwendungen. Vol. I,2*, 733–767. Teubner, Leipzig.

[Czuber 1903] CZUBER, EMANUEL: *Wahrscheinlichkeitsrechnung und ihre Anwendung auf Fehlerausgleichung, Statistik und Lebensversicherung.* Teubner, Leipzig.

[Czuber 1908] CZUBER, EMANUEL: *Wahrscheinlichkeitsrechnung und ihre Anwendung auf Fehlerausgleichung, Statistik und Lebensversicherung.* Zweite, erweiterte Auflage, Vol. I, Teubner, Leipzig und Berlin.

[Dathe 2007] DATHE, UWE: „Philosophie als eigene Antwort auf die Frage Welt“. Briefe Felix Hausdorffs an Franz Meyer. *NTM–Zeitschrift für Geschichte der Naturwissenschaften, Technik und Medizin, Neue Serie* 15: 137–147. The letters are reprinted in *HGW* IX, 501–513.

[Dehn 1905] DEHN, MAX: Vahlen, Abstrakte Geometrie. *Jahresbericht der DMV* 14: 535–537.

[Deligne 1970] DELIGNE, PIERRE: Equations différentielles à points singuliers réguliers. *Lecture Notes in Mathematics 163.* Springer, Berlin Heidelberg New York.

[Dierkesmann 1967] DIERKESMANN, MAGDA: Felix Hausdorff. Ein Lebensbild. *Jahresbericht der DMV* 69: 551–554.

[Dieudonné 1942] DIEUDONNÉ, JEAN: La dualité dans les espaces vectoriels topologiques. *Annales Sci. École Normale Supérieure (3)* 59: 107–139.

[Donnelly 1888] DONNELLY, IGNATIUS: *The Great Cryptogram. Francis Bacon's Cipher in the So-Called Shakespeare Plays.* R.S. Peale & Co., Chicago, New York, London.

[Du Bois-Reymond 1877] DU BOIS-REYMOND, PAUL: Über die Paradoxen des Infinitärcalcüls. *Mathematische Annalen* 6: 149–167.

[Duda 1996] DUDA, ROMAN: *Fundamenta Mathematicae and the Warsaw School of Mathematics.* In [Goldstein et al. 1996], 479–498.

[Dutta/Rhoades 2016] DUTTA, HEMEN; RHOADES, BILLY E.: *Current Topics in Summability Theory and Applications.* Springer Science+Business Media, Singapore.

[Dvoretzky 1948] DVORETZKY, ARYEH: A note on Hausdorff dimension functions. *Proc. Cambridge Philosophical Soc.* 44: 13–16.

[Dynkin 1947] DYNKIN, E.B.: Wytschislenije Koeffizientov v Formulje Campbella-Hausdorffa. (Berechnung der Koeffizienten in der Formel von Campbell-Hausdorff). *Doklady Akademii Nauk SSSR (1947)*, 323–326.

[Ebbinghaus 2007a] EBBINGHAUS, HEINZ-DIETER: Zermelo and the Heidelberg Congress 1904. *Historia Mathematica* 34: 428-432.

[Ebbinghaus 2007b] EBBINGHAUS, HEINZ-DIETER: *Ernst Zermelo. An Approach to His Life and Work.* Springer, Heidelberg.

[Eggebrecht 1935] EGGEBRECHT, ERNST: *Vom Jungsein und Altern.* Selbstverlag.

[Eichhorn/Thiele 1994] EICHHORN, EUGEN; THIELE, ERNST-JOCHEN: *Vorlesungen zum Gedenken an Felix Hausdorff.* Heldermann Verlag, Berlin.

[Einstein 1905] EINSTEIN, ALBERT: Über die von der molekularkinetischen Theorie der Wärme geforderte Bewegung von in ruhenden Flüssigkeiten suspendierten Teilchen. *Annalen der Physik* 322: 549–560.

[Enders 1956] ENDERS, CARL (Ed.): *Wachstum und Wandel. Lebenserinnerungen von Oskar Walzel.* Erich Schmidt Verlag, Berlin.

[Engelking 1998] ENGELKING, RYSZARD: *Kazimierz Kuratowski (1896–1980). His Life and Work in Topology.* In [Aull/Lowen 1998, 431–452].

[Epple 2006] EPPLE, MORITZ: *Felix Hausdorff's Considered Empiricism.* In JOSÉ FERREIROS, JEREMY J. GRAY (Eds.): *The Architecture of Modern Mathematics: Essays in History and Philosophy*, 263–289. Oxford Univ. Press, Oxford.

[Evans 2004] EVANS, RICHARD J.: *The Coming of the Third Reich: How the Nazis Destroyed Democracy and Seized Power in Germany.* Penguin Group, New York. Deutsche Ausgabe: DVA München.

[Falconer 1990] FALCONER, KENNETH J.: *Fractal Geometry–Mathematical Foundations and Applications.* John Wiley, Chichester.

[Federer 1969] FEDERER, HERBERT: *Geometric Measure Theory.* Springer, Berlin Heidelberg.

[Felgner 2002] FELGNER, ULRICH: *Die Hausdorffsche Theorie der* η_α*-Mengen und ihre Wirkungsgeschichte. HGW* II, 645–674.

[Feller 1966] FELLER, WILLIAM: *An Introduction to Probability Theory and Its Applications.* Vol. 2. Wiley and Sons, New York London Sydney.

[Ferreirós 2007] Ferreirós, José: *Labyrinth of Thought. A History of Set Theory and Its Role in Modern Mathematics.* 2th revised edition, Birkhäuser, Basel.

[Fichtenholz/Kantorovitch 1935] Fichtenholz, Grigory; Kantorovitch, Leonid: Sur les opérations linéaires dans l'espace des fonctions bornées. *Studia Mathematica* 5: 69–98.

[Fischer 1907] Fischer, Ernst: Sur la convergence en moyenne. *Comptes Rendus Acad. Sci. Paris* 144: 1022–1024.

[Fischer 2000] Fischer, Jens Malte: *Richard Wagners ‚Das Judentum in der Musik'. Eine kritische Dokumentation als Beitrag zur Geschichte des europäischen Antisemitismus.* Insel Verlag, Frankfurt/Main.

[Fraenkel 1919] Fraenkel, Abraham A.: *Einleitung in die Mengenlehre. Eine allgemeinverständliche Einführung in das Reich der unendlichen Größen.* Springer, Berlin.

[Fraenkel 1922] Fraenkel, Abraham A.: Zu den Grundlagen der Cantor-Zermeloschen Mengenlehre. *Mathematische Annalen* 86: 230–237.

[Fraenkel 1923] Fraenkel, Abraham A.: *Einleitung in die Mengenlehre.* Zweite erweiterte Auflage. Springer, Berlin.

[Fraenkel 1953] Fraenkel, Abraham A.: *Abstract Set Theory.* North-Holland, Amsterdam.

[Fréchet 1906] Fréchet, Maurice: Sur quelque points du calcul fonctionnel. *Rendiconti del Circolo Matematico di Palermo* 22: 1–74.

[Fréchet 1921] Fréchet, Maurice: Sur les ensembles abstraits. *Annales de l'École Normale Sup.* 38: 341–388.

[Fréchet 1928a] Fréchet, Maurice: *Les espaces abstraits.* Gauthier-Villars, Paris.

[Fréchet 1928b] Fréchet, Maurice: Sur l'hypothèse de l'additivité des erreurs partielles. *Bulletin des Sciences Mathématiques* 52: 203–216.

[Fritsch 1907] Fritsch, Theodor: *Handbuch der Judenfrage : die wichtigsten Tatsachen zur Beurteilung des jüdischen Volkes*, 26. Aufl. Hammer, Leipzig.

[Fritzsche 1991] Fritzsche, Bernd: Einige Anmerkungen zu Sophus Lies Krankheit. *Historia Mathematica* 18: 247–252.

[Frobenius 1912] Frobenius, Georg: Über den Stridsbergschen Beweis des Waringschen Satzes. *Sitzungsberichte der Preussischen Akademie der Wissenschaften (1912)*, 666–670.

[Frostman 1935] Frostman, Otto: Potentiel d'équilibre et capacité des ensembles avec quelques applications à la théorie des fonctions. *Meddelanden fran Lunds Universitets Matematiska Seminarium* 3: 1–118.

[Gallwitz 1898] GALLWITZ, HANS: Review of Sant' Ilario. Gedanken aus der Landschaft Zarathustras. *Preussische Jahrbücher* 91: 555–562.

[Giaquinta 1983] GIAQUINTA, MARIANO: Multiple Integrals in the Calculus of Variations and Nonlinear Elliptic Systems. *Annals of Math. Studies 105*, Princeton University Press, Princeton.

[Girlich 1996] GIRLICH, HANS-JOACHIM: Hausdorffs Beiträge zur Wahrscheinlichkeitstheorie. In [Brieskorn 1996], 31–70.

[Gödel 1930] GÖDEL, KURT: Die Vollständigkeit der Axiome des logischen Funktionenkalküls. *Monatshefte für Mathematik und Physik* 37: 349–360.

[Gödel 1931] GÖDEL, KURT: Über formal unentscheidbare Sätze der Principia mathematica und verwandter Systeme I. *Monatshefte für Mathematik und Physik* 38: 173–198.

[Gödel 1938] GÖDEL, KURT: The consistency of the axiom of choice and of the generalized continuum hypothesis. *Proceedings of the National Academy of Sciences, U.S.A.* 24: 556–557.

[Gödel 1939] GÖDEL, KURT: Consistency proof for the generalized continuum hypothesis. *Proceedings of the National Academy of Sciences, U.S.A.* 25: 220–224.

[Gödel 1940] GÖDEL, KURT: The consistency of the axiom of choice and of the generalized continuum hypothesis with the axioms of set theory. *Annals of mathematics studies, Vol. 3*, Princeton University Press, Princeton.

[Gödel 1986–1995] FEFERMAN, SOLOMON et al. (Eds.): *Kurt Gödel. Collected Works.* Volume I (1986) *Publications 1929–1936*, Volume II (1990) *Publications 1938–1974*, Volume III (1995) *Unpublished essays and lectures.* Clarendon Press, Oxford.

[Goldschmidt 1892] GOLDSCHMIDT, SIGISMUND: *Der Kurort Bad Reichenhall und seine Umgebung.* W. Braumüller, Leipzig und Wien.

[Goldstein et al. 1996] GOLDSTEIN, CATHERINE; GRAY, JEREMY; RITTER, JIM (Eds.): *Mathematical Europe–History, Myth, Identity.* Éditions de la Maison des sciences de l'homme, Paris.

[Golomb 1997] GOLOMB, JACOB (Ed.): *Nietzsche and Jewish Culture.* Routledge, London.

[Gottwald/Kreiser 1984] GOTTWALD, SIEGFRIED; KREISER, LOTHAR: Paul Mahlo–Leben und Werk. *NTM Schriftenreihe für Geschichte der Naturwissenschaften, Technik und Medizin* 21: 1–22.

[Gottwald/Schulle 2007] GOTTWALD, ALFRED; SCHULLE, DIANA: *Die ‚Judendeportationen' aus dem Deutschen Reich 1941–1945.* Marixverlag, Wiesbaden.

[Gray 1984] GRAY, JEREMY J.: Fuchs and the theory of differential equations. *Bulletin of the AMS (New Series)* 10: 1–26.

[Gray 2000] GRAY, JEREMY J.: *The Hilbert Challenge.* Oxford University Press, Oxford.

[Gromov 1999] GROMOV, MICHAIL: *Metric structures for Riemannian and non-Riemannian spaces.* Birkhäuser, Basel.

[Handlbauer 2004] HANDLBAUER, BERNHARD (Ed.): *Else Pappenheim. Hölderlin, Feuchtersleben, Freud–Beiträge zur Geschichte der Psychoanalyse, der Psychiatrie und Neurologie.* Nausner & Nausner, Graz.

[Hankel 1867] HANKEL, HERMANN: *Theorie der complexen Zahlensysteme.* Voss, Leipzig.

[Hardy/Heilbronn 1938] HARDY, GODFREY HAROLD; HEILBRONN, HANS: Edmund Landau. *The Journal of the London Mathematical Society* 13: 302–310.

[Hartleben 1887] HARTLEBEN, OTTO ERICH: *Studenten-Tagebuch. 1885–1886.* Verlags-Magazin (J.Schabelitz), Zürich.

[Hartleben 1906] HARTLEBEN, OTTO ERICH: *Tagebuch. Fragment eines Lebens.* Albert Langen, München.

[Harzer 1922–1924] HARZER, PAUL: *Berechnung der Ablenkung der Lichtstrahlen in der Atmosphäre der Erde auf rein meteorologisch-physikalischer Grundlage.* Publikationen der Universitätssternwarte in Kiel.

[Hawkins 1975] HAWKINS, THOMAS: *Lebesgue's Theory of Integration. Its Origins and Development.* Chelsea, New York.

[Hayman 1980] HAYMAN, RONALD: *Nietzsche: A Critical Life.* Oxford University Press, Oxford.

[Helmert 1872] HELMERT, FRIEDRICH ROBERT: *Die Ausgleichsrechnung nach der Methode der kleinsten Quadrate: mit Anwendungen auf die Geodäsie und die Theorie der Messinstrumente.* Teubner, Leipzig.

[Helmholtz 1868] HELMHOLTZ, HERMANN VON: Über die Thatsachen, die der Geometrie zum Grunde liegen. *Nachrichten von der Königlichen Gesellschaft der Wissenschaften zu Göttingen, Jahrgang 1868, (9)*, 193–221.

[Helmholtz 1870] HELMHOLTZ, HERMANN VON: *Über den Ursprung und die Bedeutung der geometrischen Axiome. Vortrag, gehalten im Docentenverein zu Heidelberg im Jahre 1870.* In HERMANN VON HELMHOLTZ: *Vorträge und Reden*, Vol. 2, Vieweg, Braunschweig 1884, 2–34.

[Hessenberg 1906] HESSENBERG, GERHARD: Grundbegriffe der Mengenlehre. *Abhandlungen der Friesschen Schule* 1: 478–706.

[Hevesi 1906] HEVESI, LUDWIG: *Acht Jahre Sezession (März 1897–Juni 1905). Kritik–Polemik–Chronik.* Carl Konegen, Wien.

[Hilbert 1899] HILBERT, DAVID: *Grundlagen der Geometrie.* Teubner, Leipzig.

[Hilbert 1900] HILBERT, DAVID: Mathematische Probleme. Vortrag, gehalten auf dem Internationalen Mathematikerkongreß zu Paris 1900. *Göttinger Nachrichten, Jahrgang 1900*, 253–297.

[Hilbert 1903] HILBERT, DAVID: *Grundlagen der Geometrie. Zweite, durch Zusätze vermehrte und mit fünf Anhängen versehene Auflage.* Teubner, Leipzig.

[Hilbert 1909] HILBERT, DAVID: Beweis für die Darstellbarkeit der ganzen Zahlen durch eine feste Anzahl n^{ter} Potenzen (Waringsches Problem). *Mathematische Annalen* 67: 281–300.

[Hölder 1882] HÖLDER, OTTO: Grenzwerte von Reihen an der Convergenzgrenze. *Mathematische Annalen* 20: 535–549.

[Hölder 1887] HÖLDER, OTTO: Ueber die Eigenschaft der Gammafunktion keiner algebraischen Differentialgleichung zu genügen. *Mathematische Annalen* 28: 1–13.

[Hoerlin 2011] HOERLIN, BETTINA: *Steps of Courage. My Parents' Journey from Nazi Germany to America.* Author House, Bloomington, IN.

[Hoffmann 1991] HOFFMANN, DAVID MARC: *Zur Geschichte des Nietzsche-Archivs. Elisabeth Förster-Nietzsche · Fritz Kögel · Rudolf Steiner · Gustav Naumann · Josef Hofmiller. Chronik, Studien und Dokumente.* Walter de Gruyter, Berlin.

[Hoffmann et al. 1998] HOFFMANN, DAVID MARC; PETER, NIKLAUS; SALFINGER, THEO (Eds.): *Franz Overbeck, Heinrich Köselitz [Peter Gast]: Briefwechsel.* Walter de Gruyter, Berlin.

[Holden 1986] HOLDEN, ARUN V. (Ed.): *Chaos.* Manchester Univ. Press, Manchester.

[Holmes 1866] HOLMES, NATHANIEL: *The Autorship of Shakespeare.* Hurd and Houghton, New York.

[Horneffer 1900] HORNEFFER, ERNST: Die Nietzsche-Ausgabe. *Die Zeit*, Wien, 28. Juli 1900, 58–59.

[Huschke 1938] HUSCHKE, KONRAD: Max Klinger und die Musik, Teil 4, Klinger und Reger. *Zeitschrift für Musik* 105: 861–866.

[Hušek 1992] HUŠEK, MIREK: *History and development of Hausdorff's work in extension in metric spaces.* In WERNER GÄHLER, HORST HERRLICH, GERHARD PREUSS (Eds.): *Recent Developments of General Topology and its Applications*, 160–169. Akademie-Verlag, Berlin.

[Ilgauds 1985] ILGAUDS, HANS JOACHIM: Zur Biographie von Felix Hausdorff. *Mitteilungen der Mathematischen Gesellschaft der DDR*, Jahrgang 1985, Heft 2–3, 59–69.

[Jammer 1960] JAMMER, MAX: *Das Problem des Raumes. Die Entwicklung der Raumtheorien.* Wissenschaftliche Buchgesellschaft, Darmstadt.

[Jarník 1929] JARNÍK, VOJTĚCH: Diophantische Approximationen und Hausdorffsches Mass. *Matematitscheskij Sbornik* 36: 371–382.

[Jarník 1931] JARNÍK, VOJTĚCH: Über die simultanen diophantischen Approximationen. *Mathematische Zeitschrift* 33: 505–543.

[John 1981] JOHN, HARTMUT: *Das Reserveoffizierkorps im Deutschen Kaiserreich 1890–1914. Ein sozialgeschichtlicher Beitrag zur Untersuchung der gesellschaftlichen Militarisierung im Wilhelminischen Deutschland.* Campus Verlag, Frankfurt New York.

[Johnson 1979/1981] JOHNSON, J.M.: The Problem of the Invariance of Dimension in the Growth of Modern Topology. *Archive for History of Exact Sciences* 20: 97–188; 25: 85–267.

[Jónsson 1956] JÓNSSON, BJARNI: Universal relational systems. *Mathematica Scandinavica* 4: 193–208.

[Jónsson 1960] JÓNSSON, BJARNI: Homogeneous universal relational systems. *Mathematica Scandinavica* 8: 137–142.

[Julia 1918] JULIA, GASTON: Mémoire sur l'itération des fonctions rationnelles. *Journal de Math. Pures et Appl.* (VII)4: 47–245.

[Kahle/Bleek 2003/2006] KAHLE, JOHN H.; BLEEK, WILHELM (Ed.): Marie Kahle: *Was hätten Sie getan? Die Flucht der Familie Kahle aus Nazi-Deutschland.* Paul Kahle: *Die Universität Bonn vor und während der Nazi-Zeit (1923–1939).* Bouvier Verlag, Bonn.

[Kameda 1915/16] KAMEDA, TOYOJIRO: Theorie der erzeugenden Funktion und ihre Anwendung auf die Wahrscheinlichkeitsrechnung. *Proceedings of the Tôkyô Mathematico-Physical Society, Ser. 2*, 8: 262–295, 336–360.

[Kampe 1988] KAMPE, NORBERT: *Studenten und „Judenfrage" im Deutschen Kaiserreich.* Vandenhoeck & Ruprecht, Göttingen.

[Kanamori 1994/2005] KANAMORI, AKIHIRO: *The Higher Infinite. Large Cardinals in Set Theory from Their Beginnings.* Springer, Berlin.

[Kanamori 2007] KANAMORI, AKIHIRO: Gödel and set theory. *Bulletin of Symbolic Logic* 13: 153–188.

[Kanovei 2013] KANOVEI, VLADIMIR: *Gaps in partially ordered sets and related problems. Commentary to* [H 1909a] *and* [H 1936b]. In *HGW* IA, 367–405.

[Kantorovitch/Livenson 1932] Kantorovitch, Leonid; Livenson, E.M.: Memoir on the analytical operations and projective sets I. *Fundamenta Mathematicae* 18: 214–279.

[Kechris 1995] Kechris, Alexander S.: *Classical Descriptive Set Theory.* Springer, New York.

[Kempf 1896] Kempf, Paul: Review of F. Hausdorff: Über die Absorption des Lichtes in der Atmosphäre. *Vierteljahresschrift der astronomischen Gesellschaft* 31: 2–28.

[Kloss 1919] Kloss, Erich (Ed.): *Briefwechsel zwischen Wagner und Liszt.* 4. Aufl., Breitkopf & Härtel, Leipzig.

[Knopf 2003] Knopf, Sabine: Gustav Kirstein–ein jüdischer Verleger, Bibliophile und Kunstsammler. *Imprimatur. Ein Jahrbuch für Bücherfreunde. Neue Folge* XVIII: 289–312.

[Koenigsberger 1919] Koenigsberger, Leo: *Mein Leben.* Winters Universitätsbuchhandlung, Heidelberg.

[Köthe 1937] Köthe, Gottfried: Die Theorie der Verbände, ein neuer Versuch zur Grundlegung der Algebra und der projektiven Geometrie. *Jahresbericht der DMV* 47: 125–144.

[Köthe 1969] Köthe, Gottfried: *Topological Vector Spaces I.* Springer, Berlin.

[Kolmogoroff 1933] Kolmogoroff, Andrej Nikolajewitsch: *Grundbegriffe der Wahrscheinlichkeitsrechnung.* Springer, Berlin.

[Kowalewski 1950] Kowalewski, Gerhard: *Bestand und Wandel.* Oldenbourg, München.

[Krazer 1905] Krazer, Adolf (Ed.): *Verhandlungen des dritten Internationalen Mathematiker-Kongresses in Heidelberg vom 8. bis zum 13. August 1904.* Teubner, Leipzig.

[Küttner 1906] Küttner, W. *Das Risiko der Lebensversicherungs-Anstalten und Unterstützungskassen.* Mittler & Sohn, Berlin.

[Kuratowski 1933] Kuratowski, Casimir: *Topologie I.* Monografie Matematyczne, Vol. III, Warszawa.

[Kuratowski 1934] Kuratowski, Casimir: Sur une généralisation de la notion d'homéomorphie. *Fundamenta Mathematicae* 22: 206–220.

[Landauer 1903] Landauer, Gustav: *Skepsis und Mystik. Versuche im Anschluß an Mauthners Sprachkritik.* Egon Fleischel & Co., Berlin.

[Lange 2016] Lange, Bernd-Lutz: *Das Leben ist ein Purzelbaum. Von der Heiterkeit des Seins.* 2. Auflage, Aufbau Verlag, Berlin.

[Lebesgue 1902] LEBESGUE, HENRI: Intégrale, longueur, aire. Thèse, Paris 1902. *Annali di Mat. (3)* 7: 231–359.

[Lebesgue 1904] LEBESGUE, HENRI: *Leçons sur l'intégration et la recherche de fonctions primitives.* Gauthier-Villars, Paris.

[Lebesgue 1905] LEBESGUE, HENRI: Sur les fonctions représentables analytiquement. *Journal de Math. (Ser. 6)* 1: 139–216.

[Lévy 1925] LÉVY, PAUL: *Calcul des Probabilités.* Gauthier-Villars, Paris.

[Lévy 1934] LÉVY, PAUL: Sur les intégrales dont les éléments sont des variables aléatoires indépendantes. *Annali Scuola norm. sup. Pisa (II)* 3: 337–366.

[Lévy 1935] LÉVY, PAUL: Propriétés asymptotiques des sommes de variables aléatoires indépendantes ou enchainées. *Journal de Math. pures et appl. (IX)* 14: 347–402.

[Lévy 1970] LÉVY, PAUL: *Quelques aspects de la pensée d'un mathématicien.* Albert Blanchard, Paris.

[Liapounoff 1900] LIAPOUNOFF, ALEXANDER: Sur une proposition de la théorie des probabilités. *Bulletin de l'Academie Imperiale des Sciences de St. Pétersbourg, Série 5* 13: 359–386.

[Liapounoff 1901] LIAPOUNOFF, ALEXANDER: Nouvelle forme du théorème sur la limite de probabilité. *Mémoires de l'Academie Imperiale des Sciences de St. Pétersbourg* 12: 1–24.

[Liebmann 1876] LIEBMANN, OTTO: *Zur Analysis der Wirklichkeit–Philosophische Untersuchungen.* Verlag von Karl J. Trübner, Straßburg.

[Liebmann 1905/1912] LIEBMANN, HEINRICH: *Nichteuklidische Geometrie.* Göschen, Leipzig.

[Lorentz 2001] LORENTZ, GEORGE G.: Who Discovered Analytic Sets? *Mathematical Intelligencer* 23: 28–32.

[Lorenz 1963] LORENZ, EDWARD N.: Deterministic nonperiodic flow. *Journal of the Atmospheric Sciences* 20: 130–141.

[Lusin 1930] LUSIN, NIKOLAI N.: *Leçons sur les ensembles analytiques et leurs applications.* Gauthier-Villars, Paris.

[Mahlo 1909] MAHLO, PAUL: Über homogene Teilmengen des Kontinuums. *Berichte über die Verhandlungen der Königl. Sächs. Ges. der Wiss. zu Leipzig. Math.-phys. Klasse* 61: 121–124.

[Mandelbrot 1977] MANDELBROT, BENOÎT: *Fractals: Form, Chance and Dimension.* Freeman, San Francisco.

[Mandelbrot 1982] MANDELBROT, BENOÎT: *The Fractal Geometry of Nature.* Freeman, San Francisco.

[Markoff 1912] MARKOFF, ANDREJ ANDREJEWITSCH: *Wahrscheinlichkeitsrechnung.* Teubner, Leipzig und Berlin (translated by Heinrich Liebmann from the second Russian edition, St. Petersburg 1908).

[Marr 1879] MARR, WILHELM: *Der Sieg des Judenthums über das Germanenthum : vom nicht confessionellen Standpunkt aus betrachtet.* Costenoble, Bern.

[Mazurkiewicz 1932] MAZURKIEWICZ, STEFAN: Sur les transformations intérieures. *Fundamenta Mathematicae* 19: 198–204.

[Mehmel 1996] MEHMEL, ASTRID: *Dora Philippson.* In ANNETTE KUHN et al. (Ed.): *100 Jahre Frauenstudium: Frauen der Rheinischen Friedrich-Wilhelms-Universität Bonn,* 200–204. Edition Ebersbach, Dortmund.

[Mehmel 2014] MEHMEL, ASTRID: *Alfred Philippson–Bürger auf Widerruf.* In REINHOLD BOSCHKI, RENÉ BUCHHOLZ (Ed.): *Das Judentum kann nicht definiert werden. Beiträge zur jüdischen Geschichte und Kultur,* 173–202. LIT Verlag, Berlin.

[Mehrtens 1979] MEHRTENS, HERBERT: *Die Entstehung der Verbandstheorie.* Gerstenberg, Hildesheim.

[Mehrtens 1987] MEHRTENS, HERBERT: Ludwig Bieberbach and "Deutsche Mathematik," in *Studies in the History of Mathematics,* Esther Phillips, ed., Washington: The Mathematical Association of America, pp. 195–241.

[Menger 1932] MENGER, KARL: *Kurventheorie.* Teubner, Leipzig Berlin.

[Menzel 1897] MENZEL, MAX: *Der Einjährig-Freiwillige und Offizier des Beurlaubtenstandes der Infanterie. Seine Ausbildung und Doppelstellung im Heer und Staat.* Verlag R.Eisenschmidt, Berlin.

[Meschkowski 1967] MESCHKOWSKI, HERBERT: *Probleme des Unendlichen. Werk und Leben Georg Cantors.* Vieweg, Braunschweig.

[Mises 1919] MISES, RICHARD VON: Fundamentalsätze der Wahrscheinlichkeitsrechnung. *Mathematische Zeitschrift* 4: 1–97.

[Moeli 1888] MOELI, CARL: *Ueber irre Verbrecher. I. Krankengeschichten. II. Ueber den Zusammenhang von Geistesstörung und Verbrechen. III. Ueber Feststellung des Geisteszustandes. IV. Die Simulation von Geisteskrankheiten. V. Behandlung und Unterbringung irrer Verbrecher.* Fischer, Berlin.

[Moore 1897] MOORE, ELIAKIM HASTINGS: Concerning transcendentally transcendental functions. *Mathematische Annalen* 48: 49–74.

[Moore 1982] MOORE, GREGORY: *Zermelo's Axiom of Choice. Its Origins, Development and Influence.* Springer, New York Heidelberg Berlin.

[Moore 1989] MOORE, GREGORY: *Towards a History of Cantor's Continuum Problem.* In DAVID E. ROWE, JOHN MCCLEARY (Eds.): *The History of Modern Mathematics,* Volume I., 78–121. Academic Press, Boston.

[Naumann 1898] NAUMANN, GUSTAV: *Antimoralisches Bilderbuch. Ein Beitrag zu einer vergleichenden Moralgeschichte.* Verlag von H.Haessel, Leipzig.

[Neuenschwander 1996] NEUENSCHWANDER, ERWIN: *Felix Hausdorffs letzte Lebensjahre nach Dokumenten aus dem Bessel-Hagen-Nachlaß.* In [Brieskorn 1996], 253–270.

[Neumann 1922] NEUMANN, JOHN VON: Zur Einführung der transfiniten Zahlen. *Acta litterarum ac scientiarum Regiae Universitatis Hungaricae*, 199–208.

[Neumann 1929] NEUMANN, JOHN VON: Zur allgemeinen Theorie des Masses. *Fundamenta Mathematicae* 13: 73–116.

[Niemytzki/Tychonoff 1928] NIEMYTZKI, VIKTOR; TYCHONOFF, ANDREI: Beweis des Satzes, dass ein metrisierbarer Raum dann und nur dann kompakt ist, wenn er in jeder Metrik vollständig ist. *Fundamenta Mathematicae* 12: 118–120.

[Nietzsche KSA] NIETZSCHE, FRIEDRICH: Sämtliche Werke. *Kritische Studienausgabe in 15 Bänden.* Herausgegeben von Giorgio Colli und Mazzino Montinari. 2. Aufl., Walter de Gruyter, Berlin.

[Ostrowski 1919] OSTROWSKI, ALEXANDER: Neuer Beweis des Hölderschen Satzes, daß die Gammafunktion keiner algebraischen Differentialgleichung genügt. *Mathematische Annalen* 79: 286–288.

[Owen 1893–1895] OWEN, ORVILLE WARD: *Sir Francis Bacon's Cipher Story.* Fünf Bände. Howard Publishing Company, Detroit New York.

[Pasch 1882] PASCH, MORITZ: *Vorlesungen über neuere Geometrie.* Teubner, Leipzig.

[Paul 1873] PAUL, OSCAR: *Handlexikon der Tonkunst.* Vol. 2. Verlag Heinrich Schmidt, Leipzig.

[Peitgen/Richter 1986] PEITGEN, HEINZ OTTO; RICHTER, PETER H.: *The Beauty of Fractals. Images of Complex Dynamical Systems.* Springer, Heidelberg.

[Peitgen et al. 1992] PEITGEN, HEINZ-OTTO; JÜRGENS, HARTMUT; SAUPE, DIETMAR: *Chaos and Fractals. New Frontiers of Science.* Springer, New York.

[Perrin 1908] PERRIN, JEAN BAPTISTE: Mouvement brownien et réalité moléculaire. *Annales de chimie et de physique* VIII 18: 5–114.

[Peters 2009] PETERS, LEO (Ed.): *Eine jüdische Kindheit am Niederrhein. Die Erinnerungen des Julius Grunewald (1860 bis 1929)*, Böhlau, Köln.

[Plotkin 2005] PLOTKIN, JACOB M. (Ed.): *Hausdorff on Ordered Sets.* American Mathematical Society Publications, Providence (Rhode Island).

[Pólya 1916] PÓLYA, GEORGE: Über Potenzreihen mit ganzzahligen Koefizienten. *Mathematische Annalen* 77: 497–513.

[Pólya 1918] PÓLYA, GEORGE: Über die Potenzreihen, deren Konvergenzkreis natürliche Grenze ist. *Acta Mathematica* 41: 99–118.

[Purkert 1983] PURKERT, WALTER: Die Bedeutung von A. Einsteins Arbeit über Brownsche Bewegung für die Entwicklung der modernen Wahrscheinlichkeitstheorie. *Mitteilungen der Mathematischen Gesellschaft der DDR*, Jahrgang 1983, Heft 3, 41–49.

[Purkert 1984] PURKERT, WALTER: Zum Verhältnis von Sophus Lie und Friedrich Engel. *Wissenschaftliche Zeitschrift der Ernst-Moritz-Arndt-Universität Greifswald, Math.-naturwissenschaftliche Reihe* 33: 29–34.

[Purkert 1986] PURKERT, WALTER: Georg Cantor und die Antinomien der Mengenlehre. *Bulletin de la Société Mathématique de Belgique*, Ser. A, 38 (1986), 313–327.

[Purkert/Ilgauds 1987] PURKERT, WALTER; ILGAUDS, HANS JOACHIM: *Georg Cantor 1845–1918.* Birkhäuser, Basel.

[Purkert 2015a] PURKERT, WALTER: *Felix Hausdorff–ein deutscher Jude als „Gründungsvater" der Moskauer topologischen Schule.* In INGRID KÄSTNER, JÜRGEN KIEFER (Ed.): *Von Maimonides bis Einstein–Jüdische Gelehrte und Wissenschaftler in Europa,* 349–362. Shaker Verlag, Aachen.

[Purkert 2015b] PURKERT, WALTER: *On Cantor's Continuum Problem and Well Ordering: What really happened at the 1904 International Congress of Mathematicians in Heidelberg.* In DAVID E. ROWE, WANN-SHENG HORNG (Eds.): *A Delicate Balance: Global Perspectives on Innovation and Tradition in the History of Mathematics. A Festschrift in Honor of Joseph W. Dauben,* 3–24. Birkhäuser, Basel.

[Rademacher/Toeplitz 1930] RADEMACHER, HANS; TOEPLITZ, OTTO: *Von Zahlen und Figuren. Proben mathematischen Denkens für Liebhaber der Mathematik.* Springer, Berlin.

[Rademacher/Toeplitz 1966] RADEMACHER, HANS; TOEPLITZ, OTTO: *The Enjoyment of Math.* Princeton University Press, Princeton.

[Rapp 2011] RAPP, JEANETTE: *Von Jüdin für Jüdin. Die soziale Arbeit der Leipziger Ortsgruppe des Jüdischen Frauenbundes und ihrer Mitgliedsorganisationen bis zum Ende der Weimarer Republik.* Dissertation am Fachbereich Erziehungswissenschaften und Psychologie der FU Berlin.

[Reibnitz/Stauffacher 1995] REIBNITZ, BARBARA VON; STAUFFACHER-SCHAUB, MARIANNE (Eds.): *Franz Overbeck: Werke und Nachlass.* Vol. 5: *Kirchenlexikon Texte, Ausgewählte Artikel J–Z.* Verlag J.B.Metzler, Stuttgart Weimar.

[Reid 1979] REID, CONSTANCE: *Richard Courant 1888–1972. Der Mathematiker als Zeitgenosse* Springer, Berlin Heidelberg.

[Remak 1912] REMAK, ROBERT: Bemerkung zu Herrn Stridsbergs Beweis des Waringschen Theorems. *Mathematische Annalen* 72: 153–156.

[Rey 1994] REY, MANFRED VON: *Die Vernichtung der Juden in Bonn.* In [Eichhorn/ Thiele 1994], 227–250.

[Riemann 1868] RIEMANN, BERNHARD: Ueber die Hypothesen, welche der Geometrie zu Grunde liegen. Aus dem Nachlaß des Verfassers mitgetheilt durch R.Dedekind. *Abhandlungen der mathematischen Classe der Königlichen Gesellschaft der Wissenschaften zu Göttingen* 13: 133–155.

[Riesz 1907] RIESZ, FRIEDRICH: Sur les systèmes orthogonaux de fonctions. *Comptes Rendus Acad. Sci. Paris* 144: 615–619.

[Riesz 1910] RIESZ, FRIEDRICH: Untersuchungen über Systeme integrierbarer Funktionen. *Mathematische Annalen* 69: 449–497.

[Riesz 1923] RIESZ, FRIEDRICH: Über eine Verallgemeinerung der Parsevalschen Formel. *Mathematische Zeitschrift* 18: 117–124.

[Riesz 1921/1922/1923] RIESZ, MARCEL: Sur le Problème des moments. Première note. *Arkiv för matematik, astronomi och fysik* 16(12): 1-23. Deuxième note 16(19): 1–21. Troisième note 17(16): 1–52.

[Rosenthal 1928] ROSENTHAL, ARTHUR: Felix Hausdorff, Mengenlehre. *Deutsche Literaturzeitung* 49: 294–295.

[Rowe 2021] ROWE, DAVID: *Emmy Noether–Mathematician Extraordinaire.* Springer, London.

[Ruelle/Takens 1971] RUELLE, DAVID; TAKENS, FLORIS: On the nature of turbulence. *Communications in Mathematical Physics* 20: 167–192.

[Ruelle 1989] RUELLE, DAVID: *Chaotic evolution and strange attractors.* Cambridge University Press, Cambridge.

[Schappacher 1987] SCHAPPACHER, NORBERT: *Das Mathematische Institut der Universität Göttingen 1929–1950.* In HEINRICH BECKER, HANS-JOACHIM DAHMS, CORNELIA WEGELER (Eds.): *Die Universität Göttingen unter dem Nationalsozialismus,* 345–373. K.G. Saur, München.

[Schappacher/Scholz 1992] SCHAPPACHER, NORBERT; SCHOLZ, ERHARD: Oswald Teichmüller–Leben und Werk. *Jahresbericht der DMV* 94: 1–39.

[Scheepers 1993] SCHEEPERS, MARION: *Gaps in* ω^ω. In HAIM JUDAH (Ed.): *Israel Math. Conference Proceedings, Nr. 6,* 439–561. Bar-Ilan University.

[Scheffers 1891] SCHEFFERS, GEORG: Zurückführung complexer Zahlensysteme auf typische Formen. *Mathematische Annalen* 41: 293–390.

[Schlick 1917a] SCHLICK, MORITZ: Raum und Zeit in der gegenwärtigen Physik. Zur Einführung in das Verständnis der allgemeinen Relativitätstheorie. *Die Naturwissenschaften* 5: 161–167, 177–186.

[Schlick 1917b] SCHLICK, MORITZ: *Raum und Zeit in der gegenwärtigen Physik. Zur Einführung in das Verständnis der allgemeinen Relativitätstheorie.* Springer, Berlin.

[Schlick 1918] SCHLICK, MORITZ: *Allgemeine Erkenntnislehre.* Springer, Berlin.

[Schlick 1920] SCHLICK, MORITZ: *Raum und Zeit in der gegenwärtigen Physik. Zur Einführung in das Verständnis der Relativitäts- und Gravitationstheorie.* Third edition of [Schlick 1917b]. Springer, Berlin.

[Schlick 1925] SCHLICK, MORITZ: *Allgemeine Erkenntnislehre.* Second edition. Springer, Berlin.

[Schneider 1989] SCHNEIDER, IVO: *Die Entwicklung der Wahrscheinlichkeitstheorie von den Anfängen bis 1933.* Akademie-Verlag, Berlin.

[Schneider 1997] SCHNEIDER, ANATOL: *Nietzscheanismus. Zur Geschichte eines Begriffs.* Königshausen & Neumann, Würzburg.

[Schneider 2011] SCHNEIDER, THOMAS: Ägyptologen im Dritten Reich. Biographische Notizen anhand der sogenannten „Steindorff-Liste". *Journal of Egyptian History* 4: 120–247.

[Schoenflies 1900] SCHOENFLIES, ARTHUR: *Entwickelung der Lehre von den Punktmannigfaltigkeiten. Bericht, erstattet der Deutschen Mathematiker-Vereinigung. Erster Teil.* Teubner, Leipzig.

[Schoenflies 1908] SCHOENFLIES, ARTHUR: *Entwickelung der Lehre von den Punktmannigfaltigkeiten. Bericht, erstattet der Deutschen Mathematiker-Vereinigung. Zweiter Teil.* Teubner, Leipzig.

[Schoenflies 1913] SCHOENFLIES, ARTHUR: *Entwickelung der Mengenlehre und ihrer Anwendungen. Erste Hälfte: Allgemeine Theorie der unendlichen Mengen und Theorie der Punktmengen.* (Revised version of [Schoenflies 1900]). Teubner, Leipzig Berlin.

[Schoenflies 1922] SCHOENFLIES, ARTHUR: Zur Erinnerung an Georg Cantor. *Jahresbericht der DMV* 31: 97–106.

[Schopenhauer 1999] SCHOPENHAUER, ARTHUR: *Die Welt als Wille und Vorstellung,* Vol. I. Herausgegeben von Ludger Lütkehaus nach der Ausgabe letzter Hand. Haffmanns Verlag, Zürich.

[Schubring 1985] SCHUBRING, GERT: Die Entwicklung des Mathematischen Seminars der Universität Bonn 1864–1929. *Jahresbericht der DMV* 87: 139–163.

[Segal 2003] SEGAL, SANFORD L.: *Mathematics under the Nazis.* Princeton University Press, Princeton.

[Siebe/Prüfer 1922] SIEBE, JOSEPHINE; PRÜFER, JOHANNES: *Henriette Goldschmidt. Ihr Leben und Schaffen.* Akademische Verlagsgesellschaft, Leipzig.

[Siegmund-Schultze 1984] SIEGMUND-SCHULTZE, REINHARD: Theodor Vahlen–zum Schuldanteil eines deutschen Mathematikers am faschistischen Mißbrauch der Wissenschaft. *NTM–Schriftenreihe für Geschichte der Naturwissenschaften, Technik und Medizin* 21: 17–32.

[Siegmund-Schultze 2009] SIEGMUND-SCHULTZE, REINHARD: *Mathematicians Fleeing from Nazi Germany. Individual Fates and Global Impact*, Princeton University Press, Princeton.

[Sierpiński 1928/1934] SIERPIŃSKI, WACŁAW: *Topologia ogólna* (Allgemeine Topologie), Warszawa. English translation: *Introduction to General Topology.* Univ. of Toronto Press, Toronto.

[Skolem 1922] SKOLEM, THORALF: *Einige Bemerkungen zur axiomatischen Begründung der Mengenlehre.* In *Wissenschaftliche Vorträge, gehalten auf dem fünften Kongreß der skandinavischen Mathematiker in Helsingfors 1922,* 217–232. Akademische Buchhandlung, Helsingfors.

[Sparrow 1982] SPARROW, COLIN: *The Lorenz Equations: Bifurcation, Chaos, and Strange Attractors.* Springer, New York.

[Steffen 1996] STEFFEN, KLAUS: *Hausdorff-Dimension, reguläre Mengen und total irreguläre Mengen.* In [Brieskorn 1996], 185–227.

[Stegmaier/Krochmalnik 1997] STEGMAIER, WERNER; KROCHMALNIK, DANIEL (Ed.): *Jüdischer Nietzscheanismus.* Walter de Gruyter, Berlin.

[Steiner 1900] STEINER, RUDOLF: Das Chaos. *Das Magazin für Litteratur* 69: 570–575.

[Stirner 1893] STIRNER, MAX: *Der Einzige und sein Eigentum.* Herausgegeben und mit einer Einführung versehen von Paul Lauterbach. Verlag von Philipp Reclam jun., Leipzig.

[Stridsberg 1912] STRIDSBERG, ERIK: Sur la démonstration de M. Hilbert du théorème de Waring. *Mathematische Annalen* 72: 145–152.

[Stubhaug 2002] STUBHAUG, ARILD: *The Mathematician Sophus Lie.* Springer, Berlin Heidelberg.

[Stubhaug 2003] STUBHAUG, ARILD: *Es war die Kühnheit meiner Gedanken. Der Mathematiker Sophus Lie.* (Translation of [Stubhaug 2002]). Springer, Berlin Heidelberg.

[Study 1898] STUDY, EDUARD: *Theorie der Gemeinen und Höheren complexen Zahlen.* Encyclopädie der Math. Wiss., Vol. I, 1. Teubner, Leipzig.

[Study 1918] STUDY, EDUARD: Franz London. *Jahresbericht der DMV* 26: 153–157.

[Study 1926] STUDY, EDUARD: Über einige Arbeiten des Herrn H. Beck. *Jahresbericht der DMV* 35: 295–298.

[Szpilrajn 1937] SZPILRAJN, EDWARD: La dimension et la mesure. *Fundamenta Mathematicae* 28: 81–89.

[Tack 2004] TACK, LIEVEN: *Transfert et Traduction de 'Pierrot lunaire': une description sociosémiotique.* In MARK DELAERE, JAN HERMAN (Eds.): *Pierrot lunaire: Albert Giraud, Otto Erich Hartleben, Arnold Schoenberg: une collection d'études musico-littéraires.* Peeters, Leuven.

[Tamari 2007] TAMARI, DOV: *Moritz Pasch–Vater der modernen Axiomatik.* Shaker Verlag, Aachen.

[Taylor 1985] TAYLOR, ANGUS: A Study of Maurice Fréchet: II. Mainly about his Work on General Topology, 1909–1928. *Archive for History of Exact Sciences* 34: 279–380.

[Tetz 1962] TETZ, MARTIN: *Overbeckiana. Übersicht über den Franz-Overbeck-Nachlaß der Universitätsbibliothek Basel. Teil II: Der wissenschaftliche Nachlaß Franz Overbecks.* Universitätsbibliothek Basel, Basel.

[Tevenar 2007] TEVENAR, GUDRUN VON (Ed.): *Nietzsche and Ethics.* Peter Lang, Bern.

[Tietze 1915] TIETZE, HEINRICH: Über Funktionen, die auf einer abgeschlossenen Menge stetig sind. *Journal für die reine und angewandte Mathematik* 145: 9–14.

[Tietze 1923] TIETZE, HEINRICH: Beiträge zur allgemeinen Topologie I. *Mathematische Annalen* 88: 290–312.

[Tietze 1924] TIETZE, HEINRICH: Beiträge zur allgemeinen Topologie II. Über die Einführung uneigentlicher Elemente. *Mathematische Annalen* 91: 210–224.

[Toeplitz 1911] TOEPLITZ, OTTO: Über allgemeine lineare Mittelbildungen. *Prace Matematyczno-Fizyczne* 22: 113–119.

[Toeplitz 1918] TOEPLITZ, OTTO: Das algebraische Analogon zu einem Satze von Fejér. *Mathematische Zeitschrift* 2: 187–197.

[Toeplitz 1949] TOEPLITZ, OTTO: *Die Entwicklung der Infinitesimalrechnung. Eine Einleitung in die Infinitesimalrechnung nach der Genetischen Methode.* Springer, Berlin.

[Toeplitz 1963/2007] TOEPLITZ, OTTO: *The Calculus: A Genetic Approach.* University of Chicago Press, Chicago.

[Treitschke 1879] TREITSCHKE, HEINRICH VON: Unsere Aussichten. *Preußische Jahrbücher* 44: 559–576.

[Tricot 1982] TRICOT, C.: Two definitions of fractal dimension. *Mathematical Proceedings of the Cambridge Philosophical Society* 91: 57–74.

[Tychonoff 1930] TYCHONOFF, ANDREI: Über die topologische Erweiterung von Räumen. *Mathematische Annalen* 102: 544–561.

[Ullrich 2004] ULLRICH, PETER: *„Lieber Freund und Schutzengel!... Herzlich gruesst Dein Deibel.“ – Über die Korrespondenz zwischen Friedrich Engel und Eduard Study.* In HARTMUT ROLOFF, MANFRED WEIDAUER (Ed.): *Wege zu Adam Ries.* Algorismus, Vol. 43, 389–403. Rauner Verlag, Augsburg.

[Urysohn 1925/1926] URYSOHN, PAUL: Mémoire sur les multiplicités Cantoriennes. *Fundamenta Mathematicae* 7: 30–137, 8: 225–351.

[Vaihinger 1898] VAIHINGER, HANS: Literaturbericht. Sant' Ilario. Gedanken aus der Landschaft Zarathustras. *Kant-Studien* 3: 201.

[Vaihinger 1899] VAIHINGER, HANS: Mongré, Paul. Das unreinliche Jahrhundert. *Kant-Studien* 3: 201.

[Vietoris 1921] VIETORIS, LEOPOLD: Stetige Mengen. *Monatshefte für Mathematik* 31: 173–204.

[Volkert 2015] VOLKERT, KLAUS, ED.: *David Hilbert, Grundlagen der Geometrie (Festschrift 1899).* Springer, Heidelberg.

[Wagon 1993] WAGON, STAN: *The Banach-Tarski Paradox.* Cambridge University Press, Cambridge.

[Walk 1981/1996] WALK, JOSEPH (Ed.): *Das Sonderrecht für die Juden im NS-Staat.* C.F. Müller, Heidelberg.

[Walzel 1917] WALZEL, OSKAR: *Wechselseitige Erhellung der Künste. Ein Beitrag zur Würdigung kunstgeschichtlicher Begriffe.* Reuther & Reinhard, Berlin.

[Wedekind 1924] WEDEKIND, FRANK: *Gesammelte Briefe.* Vol. 2, herausgegeben von FRITZ STRICH. Georg Müller Verlag, München.

[Weierstrass 1894] WEIERSTRASS, KARL: Zur Theorie der aus n Haupteinheiten gebildeten complexen Grössen. *Göttinger Nachrichten, Jahrgang 1894*, 395–419.

[Weisskopf 2004/2012] WEISSKOPF (SCHMID), DUSCHA:: *Willi Schmid: A Life in Germany.* Private Printing, Boston.

[Whittaker/Robinson 1924] WHITTAKER, E.T.; ROBINSON, G.: *The Calculus of Observations.* Blackie and Son, London; 4th edition at Dover, 1958 at Blackie with reprints in 1962, 1965, 1966 and 1967.

[Wiener 1921] WIENER, NORBERT: The average of an analytic functional and the Brownian movement. *Proceedings of the National Academy of Sciences of the U.S.A.* 7: 294–298.

[Wiener 1923] WIENER, NORBERT: Differential-Space. *Journal of Mathematics and Physics* 2: 131–174.

[Witkowski 2010] WITKOWSKI, GEORG: *Von Menschen und Büchern. Erinnerungen 1863–1933.* Lehmstedt Verlag, Leipzig.

[Zeller/Beekmann 1970] ZELLER, KARL; BEEKMANN, WOLFGANG: *Theorie der Limitierungsverfahren.* Springer, Berlin Heidelberg.

[Zermelo 1904] ZERMELO, ERNST: Beweis, daß jede Menge wohlgeordnet werden kann. Aus einem an Herrn Hilbert gerichteten Briefe. *Mathematische Annalen* 59: 514–516.

[Zermelo 1908] ZERMELO, ERNST: Untersuchungen über die Grundlagen der Mengenlehre. I. *Mathematische Annalen* 65: 261–281.

Name Index

E. Brieskorn, W. Purkert, *Felix Hausdorff*, Vita Mathematica 21,
https://doi.org/10.1007/978-3-031-52135-5

MIX
Papier aus verantwortungsvollen Quellen
Paper from responsible sources
FSC® C105338
FSC
www.fsc.org

If you have any concerns about our products,
you can contact us on
ProductSafety@springernature.com

In case Publisher is established outside the EU,
the EU authorized representative is:
Springer Nature Customer Service Center GmbH
Europaplatz 3, 69115 Heidelberg, Germany

Printed by Libri Plureos GmbH
in Hamburg, Germany